Transport in
Nonstoichiometric
Compounds

NATO ASI Series

Advanced Science Institutes Series

A series presenting the results of activities sponsored by the NATO Science Committee, which aims at the dissemination of advanced scientific and technological knowledge, with a view to strengthening links between scientific communities.

The series is published by an international board of publishers in conjunction with the NATO Scientific Affairs Division

A	**Life Sciences**	Plenum Publishing Corporation
B	**Physics**	New York and London
C	**Mathematical and Physical Sciences**	D. Reidel Publishing Company Dordrecht, Boston, and Lancaster
D	**Behavioral and Social Sciences**	Martinus Nijhoff Publishers
E	**Engineering and Materials Sciences**	The Hague, Boston, and Lancaster
F	**Computer and Systems Sciences**	Springer-Verlag
G	**Ecological Sciences**	Berlin, Heidelberg, New York, and Tokyo

Recent Volumes in this Series

Series B: Physics

Transport in Nonstoichiometric Compounds

Edited by

George Simkovich

and

Vladimir S. Stubican

Pennsylvania State University
University Park, Pennsylvania

Plenum Press
New York and London
Published in cooperation with NATO Scientific Affairs Division

Proceedings of the Third International Conference and NATO
Advanced Research Workshop on
Transport in Nonstoichiometric Compounds,
held June 11–15, 1984,
at Pennsylvania State University, University Park, Pennsylvania

---

Library of Congress Cataloging in Publication Data

International Conference on Transport in Nonstoichiometric Compounds (3rd:
    1984: Pennsylvania State University)
    Transport in nonstoichiometric compounds.

    (NATO ASI series. Series B, Physics; v. 129)
    "Proceedings of the Third International Conference and NATO Advanced
Research Workshop on Transport in Nonstoichiometric Compounds, held
June 11–15, 1984, at Pennsylvania State University, University Park, Penn-
sylvania"—T.p. verso.
    "Published in cooperation with NATO Scientific Affairs Division."
    Includes bibliographical references and index.
    1. Crystals—Defects—Congresses. 2. Mass transfer—Congresses. I.
Simkovich, George. II. Stubican, Vladimir S. III. NATO Advanced Research
Workshop on Transport in Nonstoichiometric Compounds (1984: Pennsylvania
State University) IV. North Atlantic Treaty Organization. Scientific Affairs Divi-
sion. V. Title. VI. Series.
QD921.I5357   1984   530.4'1   85-17009
ISBN-13: 978-1-4612-9522-8   e-ISBN-13: 978-1-4613-2519-2
DOI: 10.1007/978-1-4613-2519-2

©1985 Plenum Press, New York
Softcover reprint of the hardcover 1st edition 1985

A Division of Plenum Publishing Corporation
233 Spring Street, New York, N.Y. 10013

PREFACE

Prior to the 9th International Conference on Reactivity Solids in Krakow, Poland a group of about 25 international scientists held a special conference entitled "Transport in Nonstoichiometric Compounds" in late Aug. 1980 in Mogilany, Poland (near Krakow). This conference was well received in view of the interaction between the participants, as well as the resulting publication of the proceedings (Elsevier Scientific Publishing Company, 1982, edited by J. Nowotny).

At this first conference the participants decided that it would be desirable to organize similar conferences at about two year intervals. Thus, a second meeting was held in late June, early July at Alenya, Pyrenees Orientales, France. This conference had a larger number of participants, about 50, but still managed to promote excellent interaction between all the participants. These proceedings, with editors G. Petot-Ervas, Hj. Matzke and C. Monty, have also been published by Elsevier as a special edition of the journal, Solid State Ionics, Vol. 12 (1984).

In view of the success of the initial two conferences, a third meeting was organized and held at The Pennsylvania State University, University Park, PA., 16802, U.S.A. from 11 June 84 to 15 June 84. The proceedings of this conference are presented in the following text.

It is perhaps well to note that the directors of this conference also limited the number of participants to about 50 in order to maximize mutual interaction. Additionally, because of the fact that the participants were essentially all internationally well recognized, presentation of all papers was held to about 30 minutes. This permitted sufficient interaction and discussion time during the 5 working days. At the end of the conference most participants felt that such a schedule appeared to work well.

It is a pleasure for the conference chairmen to express their appreciation to: Mrs. Anna Dorman who aided extensively in our search for financing the meeting and Mrs. Gretchen Leathers who aided much in carrying out the details of organizing the conference; to the organizing and international committee for their timely probes and aid; to Mrs. Linda Decker for her extensive typing aid; and especially to Prof. Emeritus W. O. Williamson for his thorough review of the published proceedings.

We extend our best wishes to the organizers of future "Transport in Nonstoichiometric Compounds" meetings and it is with much pleasure that we look ahead to these stimulating, scientific interactions.

George Simkovich
Professor of Materials Science

V.S. Stubican
Professor of Ceramic Science

CONTENTS

ELECTRICAL PROPERTIES

ELECTRICAL PROPERTIES

TRANSPORT IN MATERIALS CONTAINING A DISPERSED SECOND PHASE

J. Bruce Wagner, Jr.

Center for Solid State Science
Arizona State University
Tempe, AZ 85287

The purpose of this paper is to review the effects of a dispersed second phase on electrical and mass transport in solids. According to classical theories, when an insulator is added to a conductor, the electrical conductivity decreases. However, when the insulator particles are small and uniformly dispersed in a conducting matrix, just the opposite behavior is observed. Liang[1] first published an experimental study of the ionic conductivity of LiI containing a dispersion of fine Al_2O_3 particles. At room temperature, the ionic conductivity of LiI (35 - 40 m/o Al_2O_3) was increased by over an order of magnitude (see Fig. 1). Liang assembled batteries (cells) utilizing LiI (Al_2O_3) as the solid electrolyte and the open circuit voltage of the cells corresponded with the calculated from the cell reactions thus indicating a negligible electronic conduction in the LiI (Al_2O_3) composites. Jow and Wagner[2] studied the CuCl (Al_2O_3) system and found behavior similar to that for LiI (Al_2O_3) reported by Liang.

This phenomenon for composite solid electrolytes has been shown to be a general one and data are available for AgI (Al_2O_3)[3,4] AgCl (Al_2O_3)[5], LiCl (Al_2O_3)[6], CaF_2 (Al_2O_3)[7], CuBr (Al_2O_3)[8] and LiBr (Al_2O_3)[9]. Furthermore, other dispersoids have been used such as flyash and silica[10,11].

In essence, the enhancement in conductivity is proportional to the surface area of the added dispersoid. Thus for a given concentration, the smaller the particle size, the larger is the enhancement. Wagner[12] had earlier derived a quantitative expression for the enhanced conduction of a semiconductor containing a dispersion of a metallic conductor. Wagner's model involves a space charge layer surrounding each dispersoid particle.

3

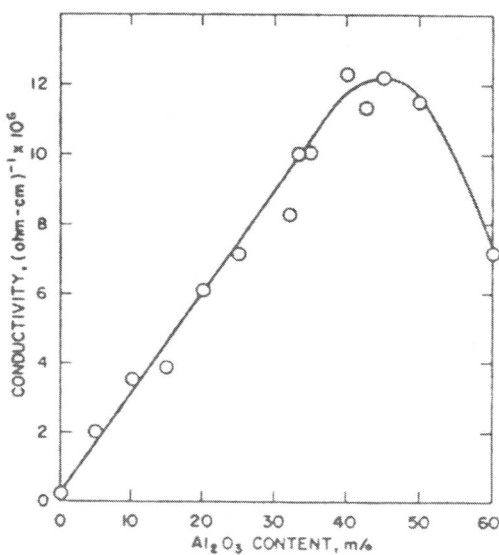

Fig. 1.
    Conductivity of LiI(Al$_2$O$_3$)
    as a function of Al$_2$O$_3$ conc-
    centration at 20°C. Liang
    [1]. Reprinted by permission
    of the publisher, the Electro-
    chemical Society.

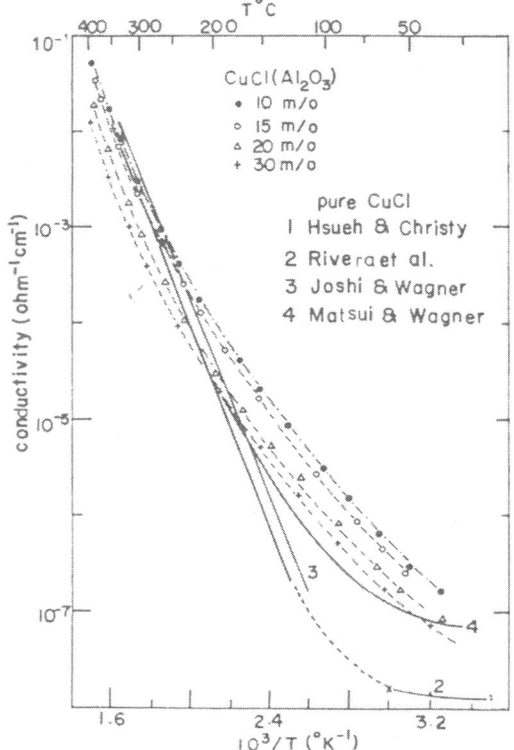

Fig. 2.
    Arrhenius plot of conductiv-
    ity of CuCl(Al$_2$O$_3$). Jow and
    Wagner [2]. Data of Hsueh
    and Christy [21]; of Rivera,
    et al. [22]; Joshi and Wagner
    [23] and Matsui and Wagner
    [24]. Reprinted by permis-
    sion of the publisher, the
    Electrochemical Society.

Jow and Wagner[2] developed a space charge model which fitted their data for CuCl (Al$_2$O$_3$) and Liang's data for LiI (Al$_2$O$_3$) in a semi-quantitative way. Their equation for the conductivity enhancement, $\Delta\sigma$, when the dispersoid concentration was low was:

$$\Delta\sigma \cong 3 \sum_i e \, \mu_i <\Delta n_i> [\lambda/r_1] \, (v/[1 - v]) \qquad (1)$$

where

$\quad$ e $\quad$ = electronic charge
$\quad \mu_i \quad$ = mobility of ith species
$\quad <\Delta n_i> \quad$ = average excess charge density within the space charge region
$\quad \lambda \quad$ = space charge layer thickness $\cong$ Debye length
$\quad$ v $\quad$ = volume fraction of dispersoid
$\quad r_1 \quad$ = radius of dispersoid

Jow and Wagner[2] tested the CuCl (Al$_2$O$_3$) system by measuring both the total a.c. conductivity and the electronic conductivity using d.c. polarization methods and Shahi and Wagner[3] performed similar experiments on AgI (Al$_2$O$_3$). It was surprising that the ionic conductivity of solid electrolytes is increased by adding a fine dispersion of insulator with no apparent change in the electronic conductivity.

The maximum enhancement occurs for samples with small concentrations of lattice defects. Thus the larger enhancements are found at low temperatures. Usually the plot of the isothermal conductivity at low temperatures versus concentration of added insulator particles yields a parabola. The maximum enhancement occurs for concentrations of insulator particles in the 15 to 50 volume percent range. At higher temperatures, the addition of insulator particles decreases the conductivity. This decrease is usually directly proportional to the volume of added insulator, i.e. the decrease appears to be a "blocking effect" caused by the volume of added insulator. Figs. 2 and 3 show typical Arrhenius plots for CuCl (Al$_2$O$_3$) and for AgI (Al$_2$O$_3$). Fig. 4 shows a plot of σ versus $(1/r_1)(v/[1-v])$ for CuCl (Al$_2$O$_3$) at 84°C. The straight line for low concentrations of alumina indicates behavior consistent with Eq. (1). Fig. 5 shows a plot of log σ versus log (particle size) for the CuCl (Al$_2$O$_3$) and for the AgI (Al$_2$O$_3$) systems (dissimilar temperatures were chosen to plot the available data on the same graph). A slope of minus one is also consistent with Eq. (1).

One of the unusual features of the electrolyte (insulator) composites is that an enhancement in σ is observed for very small additions of insulator. For example, adding 1, 2, 3, 4, ..., mole % Al$_2$O$_3$ (0.06μm) to AgI results in an enhancement from the first addition of insulator. There is no threshold concentration (percolation threshold)[4]. Secondly, there is an effect of moisture.

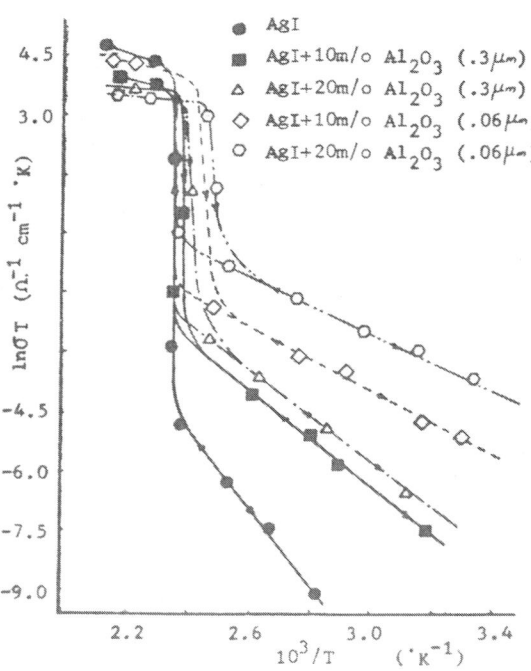

Fig. 3.
   Arrhenius plot of conductivity of AgI(Al$_2$O$_3$).
Chowdhary, Tare and Wagner
[4]. Reprinted by permis-
sion of the publisher, the
Electrochemical Society.

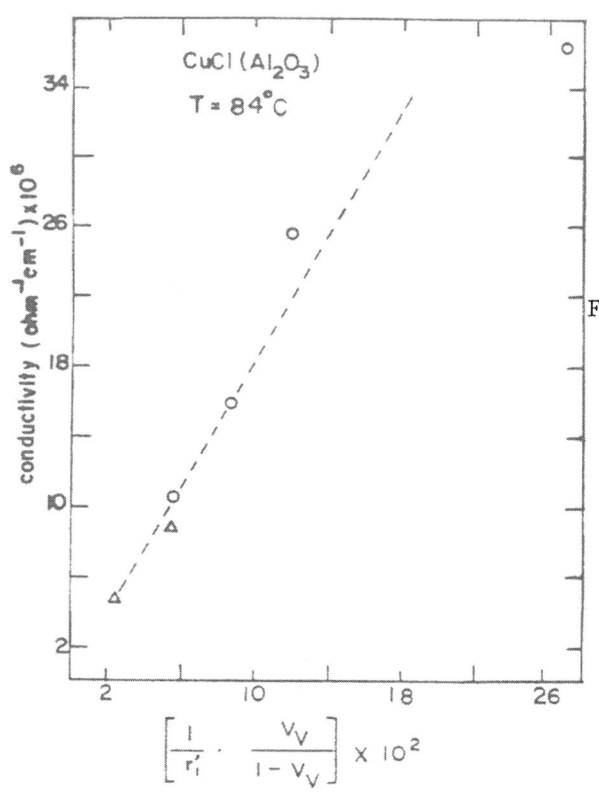

Fig. 4.
   Conductivity, σ for
CuCl(Al$_2$O$_3$) versus the
parameter (1/r$_1$)
[v/(1 - v)]. See Eq. (1).
Jow and Wagner [22].
Reprinted by permission
of the publisher, the
Electrochemical Society.

The Al_2O_3 which has been used in the as-received condition in this laboratory contains 3 to 5 weight % water. Adding Al_2O_3 in the as-received (moist) condition to AgI yields an order of magnitude greater enhancement in conductivity than adding dried Al_2O_3 of the same particle size. The drying procedure used in our laboratory is to heat the Al_2O_3 at about 750°C under a vacuum for 8 to 12 hours. While this treatment undoubtedly does not drive off all the water, it does remove the 3 to 5 weight % moisture so that comparison is useful. No explanation for the effect of moisture is available. Two suggestions are currently being tested in our laboratory. One suggestion is that the enhanced conductivity of AgI (moist Al_2O_3) may be due to transport by H^+ or some species other than Ag^+. We are currently testing the validity of Faraday's Law (a Tubandt experiment) on such composites. Secondly, the moisture may act to change the dielectric constant, ε, which appears in the expression for the Debye length. Currently we are testing pre-dried Al_2O_3 to which has been added liquids of varying dielectric constant. These materials are being mixed with AgI and the resulting conductivity measured to test for trends in σ versus dielectric constant of the liquid added to the Al_2O_3.

It has been found in our laboratory that for the same type and concentration of Al_2O_3 particles, the enhancement increases in the following order of preparation: cold pressed pellets of the mechanically mixed pwoders, hot pressed pellets of the mechanically mixed powders and pellets prepared from mixtures in which the electrolyte has been melted prior to pelletizing. These facts indicate that an intimate contact between the insulator and the matrix is important. Khandkar[5] prepared AgCl (Al_2O_3) composites by suspending the Al_2O_3 in an aqueous solution of $AgNO_3$ to which was added HCl. The AgCl precipitated onto the Al_2O_3. The resulting mixture was dried and pelletized. This composite exhibited a higher conductivity than similar AgCl (Al_2O_3) composites prepared by melting the AgCl phase. It was also shown that the grain size of the matrix, in this example, AgCl, affects the conductivity. Small particle sizes, 0.06 and 0.3 µm, of Al_2O_3 were used and the grain size of the pre-melted electrolyte matrix was of the order of 10 to 20 µm. In the case of the AgCl (Al_2O_3) formed by precipitation, the grain size of the AgCl was ~1.5 µm. A comparison of the behavior of samples prepared without Al_2O_3 showed the samples with the smaller grain size exhibited the higher conductivity. Thus in comparing data on composites from different laboratories, the grain size of the matrix material should be taken into account.

Silver iodide exhibits a Frenkel disorder. The high temperature α phase exhibits the "average structure" in which the silver ions are distributed among relatively immobile iodine ions in a bcc arrangement. The low temperature β phase has the wurtzite structure and the α-β transformation occurs at 147.5°C. When the

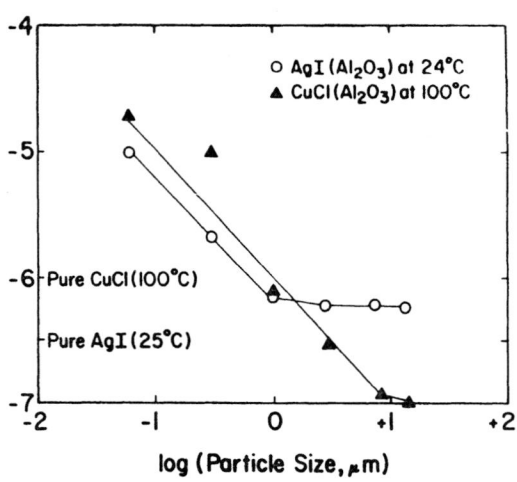

Fig. 5.
Log σ versus log (particle size) for CuCl containing 10 m/o dried Al_2O_3 at $100°C$ and for AgI containing 30 m/o of dried Al_2O_3 at $24°C$. The slope of the straight line is minus one in accord with Eq. (1). Chang, Shahi and Wagner [13]. Reprinted by permission of the publisher, The Electrochemical Society

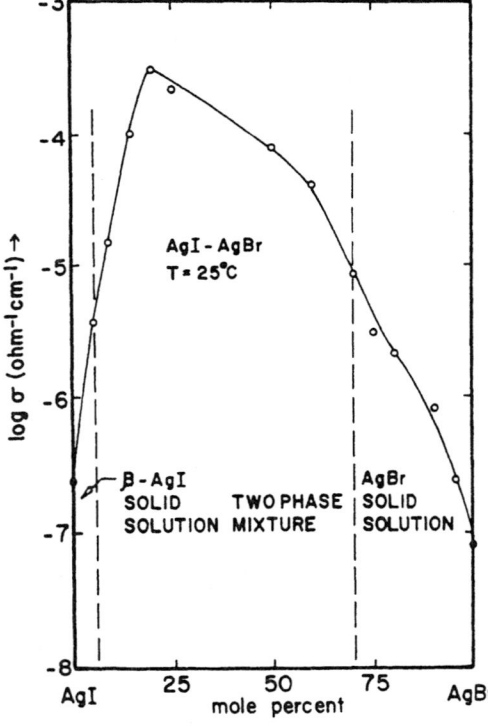

Fig. 6.
Log σ versus composition for AgI-AgBr system. Shahi and Wagner [10]. Reprinted by permission of Academic Press, publisher of J. Solid State Chem.

AgI (Al$_2$O$_3$) composite is heated from low temperatures, the β to α transition occurs at 147.5°C. However, when the composite is cooled, from above 147.5°C, the transition temperature is decreased[4]. Furthermore this decrease is proportional to the surface area of the Al$_2$O$_3$. Shahi and Wagner[3] used thermoelectric power measurements to infer cation vacancies in AgI (Al$_2$O$_3$) were the dominant current carrier at low temperatures. From an analysis of conductivity data, Khandkar[5] suggested cation vacancies were the dominant current carriers in AgCl (Al$_2$O$_3$) at low temperatures. If cation vacancies are the dominant carriers, then a working hypothesis for the conductivity changes and for the transition temperature change in AgI can be suggested as follows: at low temperatures additional cation vacancies are formed in AgI (Al$_2$O$_3$). Of course, there must be electroneutrality. At this time, the compensating species of positive charge is not known. According to Slifkin, et al.[14], and H. A. Hoyden[15] there is an excess of cation interstitials formed at a free surface. Our data which involve the Al$_2$O$_3$ - silver halide interface suggest cation vacancies are created at that interface. These conclusions are not necessarily in conflict with those of Slifkin and of Hoyden because the Al$_2$O$_3$ may act quite differently from a silver halide-vacuum or a silver halide-silver halide interface. We suggest that at low temperatures the excess cation vacancies in the Al$_2$O$_3$ composites remain localized at the interface while the bulk electrolyte has the equilibrium bulk concentration equal to the square root of the Frenkel constant. Thus on heating from room temperature to 147.5°C, the bulk halide transforms at the normal β–α temperature. Above 147.5°C, ion mobility is exceptionally high and the vacancies originally at the Al$_2$O$_3$ - AgI interface may react with cation interstitials which results in lowering the conductivity in the α phase as is observed (see Fig. 3). Next consider cooling the α - AgI (Al$_2$O$_3$). According to this model, there is a reduced concentration of cation interstitials in the bulk. According to Rice, Strassler and Tombs[16], there is a critical concentration of interstitials required to drive a phase transformation. In the case of α - AgI (Al$_2$O$_3$), the concentration of interstitials is decreased in the bulk. Accordingly, the composite must be cooled to a lower temperature, before the α–β transition occurs. This decrease in temperature on the cooling cycle is therefore proportional to the surface area of the added Al$_2$O$_3$ but on the heating cycle, the transition occurs at 147.5°C. Silver chloride has no phase transition in the solid state. However, the conductivity of AgCl (Al$_2$O$_3$) at low temperatures is enhanced while at high temperatures it is decreased relative to AgCl without Al$_2$O$_3$. The temperature at which the change from enhancement to decrease occurs depends on the surface area of Al$_2$O$_3$. This behavior is also consistent with our tentative working hypothesis that cation vacancies are introduced by the presence of Al$_2$O$_3$. They are localized at or near the interface at low temperatures but are mobile at the elevated temperatures and can annihilate cation interstitials at the elevated temperatures thus lower the electrical conductivity.

Shahi and Wagner[10] have studied the conductivity of the two phase AgI – AgBr system within the miscibility gap. No Al_2O_3 particles were added (see Fig. 6). Classically, the conductivity should be the weighted average of the conductivities of the two end terminal solid solutions, AgI saturated with AgBr and AgBr saturated with AgI. Actually, an enhancement of about a factor of fifty above the weighted average was observed (see Fig. 6). The grain size of the particles was about 15 μm. The conductivity is given by

$$\sigma = \sum_i e\, z_i\, \mu_i\, c_i \qquad (2)$$

where e is the electron charge, z_i the charge of the ith species, c_i the concentration and μ_i the mobility. To test whether only the mobility is affected owing to the large concentration of grain boundaries or whether the concentration of defects might be affected, drop calorimetric studies of the melted mixtures such as used in the conductivity studies on the one hand, and mechanical mixtures (unmelted) on the other were carried out by Khandkar, et al.[17]. Classically the difference between the enthalpies of these two should be zero, i.e. Excess Enthalpy = $[H_{518} - H_{298}]_{unmelted}$. However, an endothermic excess enthalpy was obtained as shown in Fig. 7. The maximum in the excess enthalpy occurred at about the 50 volume % concentration as would be expected from the maximum dissimilar interfaces. On the other hand the maximum in conductivity for the two phase mixtures occurs at ~25 mole % AgBr solid solution (approximately 25 volume%). This strongly suggests that the mobilities are also influenced in addition to the concentrations.

Carl Wagner[12] derived an expression for the electrical conductivity of a semiconductor containing a dispersion of a metallic conductor. V. B. Tare and Wagner[18] have studied the systems NiO containing a dispersion of Ni_3S_2 or a dispersion of nickel particles. In these cases the total conductivity is primarily electronic. The NiO is p-type according to:

$$1/2\ O_2\ (g) = O_0 + V_{Ni}^{z'} + zh\cdot$$

where a notation similar to that of Kroger and Vink[19] is used. In the NiO (Ni_3S_2) experiments, the NiO was annealed (separately) in air at $1000°C$ in order to fix the O/Ni ratios. The Ni_3S_2 is a metallic conductor[19]. Pellets were prepared. The particle sizes were ~5 μm for NiO and ~75 μm for Ni_3S_2. The conductivities of pure NiO, NiO containing sulfur in solid solution (no second phase) and NiO containing 16, 29, 50 and 67 volume % Ni_3S_2 plus pure Ni_3S_2 were studied. Fig. 8 shows Arrhenius plots for the NiO (Ni_3S_2) system. Fig. 9 shows an isothermal plot of log σ versus concen-

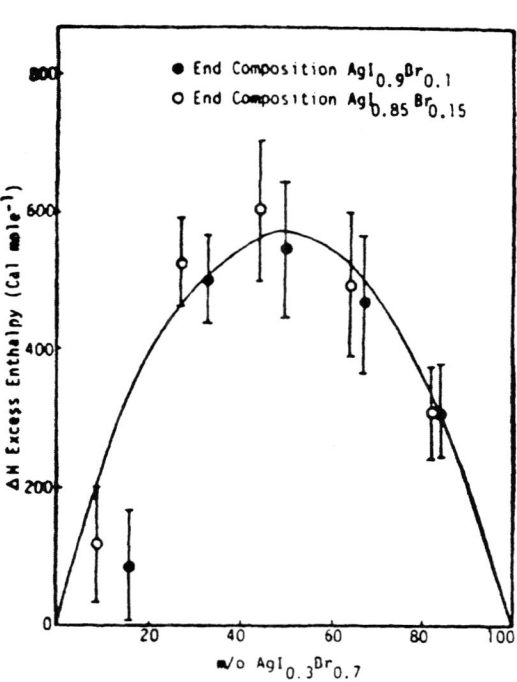

Fig. 7.
Excess Enthalpy as a function of composition for the two phase system AgI (solid solution with AgBr) and AgBr (solid solution with AgI). Khandkar, et al. [17]. Reprinted by permission of the publisher, the Electrochemical Society.

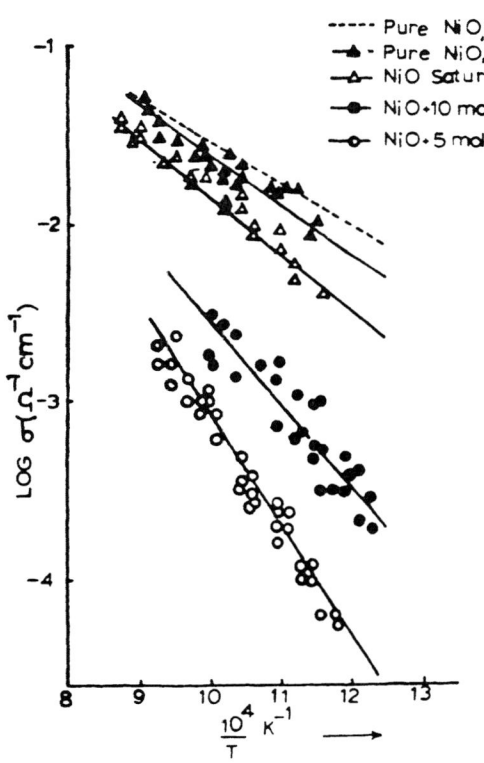

Fig. 8.
Electrical conductivity as a function of temperature for various NiO $-$ Ni_3S_2 compositions. Tare and Wagner [18]. Reprinted by permission of the American Physical Society.

tration of Ni_3S_2 at 600°C. Note that the conductivity of NiO decreases by a factor of 10^3, passes through a minimum and then increases. Seebeck measurements showed the predominant carriers for concentrations of Ni_3S_2 <16 volume % were electron holes while for higher concentrations the predominant carriers were electrons. These facts are explained as follows. Around each particle of Ni_3S_2 is a space charge layer of electrons. These electrons annihilate with the electron holes of the NiO. Consequently as Ni_3S_2 is added to NiO, the conductivity due to electron holes decreases, goes through a minimum (a virtual p to n transition at approximately 16 volume % Ni_3S_2), then increases as the concentration of electrons due to the Ni_3S_2 dominates. Eventually the NiO (Ni_3S_2) becomes a metallic conductor (~50 volume % at 600°C). A similar behavior was found for NiO containing Ni particles (see Fig. 10). In these experiments the NiO was pre-annealed at the Ni/NiO dissociation pressure. Classically the addition of small concentrations of nickel to NiO already equilibrated with nickel should result in no change in electrical conductivity.

The decrease in conductivity (see Fig. 10) is not as pronounced as in the NiO (Ni_3S_2) case because the NiO contained fewer electron holes having been pre-equilibrated at the Ni - NiO Phase boundary. The minimum in conductivity for NiO(Ni) is likewise shifted to lower concentrations of Ni than that of the NiO (Ni_3S_2).

Wagner[12] has derived an equation for the conductivity of a semiconductor containing a dispersion of metallic particles (at concentrations sufficiently high that p- to -n behavior was considered). His equation may be written as:

$$\sigma = \frac{22\varepsilon\kappa T\mu}{4\pi e} \cdot \frac{V}{r_1^2}$$

where ε is the dielectric constant of the matrix, μ the mobility of the electron, e the electron charge, V the volume fraction of metallic dispersoids, r_1 the radius of the metallic dispersoids and the other terms have their usual significance. Consequently for the case of a semiconductor containing a metallic dispersion, σ should be directly proportional to the volume fraction and inversely proportional to the square of the radius of the particles. The dependence on volume fraction for NiO(Ni) has been tested and confirmed (see Fig. 11). Tests of the dependence of conductivity on particle size are being carried out in the laboratory.

In conclusion we have summarized conductivity behavior in three types of two phase systems in which the second phase particles were small. In one type, it was shown that addition of a fine dispersion of an insulator to a solid electrolyte enhances the ionic conductivity with no apparent change in the electronic conductivity.

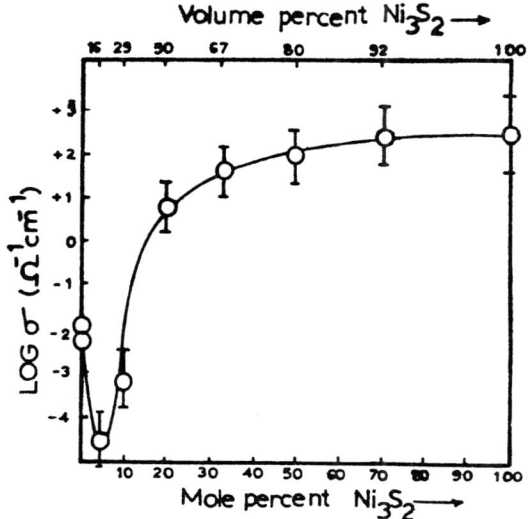

Fig. 9. Electrical conductivity as a
function of composition of $NiO-Ni_3S_2$
mixtures at 600°C. Tare and Wagner
[18]. Reprinted by permission of the
American Physical Society.

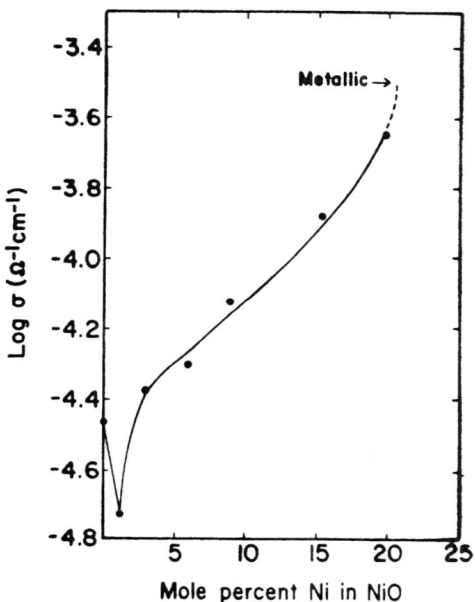

Fig. 10. Electrical conductivity of
NiO-Ni mixtures at 1000°K. Tare and
Wagner [18]. Reprinted by permission
of the American Physical Society.

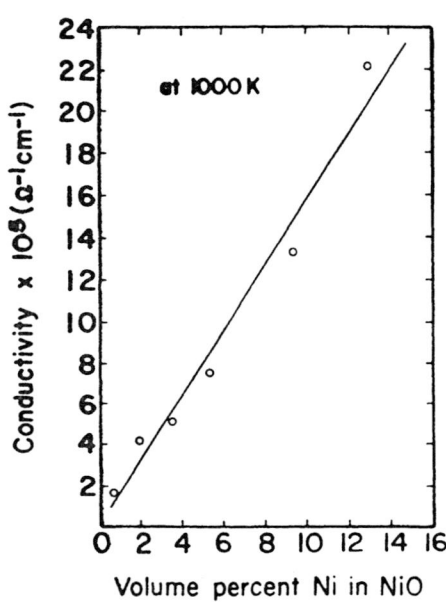

Fig. 11. Electrical conductivity of
NiO-Ni mixtures as a function
of volume percent Ni at 1000°K.
Tare and Wagner [18]. Reprinted
by permission of the American
Physical Society.

In the second, it was shown that co-existing phases within a miscibility gap may exhibit unusual enhanced conductivity when the grain sizes of the phases are small. Finally, when a metallic conductor is dispersed in a semiconductor, changes of both the carrier (electron holes or electrons) can be effected. All of these examples involve a space charge layer around the dispersoid and the magnitude of the effect depends on the surface area of the dispersoid.

REFERENCES

1. C. Liang, J. Electrochem. Soc. 120:1289 (1973).
2. T. Jow and J. B. Wagner, Jr., J. Electrochem. Soc. 126:1962 (1979).
3. K. Shahi and J. B. Wagner, Jr., J. Electrochem. Soc. 128:6 (1981).
4. P. Chowdhary, V. B. Tare and J. B. Wagner, Jr., J. Electrochem. Soc. (submitted), April, 1984.
5. A. Khandkar and J. B. Wagner, Jr., Abstr. No. 833 presented at the 163rd Meeting of The Electrochemical Society, San Francisco, CA, May 8 - 13, 1983.
6. C. Li-chuan, Z. Zong-yuan, W. Gang, L. Zi-rong, Abstract C209, Extended Abstracts of the Third International Meeting on Solid Electrolytes - Solid State Ionics and Galvanic Cells, September 15-19, 1980, Tokyo, Japan.
7. T. L. Wen, R. A. Huggins, A. Rabenau and W. Weppner, Revue de Chimie Minerale 20:643 (1983).
8. P. M. Dubec and J. B. Wagner, Jr., Materials Letters (in press).
9. A. Khandkar and J. B. Wagner, Jr., "Investigations on the LiBr-H_2O and on the LiBr (Al_2O_3) Systems," presented at the Second International Meeting on Lithium Batteries, Paris, France, April 25-27, 1984.
10. K. Shahi and J. B. Wagner, Jr., J. Solid State Chem. 42:107 (1982).
11. K. Shahi and J. B. Wagner, Jr., J. Solid State Ionics 3/4:295 (1981).
12. C. Wagner, J. Phys. Chem. Solids 33:1051 (1972).
13. M. R-W. Chang, K. Shahi and J. B. Wagner, Jr., J. Electrochem. Soc. 131:1213 (1984).
14. L. Slifkin, W. McGowan, A. Fukai and J. S. Kim, Photographic Sci. and Engn. 11:79 (1967).
15. H. A. Hoyden, paper A-1 page 1 in The Physics and Chemistry of The Silver Halide Crystal, reprints of papers at an International Conference at The University of Montreal, The Society of Photographic Scientists and Engineers (1972).
16. M. J. Rice, S. Strassler and G. A. Tombs, Phys. Rev. Lett. 32:596 (1974).

17. A. Khandkar, V. B. Tare, A. Navrotsky and J. B. Wagner, Jr., "The System AgI - AgBr: Energetic Consequences of Defect Equilibria in Single Phase and Two Phase Regions," J. Electrochem. Soc. (in press).

18. V. B. Tare and J. B. Wagner, Jr., J. Appl. Phys. 54:252 (1983); 54:6459 (1983).

19. H. J. Vink and F. A. Kroger in "Solid State Physics," Vol. 3 (1956) Ed. by F. Seitz and D. Turnbull, Academic Press, New York, p. 310.

20. H. Yagi and J. B. Wagner, Jr., Oxidation of Metals 18:41 (1983).

21. V. W. Hsueh and R. W. Cristy, J. Chem. Phys. 39:3519 (1963).

22. J. Rivera, L. A. Murray and P. A. Moss, J. Cryst. Growth 1:7 (1967).

23. A. V. Joshi and J. B. Wagner, Jr., J. Electrochem. Soc. 124:1071 (1975).

24. T. Matsui and J. B. Wagner, Jr., J. Electrochem. Soc. 124:610 (1977).

25. C. M. Osburn and R. W. Vest, J. Phys. Chem. Solids 32:1331 (1971); 32:1343 (1971).

THERMOGRAVIMETRIC AND ELECTRICAL CONDUCTIVITY STUDIES

OF Mg-DOPED $LaCrO_3$ AND La-DOPED $SrTiO_3$

B. K. Flandermeyer*, M. M. Nasrallah,
D. H. Sparlin and H. U. Anderson

Ceramic Engineering Department
University of Mo-Rolla
Rolla, MO 65401

I. INTRODUCTION AND MODELS

$SrTiO_3$ is known to be an n-type oxide (1-3) with predominantly oxygen vacancies. Fredrikse, et al. (4) report that it is a band type conductor, the band gap as determined by optical methods (5) is reported to be 3.15 ev (304 KJ/mole) at $0^{o}K$. The conduction band is believed to be composed of the 3-d orbitals of the Ti ions (6). $SrTiO_3$ shows a p-n transition due to the presence of acceptor impurities (3, 7).

X-ray diffraction studies on La doped $SrTiO_3$ indicate that it is a single phase up to 40% La additions (8); the latter is expected to act as a donor thus increasing the electron concentration and suppressing the oxygen vacancy concentration. Electrical conductivity measurements on La doped $SrTiO_3$ (9, 10), indicate the presence of strontium vacancies at high P_{O_2} with possible electronic compensation at low P_{O_2}. Thermogravimetric studies on the same system (11) indicate that $SrTiO_3$ displays a reversible weight loss of oxygen upon reduction which is sufficient to compensate the donor up to 20% whereas at high oxygen activity an uptake of stoichiometric excess of oxygen was observed. The excess oxygen absorbed may be accounted for by the presence of a crystallographic accommodation, possibly in the form of "SrO" layers interleaved within the perovskite structure leaving the oxygen sublattice intact in a manner similar to the Ruddlesdon-Popper type structure (12). Tilley (13) reports the presence of SrO layers accommodated in $SrTiO_3$ by the formation of lamellar type structures within the grains.

----

*Argonne National Laboratory, Argonne, Illinois

The model developed is based on the assumption that at high activity the absorbed oxygen reacts to form "SrO" type layers with the subsequent formation of V_{Sr}'' to compensate for the donor in the host lattice, whereas, at lower activities a controlled valency type compensation occurs by the formation of Ti^{3+} resulting in a sharp increase in electrical conductivity, according to the reaction,

$$Sr_{1-y-x}La_y^{\cdot}V_{Sr_x}''Ti_{y-2x}'Ti_{1-y+2x}O_3 + "SrO" \underset{\text{red.}}{\overset{\text{oxid.}}{\rightleftarrows}} \quad (1)$$

$$Sr_{1-y}La_y^{\cdot}Ti_y'Ti_{1-y}O_3 + \frac{x}{2}O_2$$

where y represents the dopant concentration and x the excess oxygen. This reaction can be reduced to:

$$"SrO" + 2Ti + V_{Sr}'' \underset{\text{oxid.}}{\overset{\text{red.}}{\rightleftarrows}} \quad Sr + 2Ti' + \tfrac{1}{2}O_2 \text{ (g)} \quad (2)$$

where La_{Sr}' denotes La^{3+} on a substitutional Sr^{2+} site, Ti' represent either an electron associated with Ti^{4+} or an electron in the conduction band and V_{Sr}'' is an ionized Sr vacancy. The equilibrium constant for the reaction is given by:

$$K_1 = \frac{[Sr][Ti']^2}{[SrO][Ti]^2[V_{Sr}'']} P_{O_2}^{\frac{1}{2}} \quad (3)$$

If the assumption is made that the activities can be replaced by mole fractions of the respective terms and that "SrO" is not a separate phase but is due to structural accomodation which yields not only "SrO" but Sr vacancies then the expression for the equilibrium constant becomes

$$K_1 = \frac{(1-y+x)(y-2x)^2}{x^2(1-y+2x)} P_{O_2}^{\frac{1}{2}} \quad (4)$$

which with little loss in accuracy can be approximated to

$$K_1 = \frac{(y-2x)^2}{x^2} P_{O_2}^{\frac{1}{2}} \quad (5)$$

Rearrangement of this equation yields an expression for the excess oxygen as function of K_1 and P_{O_2}

18

$$x = \frac{y\, P_{O_2}^{+\frac{1}{4}}}{(K_1)^{\frac{1}{2}} + 2P_{O_2}^{\frac{1}{4}}} \qquad (6)$$

The electrical conductivity $\sigma = e\mu n$, where the carrier concentration $n = y=2x$ which can now be related to K_1 and P_{O_2}

$$\sigma = \frac{e\mu y\, (K_1)^{\frac{1}{2}}}{(K_1)^{\frac{1}{2}} + 2P_{O_2}^{\frac{1}{4}}} \qquad (7)$$

At low P_{O_2}, $n = y$ and the amount of excess oxygen becomes

$$x = \frac{y}{(K_1)^{\frac{1}{2}}}\, P_{O_2}^{\frac{1}{4}} \qquad (8)$$

However, at high P_{O_2}, $x = y/2$, and

$$\sigma = \frac{e\mu y}{2}\, (K_1^{\frac{1}{2}})\, P_{O_2}^{-\frac{1}{4}} \qquad (9)$$

This indicates that x/y approaches $\frac{1}{2}$ and is independent of P_{O_2} at high oxygen activities (Eq. 6) whereas at low oxygen activity x/y possesses a $\frac{1}{4}$ power dependence on P_{O_2} (Eq. 8).

The model is extended to predict the electrical conductivity as expressed in Eqs. 7 and 9. However, the properties of a band type conductor need to be considered. In contrast to a small polaron type conductor where the temperature dependence is due to the energy barrier seen by the carriers, that of a band conductor is either associated with the scattering and trapping phenomena or with the excitation of carriers. In either case the carrier mobility μ is expressed as:

$$\mu = \frac{eL}{m^*V} \qquad (10)$$

where L is the carrier mean free path, e the electron charge, m^* is the carrier effective mass and V is the carrier velocity. If several mechanisms are operative, then L should be expressed in terms of such mechanisms, for example phonon scattering, impurity scattering, trapping, etc. The composite carrier mean free path is accordingly expressed as:

$$L = \frac{1}{r_p T + r_y N_g + r_{V_{Sr}} N_{V_{Sr}} + r_t N_t + D} \qquad (11)$$

where r_p, r_y, $r_{V_{Sr}}$ and r_t represent the cross section of phonons, dopant, strontium vacancies and traps (where a band electron is excited and falls into a trap) respectively. N_g, $N_{V_{Sr}}$ and N_t are the concentrations of dopant, strontium vacancies and trapping sites respectively, T is the absolute temperature and D is a constant associated with the inherent defects. Generally, L will be dominated by one of these mechanisms depending on the set of conditions provided.

Lanthanum chromite, on the other hand, is known to be a p-type oxide (14) and becomes nonstoichiometric through the formation of cation vacancies. It is one of the promising materials for high temperature electrode applications and fuel cell interconnects (15). However, volatilization and possibly corrosion impose certain limitations on its use. Several studies have been reported (16, 17, 18) relating its chemical stability and cation stoichiometry to volatilization and electrical conductivity.

The electrical conductivity of $LaCrO_3$ is essentially due to the 3d band of the Cr ions (19, 20), thus an increase in conductivity is expected due to lower valence substitution on either the La^{3+} or Cr^{3+} sites resulting in the formation of Cr^{4+}. If such a substitution is compensated by the formation of oxygen vacancies, no additional contribution to the electronic conductivity is anticipated. Karim and Aldred (21) have performed extensive electrical conductivity, Seebeck and magnetic susceptability measurements on pure and Sr-Doped $LaCrO_3$ and present convincing evidence that a small polaron conduction mechanism applies. Meadowcroft (16) observed a reversible decrease in conductivity of three orders of magnitude between the oxidized and the reduced states. Therefore, under oxidation Mg-doped $LaCrO_3$ has the high conductivity expected of a p-type material which has been acceptor doped to enhance the concentration of holes but then under reducing conditions displays an unexpected decrease in carrier concentration. Apparently a change from electronic to ionic compensation of the acceptor is occurring at low P_{O_2}'s.

In this study Mg^{2+} was used as the acceptor dopant and assumed to enter the $LaCrO_3$ structure substitutionally for Cr^{3+}. Assuming p-type disorder in nonstoichiometric undoped $LaCrO_3$ and that all defects are fully ionized and the stoichiometric A/B ratio is maintained, the defect reaction under oxidizing and reducing conditions can be represented by:

$$(12)$$

$$LaCr_{1-2y}Cr^{\cdot}_y Mg'_{Cr_y} O_3 \xrightleftharpoons[\text{oxid.}]{\text{red.}} LaCr_{1-2y+2x}Cr^{\cdot}_{y-2x}V^{\cdot\cdot}_{o_x} Mg'_{Cr_y} O_{3-x} + \frac{x}{2}O_2$$

or

$$O_O + 2Cr^{\bullet}_{Cr} \rightleftarrows 2Cr_{Cr} + V^{\bullet\bullet}_O + \tfrac{1}{2} O_2 \tag{13}$$

The equilibrium constant for this reaction is given by:

$$K_{12} = \frac{[Cr_{Cr}]^2 \, [V^{\bullet\bullet}_O]}{[Cr^{\bullet}_{Cr}]^2 \, [O_O]} P_{O_2}^{\tfrac{1}{2}} \tag{14}$$

and by substitution in terms of mole fractions

$$K_{12} = \frac{(1-2y+2x)^2 x}{(y-2x)^2 \, (3-x)} P_{O_2}^{\tfrac{1}{2}} \tag{15}$$

which can be approximated as

$$K_{12} = \frac{y}{(y-2x)^2} P_{O_2}^{-\tfrac{1}{2}} \tag{16}$$

The electrical conductivity $\sigma = e\mu p$, where the carrier concentration $p = y-2x$, which when combined with Eq. 16 becomes

$$\sigma = \frac{e\mu}{4K_{12}} P_{O_2}^{\tfrac{1}{2}} [(8yK_{12}P_{O_2}^{-\tfrac{1}{2}} + 1)^{\tfrac{1}{2}} - 1] \tag{17}$$

At the high P_{O_2} limit (x=0) Eq. 17 reduces to $\sigma = e\mu y$, whereas at low P_{O_2},

$$p = (\frac{y}{2K_{12}})^{\tfrac{1}{2}} P_{O_2}^{\tfrac{1}{4}} \tag{18}$$

or

$$\sigma = e\mu \, (\frac{y}{2K_{12}})^{\tfrac{1}{2}} P_{O_2}^{\tfrac{1}{4}}$$

II. EXPERIMENTAL

A series of La doped $SrTiO_3$ and Mg doped $LaCrO_3$ (0-20 a/o) compounds was prepared by the liquid mix process (22). In all cases stoichiometric A/B ratio was maintained, i.e., (Sr + La)/Ti or La/(Cr + Mg) = 1.

Thermogravimetric measurements were made in a TG system designed to measure weight changes on a 70-80 gm powder sample to an accuracy of ± 1 mg (± 6×10^{-5} moles oxygen). Measurements were conducted at 1000° to $1400^{\circ}C$ at an oxygen activity of $10^{-12} - 10^5$ pascals. The oxygen activity was monitored by a calibrated zirconia sensor.

Electrical conductivity measurements were conducted on rectangular bars with four embedded Pt wires welded to Pt leads and attached to a four terminal digital voltmeter. Conductivity measurements were carried out at 1000-1400°C in an oxygen activity range of 10^{-12}-10^{5} pascal.

III. RESULTS AND DISCUSSION

Figure 1 illustrates typical results for $Sr_{0.96}La_{.04}TiO_3$. As can be seen, to maintain a given level of excess oxygen, higher oxygen activities are required as the temperature increases from 1200 to 1400°C. Figure 2 shows that within our experimental range at 1350°C the amount of excess oxygen absorbed at a given activity is dependent upon the dopant concentration. The equilibrium constant for the oxidation-reduction reaction (Eq. 1) of donor doped $SrTiO_3$ is given by:

$$K_1 = 1.8 \times 10^6 \exp \frac{(-302 \pm 9 \text{ KJ/mole})}{RT} \qquad (19)$$

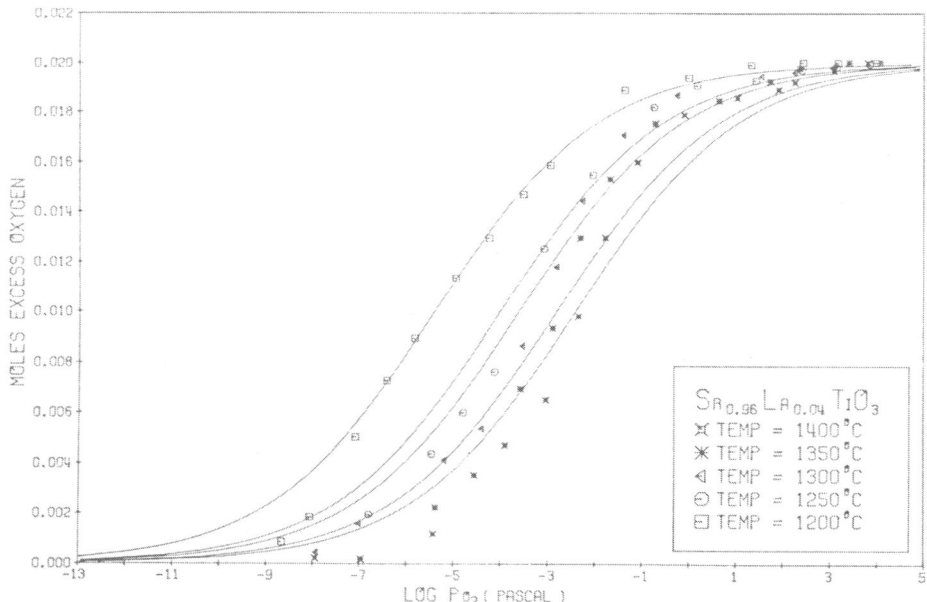

Figure 1. Plots of moles oxygen absorbed per mole sample vs. log P_{O_2} at various temperatures for 4 a% La-doped $SrTiO_3$. The curves displayed are plots of Equation (4) using experimentally determined equilibrium constants.

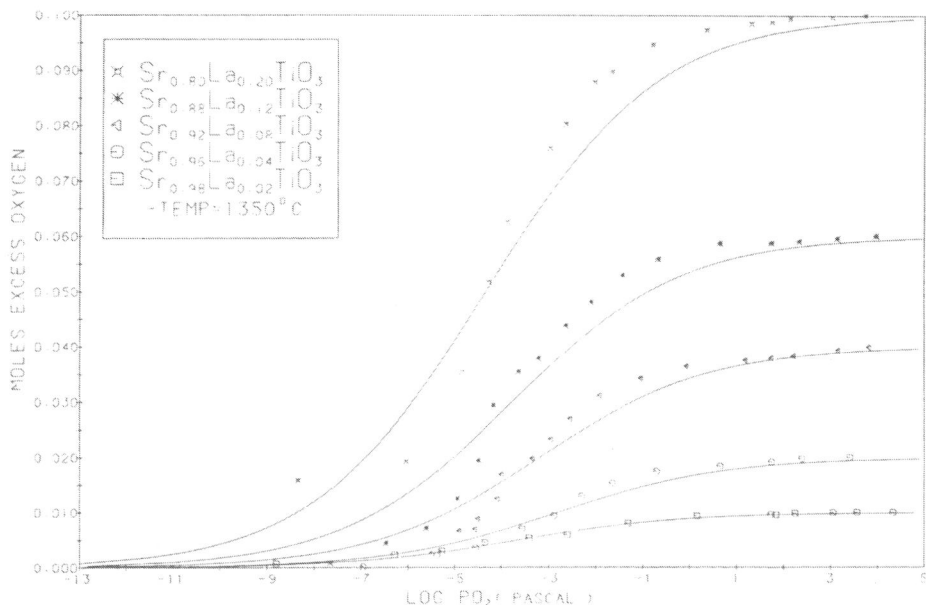

Figure 2. Plots of moles oxygen absorbed per mole sample vs. log
P_{O_2} for various La-dopant levels at 1350°C. The curves
displayed are plots of Equation (4) using experimentally
determined equilibrium constants.

Typical conductivity results are plotted as log conductivity
versus log P_{O_2} for 12 a/o La doped $SrTiO_3$ in Figure 3. The curves
displayed are plots of Equation (7) which is the conductivity predicted by the model using the K_1 values determined from the TG
experiments and assuming exact compensation of the donor with no
detectable change in carrier mobility. While these results agree
well with Balachandran and Eror (10) both in magnitude and slope,
the fit to the model using the TG determined equilibrium constants
is not good. This discrepancy is further confirmed by the conductivity versus y-2x plots shown in Figures 4 and 5 as a function
of temperature and dopant respectively. It becomes apparent that
either the carrier concentration or mobility is thermally activated
and reciprocally related to the concentration of V_{Sr}''. Activation
energies were determined to range from 48-125 KJ/mole, (0.5-1 3ev)
with the mean value of 58 KJ/mole (0.6ev). If the mobility is
thermally activated, the temperature dependence should be due to
changes in the concentration of scattering and trapping centers as
expressed in equation 11, however this does not provide a reasonable
explanation for the observed temperature dependency of the conductivity. If, on the other hand, the carrier concentration is
assumed activated, then the observed behavior can be explained if

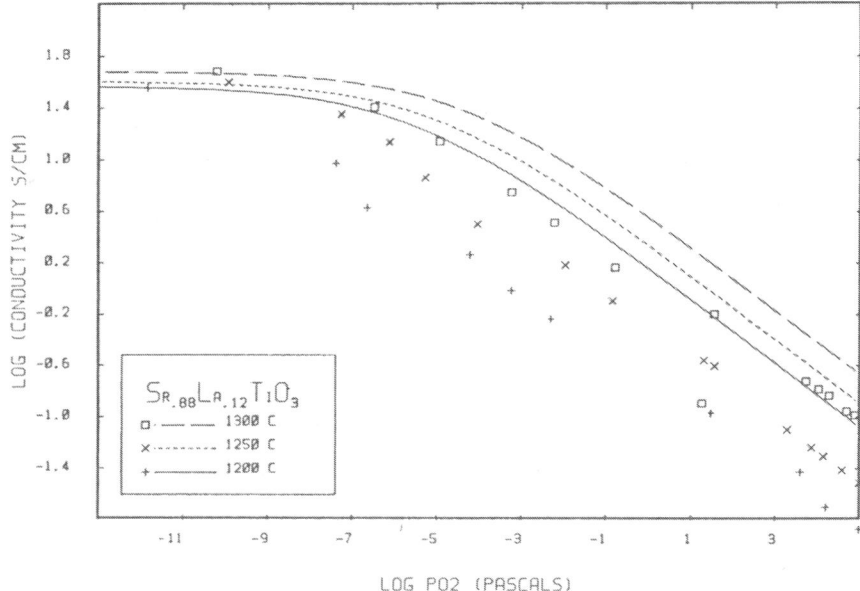

Figure 3.   Plots of log conductivity vs. log P_{O_2} at 1200°, 1250°, and 1300°C for 12 a% La-doped $SrTiO_3$. Data was taken from the oxidized to the reduced state. The curves displayed are plots of Equation (7) using the K_1 values from equation 19.

the La^{3+} donor energy levels are 58 KJ/mole (0.6ev) away from the band edge and that the carriers are provided by the thermal ionization of the donors. The V''_{Sr} could still act as scatters and the carrier concentration would then be:

$$n = (y-2x) \exp \left(\frac{-58 \text{ KJ/mole}}{RT} \right) \qquad (21)$$

This implies that $n \cong 0.01$ of that expected, and the mobility accordingly becomes $\sim 10^{-4}$ m^2/V.Sec. in agreement with published data (4). Strong evidence for the existance of V''_{Sr} has been provided by our analytical electron microscopy data (23) where inclusions of Sr rich phase were detected in La doped $SrTiO_3$. These inclusions were observed to be surrounded by a relatively high concentration of facetted cavities that appear to be due to vacancy clustering. Figure (5) generally indicates that as the dopant content increased the conductivity increased at low oxygen activity and decreased at high oxygen activity in accordance with Equation (7) and that the expected compensation was only achieved at the limits.

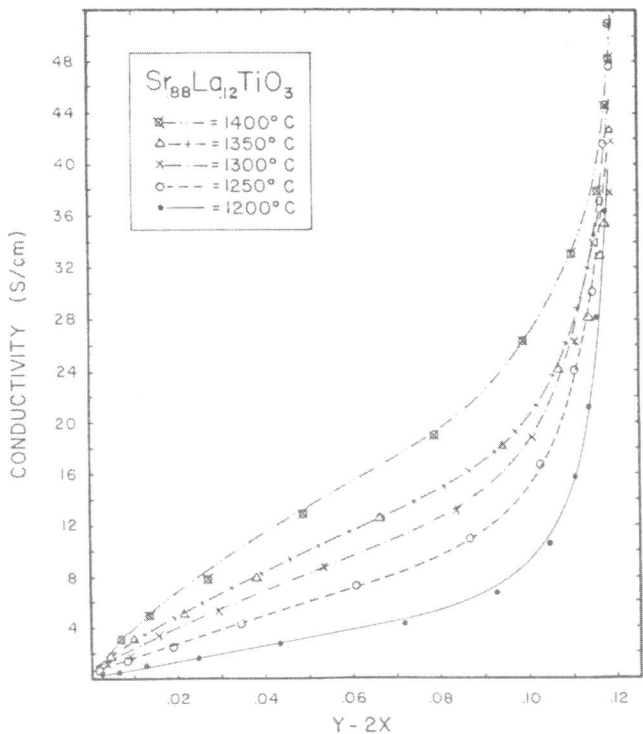

Figure 4. Plots of conductivity vs. (y-2x) at various temperatures
for 12 a% La-doped $SrTiO_3$.

Typical TG results for the p-type oxide, $LaCrO_3$, as a function
of temperature and dopant are displayed in Figures 6 and 7 respectively. The total amount of oxygen absorbed or evolved is
approximately equal to half the total moles of Mg present,
which implies that full compensation of the dopant is being achieved.
No irreversible weight loss was detected for any of the samples
within the sensitivity range of the system. The solid curves shown
are the values predicted from the proposed models.

Figure (6) illustrates the results obtained for $LaMg_{0.10}Cr_{0.90}$
O_3, equilibrated at temperatures between 1000°C to 1368°C. The
figure shows that the degree of nonstoichiometry is both temperature
and P_{O_2}, dependent. It increases with temperature at any given P_{O_2}.
Furthermore, the same degree of nonstoichiometry is achieved at
lower P_{O_2} as the temperature decreases.

The dependence of oxygen deficiency on Mg concentration and
P_{O_2} is illustrated in Figure (7) for data obtained at 1304°C. This
figure shows that the degree of nonstoichiometry increases as the

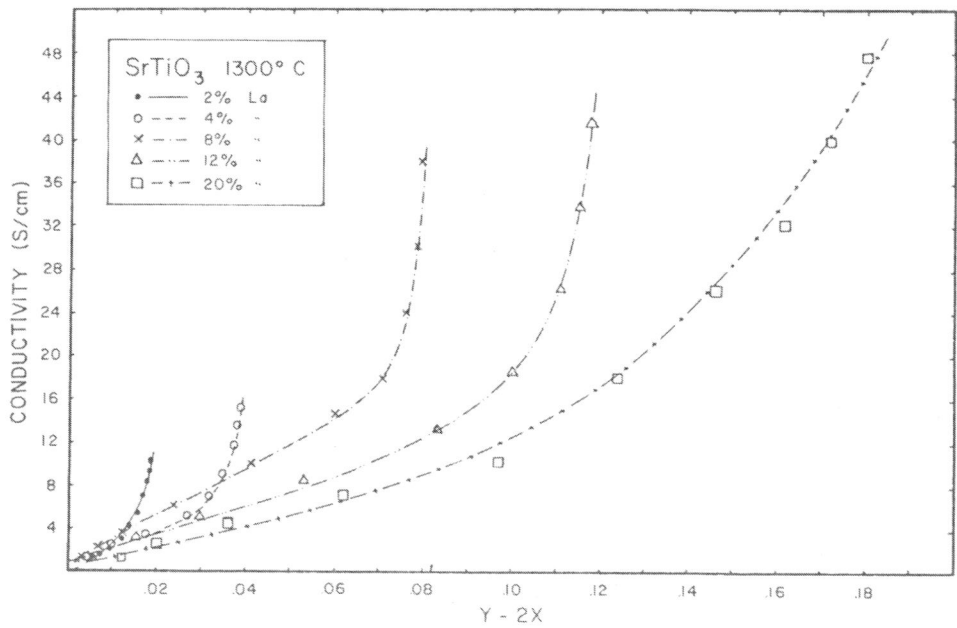

Figure 5. Plots of conductivity vs. (y-2x) for various La-dopant levels at 1300°C.

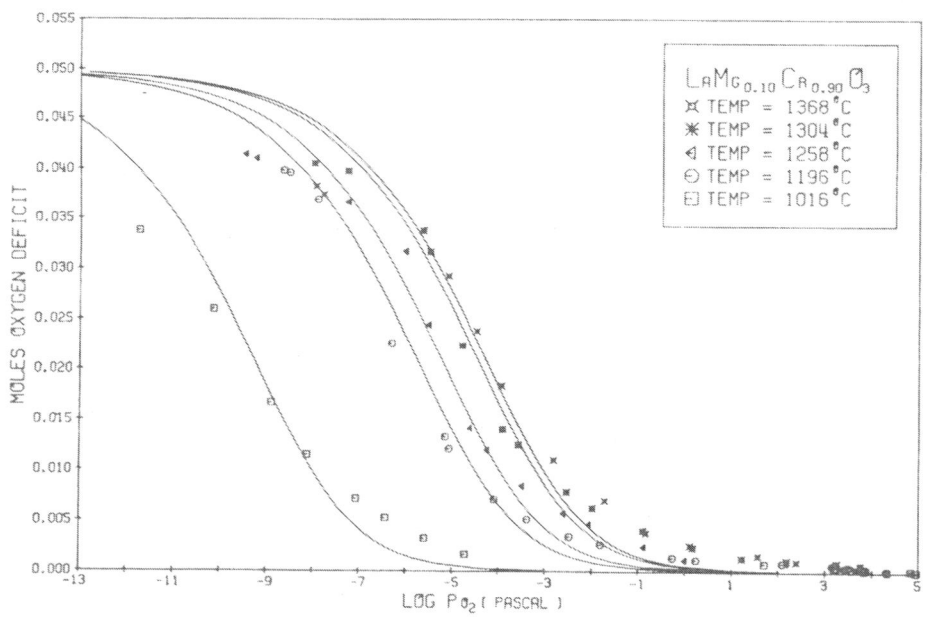

Figure 6. Moles oxygen weight loss per mole sample vs. log P_{O_2} at various temperatures for 10 a% Mg-doped $LaCrO_3$. The curves displayed are best fit solutions to Equation (16).

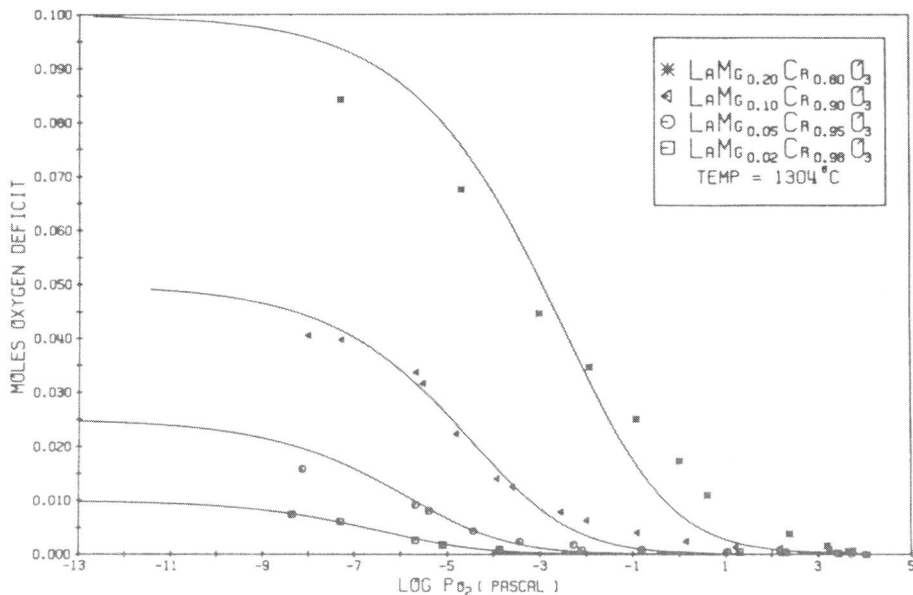

Figure 7. Moles oxygen weight loss per mole sample vs. log P_{O_2} at various dopant levels for 1304°C. The curves displayed are best fit solutions to Equation (16).

amount of dopant increases at any given oxygen activity within the experimental range of 2-10 a% Mg and 1000 to 1400°C. The equilibrium constant for the oxidation-reduction reaction (Eq. 12) of acceptor doped $LaCrO_3$ is accordingly given by:

$$K_{12} = 2.1 \times 10^4 \exp \frac{(-272 \pm 16 \text{ KJ/mole})}{RT} \qquad (22)$$

(where 272 ±16 KJ/mole is the enthalpy of formation of $V_O^{\cdot\cdot}$)

Figure (8) displays typical conductivity results plotted as log conductivity versus log P_{O_2} and compared to a plot of Equation (18), using the equilibrium constants determined by the TG data. While there is considerable scatter present and the truncation of the reaction restricts the ionically compensated region, there still is good correlation to the model throughout the entire composition range except for the 20% Mg sample where the dopant concentration appears to exceed the solubility limit (24) and deviation from the model was observed. The activation energy for the mobility was determined from Arrhenius plots of ln (σT) vs 1/T and found to be 22.3 ± 7 KJoule/mole (0.20 ± 0.07ev) which is in reasonable agreement with Karim and Aldred's (21) data. However our mobility values appear to be about an order of magnitude lower than those for the

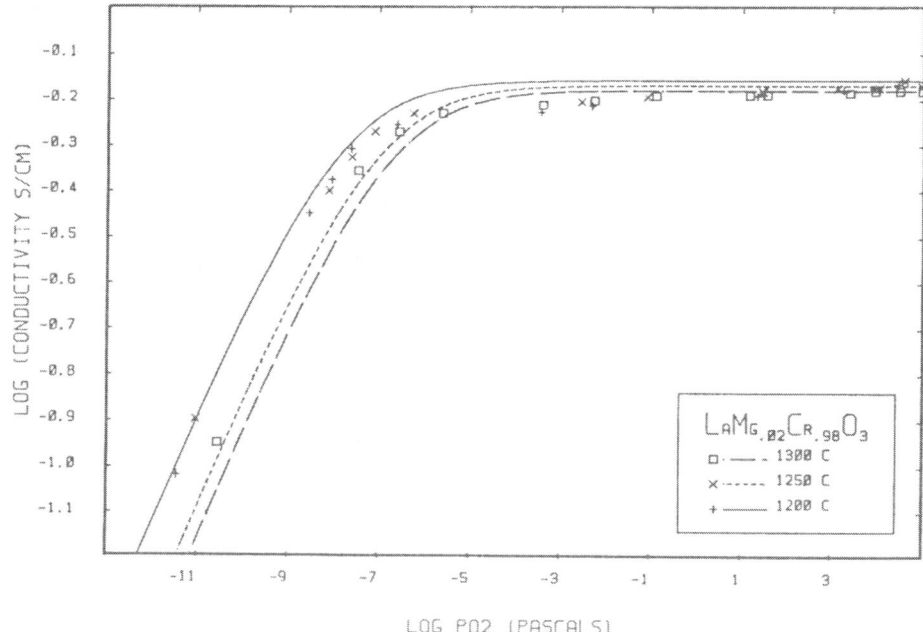

Figure 8. Plots of log conductivity vs. log P_{O_2} at 1200°, 1250°, and $1300^{\circ}C$ for 2 a% Mg-doped $LaCrO_3$. The curves displayed are plots of Equation (18) using the K_1 values from Equation 22.

Sr doped samples implying that Mg which substitutes for Cr on the "B" sites is significantly different from Sr substituting for La on the "A" sites.

A plot of carrier concentration (y-2x) versus conductivity is shown in Figure (9) which confirms the assumption that the $V_O^{..}$ directly compensates for Mg ions and that the mobility does not change under the experimental conditions. Our data show a slight depression at about 80-90% of the maximum expected carrier concentration. This might be due to a band component that exists in doped but uncompensated $LaCrO_3$ and scattering or trapping by $V_O^{..}$. Otherwise, the results show a linear dependence of carrier concentration on conductivity for y-2x values less than 0.05 regardless of the dopant level, implying that the mobility is constant and approximately equal $5 \times 10^{-6} m^2/V.sec$. which is about a factor of 5 lower than that reported by Karim and Aldred (21) ($2.5 \times 10^{-5} m^2/V.sec$. for 2% Sr). This tends to agree with the higher conductivities that they observed.

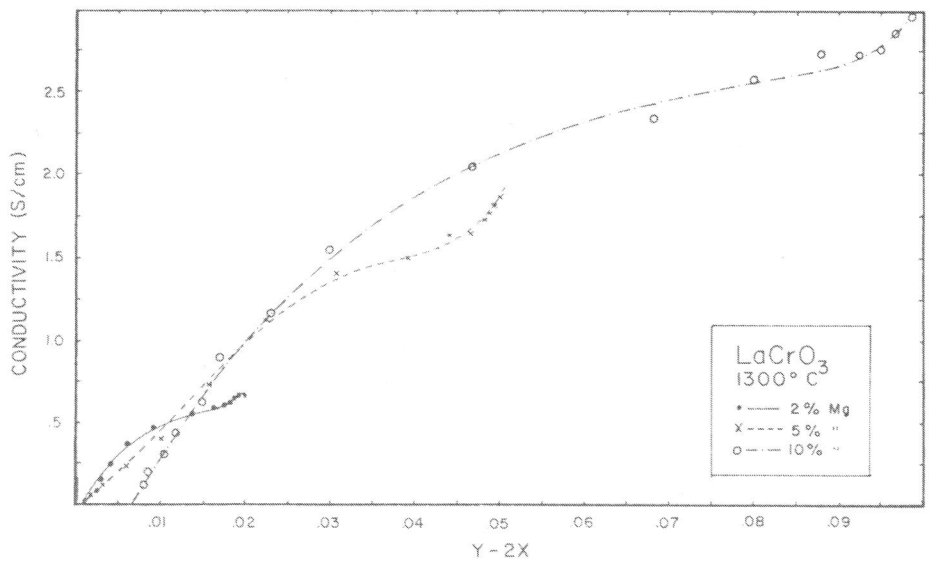

Figure 9. Plots of (y-2x) vs. conductivity for various levels of
Mg-doped LaCrO$_3$ at 1300°C.

CONCLUSIONS

In general the data obtained from TG experiments support the
models proposed for the oxidation-reduction behavior of both acceptor doped LaCrO$_3$ and donor doped SrTiO$_3$. In the case of LaCrO$_3$
the electrical conductivity data follows the model well and cor-
relates with a small polaron conduction mechanism. For SrTiO$_3$,
however, the agreement between theory and the electrical conductivity data is not nearly as good. The lack of agreement appears
to be related to the fact that SrTiO$_3$ is a band conductor and that
the La^{+3} donors are not completely ionized except at high temperature.
These results suggest that the La^{+3} donor levels may be as much as
0.6 ev below the band edge and are not fully ionized even at 1200°C.
Due to the lack of other evidence this result must be considered
only tentative.

ACKNOWLEDGEMENT

The authors wish to thank the Basic Energy Science Division
of the Department of Energy (Grant No. DEA-C02 80ER10598400) for
support of this research.

REFERENCES

1. L. C. Walters and R. E. Grace, J. Phys. Chem. Solids 28:239 (1967).
2. H. Yamada and G. R. Miller, J. of Solid State Chemistry 6:169 (1973).
3. N. H. Chan, R. K. Sharma, and D. M. Smyth, J. Electrochem. Soc. 128 (8):1762 (1981).
4. H. P. R. Frederikse, W. R. Thurber and W. R. Hosler, Phys. Rev. A, 134:442 (1964).
5. H. W. Gandy, Phys. Rev. 113:796 (1959).
6. A. H. Kahn and Leyendecker, Phys. Rev. A, 135:1321 (1964).
7. U. Balachandran and N. G. Error, J. of Solid State Chemistry 39:351 (1981).
8. B. Odekirk, U. Balachandran, N. G. Eror and J. S. Blakemore, Mat. Res. Bull. 17:199 (1982).
9. U. Balachandran and N. G. Eror, Comm. Am. Cer. Soc. C64:75 (1981).
10. U. Balachandran and N. G. Eror, J. Electrochem. Soc. 129:1021 (1982).
11. N. G. Eror and U. Balachandran, J. of Solid State Chemistry 40:85 (1981).
12. S. N. Ruddlesdon and P. Popper, Acta Cryst. 11: 54 (1958).
13. R. J. D. Tilley, J. Solid State Chemistry 21: 293 (1977).
14. D. B. Meadowcroft, Proc. Intl. Conf. on Strontium Containing Compounds, Halifax, N. S., T. Gray, Ed., Atlantic Res. Inst. p. 119 (1973).
15. D. B. Meadowcroft and J. M. Wimmer, Bull. Amer. Ceram. Soc. 58:610 (1979).
16. D. B. Meadowcroft, Brit. J. Appl. Phys., Ser. 2, 2:1225 (1969).
17. H. U. Anderson et al., "Conference of High Temperature Sciences Related to Open Cycle, Coal-fired MHD Systems," Argonne National Laboratory, April 4-6, 1977, Argonne, IL.
18. H. U. Anderson et al., "High Temperature Solid Oxide Fuel Cells," May 5-6, 1977, Brookhaven Natl. Laboratory, Upton, N.Y.
19. I. G. Austin and N. E. Mott, Adv. in Phys. 18:41 (1965).
20. J. Faber, M. Mueller, W. Procarioni, A. Aldred and H. Knott, "Conference of High Temperature Sciences Related to Open Cycle Coal-fired MHD Systems," Argonne National Laboratory, April 4-6, 1977, Argonne, IL.
21. D. P. Karim and A. T. Aldred, Phys. Rev. B, 20:2255 (1979).
22. M. Pechini, Methods of Preparing Lead and Alkaline Earth Titanates and Niobates and Coating Methods Using the Same to Form a Capacitor, U.S. Pat. 3,330,597 (1967).
23. J. Bentley, M. M. Nasrallah and H. U. Anderson, "Electron Microscopy of Donor doped $SrTiO_3$," Presented at the Am. Cer. Soc. 86th Annual Meeting, Pittsburgh, PA (1984).
24. D. Schilling, M. S. Thesis, University of Missouri-Rolla (1984).

EXPERIMENTAL STUDY OF MoO_3 ELECTRICAL CONDUCTIVITY CHANGES UNDER LOW OXYGEN PRESSURES

A. Steinbrunn, H. Reteno and J.C. Colson

Laboratoire de Recherches sur la Réactivité des Solides
C.N.R.S. LA 23 - Faculté des Sciences Mirande
B.P. 138 - 21004 Dijon Cedex (France)

INTRODUCTION

MoO_3 forms a series of non-stoichiometric oxides on heat treatment in vacuum or in reducing atmospheres, and the departure from stoichiometry is generally associated with the formation of colour centers. The basic mechanism leading to colour center formation has been suggested to be the trapping of electrons at molybdenum (Mo^{5+}) sites in the vicinity of oxygen vacancies. Owing to the simple description of the MoO_3 structure (MoO_6 octahedra linked by sharing corners) Kihlborg, Magneli, Bursill[1-3] have shown that the sub-oxides structures can be deduced from that of MoO_3 by shear transformations. These latter can explain the discrete values of the composition range MoO_{3-x} such as Mo_nO_{3n-1} (Mo_4O_{11}, Mo_5O_{14} ...) Mo_nO_{3n-2} ($Mo_{18}O_{52}$) and Mo_nO_{3n-m+1} ($Mo_{17}O_{47}$). Nevertheless, the exact mechanism of reduction is still under discussion (4,5). As for the electrical conduction properties of MoO_3, the analysis of previous work shows that the experimental study is not easy to carry out from thin films as well as from single crystals. The MoO_3 electrical conductivity values lie in the 10^{-15} to 10^{-8} Ω^{-1} cm^{-1} range respectively for temperatures equal to $-60°C$ and $20°C$. The activation energies (from 0.29 to 1.83 eV) depend on the atmosphere of crystal growing and heating. In this work, we made more detailed electrical measurements from a great number of crystals. Their crystallization was carefully carried out under reproductible conditions (T and P_{O_2}). A twinned layer model is proposed to explain the time dependence of the electrical conductivity. The platelet crystal loses lattice oxygen continuously through its largest face (010). A subsequent non-stoichiometric layer is built up by an oxygen diffusion process. An electron hopping conduction model between localized sites (Mo^{5+}) seems the more suitable to ex-

31

plain our results in agreement with that established by Sayer
et al. (6).

EXPERIMENTAL PROCEDURE

Preparation and characterization of MoO_3 crystals

MoO_3 crystals were grown by physical vapor transport and chemical reactive sublimation in an oxygen atmosphere. MoO_3 sublimation
is very easy in vacuum at low temperatures (T < 800°C), but preparing crystals in the stoichiometric composition range needs to rea-
lise the oxide vapor transport in an oxygen atmosphere (typically
p_{O_2} = 350 Torr, 1 Torr = 133.3 Pa). The crystal growing apparatus
consists of a vertical quartz tube (ϕ = 50 mm) evacuated by a classical pumping system (limit 10^{-6} Torr) and it is connected to a gas
manifold via a leak valve. Pressure is monitored by a cold cathode
gauge and a manometer (Bourdon type). A temperature gradient is ensured by two superimposed annular furnaces between $T_1 \cong 780°C$ and
$T_2 \cong 680°C$. MoO_3 powder (Merck for analysis) was deposited into a
quartz crucible positioned in the 780°C region. MoO_3 crystal
growth occurs on the tube walls. The source to crystal distance was
30 cm. The thin plate-like crystals have a translucent yellow colour. The X-Ray diffraction and I.R. spectroscopy analyses give
spectra which are characteristic of the orthorhombic form of MoO_3
(A.S.T.M. 5-0508). The crystals selected for conductivity measurements have an average thickness of 0.05 mm. They are between 5 and
15 mm long and about 5 mm wide. Their largest face is parallel
to the (010) crystallographic plane (figure 1).

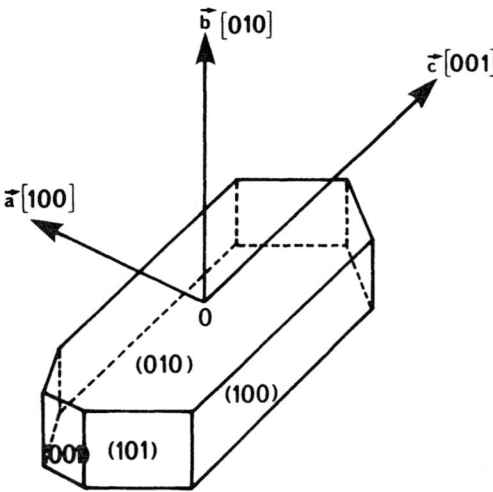

Fig. 1. Characteristic habit of orthorhombic sublimed molybdenum
        trioxide crystals.

32

Repeated Auger Electron Spectroscopy (A.E.S.) analyses of (010) surface of a crude MoO_3 crystal show the expected Auger peaks of Mo and O. No impurity, except carbon, is detected. This impurity is removed by heating the surface in vacuum at 350°C during 20 min. As by further heating, carbon is no more observed, this element can be regarded as a typical superficial contaminant. Within the experimental uncertainty of our A.E.S. analyses, one can consider that pure MoO_3 single crystals have been prepared. Their purity is at least equal to the source material i.e. :

Cl   : max 0.002 %
SO_4 : max 0.01   %
Pb   : max 0.001 %
Fe   : max 0.0005 %
NH_4 : max 0.005 %
P, As, Si : max 0.002 %

Reflexion High Energy Electron Diffraction (R.H.E.E.D.) analysis was performed from the MoO_3 (010) surface (figure 2). A simple rectangular expected unit cell is obtained i.e. : a = 3.96 Å, c : 3.69 Å. The modulation intensity along the R.H.E.E.D. diffraction streaks indicates that the surface is smooth. Some Kikuchi lines are visible on the original patterns which reveal the good crystallinity of the prepared MoO_3 crystals.

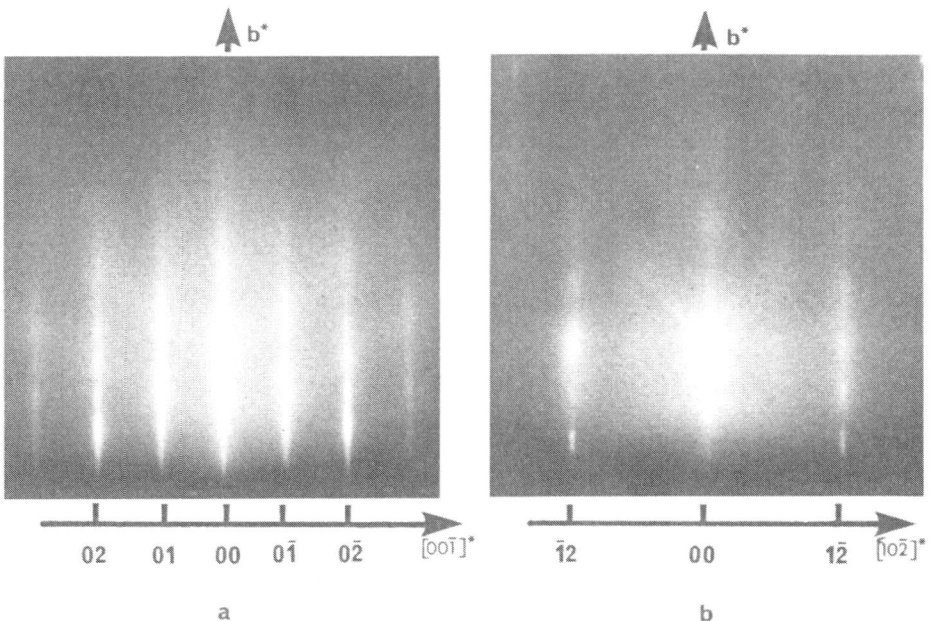

Fig. 2. R.H.E.E.D. patterns of MoO_3 surface.
(a) observation azimuth [100]
(b) observation azimuth [201]

Measurements of the electrical conductivity

Electrical measurements were made by the direct current (D.C.) two probe method. Figure 3 shows the electrical conductivity cell especially designed for that study. The electrical contacts to crystals were made by using an evaporated gold thin film ($\sim$ 1 μm thick) on each side of the monocrystal (figure 4).

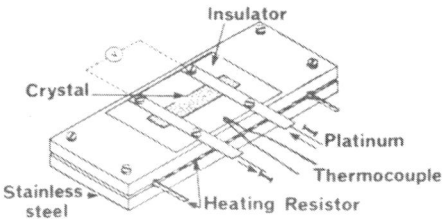

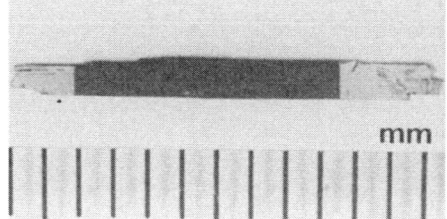

Fig. 3. Schematic view of the electrical conductivity cell.

Fig. 4. MoO_3 crystal with its evaporated gold contacts.

Smooth platinum electrodes tightly bound to the thin plate-like sample ensure that electrical and mechanical contacts are maintained. The whole support was heated by means of a thermocoax resistor. Temperature was monitored using a Pt/PtRh 10 % thermocouple which was positionned as close to the sample as possible. D.C. conductivity measurements were carried out by using a simple circuit by application of Ohm's law. The stabilized current I was delivered by a Keithley 225 source supply, while the potentiel V was continuously recorded by a Keithley 616 electrometer. The samples are maintained in a quartz vessel in which gas can be streamed by means of an adjustable leak valve in the pressure range 10^{-1} to 10^{-6} Torr monitored by a Bayard Alpert ionization gauge.

RESULTS

The electrical conductivity of the MoO_3 single crystals is found to be unstable, especially on heating in vacuum. The gas phase composition strongly affects the σ measurements - we chose two reproducible, but quite different atmospheres to study the time dependence of σ : on the one hand, a high oxygen pressure atmosphere ($p_{O2} \cong 350$ Torr), and on the other hand, a very low oxygen atmosphere which consists of the residual atmosphere of our vacuum apparatus ($p \cong 10^{-6}$ Torr). For this latter, it is reasonable to admit that P_{O_2} is quite within 10^{-6} Torr as the rarefied gas-composition is essentially due to H_2, CO, CO_2 and hydrocarbons (oil DC 704 diffusion pump). We regret to have no data about the exact value of the oxygen potential surrounding the sample. All these experiments were performed for a series of nearly stoichiometric MoO_3 crystals (translucent yellow). Figure 5 shows the I-V characteristics for as-grown MoO_3 samples recorded in the 200-350°C temperature range in vacuum. The I-V curves are linear within a few percent for an applied voltage of 0.2 to 20V. This linearity strongly suggests that the ohmic regime is satisfied for our experimental conditions.

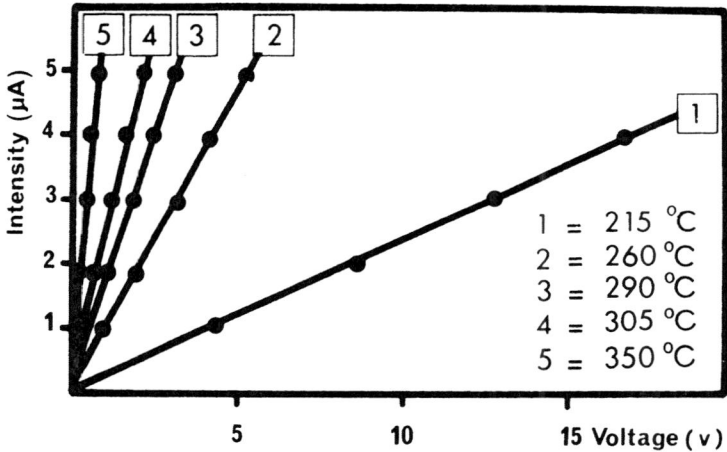

Fig. 5. (I-V) characteristics at different temperatures for as-grown MoO_3 crystals.

Oxygen pressure dependence of σ

For high oxygen pressure (350 Torr) and relatively high temperature, 340°C, the electrical conductivity σ of the MoO_3 single crystals is found to be quite stable ($\frac{\Delta\sigma}{\sigma} < 1$ % for a two hours' heating). The σ value deduced from the experimental conductance G shows that the sample is quite an insulator ($\sigma = 10^{-6}$ $\Omega^{-1}.cm^{-1}$ at room temperature). Performing measurements of oxygen pressure dependence of σ in the 10^{-5} to 10^{-3} Torr pressure range, we noticed that the time stabilization of the electrical signals is not satisfactory in agreement with other authors (7 to 9). In that case an accurate variation law of σ as a function of oxygen pressure is difficult to establish. All our measurement attempts failed. Nevertheless, for a given temperature, we noted that σ decreases when oxygen pressure increases. The n-type semi-conductor behaviour of MoO_3 was qualitatively corroborated during our preliminary measurements.

Time dependence of σ on heating in vacuum

Measurements on as-grown samples freshly mounted in vacuum show a fairly good stability which depends on the heating temperature. For a temperature within 250°C and for heating time not exceeding 2 hours, the changes of σ are stable with time. But on increasing the temperature the σ values increase with time. Figure 6 shows the isothermal variation of the relative conductance G/G_O with time, for different single crystals at 340°C. These data were continuously recorded during the heating time in the residual vacuum. The general shape of the curves obtained suggests a parabolic law for the changes of G/G_O with time. Indeed, when plotting G/G_O values against the square roots of time one gets the linear curves shown on figure 7.

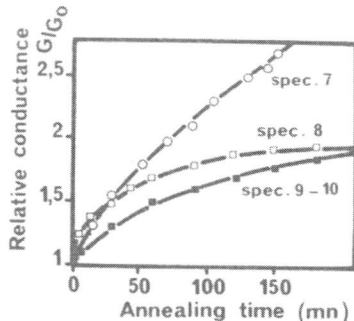

Fig. 6. Relative conductance G/G_O dependence with time for specimens 7 to 10 (table 1).

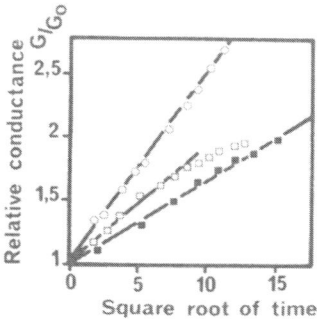

Fig. 7. G/G_O plotted against the square root of time. T = 340°C.

Temperature dependence of σ in vacuum

As the σ values taken during the increase and decrease of the temperature do not coincide, the following experimental protocol was performed in order to achieve experimental conduction activation energy measurements.

- The investigated temperature are always within the highest of the thermal cycle (i.e. 350°C).

- The as-grown samples are heated slightly from 100°C to 250°C through successive plateau temperature values. For heatings beyond 250°C the times are reduced at their maximum to prevent oxygen loss from the samples.

- When the maximal temperature T_M = 350°C is reached, the sample has undergone an approximately 3 hours' treatment in vacuum

- From this T_M, the temperature is progressively decreased.

The typical evolution of σ with the temperature is shown on figure 8.

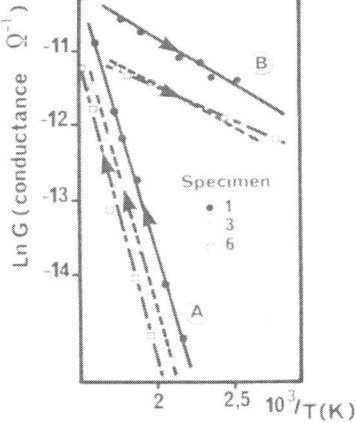

Fig. 8. Plot of the logarithm of the conductance G as a function of reciprocal temperature.
Curve A : increase of temperature.
Curve B : decrease of temperature.

Temperature increase and decrease are shown by arrows. (A) curves correspond to the increasing temperature while (B) curves to the decreasing temperature. The slopes of the curves are very different but in any case the electrical conductivity obeys the classical law :

$$\sigma = \sigma_o \, T^{\beta} \, \exp - \frac{E}{kT}$$

Our experimental results do not allow the choice between $\beta = 0$ and $\beta = -1$ because the experimental points' alignement obtained by plotting ln GT versus T^{-1} is within the same error range. The average experimental activation energy for nearly stoichiometric MoO_3 is $E_1 = (0.70 \pm 0.06)$ eV in agreement with some literature data (6, 8, 10 - 13). After heating in vacuum (curve B) the activation energy diminishes down to $E_2 = (0.10 \pm 0.03)$ eV. Table 1 summerizes our results.

Table 1. Conductivity activation energy of as-grown MoO_3 crystals before and after vacuum annealings.

Samples	Initial E_{act}	E_{act} after vacuum annealing	G/G_0 after 6h at $T \cong 300°C$	Crystal growth conditions
1	0.70 eV	0.07 eV	30.9	T. Gradient
2	0.75 eV	-		780-680°C
3	0.63 eV	0.07 eV	18.8	p_{O2} = 150 Torr
4	0.76 eV	-		
5	0.61 eV	0.11 eV	18.6	T. Gradient
6	0.66 eV	0.11 eV	34.6	750-650°C
				p_{O2} = 340 Torr
7-8	-	0.07 eV	-	T. Gradient
9-10	-	0.09 eV	-	770-700°C
11	0.76 eV	0.12 eV	-	p_{O2} = 340 Torr

Morphological modification of the single crystals during heating in vacuum

Not only do the values of σ and activation energy change during heating in vacuum but so does the colour of the crystals. Initially yellow they were bluish after a 2 hours' heating at 340°C. This change from yellow to blue is ascribed to a small deficiency in oxygen. Indeed, no difference between the X-Ray patterns could be found between the blue and yellow crystals, both having the orthorhombic structure. A superficial topographical change is observable by optical microscopy using a polarised Nomarski interferometer. The (010) surface is roughened by extended heating in vacuum. Some "cross hatched" figures are observable as well as fracture lines along the [001] crystallographic axis in agreement with those during reduction processes mentioned by different authors (3-5).

INTERPRETATION AND DISCUSSION

Despite the experimental difficulties mainly due to the high fragility of the MoO3 single crystals, these preliminary results strongly suggest that some of the weakly bound oxygen can be removed from the crystal without destroying the MoO3 lattice. A mild heating at $\cong 300°C$ in vacuum is sufficient for changing the characteristic colour of the near stoichiometric MoO3 samples. A re-equilibration is achieved essentially by a departure of oxygen into the gas phase. The subsequent formation of oxygen vacancies $V_O^{..}$ and colour centres occurs. The centres can be described as trapped electrons at Molybdenum sites (Mo^{5+}) in the vicinity of oxygen vacancies. When $V_O^{..}$ enter the lattice, two Mo^{5+} ions will be formed in order to maintain electrical neutrality. These Mo^{5+} sites can be considered as electrons bonded to Mo^{6+} sites. Thus, a correlation between colour and stoichiometry of MoO3 seems established (14).

The time dependence of the relative conductance of MoO3 crystals can be interpreted because the crystal is so thin a plate that the oxygen exchange through edge planes can be considered as negligible.

Thus, the sample can be regarded as a twinned layered monocrystal. The external layer has a conductivity value σ higher than that of the inner one σ_0, its thickness grows with heating time in vacuum. These parallel layers behave like two parallel resistances and therefore the equivalent conductance G of the whole crystal is :

$$G = G_1 + G_2$$

where $G_1 = \sigma \dfrac{x.1}{L}$ and $G_2 = \sigma_0 \dfrac{(y_0 - x)1}{L}$

y_0 = initial thickness of the crystal.

x = thickness of external layer at time t.

1 = width of the crystal.

L = length of the crystal.

Hence, the relative conductance at time t is reduced to :

$$\frac{G}{G_O} = 1 + \frac{\sigma - \sigma_O}{\sigma_0.y_0} . x \qquad (1)$$

and the ratio $\dfrac{G - G_O}{G_O}$ is proportionnal to the thickness x of the superficial substoichiometric oxide layer if one considers that $\dfrac{\sigma - \sigma_O}{\sigma_0.y_0}$ is constant with time. Therefore, the time dependence of G/G_O would show the overgrowth of a non-stoichiometric $MoO_{3-\delta}$ layer parallel to the (010) crystallographic plane, whose thickness x depends on time according to a parabolic law. This growth kinetic rate suggests that the departure of oxygen is a diffusion controlled process from the bulk towards the surface.

Indeed, for such a system, it is easy to show that the thickness of the growing layer obeys the following law :

$$x^2 - x_o^2 = 2 \zeta D (C_1 - C_o) t. \qquad (2)$$

where

x_o : layer thickness at time t = 0.

x : layer thickness at time t.

D : oxygen diffusion coefficient.

ζ : crystal volume for a MoO_3 mole.

C_1 : $V_o^{\cdot\cdot}$ concentration close to the external interface.

C_o : $V_o^{\cdot\cdot}$ concentration for nearly stoichiometric MoO_3.

t : time.

Thus, the time dependence of electrical conductivity measurements is thought to be caused by the facility with which MoO_3 can be converted to the substoichiometric $MoO_{3-\delta}$, where δ is a small fraction. The $MoO_{3-\delta}$ layer formed by the departure of weakly bound oxygen has an electrical conductivity σ higher than that of the stoichiometric MoO_3. The insertion of oxygen vacancies into the MoO_3 lattice (i.e. : Mo^{5+} sites) seems to increase the electrical conductivity. The relative conductance values (G/G_o) obtained for different crystals (Table 1) reach a maximum at about 30 after a 6 hours' vacuum annealing (for H_2 or H_2S exposure at the same temperature, G/G_o is respectively equal to 10^3 and 10^4). Using the vacancies concentrations proposed by Gai (15) : $Co = 5.7 . 10^{-4}$ and $C_1 = 2.7 . 10^{-2}$ one can estimate the $MoO_{3-\delta}$ layer thickness x. The dispersion of the literature data for the diffusion coefficient D (10^{-14} to $10^{-16}.cm^2.sec^{-1}$) (4, 16) leads to x values within 150 nm (1500 Å). Thanks to relation 1, for given σ_o and y_o, the calculation of the electrical conductivity of the $MoO_{3-\delta}$ layer is possible. We obtain σ in the 1 to 10 $\Omega^{-1}.cm^{-1}$ conductivity range for $\sigma_o = 10^{-4} \Omega^{-1}.cm^{-1}$ (340°C) and $y_o = 50$ μm. By comparing these estimated values to the reported conductivities of non stoichiometric MoO_3, we noticed that they were in fairly good agreement with the reported data of Gruber (17) for $MoO_{3-\delta}$ phases with $\delta < 0.25$. Consequently, the oxygen departure in the superficial layer, 150 nm thick, would not exceed 0.25 with respect to the stoichiometric composition while annealing a MoO_3 crystal during 3 hours at 340°C in vacuum.

For nearly stoichiometric MoO_3, the intrinsic conductivity has an activation energy of about 1.8 eV since the band gap is equal to 3.6 eV (10 - 12). As the conductivity measurements were performed in vacuum, a value slightly within 1.8 is expected, while the activation energies determined from several samples are equal only to about 0.7 eV. The interpretation of that deviation is possible if the present results are combined with previously reported data and if we admit that :

(i)   : the electrical conduction is related to the presence of Mo^{5+}.

(ii) : oxygen vacancies ($V_O^{..}$) play an important part in both conductivity and optical absorption (colour).

(iii) : the D.C. activation energy varies between 0.07 and 0.76 eV according to the defect concentration.

(iv) : the carriers are electrons which jump in between localized sites (mainly Mo).

One can consider, as well, that the crystal is modified by a non-reversible transformation on heating in vacuum. The 0.10 eV activation energy may be attributed to the extrinsic conductivity of a degenerate semi-conductor having a donor level at 0.20 eV just below the conduction band. This explanation cannot be rejected for the $MoO_{3-\delta}$ external layer as OH^- groups could be formed by interaction with the residual-atmosphere. As for the bulk nearly stoichiometric MoO_3, the previous assumptions are less able to explain the average value of 0.7 eV for the activation energy reported by many authors. The specimens prepared by various methods, have a different stoichiometry. Their Mo^{5+} sites concentrations are different and therefore the electrical conductivity data present some deviation. Nevertheless, the proposed conduction model (hopping between Mo^{5+} sites) is compatible with impurity dissolution. When their concentration is low enough, they are localized and hopping conduction between sites distributed at random is possible. The previous remarks enable one to understand the discrepancies of the literature data concerning the electrical conductivity for thin films and bulk poly-monocrystalline samples.

CONCLUSION

The conduction process in MoO_3 can be described in terms of the hopping of localized carriers between Mo sites as postulated for many transition metal oxides. For the nearly stoichiometric MoO_3, the presence of electrical carriers is linked with the existence of oxygen vacancies $V_O^{..}$. The hopping activation energy for pure MoO_3 is (0.70 ± 0.06) eV. The present results for single crystals show that oxygen can be easily removed from MoO_3. Under mild heating in vacuum, weakly bound oxygen diffuses from the bulk towards the surface. A substoichiometric $MoO_{3-\delta}$ layer, parallel to the largest crystal face (010), is built up. A twinned layer model accounts for the electrical conductivity of the $MoO_{3-\delta}$ substoichiometric layer over the nearly stoichiometric substrate layer due to a higher defect concentration ($V_O^{..}$, OH^-, ...) as well as probably to the precipitation of planar defects. As for $MoO_{3-\delta}$ with $\delta < 0.25$ the choice for the conduction model is more complicated because of the assumption about the nature of the point and planar defects :

this solid is very non-homogeneous. The conductivity activation energy is then equal to (0.10 ± 0.03) eV.

REFERENCES

1. L. Kihlborg and A. Magneli, Acta Chem. Scand. 9:471 (1955).
2. L. Kihlborg, Advan. Chem. Ser. 39:37 (1963).
3. L. A. Bursill, Proc. Roy. Soc. A 311:267 (1969).
4. W. Thoni and P. B. Hirsch, Phil. Mag. 33:4,639 (1976).
5. P. L. Gai, Phil. Mag. 43:4,841 (1981).
6. M. Sayer, A. Mansingh, J. B. Webb and J. Noad, J. Phys. C.: Solid State Phys. 11:315 (1978).
7. P. Stahelin, G. Busch, Helv. Phys. Acta 23:530 (1950).
8. V. A. Ioffe, J. B. Patrina, E. V. Zelenetskaya and V. P. Mikheeva, Phys. Stat. Sol. 35:535 (1969).
9. M. A. Khilla, Z. M. Hanafi, B. S. Farag and A. Abu-el-Sand, Thermochemica Acta, 54:35 (1982).
10. S. K. Deb and J. A. Chopoorian, J. Appl. Phys. 37:13, 4818 (1966).
11. S. K. Deb, Proc. Roy Soc. A, 304:211 (1968).
12. M. R. Tubbs, Phys. Stat. Sol. (a) 21:253 (1974).
13. R. Juriska, Phys. Stat. Sol.(b), 72:161 (1975).
14. M. J. Jagadeesh and V. Damodara Das, J. of Non-Crystalline Solids 28:327 (1978).
15. P. L. Gai, Phil. Mag. A 43:4,841 (1981).
16. Gmelings Handbuch Mo. Erg. B_1, 53:27.
17. H. Gruber and E. Krautz, Phys. Stat. Sol. a, 62:615 (1980).

ELECTRICAL PROPERTIES OF A $Fe_2(MoO_4)_3$ CATALYST

C. M. Mari

Dipartimento di Chimica Fisica ed Elettrochimica
Via C. Golgi, 19 20133 Milano (Italy)

P. Forzatti, and P. L. Villa

Dipartimento di Chimica Industriale ed Ingegneria
Chimica
Piazza L. da Vinci, 32 20133 Milano (Italy)

INTRODUCTION

Multicomponent molybdate catalysts are currently used for the industrial oxidation and ammoxidation of propylene[1-11]. Most of the studies reported in literature concern simple molybdates taken as suitable model catalysts and structural defects have been invoked as responsible for the catalytic activity of such materials. Their nature has been investigated mainly by X-ray diffraction technique [5-7]. It appears evident that a more detailed study of the structural and related bulk properties as well as systematic measurements of the collective properties of such materials could contribute to the explanation of their catalytic properties. In the frame of this view, the electrical conductivity of a "pure" $Fe_2(MoO_4)_3$ polycrystalline specimen was previously investigated[12] and in the present work, electrical conductivity measurements as a function of temperature and oxygen partial pressure have been carried out on the $Fe_2(MoO_4)_3$ catalyst.[*]

---

[*]The present compound was quoted as FeMo-O(1) in Ref. 13.

EXPERIMENTAL

$Fe_2(MoO_4)_3$ catalyst was prepared[3] by dissolving $(NH_4)_6Mo_7O_{24} \cdot 4H_2O$ in water at 80°C, adding molten $Fe(NO_3)_3 \cdot 9H_2O$ to the solution and the resulting slurry being diluted to 600 mol. The pH was raised by NH_4OH addition.

The slurry was finally spray dried and gradually heated to 510°C and then kept at this temperature for 5 hours. The X-ray powder diffraction pattern is in good agreement with literature data for monoclinic $Fe_2(MoO_4)_3$ [14-17]. No other phase was detected.

The DSC thermogram of the material differs from that of "pure" $Fe_2(MoO_4)_3$ prepared by solid state reaction in a sealed silica tube, in the presence of MoO_3 vapors, at 677°C for one week[18].

As it appears in Fig. 1, "pure" $Fe(MoO_4)_3$ shows an endothermic peak, on heating, at 510-515°C and an exothermic one, on cooling at 475-481°C, both associated with the monoclinic to orthorhombic reversible transformation[19-20]. The $Fe_2(MoO_4)_3$ catalyst presents a similar behaviour on heating and on cooling, a composite broad exothermic peak with maximum at T = 479°C and with a noticable shoulder at T = 485°C.

The electrical conductivity was measured on polycrystalline specimens obtained by pressing the oxide powder at 4000 Kg/cm^2 and sintering the pellets ($\emptyset$ = 13 mm, thickness = 1/4 mm) at 450°C, in air, for 30 hours.

No metallic film was evaporated onto the surface of the pellet in order to prevent diffusion of the conductive material into the bulk of the specimen.

Before the measurements, the pellets were equilibrated at 510°C for about 40 hours at a well defined oxygen partial pressure. To avoid decomposition of the $Fe_2(MoO_4)_3$ and to minimize MoO_3 losses, the temperature of 630°C was never exceeded and the time elapsed, for the measurements, at high temperatures, was always very short, of the order of few minutes.

A Wayne Kerr Autobalance Precision Bridge mod. B 331 MK11 operating at 1 KHz was employed.

RESULTS

Electrical conductivity measurements have been carried out in the temperature range 120 to 627°C and in the oxygen partial pressure region 1 to 10^{-3} bar.

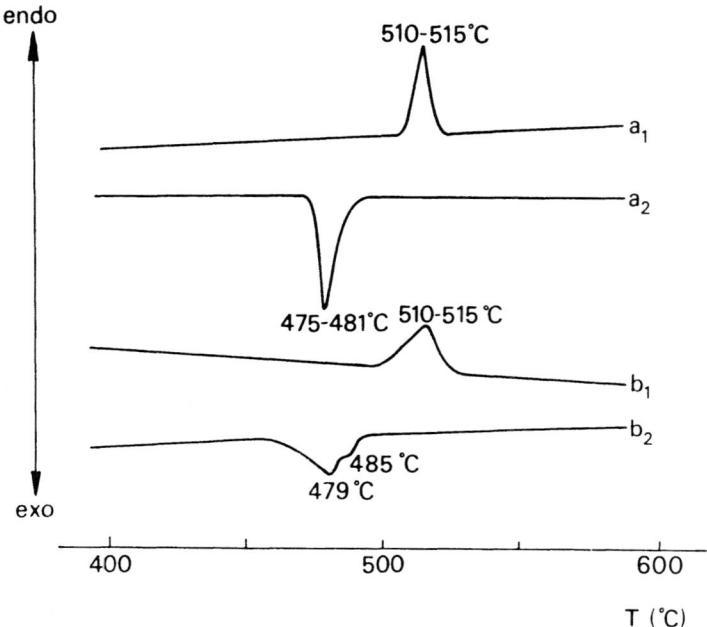

Fig. 1. DSC thermograms of "pure" $Fe_2(MoO_4)_3$ (a_1 on heating, a_2 on cooling) and $Fe_2(MoO_4)_3$ catalyst (b_1, b_2).

The results, corrected for the cell impedance, are reported in Fig. 2 as log σ vs $1/T$.

A marked discontinuity, attributed to the monoclinic $\rightleftarrows$ orthorhombic transition, was always observed at $T = 487°C$. At temperatures higher than T_c the isothermal dependence of the electrical conductivity as a function of oxygen partial pressure is described by the equation:

$$\sigma = KPo_2^{-1/n} \tag{1}$$

where $n \cong 6$ (see Fig. 3).

This behaviour can be attributed to the presence of doubly ionized oxygen vacancies according to the following reaction:

$$O_o^x \rightleftarrows V_o^{\cdot\cdot} + 2e' + 1/2\ O_2 \tag{2}$$

(using Kröger's notations) as previously reported for "pure" $Fe_2(MoO_4)_3$[18].

45

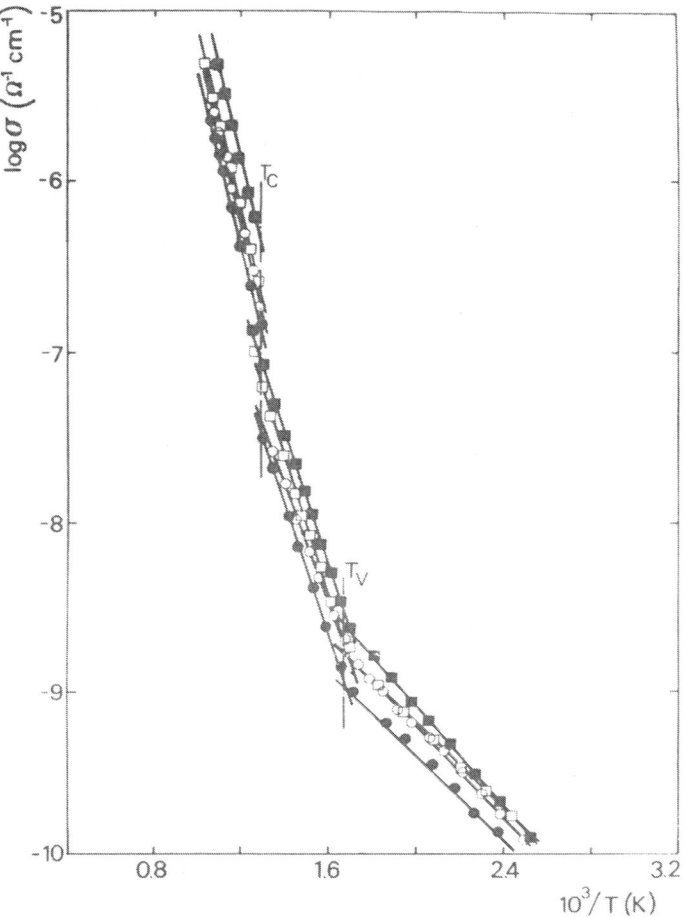

Fig. 2. Electrical conductivity of $Fe_2(MoO_4)_3$ catalyst as a
function of temperature at different oxygen partial
pressures (o = 1 bar, o = 0.21 bar, = 0.0013 bar).

The activation energy for the electrical conductivity
process seems not to depend on the oxygen partial pressure and it
is about 25 Kcal/mole.

At temperatures lower than T_c, two different electrical conduction mechanisms, depending on the temperature, were observed,
the former, between T_c and T_v (= 310°C) and the latter at
temperatures lower than T_v. In the range T_c to T_v the conduction
mechanism can be, again, attributed to the presence of doubly
ionized oxgyen vacancies according to reaction (2), since the
electrical conductivity can be described by equation (1) with

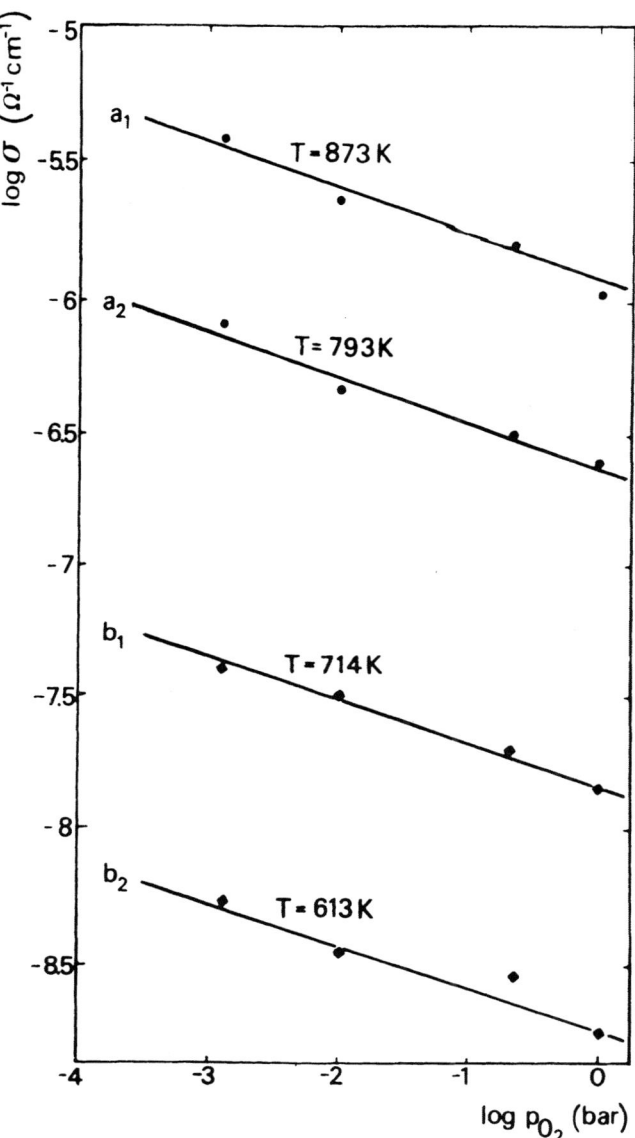

Fig. 3. Electrical conductivity of $Fe_2(MoO_4)_3$ catalyst as a function of the oxygen partial pressure (a, $T > T_c$; b, $T_c > T > T_v$).

$n \cong 6$ (see Fig. 3). The activation energy for the electrical conductivity process is about 18 Kcal/mole and does not depend on the oxygen partial pressure.

The difference in activation energy for the same proposed mechanism can be attributed to the different crystallographic phases present above and below T_c (orthorhombic and monoclinic).

At temperatures below T_V, the electrical conductivity is independent of oxygen partial pressure and, again, its value increases with the temperature. This behavior may reflect either ionic or semiconductive electrical properties. Based on the low activation energy ($\cong$ 6 Kcal/mole) calculated under these experimental conditions, the conduction mechanism is likely to be associated with the presence of a level close to the conduction band.

CONCLUSION AND DISCUSSION

The results here reported emphasize that:

i) the electrical conductivity of the orthorhombic form of the catalyst is associated with a majority defect pair constituted by $V_O^{\cdot\cdot}$ and e'

ii) a marked discontinuity occurs at the temperature (T_c = 487°C) at which the crystallographic monoclinic $\rightleftarrows$ orthorhombic transition takes place

iii) the monoclinic form of the catalyst presents two different conduction mechanisms dependent on the temperature: the former (in the temperature range $T_c > T > T_V$) attributed to the presence of doubly ionized oxygen vacancies and electrons, the latter (at $T < T_V$) associated with the presence of a level close to the conduction band.

A comparison with the results obtained in "pure" $Fe_2(MoO_4)_3$, points out a marked difference between the materials at temperatures lower than T_c.

One can recognize, in the case of "pure" $Fe_2(MoO_4)_3$, the presence of two different conduction mechanisms (at $T < T_c$) both independent of oxygen partial pressure: the former (at $P_{O_2} \leq 10^{-2}$ bar) associated with the presence of a level close to the conduction band with a activation energy of about 5 Kcal/mole and the latter (at $0.5 < P_{O_2} < 1$ bar) attributed to the presence of holes and free electrons with an activation energy of about 15 Kcal/mole[12]. The difference in the electrical conductivity behaviour could be attributed to the presence of small amounts of MoO_3, in excess the stoichiometric composition of $Fe_2(MoO_4)_3$, in the $Fe_2(MoO_4)_3$ catalyst.

The possibility of dissolving small amounts of MoO_3 in the $Fe_2(MoO_4)_3$ lattice is suggested by the structure of monoclinic iron molybdate[17], which consists of FeO_6 octahedra and MoO_4 tetrahedra sharing corners: only in such a way each oxygen atom is bound only to one Fe and one Mo atom and consequently related to a distorted $A_3B_2(SiO_4)_3$ garnet structure with the A sites unoccupied. Large voids surrounded by oxygen atoms located at the perimeter of cages with a diameter of about 5 Å are present and therefore large foreign ions can be accepted without involving significant distortion of the host structure.

Evidences of $Fe_2(MoO_4)_3$ catalysts with different amounts of Bi or Te are reported in the literature [13].

The presence of MoO_3 seems to stabilize, at every oxygen partial pressure and at temperature lower than T_v, a phase as that present at low oxygen partial pressure in the case of "pure" $Fe_2(MoO_4)_3$; the conduction mechanism appears to be same as well as the activation energy.

At temperatures ranging T_c to T_v the presence of a small excess of MoO_3 seems to influence the stability of the molybdate. A homogeneous nonstoichiometric phase appears to be more stable.

It is worth noticing that only at temperatures higher than about 300°C $Fe_2(MoO_4)_3$ catalyst becomes active in the oxidation of olefines and bulk reduction of the material takes place.

Electrical conductivity measurements seems to be a powerful technique for investigating flexible defect structures where X-ray diffraction techniques fail in detecting any structural defects. Thus, it appears proper to us to suggest extensive use of this kind of measurement to aid in contributing explanations of the catalytic properties of such catalysts.

REFERENCES

1. K. Aykan, O. Halvorson, A. W. Sleight, D. B. Rogers, J. Catal. 35:401 (1974).
2. A. W. Sleight, Crystal Chemistry and Catalytic Properties of Oxides with Scheelite Structure, in: "Advanced Materials in Catalysis," J. J. Burton, R. L. Garten, Eds., Academic Press, New York (1977).
3. A. W. Sleight K. Aykan, D. B. Rogers, J. Solid State Chem. 13:231 (1975).
4. P. L. Villa, A. Szabo, F. Trifiro, M. Carbucicchio, J. Catal. 47:122 (1977)
5. F. Veath, J. L. Callahan, J. D. Idol, E. D. Milberger, Chem. Engl. Prog. 5b (10):65 (1960).
6. J. Y. Robin, Y. Arnaud, J. Guidot, J. E. Germain, R. Acad. Sci. Ser. C 280:921 (1975).
7. W. J. Linn, A. W. Sleight, J. Catal. 41:134 (1976).

8. P. Forzatti, I. Pasquon, P. Trifiro, Tellurium as a Specific Component in New Molybdenum-Based Catalysts for the Selective Oxidation and Ammoxidation of Olefins, in: "Proceeding of the Climax 3rd International Conference on the Chemistry and Uses of Molybdenum," H. F. Barry, P. C. H. Mitchell, Eds., Ann Arbor, Mich. (1979).

9. P. Forzatti. P. Tittarelli, J. Solid State Chem. 33:421 (1980).

10. M. W. J. Wolf, P. A. Batist, J. Catal. 32:25 (1974).

11. I. Matsuura, M. W. J. Wolf, J. Catal. 37:174 (1975).

12. P. Forzatti, P. L. Villa, C. M. Mari, Mat. Chem. Phys. 10:385 (1984).

13. P. Forzatti, P. L. Villa, N. Ferlazzo, D. Jones, J. Catal. 76:188 (1982).

14. P. V. Klevtsov, R. F. Klevtsova, L. M. Kefeli, L. M. Plyasova, Inorg. Mater. 1:843 (1965).

15. L. M. Plyasova, R. F. Klevtsova, S. V. Borisov, L. M. Kefeli, Dokl. Akad. Nauk. 167:84 (1966).

16. G. Fagherazzi, N. Pernicone, J. Catal. 16:321 (1970).

17. H. Chen, Mater. Res. Bull. 14:1583 (1979).

18. A. V. Gur'ev, G. Flor, A. Marini, V. Massarotti, R. Riccardi, Z. Naturforsch. 36a:280 (1981).

19. A. W. Sleight, L. H. Brixner, J. Solid State Chem. 7:172 (1973).

20. V. Massarotti, G. Flor, A. Marini, J. Appl. Crystallogr. 14:64 (1981).

ANISOTROPY OF ELECTRICAL PROPERTIES IN PURE AND DOPED αFe_2O_3

Driss Benjelloun, Jean-Pierre Bonnet and Marc Onillon

Laboratoire de Chimie du Solide du CNRS

Université de Bordeaux I, 33405 Talence Cedex, France

ABSTRACT

The electrical conductivity and the thermoelectric power of pure and nickel-doped α Fe_2O_3, studied in the crystallographic plane (001) and along the [001] axis have shown typical anisotropic behaviour.

When the electrical conductivity is predominantly extrinsic, its anisotropy can be enhanced, the magnitude of the phenomenon depending on the nickel concentration. However an inversion of the anisotropy is observed for the atom ratio $Ni/Fe \simeq 1.4 \; 10^{-3}$. Models have been developped in order to interpret these behaviours.

In the pure material and for temperatures higher than 1050 K the electrical conductivity is essentially intrinsic.

EXPERIMENTAL STUDIES AND RESULTS

The αFe_2O_3 single crystals, either pure or doped, have been prepared using a chemical vapor transport technique [1] and analysed by atomic absorption and electron microprobe, the latter allowing control of the homogeneity. Except for nickel, no impurity was detected within the limit sensitivity of the analytical methods. The nickel content for the various crystals studied is reported in table I.

The electrical conductivity was measured using DC current, in an atmosphere of thoroughly dried argon or air ; for the electrical contacts platinum paint and leads were used [2]. All values were collected assuming that equilibrium was reached once a 10 hours stabilization has been observed.

The thermoelectric power coefficient α was measured in dried air or argon up to 1120 K using a method described by Dordor et al. [3]

Table I : Cationic fraction of nickel in the crystals studied.

	A	B	C	D	E
$[Ni] / [Fe]$	0	$9.0 \ 10^{-4}$	$1.2 \ 10^{-3}$	$1.4 \ 10^{-3}$	$1.7 \ 10^{-3}$

The variations of the electrical conductivity of pure and doped crystals of αFe_2O_3 versus temperature are reported on fig. 1 and fig. 2.

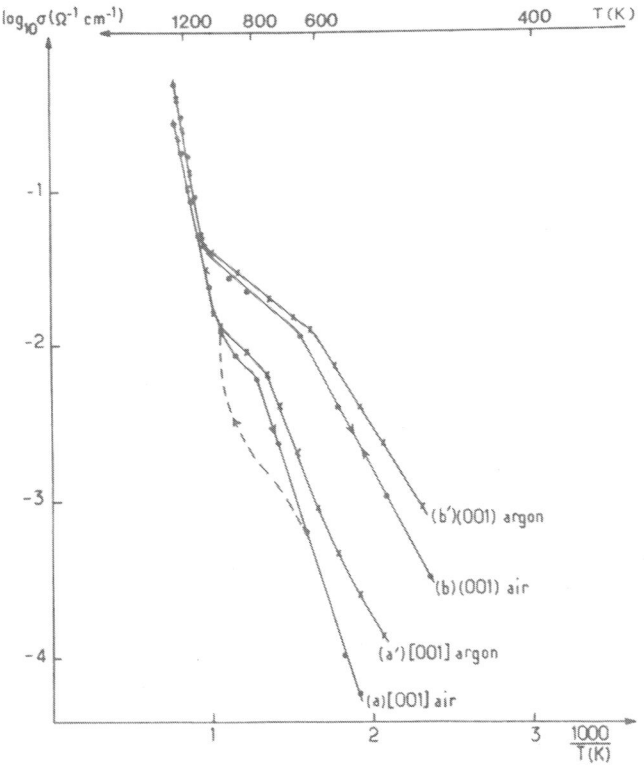

Fig. 1 . Electrical conductivity of αFe_2O_3 in air and argon as a function of temperature and crystallographic direction (crystal A, pure).

In either case, there is strong evidence for anistropy, especially below 1000 K. This anisotropy depends on the nickel concentration : when the amount of nickel is low, the electrical conductivity is higher in the (001) plane than along the [001] axis ; an opposite situation is observed for the highest concentration of nickel. The inversion occurs for the atomic ratio Ni/Fe≈1.4×10^{-3} (cf. fig. 2, crystal D).

Whatever the direction considered, there is no influence of the partial pressure of oxygen for pure crystal A above 1050 K, which can be explained by the predominance of intrinsic conductivity as reported by Kofstad [4].

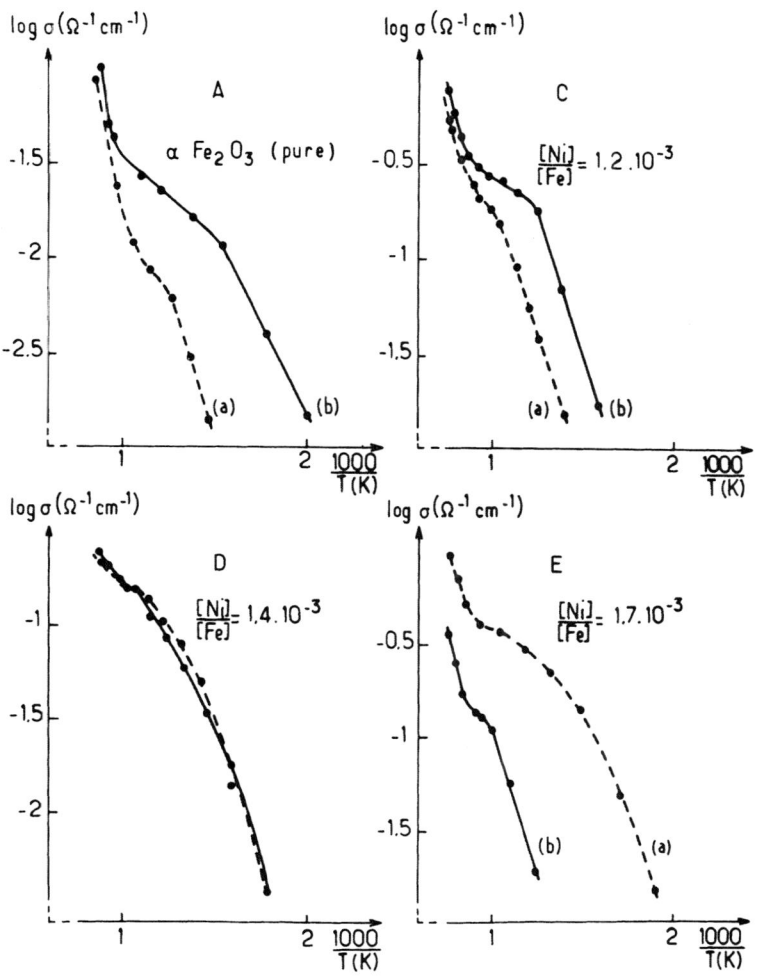

Fig. 2. Electrical conductivity of pure and doped αFe$_2$O$_3$ vs temperature in air (a): [001]; (b) (001).

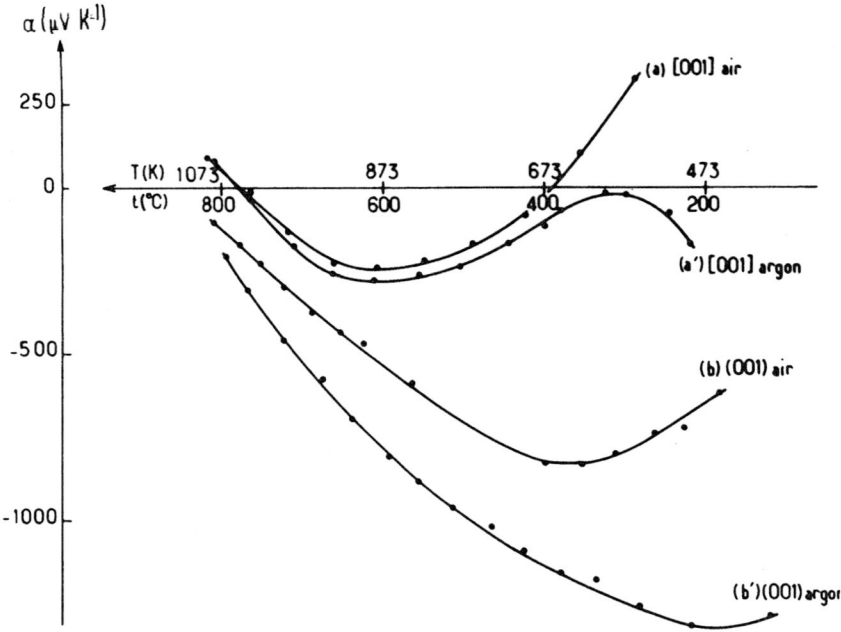

Fig. 3. Thermoelectric power of αFe_2O_3 in air and argon
as a function of temperature and crystallographic
direction.

Fig. 3 shows the values of the thermoelectric power coefficient
α versus temperature for pure crystal A : a strong anisotropy is
observed which depends on the partial pressure of oxygen especially
at low temperature. The situation is quite different for doped crystals : when the concentration of nickel is high (cf. crystal E,
fig. 4) the thermoelectric power is the same in all directions.
The values of α, measured along the $[001]$ axis for nickel doped
crystals B, D and E, are represented on fig. 5 as a function of
temperature. Their positive sign indicates a predominance of p
carriers.

DISCUSSION

For this material, a high optical absorption above 2.1 eV has
already been reported and discussed 5,6. it would originate in
an electronic transfer between neighbouring Fe^{3+} according to:

$$2\ Fe^{3+} \xrightarrow{h\nu} Fe^{2+} + Fe^{4+}$$

involving only the d levels of the iron atoms 7,8. This charge
transfer would allow the creation of intrinsic carriers 9.

54

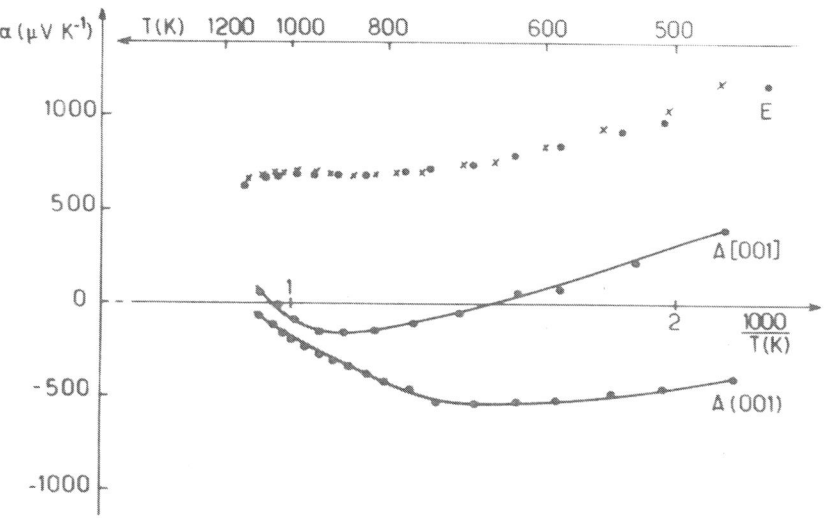

Fig. 4. Influence of doping, crystallographic direction
and temperature on the thermoelectric power of
αFe_2O_3.

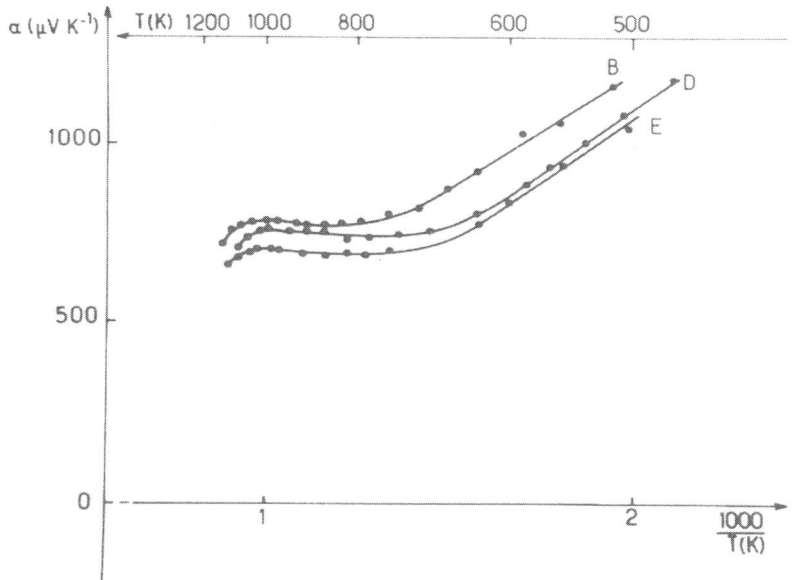

Fig. 5. Influence of doping and temperature on the
thermoelectric power of αFe_2O_3 along $[001]$.

Extrinsic behaviour

Below 720 K, the thermoelectric power coefficient is a linear function of the reciprocal of temperature in the case of nickel doped crystals (cf. fig. 5). Assuming the predominance of one type of carrier with extrinsic origin, the experimental curves are characteristic of a non-saturated semi-conductor. The theoretical variation of α can then be calculated after :

$$\alpha = \frac{k}{q} \left[\frac{E_F - E_V}{kT} + A \right] \tag{1}$$

where A is a constant related to the kinetic contribution of the electrons, $E_F - E_V$ is the difference between the Fermi level and the e_g band of iron.

The activation energy E_a = 0.41 eV calculated from the experimental curves is lower than the activation energy for the electrical conductivity (0.6 to 0.8 eV) in the same range of temperature. This difference indicates a marked influence of the temperature on the carrier's mobility.

The $E_F - E_V$ values deduced from relation (1) allow the calculation of the carrier's concentrations from:

$$p = N \exp \left(- \frac{E_F - E_V}{kT} \right) \tag{2}$$

where, for the e_g narrow band of iron, N can be taken as the number of iron atoms per cm^3, i.e. $N = 3.95 \times 10^{22} cm^{-3}$ 10 .

From our experimental results (fig. 5) the thermoelectric power coefficient α is practically constant from 720 K up to 1000 K which is typical of a narrow band saturated semi-conductor.

The carrier's concentrations thus calculated for various temperatures below and above 720 K are reported in table II for crystals B, D and E. In the saturated domain, assuming that every Ni^{2+} substitutes a Fe^{3+} creating then a p carrier, and taking A = 2 in relation (1), leads to calculated carrier concentrations in good agreement with the nickel concentrations obtained from microprobe analysis. As could be expected, the carrier concentrations in the unsaturated domain are much smaller than those calculated using the above assumption. This indicates the existence of an acceptor level located few tenths of an eV above the e_g band: this level could originate from the presence of Ni^{3+}.

Combining the calculated carriers concentrations with the experimental values of the electrical conductivity, it is possible to calculate the carrier mobilities along the $[001]$ axis and in the (001) plane. The values obtained for crystals B, D and E are reported in table II : they never exceed few hundredth of a $cm^2V^{-1}s^{-1}$. Their temperature dependance is shown on fig. 6.

Table II. Carriers concentrations and mobilities $\mu(cm^2V^{-1}s^{-1})$ for nickel doped αFe_2O_3.

T(K)	B nickel concentration: $3.56\ 10^{19}cm^{-3}$		D nickel concentration: $5.58\ 10^{19}\ cm^{-3}$			E nickel concentration: $6.67\ 10^{19}cm^{-3}$		
	$P(cm^{-3})$	$\log\mu$ $[001]$	$P(cm^{-3})$	$\log\mu$ $[001]$	$\log\mu$ (001)	$P(cm^{-3})$	$\log\mu$ $[001]$	$\log\mu$ (001)
1073	$3.76\ 10^{19}$	-1.42	$5.50\ 10^{19}$	-1.568	-1.56	$6.55\ 10^{19}$	-1.12	-1.80
993	$3.76\ 10^{19}$	-1.47	$5.50.10^{19}$	-1.638	-1.63	$6.55\ 10^{19}$	-1.35	-1.91
913	$3.76\ 10^{19}$	-1.50.	$5.50.10^{19}$	-1.668	-1.75	$6.55\ 10^{19}$	-1.38	-2.15
833	$3.76\ 10^{19}$	-1.58	$5.50.10^{19}$	-1.778	-1.87	$6.55\ 10^{19}$	-1.45	-2.45
753	$3.56\ 10^{19}$	-2.10	$5.50.10^{19}$	-1.958	-2.07	$6.55\ 10^{19}$	-1.55	-2.85
673	$1.14\ 10^{19}$	-2.60	$3.55.10^{19}$	-2.255	-2.28	$4.08\ 10^{19}$	-1.71	-3.21
593	$2.74\ 10^{18}$	-3.00	$1.10.10^{19}$	-2.466	-2.28	$1.27\ 10^{19}$	-1.60	-3.54
513	$4.65\ 10^{17}$		$1.87.10^{18}$		-2.43	$1.87\ 10^{18}$	-1.43	

Along the $[001]$ axis the mobility increases with the nickel concentration while the temperature dependance decreases. The inverse phenomenon is observed in the (001) plane.

This peculiar behaviour originates in the magnetic structure of αFe_2O_3.

Intrinsic behaviour

As reported before, pure crystals above 1050 K show evidence for intrinsic conductivity. In this range of temperature the material is paramagnetic.

The activation energy for electrical conductivity, calculated through $\sigma = \sigma_0 \exp(-\frac{E_a}{kT})$, is not the same along the $[001]$ axis, i.e

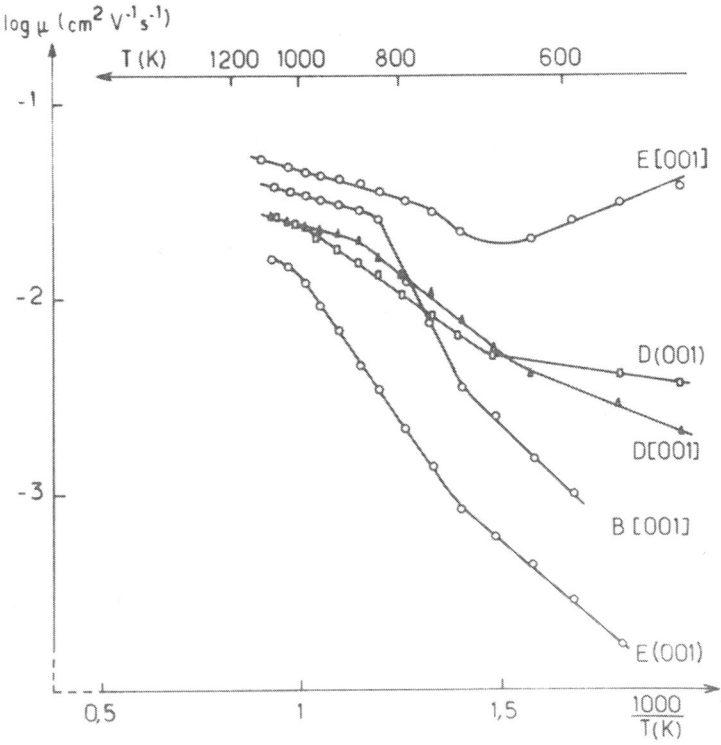

Fig. 6. Mobilities of carriers in doped αFe_2O_3.

1.08 eV and in the (001) plane, i.e. 1.22 eV.

Our optical measurements confirm a strong absorption at 2.14 eV in good agreement with the previously reported 5,6 values of 2.1 eV, but show a maximum at 2.50 eV[11]; such a maximum is also present in the spectrum of Redon et al.[11]

Analysing the photoresponse of an αFe_2O_3 anode, Hejtmaneck 8 has observed two electronic transitions at respectively 2.02 and 2.32 eV.

All these results suggest the existence of two neighbouring energy levels which could be related to the two activation energies observed for the electrical conductivity after examining the crystal structure of αFe_2O_3.

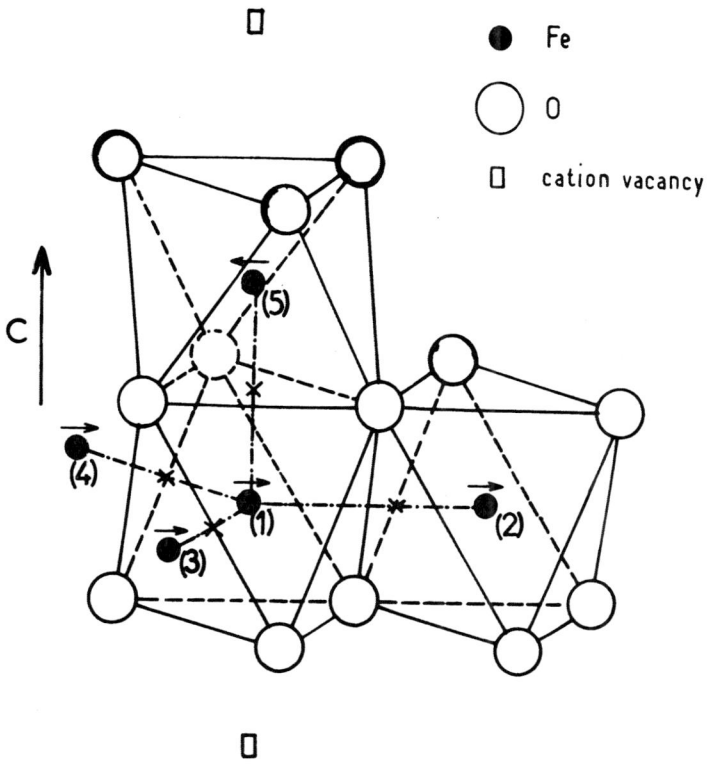

● Fe

○ O

□ cation vacancy

C

(5)

(4)

(1)

(3)

(2)

□

Fig. 7. Structural features of αFe_2O_3.

In this structure, oxygen octahedrons share a common edge in the (001) plane and a common face along the [001] axis (fig. 7). According to Goodenough [9], the resulting deformation of the crystal field then splits the t_{2g} orbitals into two e_g orbitals, directed towards the neighbouring cations in the (001) plane, and an a_{1g} orbital along the [001] axis. It is generally the case that the full orbital(a_{1g}) is likely to be more stable than the half-filled orbital (e_g) in the case of the d^6 configuration of Fe^{2+}. Thus to the charge transfer :

$$2 \ Fe^{3+} \quad \rightarrow \quad Fe^{2+} + Fe^{4+}$$

would be associated an energy of 2.12 eV when the 6th electron is in the a_{1g} orbital and 2.50 eV when it is in the e_g orbitals. Fig. 8 illustrates the splitting of the d orbitals of Fe^{2+} in a rhombohedral crystal field.

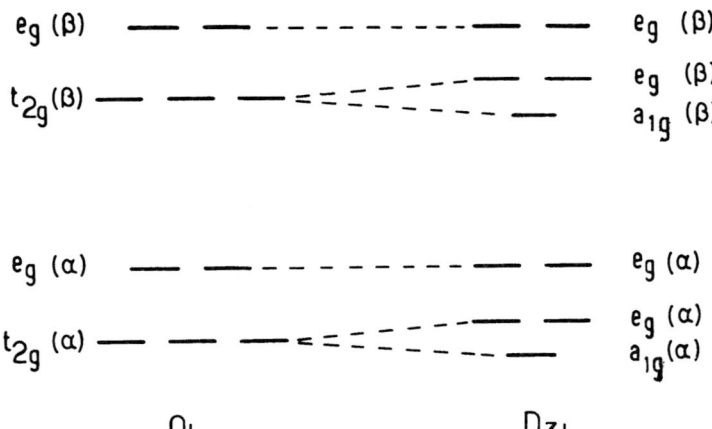

Fig. 8. Rhombohedral splitting of d orbitals in αFe_2O_3.

The anisotropy observed for the electrical conductivity and its temperature dependance can thus be explained by a corresponding anisotropy in the mobility of carriers, depending on the orbitals involved : for example a_{1g} electrons should move more easily than e_g electrons along the $[0\bar{0}1]$ axis.

REFERENCES

1. D. Benjelloun, J.P. Bonnet, J.P. Doumerc, J.C. Launay, M. Onillon and P. Hagenmuller, Mat. Chem. and Phys. (in press).
2. D. Benjelloun, J.P. Bonnet, P. Dordor, J.C. Launay, M. Onillon and P. Hagenmuller, Revue de Chimie Minérale (in press).
3. P. Dordor, E. Marquestaut and G. Villeneuve, Revue Phys. Appl., 15:1607 (1980)
4. P. Kofstad "Non Stoichiometry, Diffusion and Electrical Conductivity in Binary Metal Oxides", Wiley, New York
5. R.F.G. Gardner, F. Swett and D.W. Tanner, J. Phys. Chem. Solids, 24:1175 (1963)
6. J.H.W. De Wit, 9th International Symposium on the Reactivity of Solids, Krakow (1980).
7. S. Wittekoek, T.J.A. Popma, J.M. Roberlson and P.F.B. Bougers, Phys. Rev., B12:2777 (1975).
8. J. Hejtmaneck, Thèse d'Université, n° 134 (1982), Université de Bordeaux I, France.
9. J.B. Goodenough, "Metallic Oxides", Progress in Solid State Chemistry, H. Reiss Ed., Pergamon Press, 5:145 (1971).
10. F.J. Morin, Phys. Rev., 93:1195 (1954).
11. A.M. Redon and J. Vigneron, Solar Cells, 3:176 (1981).

ANALYSIS OF THE MOBILITY OF THE ELECTRONIC DEFECTS IN $Co_{1-\delta}O$

G. Petot-Ervas[*], P. Ochin[*], and T. O. Mason[+]

[*]C.N.R.S. Laboratoire P.M.T.M., 93430 Villetaneuse
France

[+]Department of Materials Science and Engineering
Northwestern University, Evanston, Illinois, U.S.A.

ABSTRACT

Analysis was made of the temperature and composition dependence of hole mobility in CoO via a comparative study of conductivity and thermoelectric coefficient. The conductivity at constant thermopower appears to be activated. However the magnitude ($\lesssim 0.1$ eV) does not permit unambiguous confirmation of conduction mechanism. The oxygen exponents for conductivity and thermopower are, within experimental uncertainty, identical, i.e. no detectable composition dependence of the mobility. The values of the oxygen exponents obtained are consistent with defect clustering at high p_{O_2}.

INTRODUCTION

Despite extensive investigations, the nature of the prevailing point defects in cobaltous oxide and the conduction mechanism of the charge carriers in this material are not unambiguously determined. Concerning the point defect structure, recent experiments would seem to confirm the existence of defect clustering near the Co_3O_4 phase boundary, in agreement with the predictions of Catlow and Stoneham.[1] Due to the small deviation from stoichiometry ($\delta \sim 10^{-2}$ in $Co_{1-\delta}O$) at this phase boundary, it is dubious that diffraction techniques can establish the type(s) of clusters involved. Concerning the conduction mechanism in CoO, there is good agreement about the magnitude of the hole mobility ($\mu \sim 0.3$ $cm^2s^{-1}v^{-1}$). However, there is considerable disagreement as to whether or not the mobility exhibits temperature dependence

61

and/or composition dependence. It can be shown that an exponential temperature dependence and a Co^{2+} concentration dependence of the mobility are necessary and sufficient to confirm small polaron conduction. Nowotny et al.[2] recently argued for a composition dependence of the hole mobility in CoO. Our study was undertaken to examine the temperature and composition dependences of the hole mobility to obtain information concerning the defect structure and conduction mechanism in CoO.

Of the various physical properties which have been studied (nonstoichiometry, diffusion, electrical conductivity, thermopower, etc.), the variations of conductivity as a function of temperature and oxygen partial pressure are probably known with the most precision. But a precise analysis of these results requires complementary information. The conductivity of a p-type semi-conductor is proportional to the product of the hole concentration (p) and the mobility (μ):

$$\sigma = e\,p\,\mu \tag{1}$$

where e is the charge of the electron. As we have previously shown,[3] the thermoelectric coefficient will be:

$$Q = \frac{k}{e} \ell n \frac{N_v}{p} + const. \tag{2}$$

where N_v is the density of states and k/e has the value 86.1 $\mu V \cdot K^{-1}$. Equations (1) and (2) are valid regardless of conduction mechanism as long as the material remains exclusively p-type, and the conduction mechanism does not change. It follows that a combination of thermopower and conductivity measurements can be used to determine if the mobility exhibits temperature dependence or concentration dependence.

We have previously studied lithium-doped CoO.[4] In the extrinsic regime ($P_{O_2} < 10^{-4}$ atm) where the hole concentration is fixed by the amount of doping, it was determined that the mobility obeyed the following relation:

$$\mu = 0.52 \exp - 0.07 \ (eV)/kT \tag{3}$$

A similar approach can be taken in undoped CoO by examining the conductivity at fixed thermopower, which corresponds to fixed hole concentration. At the same time the oxygen pressure dependencies of these two properties can be compared for divergence, which would suggest a variation in mobility with composition.

62

EXPERIMENTS

Measurements were performed on parallelopiped specimens
with the dimensions 2 x 3 x 12 mm for conductivity studies and
1 x 1 x 7 mm for thermoelectric studies. The conductivity specimens
were arc image furnace produced single crystals, whereas the
single crystals used for thermopower measurements were supplied
by Dieckmann from his diffusion studies.[5] The electrical contacts,
or thermocouples, were held in place with alumina holders and
platinum foils, as described elsewhere. In the Seebeck experiments,
a small auxiliary heater was employed to establish gradients
of 0 to 15°K across the specimen. In all cases the cold end
was positive (in agreement with the nature of the prevailing
electronic defects) and plots of ΔE vs. ΔT were linear with
zero intercepts within an uncertainty of $\pm$ 1°K. The slope of
these plots is the thermoelectric coefficient (Q).

The electrical conductivity results are shown in Fig. 1
as a function of oxygen partial pressure. There is good shape
agreement between these results and those displayed by Dieckmann.[5]
The small difference in absolute values is difficult to explain.

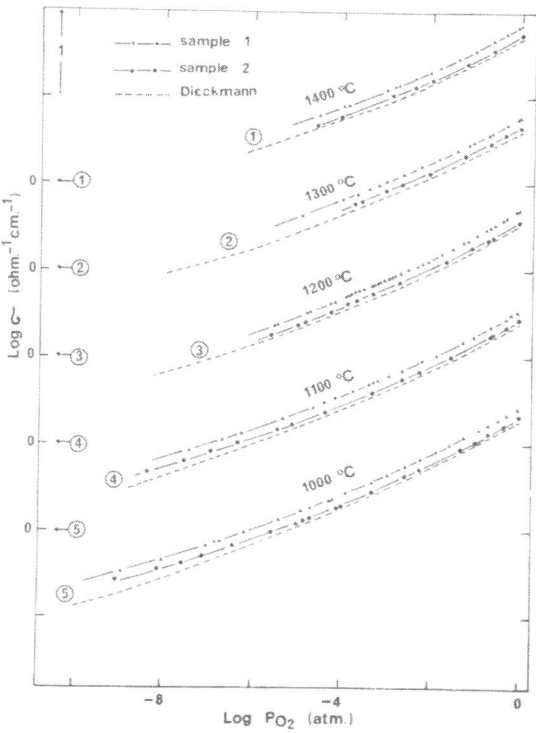

Fig. 1. Experimental conductivity results for CoO compared with
those of Ref. 5. (Reproduced from Ref. 4 with permission.

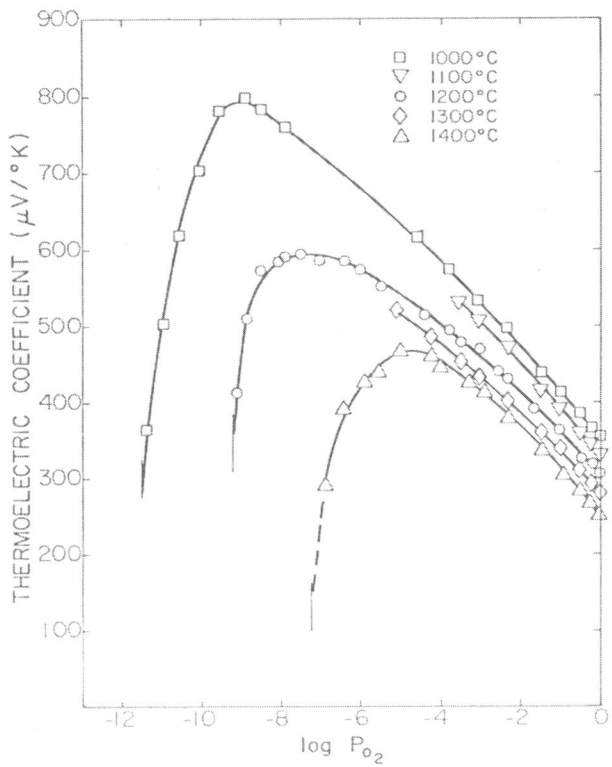

Fig. 2. Experimental Seebeck coefficient results for CoO.
(Reproduced from Ref. 3 with permission.)

It may be attributable to the crystallographic quality of the samples. The data of sample 1 will be used in the analysis below;
however, identical results would have been obtained using sample 2.

Thermoelectric coefficients obtained for CoO are displayed
in Fig. 2. Near the cobalt phase boundary a rapid decline of
the thermopower has been observed. This is attributable to
the influence of minority electrons. We have therefore limited
our analysis to high oxygen pressures ($P_{O_2} \geq 10^{-2}$ atm) where
holes are clearly dominant.

ANALYSIS

Mobility Temperature Dependence

According to Eq. (2), if Q is held constant, the ratio
$q = N_V/p$ must also be constant. Substituting for p in Eq. (1),

we obtain:

$$\sigma = e N_V q^{-1} \mu \qquad (4)$$

from which it follows that:

$$\left(\frac{\partial \ln \sigma}{\partial 1/T}\right)_Q = \left(\frac{\partial \ln \sigma}{\partial 1/T}\right)_q = \left(\frac{\partial \ln N_v}{\partial 1/T}\right)_q + \left(\frac{\partial \ln \mu}{\partial 1/T}\right)_q \qquad (5)$$

For phonon-scattered itinerant charge carriers, N_v is proportional to $T^{3/2}$ and μ is proportional to $T^{-3/2}$. Hence, no temperature dependence is expected. For small polarons, N_v should be the Co^{2+} concentration and approximately temperature independent, while μ is thermally activated. An exponential temperature dependence of σ at constant Q is therefore indicative of small polaron conduction.

   In Fig. 3 we have reported electrical conductivity in CoO at constant thermoelectric coefficient. As can be seen, these lines exhibit small negative slopes corresponding to activation energies less than 0.1 eV. It should be cautioned that whereas σ and Q were measured on different specimens and whereas such small slopes are obtained, Fig. 3 cannot be taken as conclusive evidence for small polaron conduction. Previous isothermoelectric analyses yielded zero or positive slopes.[3] These were based, however, on older and less precise measurements. Precise, simultaneous measurements of σ and Q on the same sample may resolve this issue.

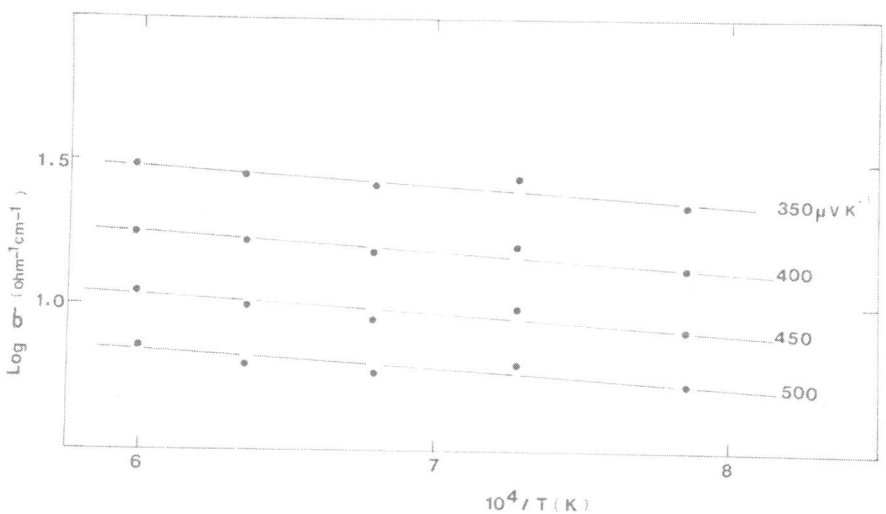

Fig. 3. Isothermoelectric conductivity plot for CoO.

Table I. Experimental Inverse Oxygen Exponents in CoO
$(10^{-2} \leq P_{O_2} \leq 1 \text{ atm})$

$T(^oC)$	n_σ	n_Q	$n_\sigma - n_Q$
1000	3.44 ± 0.09	3.35 ± 0.07	$+0.09$
1100	3.50 ± 0.04	3.28 ± 0.09	$+0.22$
1200	3.65 ± 0.06	3.72 ± 0.16	-0.07
1300	3.74 ± 0.10	3.84 ± 0.13	-0.10
1400	3.74 ± 0.07	3.64 ± 0.15	$+0.10$

It is interesting to note that the activation energies of Fig. 3 compare quite well to those of lithium-doped specimens (~ 0.07 eV).[4] Nevertheless, it should be pointed out that the thermopower of lithium-doped CoO is slightly temperature dependent,[6,7] which would not be predicted from small polaron theory.

Mobility Composition Dependence

In Table I we have reported our inverse oxygen parameters for conductivity ($n_\sigma = (\partial \ln\sigma / \partial \ln p_{O_2})_T^{-1}$) and thermopower ($n_Q = -(\partial Q\, e/k/ \partial \ln p_{O_2})_T^{-1}$). These values and their uncertainties were obtained by a least squares linear fit of the data in the oxygen partial pressure range $10^{-2}-1$ atm. Within experimental uncertainty, the slopes are identical.

From Eqs. (1) and (2) it follows that:

$$\left(\frac{\partial \ln\sigma}{\partial \ln p_{O_2}}\right)_T + \left(\frac{\partial (Q\, e/k)}{\partial \ln p_{O_2}}\right)_T = \left(\frac{\partial \ln N_v}{\partial \ln p_{O_2}}\right)_T + \left(\frac{\partial \ln \mu}{\partial \ln p_{O_2}}\right)_T \qquad (6)$$

For phonon-scattered itinerant charge carriers, $(\partial \ln N_v / \partial \ln p_{O_2})_T = 0$. Consequently, it follows from Eq. (6) and from our conductivity and thermopower results, which exhibit equal but opposite oxygen dependencies (Table I), that:

$$\left(\frac{\partial \ln \mu}{\partial \ln p_{O_2}}\right)_T \approx 0 \qquad (7)$$

This result is inconsistent with small polaron theory (see Eq. 12)). However, in the following we will demonstrate that small polaron theory leads to a composition dependence of μ and divergence of oxygen exponents n_σ and n_Q only at large enough hole concentrations.

Neglecting for the moment any defect clustering, and for the sake of simplification, we have assumed that singly-charged cobalt vacancies dominate the defect structure at high oxygen pressures:

$$1/2 \ O_2 \ (g) \rightleftarrows O_O^x + V_{Co}' + h^{\bullet} \tag{8a}$$

Assuming the holes to be localized, Eq. (8a) becomes:

$$(Co_{Co}^{2+})^* + 1/2 \ O_2 \ (g) \overset{x}{\rightleftarrows} O_O^* + \overset{x}{V_{Co}'} + \overset{x}{(Co_{Co}^{3+})^{\bullet}} \tag{8b}$$

for which the appropriate site fractions are:

$$[V_{Co}'], \ [Co_{Co}^{3+}] = x/(1 + x) \tag{9}$$

$$[Co_{Co}^{2+}] = (1 - x)/(1 + x) \tag{10}$$

For a small polaron conductor, the hole concentration is the cation site density (N) multiplied by the Co^{3+} site fraction:

$$p = N \left(\frac{x}{1+x} \right) \tag{11}$$

The mobility is given by:[8]

$$\mu = (1 - c) \ \frac{e \ a^2 \ \nu}{kT} \ \exp \left(\frac{-\Delta H_m}{kT} \right) \tag{12}$$

where e is the electron charge, a is the jump distance, ν is the vibrational frequency, ΔH_m is the hopping energy, and (1-c) is the site fraction Co^{2+} given by Eq. (10). At fixed temperature it follows that:

$$\sigma = const \cdot x \ (1 - x)/(1 + x)^2 \tag{13}$$

from which:

$$\frac{1}{n_\sigma} = \left(\frac{\partial \ \ell n \ \sigma}{\partial \ \ell n \ P_{O_2}} \right)_T = \left(\frac{\partial \ \ell n \ x}{\partial \ \ell n \ P_{O_2}} \right)_T \left[1 - \left(\frac{x}{1 - x} \right) - \left(\frac{2x}{1 + x} \right) \right] \tag{14}$$

Similarly, the thermopower will be given by:

$$Q = \frac{k}{e} \ \ell n \ \frac{(1 - x)}{x} + const. \tag{15}$$

from which it follows that:

67

$$\frac{1}{n_Q} = -\left(\frac{\partial (Q\,e/k)}{\partial \ln p_{O_2}}\right)_T = -\left(\frac{\partial \ln x}{\partial \ln p_{O_2}}\right)_T \left[1 + \left(\frac{x}{1-x}\right)\right] \qquad (16)$$

By comparing Eqs. (14) and (16), it is obvious that the magnitudes of the oxygen dependences for conductivity and thermopower will diverge when x becomes significant. In Table II are tabulated the maximum x values for NiO, CoO and MnO (assuming x equals the deviation from stoichiometry) together with predicted inverse oxygen exponents for x, σ and Q. As expected, the magnitudes of conductivity and thermopower oxygen dependencies are virtually identical at vacancy concentrations corresponding to the maximum deviation from stoichiometry in NiO. By the maximum deviation from stoichiometry in MnO, however, large differences should be obtained. At the maximum x in CoO (~1%) a difference $n_\sigma - n_Q$ ~0.20 is predicted. This is not large enough to be detected according to experimental uncertainty (Table I) and does not allow us to confirm polaron conduction in CoO. However, what the data indicates within experimental uncertainty, is no detectable mobility dependence upon concentration (Eq. 7). This is contrary to what Nowotny and co-workers[2] claimed. Their n_σ and n_Q values were 3.88 $\pm$ 0.12 and 4.20 $\pm$ 0.24, respectively, for pure CoO. There is no significant difference between these values, however, and $n_\sigma - n_Q$ has the wrong sign to be consistent with small polaron theory.

Nature of the Prevailing Defects

It follows from Eqs. (14) and (16) that the parameter n deduced from electrical property measurements is directly related to the nature of the prevailing ionized point defects in CoO. It is apparent that these values are significantly less than 4 in Table I, which confirms the non-ideal defect behavior proposed previously. Futhermore, the values of n obtained from thermogravimetric and mass transport measurements are about the

Table II. Predicted Inverse Oxygen Exponents

Oxide	x_{max}^{*}	$\left(\frac{\partial \ln x}{\partial \ln p_{O_2}}\right)_T^{-1}$	$n_\sigma = \left(\frac{\partial \ln \sigma}{\partial \ln p_{O_2}}\right)_T^{-1}$	$n_Q = -\left(\frac{\partial Q\,e/k}{\partial \ln p_{O_2}}\right)_T^{-1}$	$n_\sigma - n_Q$
NiO	0.0005	4.00	4.00	4.00	0.00
CoO	0.013	4.00	4.16	3.95	+ 0.21
MnO	0.15	4.13	8.52	3.40	+ 5.12

*As reported in Ref. 11.

68

Table III. Clusters with Oxygen Exponents Between 1/3 and 1/3.5.

$1/n$	$(v-i)$	ξ	Clusters
1/3.0	2	2	3:1, 4:2, 5:3, etc.
	4	5	5:1, 6:2, 7:3, etc.
1/3.2	5	7	6:1, 7:2, 8:3, etc.
1/3.3	3	4	4:1, 5:2, 6:3, etc.
1/3.5	4	6	5:1, 6:2, 7:3, etc.

same as for conductivity and thermopower (see, e.g., Fig. 4 in Ref. 3). Consequently, neutral cobalt vacancies are probably minority defects in CoO in agreement with the theoretical predictions of Catlow and Stoneham.[1]

To examine the influence on the electrical properties of vacancy-interstitial clusters, we can write the generalized reaction:

$$(i + \xi)\ M_M^* + \frac{(v-i)}{2}\ O_2(g) \rightleftarrows (vV_M iM_i)^{\xi'} + (v-i)O_O^* + \xi\ M_M^{\cdot} \qquad (17)$$

where v and i are the number of vacancies and interstitials, respectively, in the cluster of charge ξ. Assuming the electroneutrality condition to be:

$$[M_M^{\cdot}] = \xi\ [(vV_M iM_i)^{\xi'}] \qquad (18)$$

we obtain:

$$[(vV_M iM_i)^{\xi'}] = \frac{[M_M^{\cdot}]}{\xi} = \text{const} \cdot P_{O_2}^{\left(\frac{v-i}{2(\xi+1)}\right)} \qquad (19)$$

if $[M_M^*]$ does not vary significantly with cluster concentration.

Two experimental observations bracket the range of oxygen exponents anticipated for the cluster regime. Logothetis[9] reports ~1/3.5 at the Co_3O_4 phase boundary on the basis of thermopower and conductivity data. Bridges et al.[10] obtained ~ 1/3.1 over the range $10^{-2} < P_{O_2} < 30$ atm in parabolic oxidation studies at 950–1100°C. Table III lists clusters which should have oxygen exponents in the range 1/3 to 1/3.5 based upon Eq. (19).

CONCLUSIONS

Analysis has been made of the temperature and composition dependence of hole mobility in CoO via a comparative study of conductivity and thermoelectric coefficient. The conductivity analyzed at fixed thermopower exhibits a small ($\lesssim 0.1$ eV) activation energy. This result is consistent with the activation energy in lithium-doped CoO. Unfortunately, the value obtained for undoped CoO is not large enough, relative to experimental uncertainty, to confirm small polaron conduction unambiguously. We have also compared the oxygen pressure dependences of conductivity and thermopower to test for a composition dependent mobility, indicative of small polaron conduction. Within experimental uncertainty, the oxygen exponents are identical. This does not rule out small polaron conduction, however. Not enough defects exist in CoO at $P_{O_2} = 1$ atm to create sufficient divergence of oxygen exponents, even if small polaron conduction is operative.

The conductivity and thermopower oxygen exponents are significantly greater than 1/4 predicted on the basis of singly-charged cobalt vacancies as the dominant ionic defect. This was shown to be consistent with the existence of vacancy-interstitial clusters at high oxygen partial pressures.

ACKNOWLEDGEMENT

Professor Mason acknowledges the support of the U. S. Department of Energy under Grant No. DE-AC02-83ER45043.

70

REFERENCES

1. C.R.A. Catlow and A.M. Stoneham, J. Am. Ceram. Soc. 64:234 (1981).
2. J. Nowotny, I. Sikora, and M. Rekas, J. Electrochem. Soc. 131:94 (1984).
3. H.-C. Chen, E. Gartstein, and T. O. Mason, J. Phys. Chem. Solids 43:991 (1982).
4. G. Petot-Ervas, P. Ochin, and B. Sossa, Solid State Ionics 12:295 (1984).
5. R. Dieckmann, Z. Phys. Chem. N.F. 107:189 (1977).
6. B. Fisher and J. B. Wagner, J. Appl. Phys. 38:3838 (1967).
7. A. J. Bosman and C. Crevecoeur, J. Phys. Chem. Solids 30:1151 (1969).
8. H. L. Tuller and A. S. Nowick, J. Phys. Chem. Solids 38:859 (1977).
9. E. M. Logothetis and J. K. Park, Solid State Commun. 43:543 (1982).
10. D. W. Gridges, J. P. Baur and W. M. Fassel, J. Electrochem. Soc. 103:619 (1956).
11. N. L. Peterson and W. K. Chen, J. Phys. Chem. Solids 43:29 (1982).

A^{++}/Ti NONSTOICHIOMETRY IN ALKALINE EARTH TITANATES, $ATiO_3$

Y. H. Han, M. P. Harmer, Y. H. Hu, and D. M. Smyth

Materials Research Center #32
Lehigh University
Bethlehem, PA 18015

I. INTRODUCTION

Ternary oxides offer the possibility of variations in the ratios of the two metallic constituents as well as in the metal-nonmetal ratios. The extent of this cation-cation nonstoichiometry, and its effect on material properties, is of interest because numerous useful ferroelectric, piezoelectric, magnetic, and electro-optic materials are found among the ternary oxides. When signifi-cant cation-cation nonstoichiometry is tolerated in the single phase system, the resulting changes in ionic defect concentrations can affect charge and mass transport and oxidation-reduction equilibria. If the tolerance for nonstoichiometry is exceeded, second phases will be present that can also profoundly affect the material properties, e.g., if the second phase is liquid at the sintering temperature, densification rates and the resulting micro-structure will be strongly influenced. This paper will discuss the cation-cation nonstoichiometry of the alkaline earth titanates having the perovskite structure, $BaTiO_3$, $SrTiO_3$, and $CaTiO_3$. These compounds share the generic formula ABO_3, and the structure is characterized by having two very different types of cation sites, a large, 12-coordinate A site, occupied by the alkaline earth ions, and a smaller, 6-coordinate, octahedral B site, occupied by the titanium. In the ideally pure, stoichiometric compounds, all equivalent lattice sites are completely occupied by the appropriate species.

Cation-cation nonstoichiometry for the alkaline earth tita-nates having the formula ABO_3 corresponds to an excess of one of the constituent binary oxides, AO or BO_2. Possible incorporation reactions are as follows:

$$AO \rightarrow A_I^{\bullet\bullet} + O_I'' \qquad (1)$$

$$AO \rightarrow A_A + V_B'''' + O_O + 2V_O^{\bullet\bullet} \qquad (2)$$

$$BO_2 \rightarrow B_I^{\bullet\bullet\bullet\bullet} + 2O_I'' \qquad (3)$$

$$BO_2 \rightarrow V_A'' + B_B + 2O_O + V_O^{\bullet\bullet} \qquad (4)$$

Equations (1) and (3) involve no changes in the total number of lattice sites, while (2) and (4) involve the creation of new sites in proper stoichiometric ratio. O_I'' is a very unfavorable defect in the close-packed perovskite structure, as is $A_I^{\bullet\bullet}$, so Equations (1) and (3) are considered to be unlikely options. If cation place-exchange defects are considered, two additional options are apparent:

$$2AO \rightarrow A_A + A_B'' + 2O_O + V_O^{\bullet\bullet} \qquad (5)$$

$$2BO_2 \rightarrow B_A^{\bullet\bullet} + B_B + 3O_O + O_I'' \qquad (6)$$

Equation (6) is again considered to be unlikely because it involves O_I'', but it will be seen that Equation (5) cannot be ignored when Ca^{++} is the alkaline earth. These point defect models for incorporation of an excess of AO or BO_2 conserve the basic perovskite structure. Alternative mechanisms involving structural modifications such as defect ordering or superlattice structures are also possible in principle, and will be discussed when pertinent.

II. EXPERIMENTAL

The samples were prepared by the liquid-mix technique, starting with titanium tetraisoproxide and the alkaline earth carbonate. This technique is a modification of a process developed by Pechini[1] and has been described elsewhere[2,3]. Precise control of the A/B ratio can be achieved by this procedure. The fine-grained powders were cold-pressed and sintered in air to give polycrystalline samples. The equilibrium conductivities were measured by 4-point dc technique, mostly at 1000°C. A stabilized zirconia oxygen activity cell was used to measure the wide range of oxygen partial pressures obtained from $Ar-O_2$ and $CO-CO_2$ mixtures.

Two techniques have been used to determine the extent of solid solubility. The first is direct microscopic examination, usually of polished and etched samples by scanning electron microscopy (SEM), but occasionally by transmission electron microscopy (TEM) of thinned specimens. The SEM technique has proved to be capable of detection of second phases at the 0.1 mol % level. Secondly, all of the probable incorporation reactions, Equations (2), (4), and (5), involve the formation of oxygen vacancies, $V_o^{..}$, so a positive test for an increase in the concentration of $V_o^{..}$ would be an indication of solid solubility. A very sensitive test is available in the measurement of the equilibrium electrical conductivity. Typical results at 1000°C are shown for $BaTiO_3$ in Figure 1. The data consist of an oxygen-deficient, n-type region at low oxygen partial pressures, P_{O_2}, characterized by a reduction reaction and its mass-action expression

$$0_o \rightleftharpoons 1/20_2 + V_o^{..} + 2e' \tag{7}$$

$$[V_o^{..}]n^2 = K_n P_{O_2}^{-1/2} \tag{8}$$

and an oxygen-excess, p-type region at high P_{O_2} controlled by the reaction

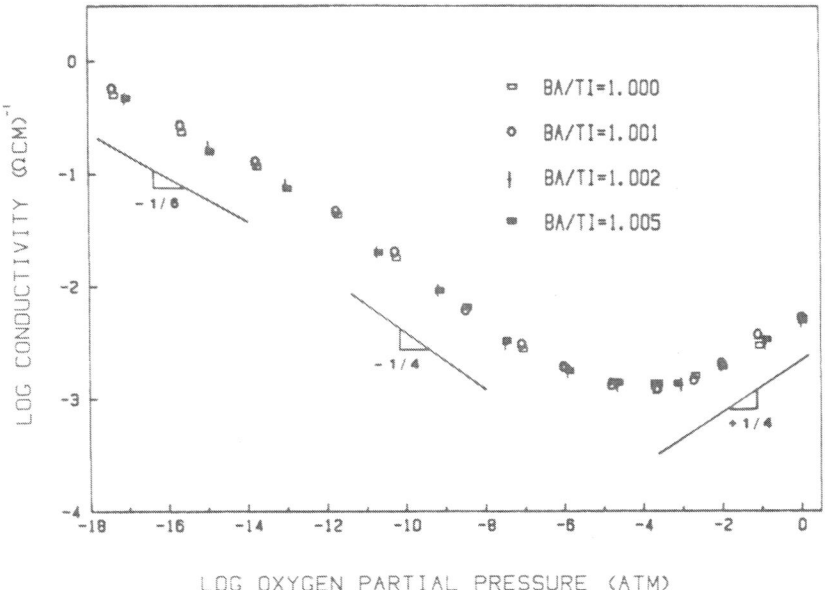

Fig. 1. Equilibrium electrical conductivity at 1000°C for $BaTiO_3$ with excess BaO.

$$V_O^{\cdot\cdot} + 1/2 O_2 \rightleftharpoons O_O + 2h^{\cdot} \tag{9}$$

$$\frac{p^2}{[V_O^{\cdot\cdot}]} = K_p \, P_{O_2}^{1/2} \tag{10}$$

The brackets denote the activity of the enclosed species, and $n \equiv [e^{\prime}]$ and $p \equiv [h^{\cdot}]$. Except under the most severely reducing conditions, i.e. highest temperature and lowest P_{O_2}, $[V_O^{\cdot\cdot}]$ in undoped $BaTiO_3$ is determined by the natural acceptor impurity content, e.g. Na^+ and K^+ substituted for Ba^{++}, or Al^{+3}, Fe^{+2}, Fe^{+3}, and Ni^{+2} substituted for Ti^{+4}.[4] Incorporation of a typical acceptor impurity oxide, Al_2O_3, can be written as

$$2BaO + Al_2O_3 \rightarrow 2Ba_{Ba} + 2Al_{Ti}^{\prime} + 5O_O + V_O^{\cdot\cdot} \tag{11}$$

The acceptor impurity oxide brings in less oxygen per cation than the oxide it replaces. These extrinsic oxygen vacancies offer an easy mechanism for incorporating a stoichiometric excess of oxygen according to Equation (9). Near the conductivity minima, the extrinsic $[V_O^{\cdot\cdot}]$ is not significantly affected by either reduction or oxidation and

$$[A^{\prime}] \approx 2[V_O^{\cdot\cdot}] \tag{12}$$

where $[A^{\prime}]$ represents the summation of all of the acceptor impurity species. The location of the conductivity minimum at $P_{O_2} = P_{O_2}^{O}$ is determined by the condition

$$\sigma_n = ne\mu_n = \sigma_p = pe\mu_p \quad (\text{at } P_{O_2}^{O}) \tag{13}$$

where σ_n and σ_p are the electron and hole conductivities and μ_n and μ_p are the respective mobilities. An expression for $P_{O_2}^{O}$ can be derived[5] from Equations (8), (10), (12), and (13)

$$P_{O_2}^{O} = \left(\frac{\mu_n}{\mu_p}\right)^2 \frac{1}{[V_O^{\cdot\cdot}]^2} \frac{K_n}{K_p} \tag{14}$$

It will be assumed for the evaluations in this paper that the mobilities, and K_n and K_p, which contain the enthalpies and entropies of the reduction and oxidation reactions, are functions only of temperature[4,5]. Thus P_{O_2} at a given temperature is a sensitive indicator of the relative concentration of extrinsic oxygen vacancies. The effect of added Al_2O_3 as an acceptor impurity has been previously described[5]. The addition causes an increase in the p-type conduction, a decrease in the n-type conduction, and the resulting decrease in $P_{O_2}^{O}$, as anticipated from Equation (14). The effect of 155 ppm of added Al^{+3}

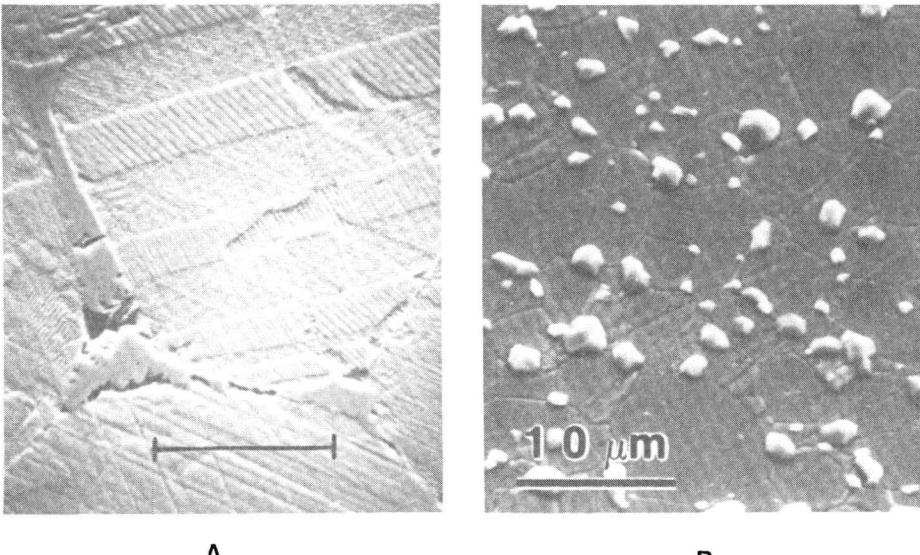

A   B

Fig. 2. A. Scanning electron micrograph of polished and acid-etched $BaTiO_3$ with Ba/Ti = 0.995. $Ba_6Ti_{17}O_{40}$ is present as a grain boundary phase. Marker=25μm. (Reprinted with permission from Ref. 5, Journal of the American Ceramic Society). B. Scanning electron micrograph of polished and thermally-etched $BaTiO_3$ with Ba/Ti = 1.04. The second phase is Ba_2TiO_4.

(expressed in terms of atomic substitution for Ti^{+4}) is clearly evident. This corresponds to an increase in $[V_O^{..}]$ by about 80 ppm, and since Al_{Ti}^{-} has been shown to be only about 30% effective in creating $V_O^{..}$, the sensitivity is significantly greater[5].

III. $BaTiO_3$

Examination of undoped $BaTiO_3$ samples by the SEM technique has disclosed the presence of second phases in samples with 0.1 mol % excess of either BaO [6] or TiO_2 [2], corresponding to Ba/Ti = 1.001 and 0.999 respectively. The second phases were identified as Ba_2TiO_4 and $Ba_6Ti_{17}O_{40}$ by X-ray diffraction patterns and by quantitative, wavelength-dispersive X-ray microanalysis. These are the adjacent phases indicated by studies of the $BaO-TiO_2$ system[7,8,9]. Examples of these second phases are shown in Figure 2. Equilibrium conductivity profiles for samples with excess BaO are shown in Figure 1. The absence of any discernable shift in the conductivity minimum indicates that the concentration of free, unassociated, extrinsic $V_O^{..}$ has been increased by <100 ppm. Conductivity measurements on $BaTiO_3$ with excess TiO_2 have given similar results. These results do not necessarily set the solubility limits, however, as it has been suggested that defect associates such as ($V_A'' V_O^{\bullet\bullet}$) could result from cation-cation nonstoichiometry of the type indicated by Equation (4), and this would keep the concentration of unassociated $V_O^{\bullet\bullet}$ at low levels (10,11). The behavior of $SrTiO_3$ and $CaTiO_3$, to be described later, indicates that such complexes are not formed in these compounds near 1000°C, however, and theoretical calculations of the association enthalpies in $BaTiO_3$ also predict their absence[12]. We do not believe that such complexes play a significant role in any of these compounds at high temperature. Thus the absence of any shift in $P_{O_2}^o$ due to the presence of excess BaO or TiO_2, sets a limit of <100 ppm on their solubility, according to Equations (2), (4), and (5).

In the case of $SrTiO_3$, large excesses of SrO can be accommodated by insertion of individual layers of SrO between blocks of perovskite structure. These structures correspond to a homologous series of compounds, $SrO \cdot nSrTiO_3$, where n corresponds to the number of perovskite layers (TiO_6 octahedra) between SrO layers. The structures have been described by Ruddlesden and Popper[13], and observed by the lattice imaging technique by Tilley[17]. The observation of Ba_2TiO_4 in $BaTiO_3$ containing an excess of only 0.1 mol % BaO indicates that accommodation of excess BaO according to a Ruddlesden-Popper type of superlattice does not occur in $BaTiO_3$. This represents a striking difference in behavior relative to $SrTiO_3$, which is otherwise extremely similar in its properties.

The absence of significant solubility for either BaO or TiO_2 in $BaTiO_3$ indicates that none of the possible incorporation reactions, Equations (1)-(6), is very favorable. Those mechanisms involving interstitial defects have already been excluded. Since $V_O^{\bullet\bullet}$ is known to be easily formed in these compounds, it can be surmised from the remaining reactions that V_{Ba}'', V_{Ti}'''', and Ba_{Ti}'' are also energetically unfavorable defects.

IV. $SrTiO_3$

SEM examination of $SrTiO_3$ having excess TiO_2 down to 0.5 mol % (Sr/Ti = 0.995) shows the presence of a Ti-rich second phase[14]. An example is shown in Figure 3. Energy dispersive analysis shows the presence of only Ti in the second phase when the particles are large compared with the volume activated by the electron beam. STEM analysis of the second phase gives the electron diffraction pattern of the rutile structure and confirms the presence of Ti as the only metallic constituent[15]. This is consistent with published

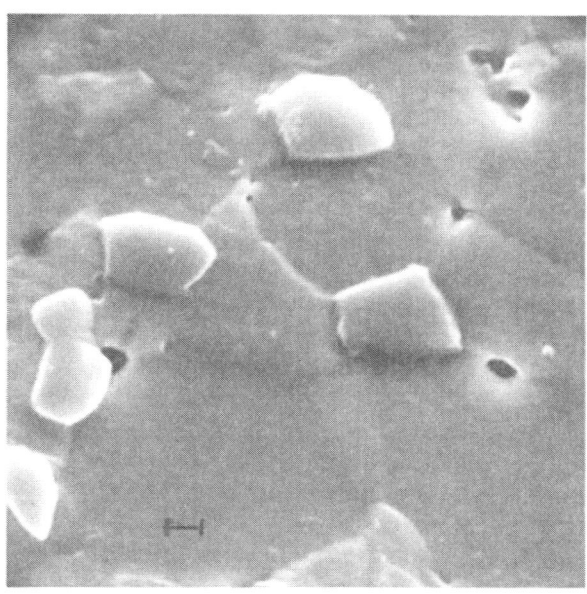

Fig. 3. Scanning electron micrograph of polished and acid-etched
SrTiO$_3$ with Sr/Ti=0.90. The second phase is TiO$_2$.
Marker=1μm. (Reprinted with permission from Ref. 14,
Journal of the American Ceramic Society).

phase diagrams that show no other phases between SrTiO$_3$ and TiO$_2$ [16].
As seen in Figure 4, the equilibrium electrical conductivity does
show a significant shift of $P_{O_2}^o$ to lower P_{O_2} in the presence of
excess TiO$_2$ [14]. For 0.5% excess TiO$_2$ (Sr/Ti=0.995) the shift corresponds to an increase in [$V_O^{\cdot\cdot}$] by about a factor of 3.9 for the slow-
cooled sample, according to Eq. (14), while the increase is by a
factor of 6.5 in the quenched sample, indicating that more excess
TiO$_2$ was retained in solid solution by the more rapid cooling[14].
The absence of any effect of cooling rate on the position of $P_{O_2}^o$
for the sample with Sr/Ti=1.000, as shown in Figure 4, shows that
the cooling rate by itself does not cause a shift; excess TiO$_2$ must
be present. If the concentration of extrinsic $V_O^{\cdot\cdot}$ is about 50 ppm
in the undoped sample, i.e. [A'] ∿ 100 ppm, which seems typical
for these materials[4,5], [$V_O^{\cdot\cdot}$] in the TiO$_2$-excess sample has been
raised to 200 and 320 ppm for the slow-cooled and quenched samples,
respectively. This is consistent with the observation of TiO$_2$ as a
second phase in samples of this composition (Sr/Ti = 0.995). It
is clear that SrTiO$_3$ has a greater tolerance than BaTiO$_3$ for excess
TiO$_2$, but the solubility seems to be limited to the order of a
few hundred ppm.

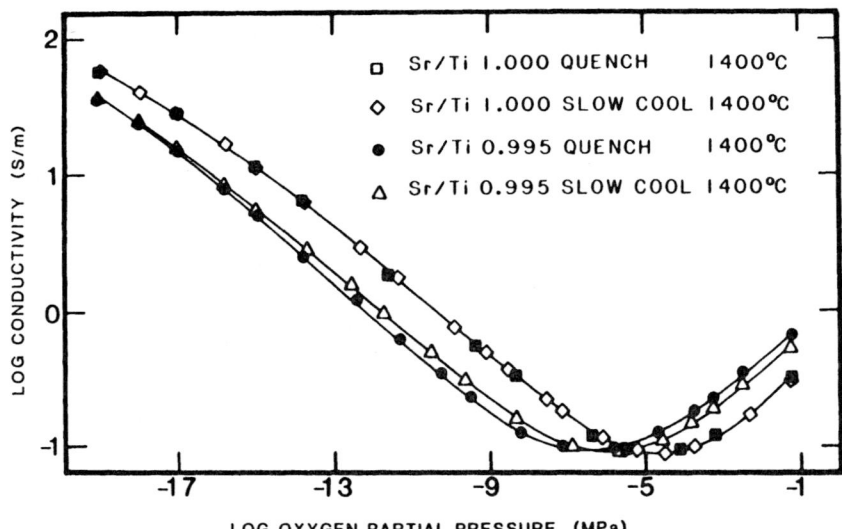

Fig. 4. Equilibrium electrical conductivity at 1000°C of quenched and slow-cooled SrTiO₃ with Sr/Ti=1.000 and 0.995. (Reprinted with permission from Ref. 14, Journal of the American Ceramic Society.)

A slight shift in $P_{O_2}^o$ to lower values has also been noted for samples with excess SrO[14]. This may represent a very small solubility according to Equations (2) or (5), involving either $V_{Ti}^{\prime\prime\prime\prime}$ or $Sr_{Ti}^{\prime\prime}$, respectively. It is known that larger amounts of excess SrO lead to superstructure ordering as described by Ruddlesden and Popper[13].

The movement of $P_{O_2}^o$ to lower values for excesses of either SrO or TiO₂ is consistent with an increase in $[V_O^{\cdot\cdot}]$ for either direction of nonstoichiometry, as indicated by Eqs (2) and (5) for excess AO and Eq (4) for excess TiO₂.

V. CaTiO₃

For CaTiO₃, the trend toward increasing solubility for excess TiO₂ with decreasing atomic weight of the alkaline earth continues. In addition, a substantial solubility for excess CaO is also apparent. Equilibrium conductivity results for CaTiO₃ with excess TiO₂ are shown in Figure 5. Not only is there a very substantial shift of $P_{O_2}^o$ to lower pressures, there is also a very pronounced raising and flattening of the minimum similar to that reported previously to result from any significant increase in $[V_O^{\cdot\cdot}]$ in BaTiO₃ [5] and SrTiO₃ [3]. This phenomenon is seen in SrTiO₃ as shown in Figure 4, but to much less extent because of the much lower solubility limit. The effect has been attributed to ionic conduction

80

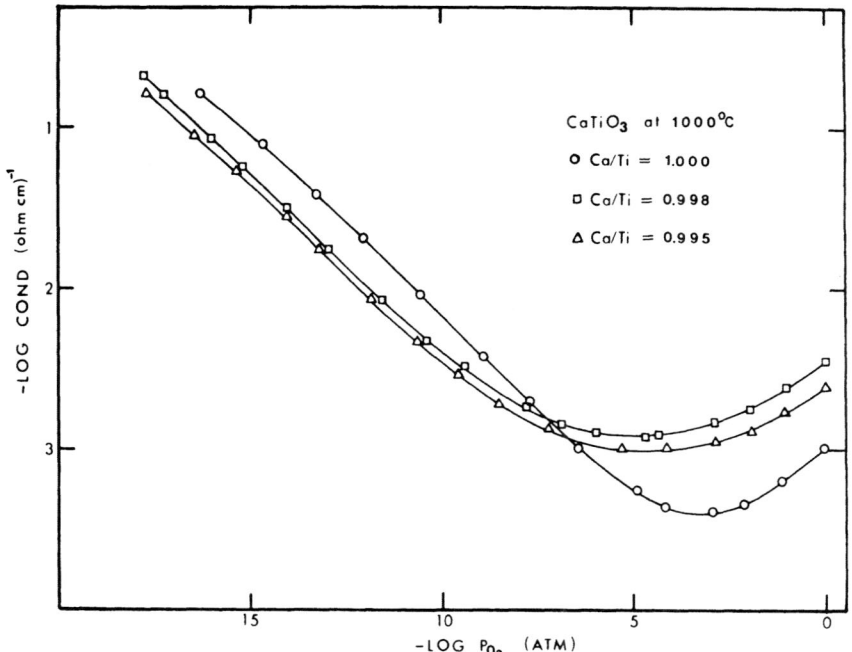

Fig. 5. Equilibrium electrical conductivity at 1000°C of CaTiO$_3$ with excess TiO$_2$.

by oxygen vacancies, and this hypothesis is supported by transport number measurements made by the oxygen concentration cell technique by Takakashi on very heavily acceptor-doped samples, e.g. SrAl$_{0.1}$Ti$_{0.9}$O$_{2.95}$ or CaAl$_{0.3}$Ti$_{0.7}$O$_{2.85}$ [19]. The effect is thus related to the more familiar case of acceptor-doped ZrO$_2$, i.e., calcia- or yttria-stabilized zirconia. It serves as additional evidence for a substantial increase in [V$_o^{\cdot\cdot}$], and, in the case of CaTiO$_3$ supports the observed shift in P$_{O_2}^o$ as an indication of substantial solubility of excess TiO$_2$ in CaTiO$_3$.

There is no apparent difference in the position of P$_{O_2}^o$ for the samples with Ca/Ti = 0.998 and 0.995 in Figure 5, but rather a general lowering of the entire conductivity curve for the latter sample. This generally downward movement persists for even larger additions of excess and may result from interference with the conduction processes by increasing amounts of second phase. The maximum shift in P$_{O_2}^o$ indicates that the concentration of V$_o^{\cdot\cdot}$ has been increased by about an order of magnitude by the solubility of excess TiO$_2$.

Only in the case of CaTiO$_3$ is there a major shift in P$_{O_2}^o$ in the presence of excess alkaline earth oxide. As shown in Figure 6 P$_{O_2}^o$ moves the lower values and the conductivity minima become higher

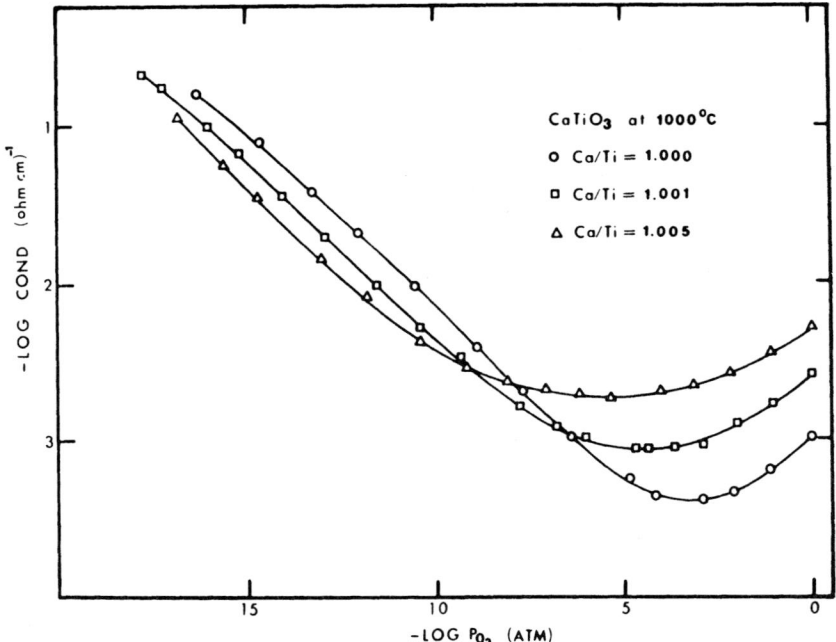

Fig. 6. Equilibrium electrical conductivity at 1000°C of CaTiO$_3$
         with excess CaO.

and flatter with increasing CaO, all indicative of increased $[V_o^{..}]$.
Equations (2) and (5) have been suggested as the most likely
incorporation reactions for excess alkaline earth oxide. The very
limited extent of solid solution in BaTiO$_3$ and SrTiO$_3$ indicates that
$V_{Ti}^{''''}$, $Ba_{Ti}^{''}$, and $Sr_{Ti}^{''}$ are not very favorable defects in the perovskite
lattice. There is substantial evidence for solubility of excess
CaO in BaTiO$_3$ with the formation of $Ca_{Ti}^{''}$, however, according to
the incorporation reaction

$$BaO + CaO \rightarrow Ba_{Ba} + Ca_{Ti}^{''} + 2O_o + V_o^{..} \qquad (15)$$

or, written alternatively in the sense of addition of CaO to
stoichiometric BaTiO$_3$

$$2CaO \rightarrow Ca_{Ba} + Ca_{Ti}^{''} + 2O_o + V_o^{..} \qquad (16)$$

This evidence includes a shift of $P_{O_2}^o$ to lower pressures in CaTiO$_3$-
BaTiO$_3$ solid solutions when (Ca+Ba)/Ti > 1, i.e. when there is insufficient room on the A-sites for the Ca[20]. The shift is
accompanied by a pronounced raising and flattening of the conductivity minima characteristic of increased $[V_o^{..}]$. The technique of
channel-enhanced electron microscopy also gives a positive indication of Ca on Ti-sites in such samples[21]. Donor impurities in

$BaTiO_3$ can be compensated by Ca when there is insufficient room for it on A-sites[22]. Finally, Ca on Ti-sites has been found to reduce the ferroelectric transition temperature substantially in $BaTiO_3$ [23], while Ca substituted for Ba is known to have no such effect. The solubility limit for Ca_{Ti}'' is about 2-3%. There is no apparent reason to expect V_{Ti}'''' to be any more favorable in $CaTiO_3$ than in $BaTiO_3$ and $SrTiO_3$, while the evidence for Ca_{Ti}'' in $BaTiO_3$ strongly suggests that it can be formed in $CaTiO_3$. Thus we propose that excess CaO is incorporated into $CaTiO_3$ by partial occupation of Ti sites, analogous to Eqs. (5) and (16). The solubility limit is similar to that for excess TiO_2 in $CaTiO_3$. The symmetrical movement of P_{O_2} for excesses of either CaO or TiO_2 is again indicative of the formation of $V_O^{..}$ by the incorporation of either species.

VI. CONCLUSIONS

The oxygen vacancy is known to be formed easily in the perovskite structure oxides, so the solubility of an excess of either binary oxide constituent is an indication of the ease of formation of the other defect involved in the incorporation reaction, i.e. V_A'' in the case of excess TiO_2 and V_{Ti}'''' or A_{Ti}'' in the case of excess AO. The increasing solubility of excess TiO_2 from essentially zero in $BaTiO_3$, to thousands of ppm in $CaTiO_3$ indicates that the A-site vacancy is more easily formed as the A ion becomes smaller. The lattice constants in this sequence change from 4.012 Å for $BaTiO_3$ (at 200°C), to 3.905Å for $SrTiO_3$, to $\simeq$3.82 Å for (pseudocubic) $CaTiO_3$ [24]. Thus while the three-dimensional, corner-sharing array of TiO_6 octahedra is the main factor in determining the lattice dimension, the structure does contract slightly around A ions of decreasing size. The "empty space" left by a missing A ion therefore also decreases somewhat. Moreover, the ionicity should decrease with decreasing size of the A ion so that the absolute charge on the A vacancy will also decrease. These factors may contribute to the relative stabilities of A-site vacancies in the three compounds.

The striking increase in solubility of excess AO for the case of CaO in $CaTiO_3$ is attributed to the ability of Ca^{++} to occupy the octahedral B-site to a significant degree. The alternative incorporation reaction involves V_{Ti}'''', Eq (2), and it is unlikely that this defect is much more favorable in $CaTiO_3$ than in the other two compounds. There is abundant evidence for Ca_{Ti}'' in $BaTiO_3$, and since the ionic radii of the A^{++} ions change more than the lattice parameter of the corresponding ABO_3 compounds ($r_{A^{++}}$ decreases 14% from Ba^{++} to Ca^{++} while a_o decreases only 4.8% from $BaTiO_3$ to $CaTiO_3$), it seems reasonable that Ca_{Ti}'' can also be a significant defect in $CaTiO_3$. Of course the fact that $CaTiO_3$ is slightly distorted from purely cubic symmetry may also be a contributing factor.

The slight shift in the conductivity minimum toward lower P_{O_2} for $SrTiO_3$ containing an excess of SrO may indicate a very small range of solid solubility. This is supported by the absence of such a shift in the $BaTiO_3$-BaO system. There is also evidence for random substitution of Sr^{++} on the B-sites of a perovskite-related structure in the work by Lecomte et al.[25] on the system $Sr(Sr_{1/3+x}Nb_{2/3-x})O_{3-3/2x}$. For $0 < x < 1/6$, the B-sites are arranged in alternate layers perpendicular to <111> with one set of layers containing only Nb, while the other contains varying amounts of both Sr and Nb. The series does not extend to very small fractional occupation of B-sites by Sr, however.

The presence of Ba_2TiO_4 as a second phase in all samples of $BaTiO_3$ with Ba/Ti $\gtrsim$ 1.001 indicates that incorporation of excess BaO by a Ruddlesden-Popper type of superlattice does not occur. This is a major difference between the $BaTiO_3$ and $SrTiO_3$ systems.

VII. ACKNOWLEDGEMENT

This work was supported by the Ceramics Programs of the Divisions of Materials Research of the National Science Foundation and the Office of Naval Research.

REFERENCES

1. M. Pechini, U.S. pat. 3,330,697 (1967).
2. R. K. Sharma. N.-H. Chan, and D. M. Smyth, J. Am. Ceram. Soc. 64:448 (1981).
3. N.-H. Chan, R. K. Sharma, and D. M. Smyth, J. Electrochem. Soc. 128:1762 (1981).
4. N.-H. Chan, R. K. Sharma, and D. M. Smyth, J. Am. Ceram. Soc. 64:556 (1981).
5. eidem, ibid. 65:167 (1982).
6. Y. H. Hu, M. P. Harmer, and D. M. Smyth, to be presented at the 37th Pacific Coast Regional Meeting of the American Ceramic Society, San Francisco, Oct. 28-31 (1984).
7. D. E. Rase and R. Roy, J. Am. Ceram. Soc. 38:102 (1955).
8. H. M. O'Bryan, Jr. and J. Thompson, Jr., J. Am. Ceram. Soc., 57:522 (1974).
9. T. Negas, R. S. Roth, H. S. Parker and D. Minor, J. Solid State Chem. 9:297 (1974).
10. N. G. Eror and D. M. Smyth, J. Solid State Chem., 24:235 (1978).
11. N.-H. Chan and D. M. Smyth, J. Electrochem. Soc., 123:1584 (1976).
12. G. V. Lewis and C. R. A. Catlow, Radiat. Eff., 73:307 (1983).
13. S. N. Ruddlesden and P. Popper, Acta Crystallogr., 11:54 (1958).

14. S. Witek, D. M. Smyth, and H. Pickup, J. Am. Ceram. Soc. 67:372 (1984).

15. N. Stenton and M. P. Harmer, pp. 156-65 in Advances in Ceramics: VII, Additives and Interfaces in Electronic Ceramics. Edited by M. F. Yan and A. H. Heuer. The American Ceramic Society, Columbus, OH 1984.

16. E. M. Levin, C. R. Robbins, and H. F. McMurdie, Phase Diagrams for Ceramists, 1964 (Figs. 297 and 298) and 1969 (Fig. 2334). Edited by M. K. Reser. The American Ceramic Society, Columbus, Ohio.

17. R. J. D. Tilley, J. Solid State Chem., 21:293 (1977).

18. X. W. Zhang, M. Lal, and D. M. Smyth, Paper 122-E-84 presented at the 86th Annual Meeting of the American Ceramic Society, Pittsburgh, May 3 (1984).

19. T. Takahashi, in "Physics of Electrolytes," Vol. 2, J. Hladik, Editor, Chap. 24, Academic Press, New York (1972).

20. J. Appleby, Y. H. Han, and D. M. Smyth, paper 13-E-83, presented at the 85th Annual Meeting of the Am. Ceram. Soc., Chicago, April 25 (1983).

21. H. M. Chan, M. P. Harmer, M. Lal, and D. M. Smyth, to appear in the Proceedings of the Symposium on Electron Microscopy of Materials, Annual Meeting of the Materials Research Society, Boston, November (1983).

22. Y. Sakabe, J. Appleby, Y. H. Han, D. Wintergrass, and D. M. Smyth, Paper 123-E-84 presented at the 86th Annual Meeting of the American Ceramic Society, Pittsburgh, May 3 (1984).

23. M. Lal, Y. H. Hu, M. P. Harmer, and D. M. Smyth, to be presented at the 37th Pacific Coast Regional Meeting of the American Ceramic Society, San Francisco, Oct. 28-31 (1984).

24. O. Muller and R. Roy, "The Major Ternary Structural Families", Springer, New York (1974).

25. J. Lecomte, J. P. Loup, M. Hervieu, and B. Raveau, Phys. Stat. Sol. 66:551 (1981).

DEFECT STRUCTURE AND TRANSPORT IN

OXYGEN EXCESS CERIUM OXIDE-URANIUM OXIDE SOLID SOLUTIONS

H. L. Tuller and T. S. Stratton*

Department of Materials Science and Engineering
Massachusetts Institute of Technology
Cambridge, MA 02139, U.S.A.
*Presently at Raychem Corp., Menlo Park, CA

INTRODUCTION

Given the ease with which the binary oxides CeO_2 and UO_2 deviate from stoichiometry, these systems represent the paradigms for non-stoichiometry in the fluorite-related oxides with CeO_2 deviating in the direction of oxygen deficiency and UO_2 in the direction of oxygen excess. Nevertheless, important information concerning intrinsic ionic disorder in both systems and the defect structure of oxygen excess CeO_2 or oxygen deficient UO_2 is sorely lacking. In this study we report results of electrical measurements and defect modeling on cerium oxide-uranium oxide solid solutions which traverse the composition range from oxygen excess to oxygen deficiency for a series of isotherms ranging from $\sim$450-1200°C. The following features will be emphasized. Cerium dioxide may be induced to become oxygen excess under accessible experimental conditions by the addition of as little as 0.1% uranium. The Frenkel product is obtained for this system for the first time experimentally. Values for the electron mobility are obtained in a relatively straightforward manner. Subsequent articles by the authors will deal with (a) results obtained for a much broader range of compositions (including nearly pure UO_2), (b) implications of defect ordering on this type of analysis and (c) analysis of electronic carrier transport based on combined electrical conductivity and thermoelectric power measurements.

The compound cerium dioxide is of interest for a number of scientific and technological reasons. First, it deviates readily from stoichiometry in the direction of oxygen deficiency at elevated temperatures and reduced oxygen partial pressures, P_{O_2}. This transforms CeO_2 from an electrical insulator to an n-type semiconductor with an

87

exceptionally large oxygen vacancy density at the fluorite phase limit ($x \sim 0.3$ for CeO_2-x at 1000°C) [1]. The details of the defect equilibria have been interpreted in terms of the generation of doubly and singly ionized oxygen vacancies which serve as donor centers with the former dominating at smaller values of x [2]. Electrons move through the lattice via a small polaron hopping process within the narrow Ce 4f conduction band with a hopping energy of ~ 0.3 eV at low values of x which increases to nearly 0.5 eV for high values of x [3].

Large concentrations of oxygen vacancies may also be introduced by acceptor doping, e.g. Ca, Y, rendering the compound an oxygen ion conducting solid electrolyte at high P_{O_2} and a mixed ionic-electronic conductor under reducing conditions [4].

Although a good deal is now known concerning the defect equilibria and transport properties of oxygen deficient CeO_2, little or no information has been available concerning intrinsic ionic disorder at stoichiometry or at oxygen excess conditions. Since it is well known that UO_2, with the same crystal structure as CeO_2, deviates extensively in the direction of oxygen excess by the formation of oxygen interstitials [5], it might be expected that CeO_{2+x} could also be formed under the right circumstances. Examining the relative oxidation states that Ce and U readily acquire, i.e. Ce (3,4+) and U (4,5,6+) it becomes clear that CeO_2 does not readily oxidize given the stability of the 4+ valence state even under highly oxidizing conditions. On the basis of some preliminary calculations, we concluded that this oxidation state limitation could be lifted by the addition of relatively small concentrations of higher value donor impurities. Below, we summarize the essence of these calculations and later apply them to the interpretation of electrical conductivity measurements obtained for a series of U-doped CeO_2 solid solutions.

DEFECT CALCULATIONS

To simplify the present analysis, we consider only those defects which are likely to be of importance in the defect regimes both within and bordering the stoichiometric composition. These include doubly ionized oxygen vacancies ($V_O^{\cdot\cdot}$) and interstitials (O_i''), singly ionized uranium ions sitting substitutionally on cerium sites ($U_{Ce}^{\cdot}$) and electrons (e') and holes (h°). Based on an analysis of the electronic band structure of CeO_2 and UO_2 [6] we assume the uranium donor lies above the cerium conduction band and is thus ionized at all temperatures.

The defect relations which need be considered for this analysis are listed in Table 1. The first and second equations correspond to intrinsic thermal generation of oxygen Frenkel and electron-hole pairs respectively. Next, the reduction reaction in which oxygen leaves the

Table 1.   Defect Reactions Appropriate to CeO_2-UO_2 System

$$O_O \rightarrow O_i'' + V_O^{\cdot\cdot} \qquad\qquad [O_i''] \; [V_O^{\cdot\cdot}] = K_f(T) \qquad\qquad (1)$$

$$null \rightarrow e' + h^{\cdot} \qquad\qquad n \; p = K_e(T) \qquad\qquad (2)$$

$$O_O \rightarrow V_O^{\cdot\cdot} + 2e + \tfrac{1}{2}O_2 \qquad [V_O^{\cdot\cdot}]n^2 P_{O_2}^{\tfrac{1}{2}} = K_r(T) \qquad\qquad (3)$$

$$n + 2[O_i''] = p + 2[V_O^{\cdot\cdot}] + [U_{Ce}^{\cdot}] \qquad\qquad (4)$$

lattice for the gas phase leaving behind a vacant site and two electrons is given. Last is the electroneutrality relation which equates
the total net negative and positive charge. Square brackets indicate
concentrations in $\#/cm^{-3}$ while the K values correspond to the thermally activated equilibrium constants appropriate for reactions
(1) – (3).

Solutions to the four simultaneous equations give the four unknown defect concentrations in terms of the equilibrium constants,
the P_{O_2} and the uranium content. To simplify matters, the equations
of Table 1 are commonly solved in a piecewise linear fashion by eli-
minating all but the two defects with the highest concentrations for
chosen conditions of temperature and atmosphere. Solutions obtained
in this manner are listed in Table 2, and schematically plotted as a
function of P_{O_2} for a single isotherm in Figure 1.

The utility of these relations can be seen by examining the
electron concentration, n, for example. At low oxygen partial pressures, region I, n is proportional to $P_{O_2}^{-1/6}$, in region II n is
P_{O_2}-independent and in region III n is proportional to $P_{O_2}^{-1/4}$.
Thus the controlling defect regime, under given conditions, can be
established from the n-type electronic conductivity. Once the defect
region is identified, the equilibrium constants can in principle be
derived from examination of the defect densities as a function of
temperature at constant P_{O_2}.

Alternatively, one may solve for the P_{O_2}'s specifying the transition between any two adjacent defect regimes as given in Table 3:
these are particularly useful since they enable determination of the
equilibrium constants without a-priori knowledge of, in this case,
the electron mobility.

Additional key features of this analysis may be noted by further
examination of Figure 1. First, in region II the P_{O_2}-independent
electron density is constant and equal to the uranium density. This
therefore serves as a temperature independent reference point. This
has two important implications.

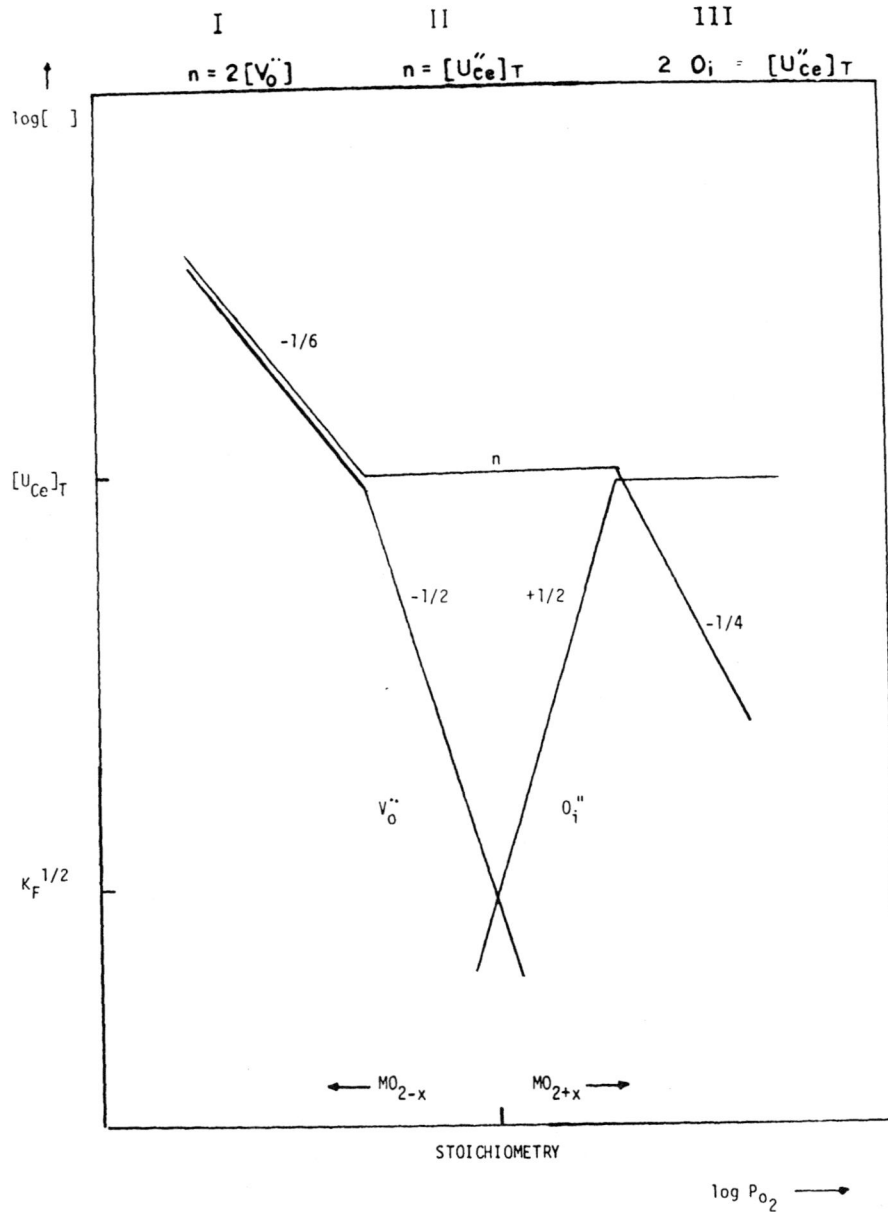

Figure 1. Schematic drawing of concentration dependence of defects on oxygen partial pressure.

TABLE 2

Region	I	II	III
Electroneutrality relation	$n = 2[V_O^{\bullet\bullet}]$	$n = [U_{Ce}^\bullet]$	$2[O_i''] = [U_{Ce}^\bullet]$
P_{O_2} range	low	intermediate	high
$[V_O^{\bullet\bullet}]$	$2^{-2/3} K_r^{1/3} P_{O_2}^{-1/6}$	$K_r [U_{Ce}^\bullet]^{-2} P_{O_2}^{-1/2}$	$2K_f [U_{Ce}^\bullet]^{-1}$
n	$2^{1/3} K_r^{1/3} P_{O_2}^{-1/6}$	$[U_{Ce}^\bullet]$	$2^{-1/2} K_r^{1/2} K_f^{-1/2} [U_{Ce}^\bullet]^{1/2} P_{O_2}^{-1/4}$
p	$2^{-1/3} K_f K_r^{-1/3} P_{O_2}^{1/6}$	$K_e [U_{Ce}^\bullet]^{-1}$	$2^{1/2} K_f^{1/2} K_e K_r^{-1/2} [U_{Ce}^\bullet]^{1/2} P_{O_2}^{1/4}$
$[O_i'']$	$2^{2/3} K_f K_r^{-1/3} P_{O_2}^{1/6}$	$K_f K_r^{-1} [U_{Ce}^\bullet]^2 P_{O_2}^{1/2}$	$2^{-1} [U_{Ce}^\bullet]$

(a) Measurement of the electrical conductivity as a function of temperature can easily establish whether the electron mobility is thermally activated since

$$\sigma = ne\mu = [U_{Ce}^{\cdot}]e\mu = \text{constant} \times \mu.$$

We later use this feature to show that μ is in fact activated.

(b) The dominant ionic defects, $V_O^{\cdot\cdot}$ and $O_i^{\prime\prime}$, cross precisely at the center of this region. This allows us to specify the P_{O_2} corresponding to stoichiometry and more importantly the magnitude of the Frenkel product K_f since $V_O^{\cdot\cdot}$ and $O_i^{\prime\prime}$ follow $P_{O_2}^{-1/2}$ and $P_{O_2}^{+1/2}$ power laws respectively with the $[U_{Ce}^{\cdot}]$ level as a reference at the boundaries of region II.

Second, in region III the oxygen excess is held constant and equal to $\frac{1}{2}[U_{Ce}^{\cdot}]$. Thus oxygen diffusion measurements in this region should enable determination of the migration energy of oxygen interstitials. Such measurements are now proceeding by collaboration with colleagues at Imperial College, London.

The sharp change in slopes at the boundaries between defect regimes, as illustrated in Figure 1, is an artifact of the approximations made above in the neutrality relations. Actual experimental data exhibit smooth transitions from one defect regime to the next. When the data do not extend well beyond such boundaries, an additional charge species should be included in the electroneutrality relation. Thus, for example, in the vicinity of the I-II boundary, the neutrality condition becomes

$$n = 2[V_O^{\cdot\cdot}] + [U_{Ce}^{\cdot}] \tag{5}$$

Substituting into equation (3) one obtains a cubic equation in n given by

$$n^3 - [U_{Ce}^{\cdot}]n^2 - 2K_r P_{O_2}^{-\frac{1}{2}} = 0 \tag{6}$$

which may be transformed to an equation in terms of conductivity σ by multiplying n through by $q\mu$ where q is the charge on an electron and μ the mobility. Fitting of the cubic equation to the data may then be used to extract K_r.

Similarly at the II-III boundary one obtains a similar expression after combining the neutrality equation

$$n + 2[O''_i] = [U^{\bullet}_{Ce}]$$

with equations (1) and (3). Here one may obtain K_f after fitting the data.

EXPERIMENTAL

Three compositions of U-doped CeO_2 are examined in this paper, i.e. 0.1, 1.0 and 5.0 mol % UO_2 in CeO_2. These were prepared by co-precipitation of a nitrate solution by NH_4 followed by rinses in deionized water, acetone-toluene and acetone. The resulting cakes were ground and fired in hydrogen at 840°C for one hour. These were then die pressed at 3,000 psi followed by isostatic pressing at 40,000 psi. The compacts were then fired in $CO/CO_2 = 1/10$ at 1515°C for 48 hours. Densities varied from 80-92% of theoretical density.

Electrical measurements were performed by the conventional four probe dc technique. Spot checks with a Wayne Kerr AC Impedance Bridge (model B634) yielded conductivities in excellent agreement with dc values.

The oxygen partial pressure was controlled by mixing O_2/Ar and CO/CO_2 gases in appropriate proportions. A stabilized zirconic cell was used to monitor the actual P_{O_2} in the sample chamber. Further experimental details may be found in [6].

RESULTS AND DISCUSSIONS

A key feature of the predictions illustrated in Figure 1 is the expected P_{O_2} dependence of the n-type electronic conductivity. Specifically, one expects a P_{O_2}-independent plateau bounded on one side by a more highly conducting oxygen deficient region characterized by a $P_{O_2}^{-1/6}$ dependence and bounded on the other side by a less highly conducting oxygen excess region characterized by a $P_{O_2}^{-1/4}$ dependence. All of these features are in fact observed in the electrical conductivity data presented in Figure 2. Here measurements of electrical conductivity, made for the three compositions examined in this study at 666°C, are plotted as a function of P_{O_2}.

Although not all the features are observed for any one composition over the 22 orders of magnitude in P_{O_2}, all of the elements are found within the group as a whole. For example, for the 5% dopant level the conductivity plateau is bounded at high P_{O_2} by a $\sim P_{O_2}^{-1/4}$ power law while for the 0.1% dopant level it is bounded by a $\sim P_{O_2}^{-1/6}$ power law at low P_{O_2}. These observations are consistent with the predictions of Table 3 which requires that the positions of both the I-II and II-III boundaries shift to lower P_{O_2} with increasing $[U^{\bullet}_{Ce}]$

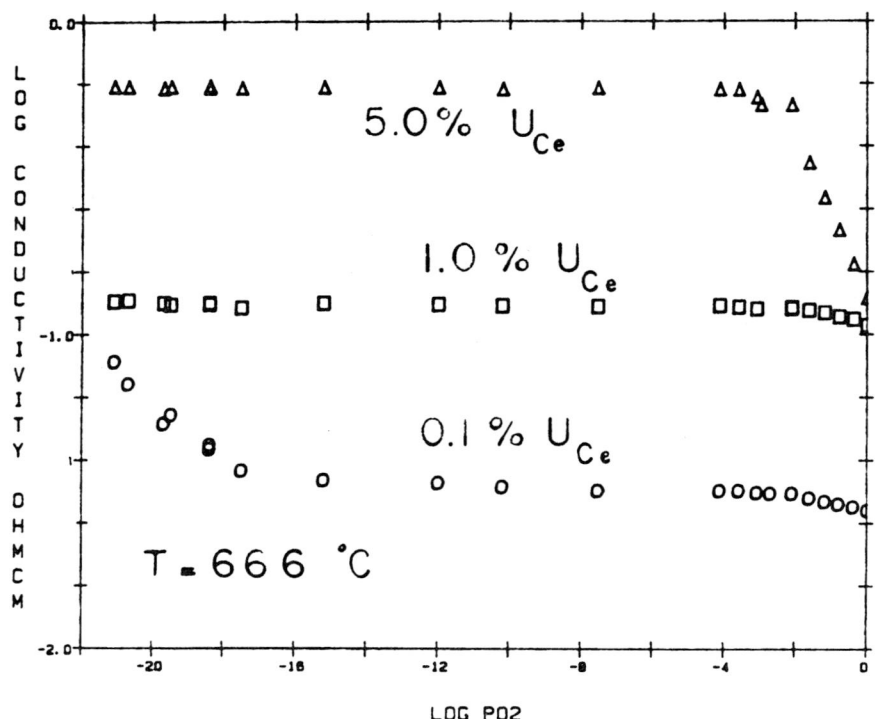

Figure 2. log σ versus log P_{O_2} at 666°C for three specimens of U-doped CeO_2 with varying U content.

at a given temperature, T.

In Figure 3 we present the measurements of electrical conductivity obtained for the 1 mol % dopant level as a function of P_{O_2} for a series of isotherms ranging from 461-1275°C. Similar data not shown was obtained for the other dopant levels. Here we find that as we increase the temperature the flat appearance of the 666°C isotherm begins to disappear (also illustrated in Figure 2) and shows a successively stronger P_{O_2} dependence with increasing T which begins at successively higher P_{O_2}'s. This is again consistent with the predictions of Table 3.

Values of K_r and K_f were obtained from the data for 0.1, 1.0 and 5.0 mol % U-doped CeO_2 by the fitting procedure described above. These values are plotted as a function of reciprocal temperature in Figures 4 and 5 respectively. The lines through the data represent linear least squares fits to the values of K_r and K_f evaluated at

Table 3. Boundaries for Defect Regions Described in Table 2

Boundaries	I-II	II-III
Defects Equal at Boundaries	$2[V_o^{\cdot\cdot}] = [U_{Ce}^{\cdot}]$	$n = 2[O_i'']$
P_{O_2} at Boundary	$2^2 K_r^2 [U_{Ce}^{\cdot}]^{-6}$	$2^{-1} K_f^{-2} K_r^2 [U_{Ce}^{\cdot}]^{-2}$

each temperature. The values of K_r are compared with earlier data derived by Tuller and Nowick [2] for undoped CeO_2. The data is seen to be in good agreement, showing no systematic dependence on uranium content up to 5 mol %.

The results for the frenkel product are of particular interest. For the 5 mol % dopant level we estimate an intrinsic carrier density of $\sim 2.5 \times 10^{17}$ cm^{-3} at 1273°K, a temperature less than one-half of the melting temperature of CeO_2. This therefore corresponds to an exceptionally high intrinsic ionic disorder level. This is, however, consistent with the large deviations from stoichiometry attained at these temperatures under reducing conditions fol-

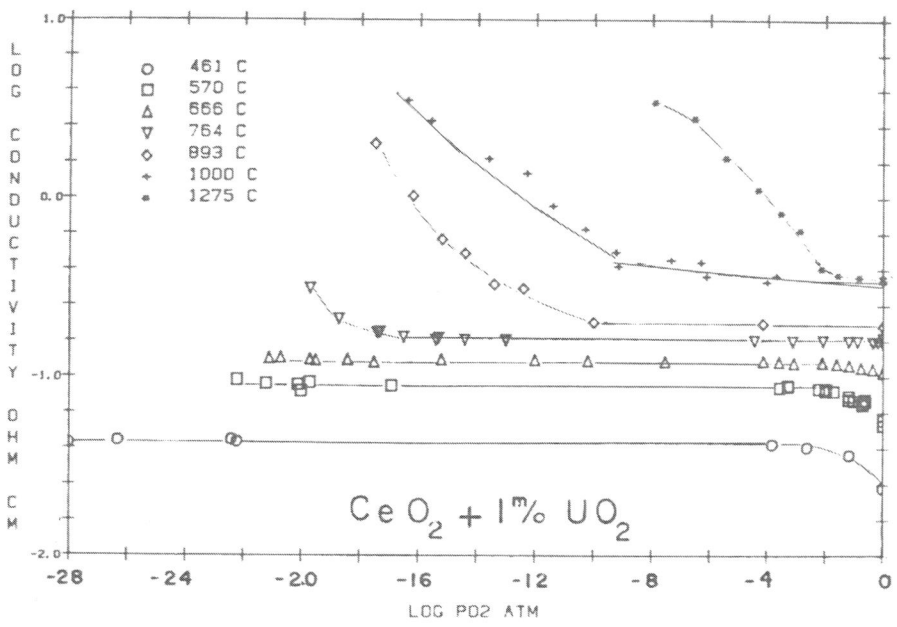

Figure 3. log σ versus log P_{O_2} for a series of isotherms for U-doped CeO_2 with 1 mol % UO_2.

95

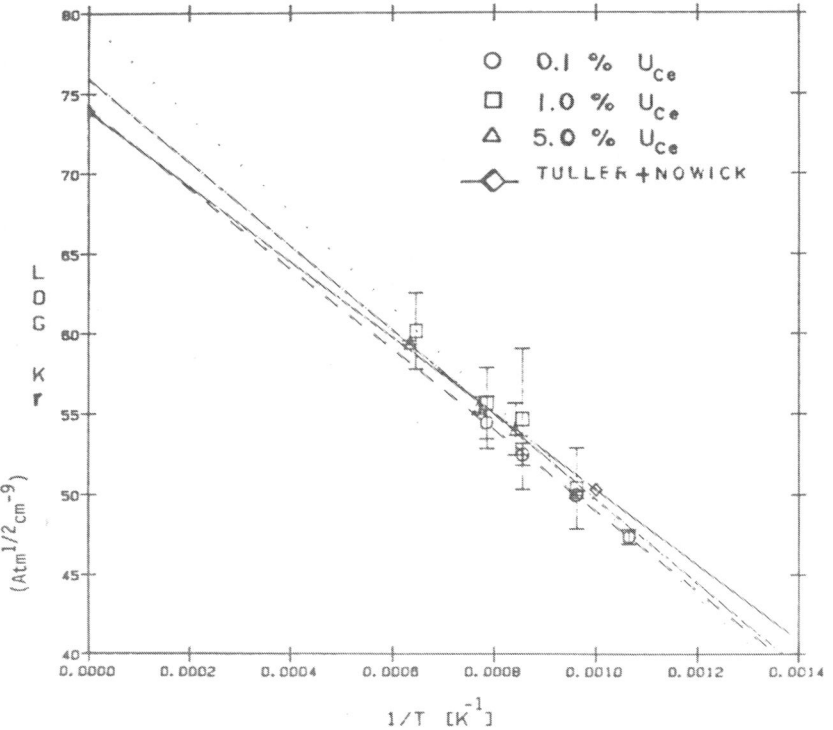

Figure 4. log K_r versus 1/T. Values of K_r obtained in this study are compared with those of Tuller and Nowick [3] obtained for undoped CeO_2.

lowing the theory developed originally by Anderson [7]. A more detailed analysis of these data will be presented in a subsequent publication after a more precise computer fitting routine is completed.

Lastly, we present measurements of the electronic conductivity obtained in the plateau region, where the carrier density is constant, as a function of temperature. These results are shown plotted in Figure 6 as log σT versus 1/T for the three compositions treated. All three systems are observed to be characterized by a thermally activated mobility with activation energies in the vicinity of 0.3 eV, a value in good agreement with previous results obtained for nearly stoichiometric CeO_2 [3]. Although σ increases with $[U_{ce}^{\cdot}]$ as expected it does not follow a linear dependence. This feature has been examined in some detail in related studies in this laboratory

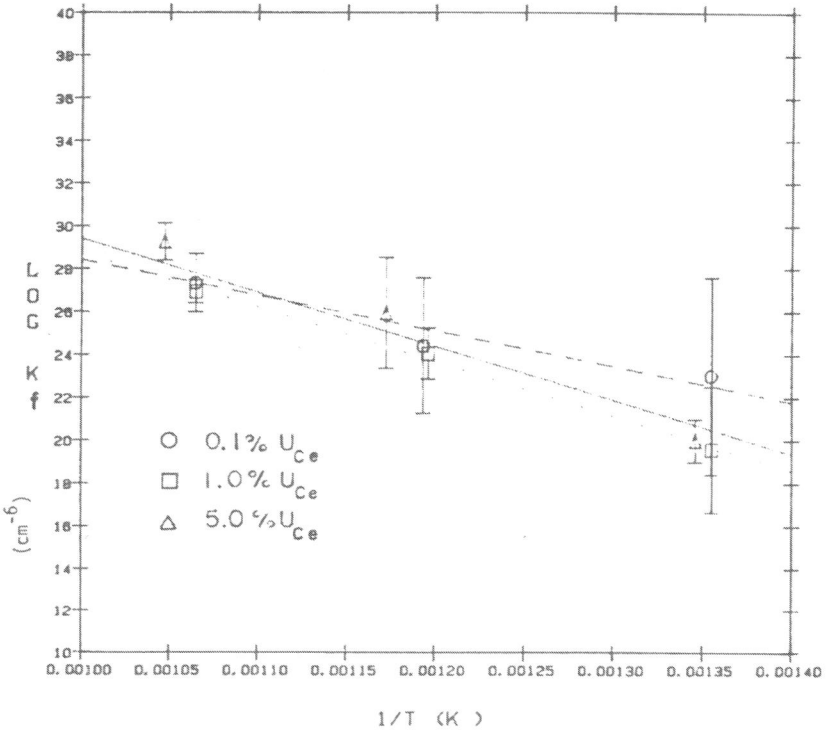

Figure 5. log K_r versus 1/T.

which include thermoelectric power measurements and are discussed in an upcoming publication.

SUMMARY

A defect model has been presented for U-doped CeO_2 which speci-fies the criteria which need be satisfied to confirm that one has reached (a) the condition of oxygen deficiency; (b) the condition of stoichiometry; and (c) the condition of oxygen excess. Further, once these criteria are met, relations are presented which enable the derivation of a number of key parameters including K_r, K_f and μ, the equilibrium constants for reduction and frenkel defect pair formation and the electron mobility respectively.

Electrical conductivity measurements performed on U-doped CeO_2 (0.1, 1.0, and 5.0 mol %) as a function of temperature and P_{O_2} have

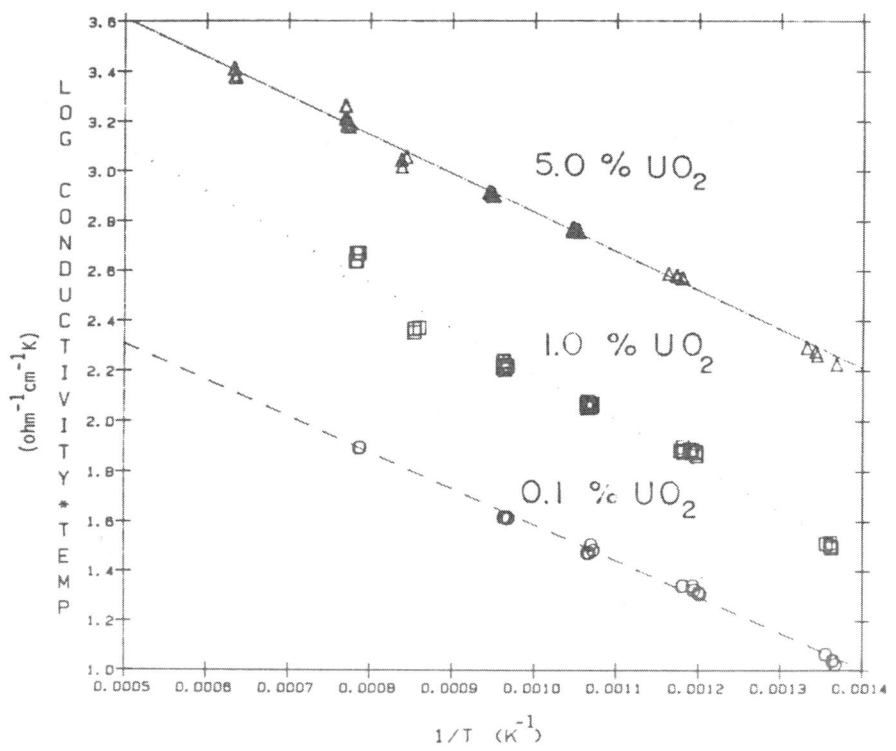

Figure 6. log σT versus 1/T. Measurements of electronic
conductivity were obtained in region II where
n is a constant.

been found to be in good agreement with the predictions of the defect
model and have confirmed the ability to achieve oxygen excess CeO_2.
Fits of the data to the model allowed derivation of the parameters
K_r, K_f and μ for this system. Of special interest is the high
intrinsic Frenkel disorder obtained at temperatures far below the
melting point of CeO_2 and which explain in part the high levels of
non-stoichiometry achieved in CeO_{2-x} under reducing conditions.

Acknowledgements

Support by the National Science Foundation for this research
under Grant No. DMR-82-03697 is gratefully acknowledged. Initial
support for this work by the Department of Energy under Grant No.
DE-AC02-77ET04436 is also appreciated.

REFERENCES

1. D. J. M. Bevan and J. Kordis, <u>J. Inorg. Nucl. Chem.</u> 26:1509 (1964).
2. H. L. Tuller and A. S. Nowick, <u>J. Electrochem. Soc.</u> 126:209 (1979).
3. H. L. Tuller and A. S. Nowick, <u>J. Phys. Chem. Sol.</u> 38:859 (1977).
4. H. L. Tuller and A. S. Nowick, <u>J. Electrochem. Soc.</u> 122:255 (1975).
5. B. T. M. Willis, <u>Nature</u> 197:755 (1963).
6. T. G. Stratton, Electrical Properties and Defect Structure of UO_2-CeO_2 Solid Solutions, Ph.D. Thesis, Department of Materials Science and Engineering, M.I.T. (1984).
7. J. S. Anderson, <u>Proc. Royal Soc.</u> A185:69 (1946).

TRANSPORT IN ANION DEFICIENT FLUORITE OXIDES

A.N. Cormack and C.R.A. Catlow

Department of Chemistry
University Collge London
20 Gordon Street, London,   WC1H OAJ, UK

INTRODUCTION

The enhanced oxygen transport in anion deficient oxides based on the fluorite structure has been widely exploited technologically in recent years, finding application in such areas as energy storage (secondary batteries) and gas sensors devices.

The oxygen vacancy concentration may be accommodated nonstoichiometrically either through reduction of the cations, as in CeO_{2-x} PrO_{2-x} or TbO_{2-x}, or through doping with an aliovalent cation such as Mg, Ca or Y, Yb, for example [1]. Obviously, transport dominated processes such as ionic conductivity and diffusion will be significantly influenced by interactions between the dopant, or reduced cations and the oxygen vacancies which are introduced as charge compensators.

Experimental studies indicate that there is an associative interaction to form defect pairs, comprising a dopant cation and an oxygen vacancy, at dilute concentrations of dopant (< 1%). Interestingly, heavily doped samples seem to behave in a similar manner (see the paper by Nowick in these proceedings), although the defect structure must be considerably more complex and cannot be simply described in terms of non-interacting defect pairs.

In this presentation we will describe the use of a theoretical approach, based on computer simulations, towards understanding the

101

defect energetics of defect association and migration. This approach
is now well established as a reliable tool for the calculation of
defect energies; its limitations are essentially constrained by the
choice of interatomic potentials used in the model. This aspect will
be further considered below, but for detailed exposition of the computational methods, reference should be made elsewhere. (2) (3)

Here we will concentrate on two aspects of transport in anion
deficient fluorites. Firstly we will take the pyrochlore $Gd_2Zr_2O_7$, in
which the vacancies are regularly ordered to give a superlattice
structure whose unit cell contains eight fluorite cells (a-pyro = 2a-
fluorite). Secondly, we will consider the tetragonal phase of ZrO_2,
in particular contrasting its behaviour with the fully stabilised
cubic phase.

$Gd_2Zr_2O_7$ PYROCHLORE

The prospect of oxygen conductivity in this material immediately
raised two structurally related problems. One, do the vacancies (on
the fluorite sublattice) in the structure contribute to, or take part
in, oxygen migration, and, two, which of the possible jump mechanisms
is involved. A projection of the structure is depicted in figure (1)
from which it can be seen that there are three possible simple
vacancy jumps between the two kinds of oxygen site.

From the theoretical viewpoint this is also an interesting material
since the question arises as to whether interatomic potentials derived
for binary oxides - ZrO_2 and Gd_2O_3 - can be transferred to ternary
or mixed metal oxide, We were able to conclude that this was,
in fact, possible, providing care was taken over the choice of shell
parameters used to model the ionic polarisabilities. A fuller
discussion of this topic is not suited to this presentation and is
elaborated in greater detail elsewhere(4).

We record in table (1) the interatomic potential model used, together
with a comparison of the observed and calculated physical properties,
from which the efficacy of the model may be predicated.

The activation energies for the possible vacancy jumps were obtained
by calculating the potential surface between the two sites. These
activation energies, which may be compared with the expermental data
of van Dijk de Vries, and Burggraaf (5), are listed in table (2) and
enable one to determine that the likely mechanism is a vacancy
jumping along <100> between two O 48(f) sites. Note that <110> jumps

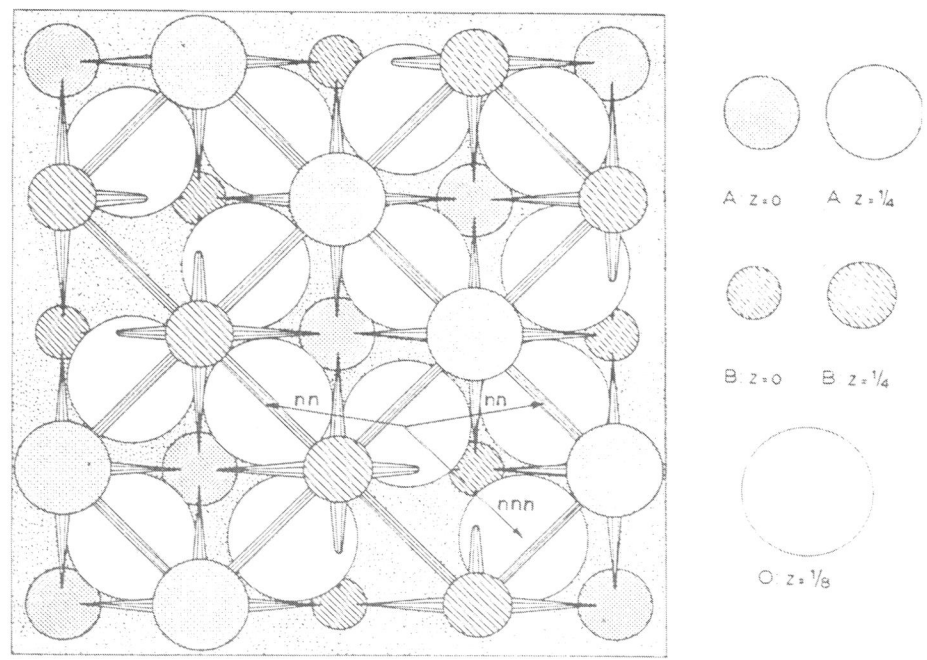

Figure 1   Projection of the pyrochlore structure on (100).   A
represents Gd, B represents Zr and the O48(f) positions are
adjacent to the vacant sites.

TABLE 1(a)   Interatomic Potential Model for Pyrochlore, $Gd_2Zr_2O_7$

Potential form: $V(r) = A \exp(-r/\rho) - Cr^{-6}$

Interaction	A	ρ	C
Gd ... O	1336.76	0.3551	0.0
Zr ... O	1453.80	0.3500	0.0
O ... O	22764.30	0.1490	27.89

Shell model parameters	Shell Charge	Spring Constant
Gd	-0.25000	145.000
Zr	0.61100	250.000
O	-2.77739	26.793

TABLE 1(b)   Observed and Calculated Parameters for $Gd_2Zr_2O_7$

Parameter	Observed	Calculated	
Lattice constant/Å	10.532	10.532	
Oxygen O48(f) parameter	0.410	0.4096	(for core
Bulk modulus, K/GPa	178	210	positions)
Dielectric Constant, ε_o	33	24.8	

TABLE 2   Activation Energies for Oxygen Vacancy Jumps in $Gd_2Zr_2O_7$

jump	energy/eV
O 8(a) ... O48(f)	2.5
O48(f) ... O48(f)*	0.9
experimental ΔH	0.8
(from van Dijk et al (5))	

*For jumps along <100>. Note that <110> jumps cannot lead to diffusion.

TABLE 3   Structural Properties of Tetragonal ZrO_2

	observed*	calculated
lattice parameter, $a_o/\overset{\circ}{A}$	5.11	5.11
axial ratio c/a	1.024	1.028
oxygen parameter	0.185	0.190
(¼ in fluorite structure)		

*From Teufer (11).

TABLE 4(a)   Calculated Activation Energies (in eV) for Oxygen Vacancy Migration in ZrO_2

	cubic	tetragonal*
no dopants	0.005	0.30
one dopant (Y)	0.21	1.07

*Calculated for vacancy jump along a.

TABLE 4(b)   Association Energies for Single Dopant-Vacancy Defect Pair in ZrO_2

Dopant	cubic	tetragonal
Sc^{3+}	-0.04	-0.43
Y^{3+}	0.19	-0.29
Gd^{3+}	0.21	-0.29
La^{3+}	0.22	-0.33

between two O 48 (f) sites cannot lead to a continuous diffusion path, and so in spite of its low activation energy, must be discounted. It is possible that this jump may form part of a more complicated migration mechanism, but this has not been examined at this stage.

To test whether the vacant O 8(b) sites (which would be normally occupied in the fluorite cell) may take part in the transport of oxygen the energy of swapping an oxygen ion from a 48(f) site to an adjacent 8(b) site was calculated, a process that would be necessary if the 8(b) sites were to be involved. It was found that this configuration was, in fact, unstable: the oxygen ion displaced from the 48(f) site into the neighbouring 8(b) site, moved straight back into the 48(f) site. This clearly indicates that the 8(b) sites do not want to be embroiled in the mechanics of oxygen ion transport and that it is not really satisfactory to think of them as oxygen vacancies. Rather, they should more properly be considered as an integral feature of the pyrochlore structure than as a set of defect species. This conclusion is also consistent with the observations of Ando and Oishi (these proceedings) that in Y_2O_3, the 'vacant' fluorite sites in the structure also do not contribute to the oxygen conductivity.

IONIC CONDUCTIVITY IN TETRAGONAL ZrO_2

Introduction

As is now well known, pure ZrO_2 can exist in one of three modifications depending on the temperature regime, but will always invert to the monoclinic form on cooling, the higher temperature phases being apparently unquenchable (6). It is, however, possible to stabilise the other (cubic or tetragonal) phases by doping with some aliovalent cations, notably Ca^{2+} or Y^{3+} (6). Although most studies have been centred on the cubic phase, it has recently been realised that the tetragonal phase offers comparable ionic conductivity and may have better mechanical properties (7). An important property of the tetragonal phase is that it requires a considerably lower dopant concentration to stabilise it; the significance of this will be realised later. (Note, of course, that one can dope to higher levels than the minimum needed to stabilise the structure.)

Bonanos et al.(8) found that $\sigma_{tet} > \sigma_{cub} > \sigma_{mono}$ for a polyphasic mixture containing 8 mol % Y_2O_3, using complex impedance measurements which enabled them to separate out the conductivities of the three phases. Experiments on single crystals provided them with an activation energy, $\Delta H\sigma$, of ~0.9 eV for conductivity in tetragonal phase, compared to ~1.1 eV in the cubic phase (8).

In cases where the vacancies contributing to the conductivity are
present as compensators for aliovalent dopant cations, such as here,
two separate contributions to the activation energy may be distinguished. Firstly, there is the vacancy migration energy, which may be
recognised as the barrier to migration of a free vacancy in the pure
(undoped) material, and secondly, there may be an association energy
arising from interactions between the vacancies and the dopant
cation. These two terms, separately, may be calculated, from theory
(9), and, in favourable cases, may also be deduced from experiment
(10).

We have applied our theoretical techniques to determine how far the
difference in ΔH_σ between the tetragonal and cubic phases is due to
differences in the two components ΔH_m and ΔH_a, caused by differences in
the two structures.

Now, whilst the migration energy ΔH_m is adequately defined, there may
be some complication with the quantity ΔH_a. This really refers to the
energy required to create a 'free' vacancy from whatever aggregates of
defects there may exist in the crystal. This causes no confusion if
(i) the only defect complex is a pair: a vacancy bound to a dopant,
such as has been characterised in very lightly doped CeO_2 (10) and
(ii) these defect pairs do not themselves interact. Obviously, these
conditions are only true at low concentrations of dopants.

At higher concentrations, there are certain to be more complex defect
aggregates; there are also likely to be a number of different aggregates coexisting. These predications stem from a consideration of
the possible distribution of dopants in the host as function of
concentration.

We are undertaking a systematic study of the energetics of defect
aggregation in the fluorite oxides, in order to throw light on the
behaviour of the heavily doped structures, but here we will just report
on the simplest situation, that of an oxygen vacancy interacting with
a single dopant cation.

Potential Model

Since differences in the structures of the tetragonal and cubic phases
are expected to important, it is obviously crucial that our theoretical
model reproduces them. The tetragonal distortions producing the departure from cubic symmetry are depicted in figure(2), from which it
can be seen that small displacements of the oxygen ions are mainly

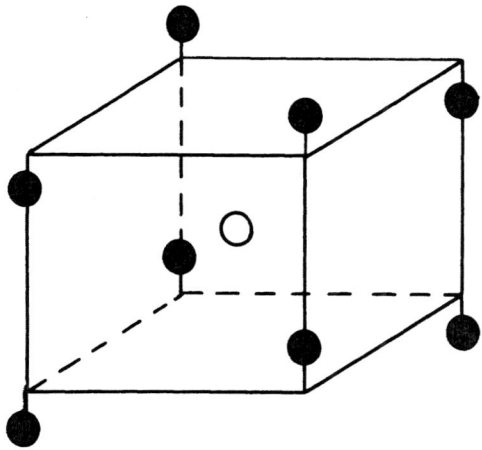

Figure 2 Structure of tetragonal ZrO_2, showing how displacements of oxygens from fluorite sites lead to lowering of symmetry.

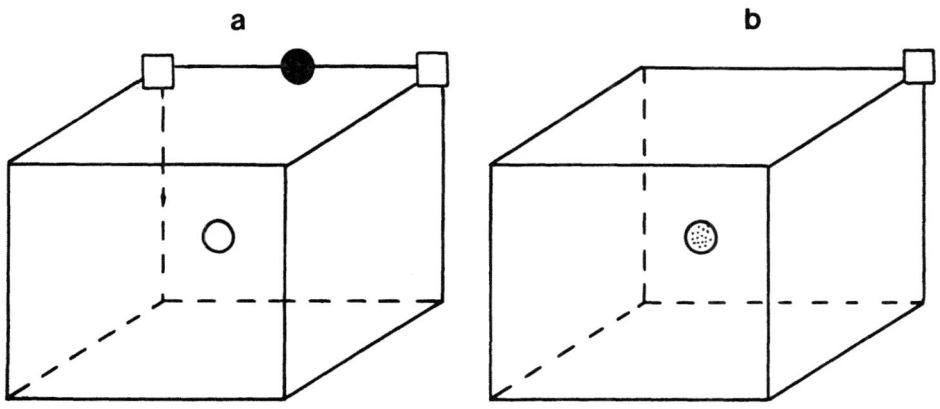

a   **b**

Figure 3 (a) Saddlepoint configuration for oxygen vacancy migration in the fluorite structure. The dopant (stippled) sits substitutionally on a normal cation site.

(b) Oxygen vacancy - dopant cation defect pair in the fluorite structure. The dopant (stippled) sits substitutionally on a normal cation site.

responsible, although there is also a slight change in the c/a ratio from unity to 1.024. It will be realised that if the oxygen displacements are not correctly modelled (there being residual strains in the structure),then when the structure is allowed to relax, these ions will want to move back onto the cubic fluorite lattice sites - a situation which would occur, for example, if the potential model had been fitted to a cubic structure. Thus, when the relaxation of the structure around a defect is calculated, there will be an erroneous contribution to the defect energy,as the model will be reverting to cubic symmetry as well as responding to the presence of the defect.

We have avoided this difficulty, by fitting the potential model to the tetragonal phase. Our calculated structural properties are compared in table (3) with the experimental structural data; the excellent agreement establishes the suitability of the potential model. The model also predicts the tetragonal phase to be more stable than the cubic phase by 0.2 eV per formula unit. This is consistent with the observed phase stabilities.

Results

In figure (3a) we illustrate schematically a saddlepoint configuration for a simple oxygen vacancy jump and in figure (3b) the structure the simple defect pair is sketched.

The activation energies corresponding to figure (3) are given in table (4a) and the binding energies of the defect pair for some trivalent dopant cations are presented in table (4b). Although these results may seem a bit surprising, in the light of some other examples (e.g. CeO_2, see the paper by Nowick in these proceedings), they do have a natural explanation, and are, indeed, consistent with the available experimental data.

Firstly, the apparent absence of a barrier to oxygen vacancy migration in the cubic structure, in the absence of any dopant cations, merely reflects the observation that this structure is not stable except at high temperatures, and, in fact, by introducing the vacancy and perturbing the arrangement of the oxygen sublattice, we are allowing the zirconium to be 7-coordinated. This is the coordination found in the monoclinic structure (12) and hence the absence of an activation energy implies that the oxygens would spontaneously rearrange towards the seven fold coordination of the monoclinic form.

The second point to emerge from the calculations concerns the

108

the association energies of the dopant vacancy pair. The positive values for the cubic structure mean that these defect aggregates will not be present, as isolated non-interacting pairs, in contrast to the tetragonal phase in which the distortions from the cubic symmetry allow them to form. This, again, is entirely consistent with experimental observations. The level of dopant concentration required to stabilise the cubic structure is so high (> 8 $^{o}/o$ M_2O_3)that the infinitely dilute solution approximation cannot possibly be valid: the probability of finding (at least) pairs of dopants on neighbouring cation sites is great enough to warrant the consideration of larger, more complex, defect aggregates. A natural extension of this sort of short-range ordering would, of course, be the formation of microdomains, itself the subject of some debate (13,14). We will not discuss it further here.

In contrast to the dopant levels found in the cubic phase, the tetragonal phase can, apparently, be stablised with quite small dopant concentrations, 2/3 mol $^{o}/o$, or lower (1). This is reflected in the stability of the individual defect pairs in the tetragonal structure, which ispresumably brought about by the displacements of the oxygen sublattice from the ideal fluorite sites.

Thus, in conclusion, at this stage, our theory predicts that whilst dopant-vacancy pairs may be stable (and we have not investigated interactions between pairs) in the tetragonal structure, this will not be the case for the cubic structure in which we may suppose that more complex defect clusters will be present, as a result of the high dopant concentrations.

ACKNOWLEDGEMENTS

We are indebted to AERE Harwell for financial support and to The University of London for the provision of computing facilities   We thank the sponsors of the conference for meeting our   financial requirements.

REFERENCES

1. O. T. Sorensen, ed.,"Non-Stoichiometric Oxides", Academic Press, London and New York (1981).
2. C.R.A. Catlow, R. James, W. C. Mackrodt and R. F. Stewart, Phys. Rev. B25:1006 (1982).
3. C.R.A. Catlow and W. C. Mackrodt, ed., "Computer Simulation of Solids", Lecture Notes in Physics, 166, Springer-Verlag, Berlin.

4. A. N. Cormack and M. P. Van Dijk, in preparation.
5. M. P. Van Dijk, J. J. de Vries and A. J. Burggraaf, Solid State Ionics, 9,10:913 (1983).
6. Subbarao, E. C., in "Advances in Ceramics: Science and Technology of Zirconia" (A. H. Heuer and L. W. Hobbs, eds.) Vol. p1, The American Ceramic Society, Columbus, Ohio (1981).
7. N. Bonanos, R. K. Stotwinski, B. C. H. Steele and E. P. Butler, J. Mater. Sci. 19:785 (1984).
8. N. Bonanos, R. D. Stlowinski, B. C. H. Steele, and E. P. Butler, J. Mater. Sci. Letters 3:245 (1984).
9. V. Butler, C. R. A. Catlow, B. E. F. Fender and J. H. Harding, Solid State Ionics 8:109 (1983).
10. R. Gerhardt-Anderson and A. S. Nowick, Solid State Ionics 5:547 (1981).
11. G. Teufer, Acta Crystallogr. 15:1187 (1962).
12. D. K. Smith and H. W. Newkirk, Acta Crystallogr. 18:983 (1965).
13. J. G. Allpress and H. J. Rossell, J. Solid State Chem. 15:68 (1975).
14. M. Morinaga and J. B. Cohen, Acta Crystallogr. A36:520 (1980).

THE ROLE OF DOPANT IONIC RADIUS IN

O^{2-} - CONDUCTING SOLID ELECTROLYTES

R. Gerhardt-Anderson and A.S. Nowick

Henry Krumb School of Mines
Columbia University
New York, N.Y. 10027

I. INTRODUCTION

The oxides of the fluorite structure (ZrO_2, CeO_2, ThO_2, HfO_2) have proved to be attractive systems for the study of mass transport. The focus has been primarily on transport on the oxygen-ion sublattice, for the cations are known to migrate much more slowly and, in fact, the entire cation sublattice is generally frozen-in below ~ 1000 ^oC.[1] Migration on the oxygen-ion sublattice can be enhanced by the introduction of oxygen vacancies, $V_O^{..}$, through doping with lower valent cations. The study of such ionic transport is stimulated, in part, by the actual and potential applications of such oxides as oxygen sensors and oxygen-ion conductors in high-temperature fuel cells.[2] Stabilized zirconia has received the most attention, but that material requires ~ 8 mole % of dopant to stabilize the fluorite structure. Ceria and thoria, on the other hand, are more versatile for fundamental studies since they possess the fluorite structure in the pure state; thus it is possible to study the entire range, from very low to high dopant concentrations.

This paper will focus primarily on M_2O_3-doped ceria, where M represents ions such as Y^{3+} and trivalent rare earths. Of principal interest here is the question of how the radius of the M^{3+} ion affects the transport properties. For every molecule of M_2O_3 introduced, there must be one $V_O^{..}$ (oxygen-ion vacancy) for charge and lattice-site conservation, i.e. one $V_O^{..}$ for every two M_{Ce}', in Kroger-Vink notation. In view of the immobility of the cations, one expects the distribution of M^{3+} ions to be more or less random, so that the completely compensated $(M_2V_O)^x$ defect cannot form. Nevertheless, the existence of a coulombic attraction between M_{Ce}' and the mobile $V_O^{..}$, should give rise to a favorable binding energy, and the formation of $(MV_o)^{\bullet}$ pairs. One such pair is illustrated in Fig. 1, which shows one-half the unit cell of the fluorite structure. Clearly the pair possesses eight equivalent orientations, lying along the various <111> directions. The reaction may be written as:

$$M_2O_3 \rightarrow 2M'_{Ce} + V_o^{\bullet\bullet} \underset{\leftarrow}{\rightarrow} (M_{Ce}V_o)^{\bullet} + M'_{Ce} \qquad (1)$$

The corresponding mass-action equation may be expressed in the form:

$$C_V C_M / C_P = K_A = 1/8 \exp(-h_A/kT) \qquad (2)$$

where C_V, C_M and C_P are the concentrations (mole fractions) of vacancies, M'_{Ce} and pairs, respectively, and h_A is the association enthalpy of the pair (taken as positive when association is favored). In the low temperature range where association is nearly complete, we should then have primarily an array of equal numbers of alternately charged defects, $(MV_o)^{\bullet}$ and M'_{Ce}, while the relatively small concentration of free vacancies will be given by

$$C_V \cong K_A \qquad (3)$$

i.e. C_V is independent of the dopant concentration. The electrical conductivity σ of these materials, which is ionic and due to the $V_o^{\bullet\bullet}$ defects, is given by

$$\sigma = C_V N_o e \mu_V \qquad (4)$$

Here N_o is the number of CeO_2 molecules per unit volume, e the electronic charge, and μ_V the vacancy mobility, given by[1]

$$\mu_V \propto \frac{1}{T} \exp(-h_m/kT) \qquad (5)$$

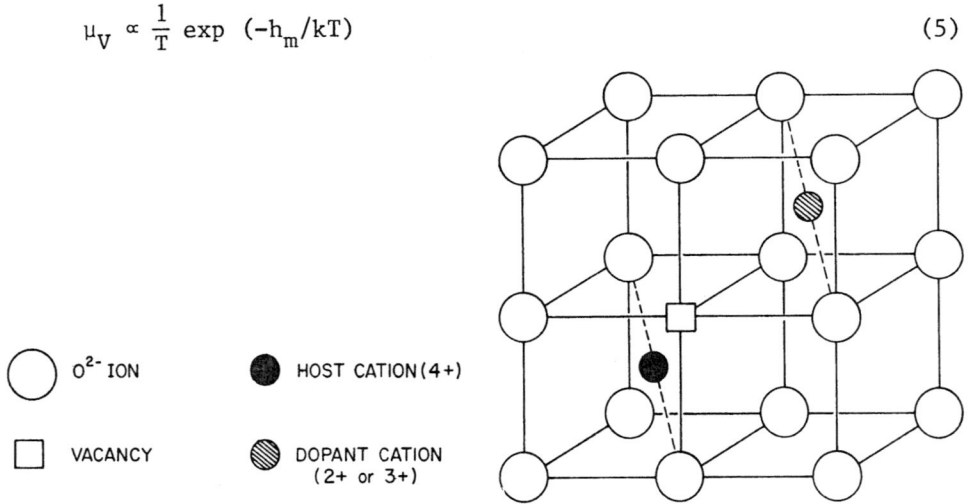

○ O^{2-} ION   ● HOST CATION(4+)

□ VACANCY   ◍ DOPANT CATION (2+ or 3+)

Fig. 1. Diagram of one-half the unit cell of a fluorite-type oxide showing an MV_o pair in a nearest-neighbor (nn) position.

where h_m is the activation energy for migration of a free vacancy. Thus in the low temperature range, in which Eq. (3) applies, σ obeys the usual Arrhenius relation

$$\sigma T = A \exp (-h_\sigma/kT) \tag{6}$$

with

$$h_\sigma = h_m + h_A \tag{7}$$

i.e. the conductivity enthalpy h_σ is the sum of a motion and an association enthalpy.

The $(MV_o)^{\cdot}$ pair, shown in Fig. 1, is both an electric and elastic dipole, viz., it can be preferentially aligned among its 8 orientations by either an electric or elastic stress field, and under alternating fields it can give rise, respectively, to dielectric or anelastic relaxation. For the simple defect pair, anelastic relaxation takes the form:[3]

$$\tan \phi = \Delta \cdot \omega\tau/(1 + \omega^2\tau^2) \tag{8}$$

where $\tan \phi$ is the phase angle by which the strain lags behind the stress (also called the "internal friction", Q^{-1}), Δ is the relaxation strength, ω is the circular frequency and τ is the relaxation time, given by

$$\tau^{-1} = \nu_o' \exp (-h_r/kT) \tag{9}$$

Here h_r is the activation enthalpy for relaxation i.e., for the jump of a $V_o^{\cdot\cdot}$ while remaining associated to the M^{3+} ion, while ν_o' is the frequency factor (which includes the activation entropy). Equation (8) gives a symmetric peak (called a Debye peak) about $\omega\tau = 1$, either in a plot of $\tan \phi$ vs. $\log \omega$ at constant T, or more conveniently, as a function of $1/T$ at constant ω.

The results for dielectric relaxation are analogous, but the τ-values in the two cases are not necessarily the same for the same defect. In fact, for the defect of Fig. 1, it can be shown[4] that

$$\tau_{diel}/\tau_{anel} = 2 \tag{10}$$

II. REVIEW OF PREVIOUS WORK

The most complete study of both conductivity and relaxation has been conducted on Y_2O_3-doped ceria. It is advantageous to begin by discussing the relaxation, since relaxation is a spectrographic property, i.e. different peaks are obtained for different defects,

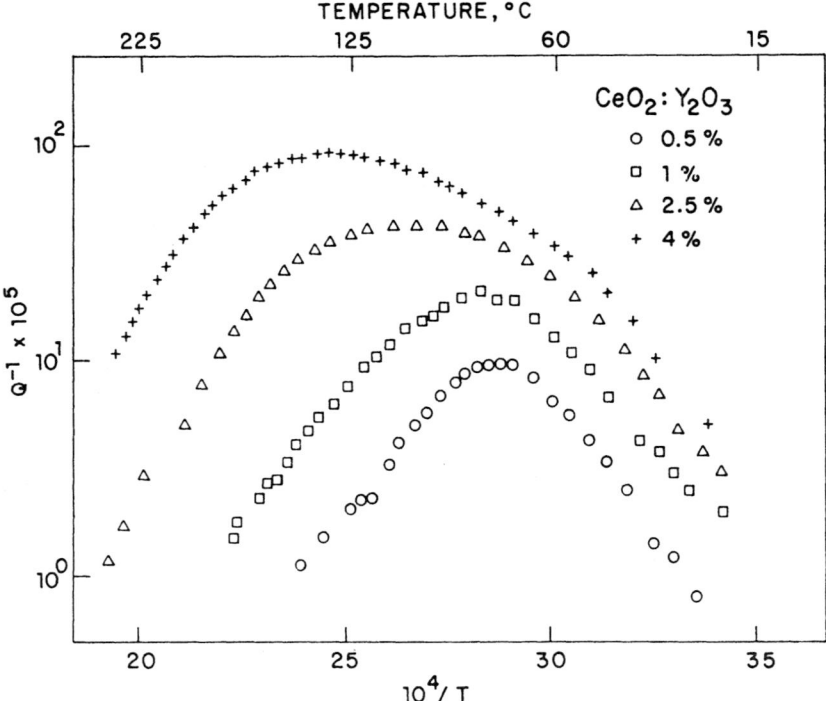

Fig. 2. Internal friction (on a logarithmic scale) as a function of 1/T for four different Y_2O_3-doped ceria solid solutions. Measurement frequency $\sim$ 8 kHz. From Anderson.[5]

while conductivity is an integrated property. Both dielectric and anelastic relaxation have been studied on this system. Results of the anelastic study by Anderson [5,6] are given in Fig. 2 for four compositions. For 0.5% Y_2O_3 the peak is very close to a Debye peak given by combining Eqs. (8) and (9). The peak is only slightly broadened toward higher temperatures for 1% Y_2O_3; but by 4% Y_2O_3 it is so considerably broadened that even resolution into 2 or 3 discrete peaks is not practical. The broadening has been attributed to higher order defect clusters present at the higher concentrations. These results lead us to conclude that a true "dilute range", where the defects are simple, exists only for $\lesssim$ 1 mole per cent of Y_2O_3.

Dielectric relaxation has also been observed,[7] and for the dilute range a ratio $\tau_{diel}/\tau_{anel} = 1.94$ is obtained, which is in excellent agreement with the expected value of 2.0 [Eq. (10)] for the nn $(MV_O)^{\cdot}$ model of Fig. 1. An additional large dielectric loss peak is observed, however, which has been shown to be due to relaxation of the whole array of charges, $(MV_O)^{\cdot} + M'_{Ce}$, through the

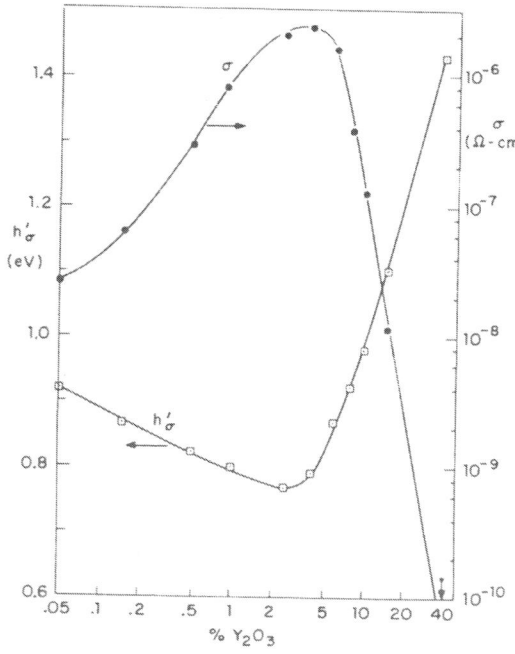

Fig. 3. Variation of activation enthalpy, h_σ, and of conductivity at 182 °C, σ, with mole % Y_2O_3 for various CeO_2:Y_2O_3 solid solutions. From reference 9.

migration of $V_O^{\cdot\cdot}$.[7] The presence of this additional effect means that the dielectric relaxation behavior at the higher concentrations is not as clean as the anelastic behavior of Fig. 2.

Turning now to the conductivity, we have studied the ceria-yttria system from the very dilute to the high concentration range. In all cases, the use of measurements over a wide frequency range, and complex-impedance analysis is essential because of the presence of an extra arc due to blocking at grain boundaries.[8,9] All of the conductivity results quoted herein will refer to the "bulk" or "lattice" conductivity obtained from the highest frequency arc. Figure 3 is a summary of results showing both conductivity at one temperature (182 °C) and activation enthalpy, h_σ, as a function of Y_2O_3 content. A logarithmic scale of concentration is used in order to amplify the low concentration range. Clearly, the prediction of Eq. (3) that c_V, and therefore σ, is a constant in the dilute range ($\lesssim$ 1% Y_2O_3) is not borne out. Rather σ increases by a factor $\sim$ 30 in this range. At the same time, h_σ is not a constant, but decreases. The preexponential, A, remains constant over this range; therefore, the rise in σ can be attributed entirely to the decrease in h_σ. It has been shown[9] that this effect is due to coulombic interactions of the migrating vacancy with the charge array. In this way a quantitative accounting for the decrease in h_σ and the increase in σ with % Y_2O_3 has been given.

Having established an understanding of the dilute range for Y_2O_3 doping, we turn to other dopants of different ionic radii: Gd^{3+}

115

and La^{3+} which are larger than Ce^{4+}, and Sc^{3+} which is smaller. Conductivity data obtained for 1 mole % solutions of each of these dopants, showed that Gd^{3+} doping gave the highest conductivity, while Sc^{3+} doping produced extremely low values.[10] From independent evidences,[9] it was deduced that the vacancy migration enthalpy h_m is close to 0.61eV; therefore, from Eq. (7) and the experimental values of h_σ, we can obtain h_A by difference. Table I summarizes the results for the four dopants studied. At the same time, it compares the ionic radii of these ions to the fractional change in lattice parameter, $C_o^{-1}\Delta a/a_o$, of doped ceria, showing that for a radius ratio <1, this lattice parameter variation is negative, as might be expected. The values of h_A for 1 % M$_2$O$_3$ are plotted versus dopant ionic radius in Fig. 4. (The values for 6% M$_2$O$_3$ will be considered later.) The figure shows a minimum close to the Ce^{4+} radius, with a sharper rise for dopant of small radius than for those of large radius. Also shown in Fig. 4 are the values calculated by Butler et al.[11] using computer simulation techniques. It is gratifying that the trends shown in the experimental values are also brought out by these calculations. Similar calculated results have been obtained for other systems,[12,13] showing that elastic as well as coulombic interactions are important in determining the association energy. The greatly enhanced bonding of small ions to the vacancy is apparently due to the fact that expansion of the lattice about the vacancy is compensated by contraction around the small dopant ion.[14]

III. CONDUCTIVITY AT HIGH CONCENTRATIONS

From Section II, it can be stated that we have achieved a

Table 1. Association Enthalpies, h_A, for 1% and 6% M$_2$O$_3$ Solid Solutions for Four Dopants of Different Ionic Radii in CeO$_2$

Dopant	$r_{M^{3+}}/r_{Ce^{4+}}$	$C_o^{-1}\Delta a/a_o$ (%)	h_A (eV) For 1% M$_2$O$_3$	h_A (eV) For 6% M$_2$O$_3$
La^{3+}	1.21	5.5	0.14	0.18
Gd^{3+}	1.03	0.9	0.12	0.16
Y^{3+}	0.98	−0.9	0.21	0.26
Sc^{3+}	0.86	−3.7	0.67	−

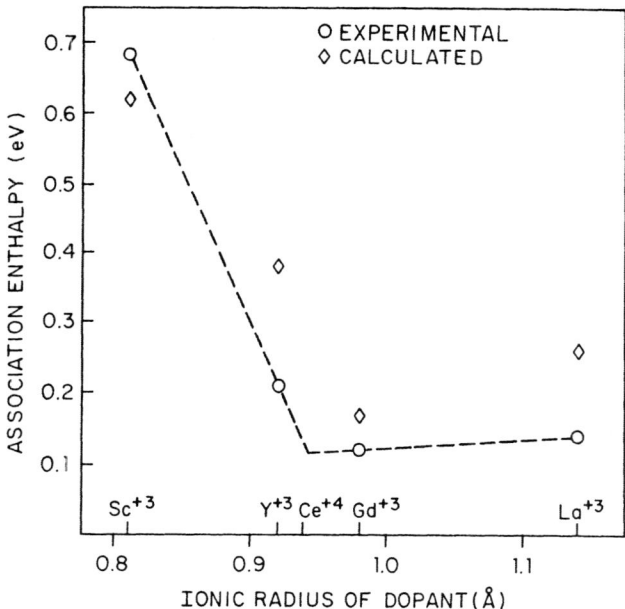

Fig. 4. Variation of association enthalpy, h_A, with ionic radius of dopant for various M^{3+} cation dopants: as obtained from conductivity of 1% solid solutions in ceria (circles),[10] and as calculated by computer simulation (diamonds).[11]

reasonable understanding of the defects present in the dilute range, and of the way that σ and h_A vary with ionic radius in this range. In considering the extension into the high-concentration range, we have already noted the behavior of σ and h_A for Y_2O_3-doped ceria (Fig. 3). It remains, therefore, to consider how conductivity varies with dopant radius in the high concentration range. Figure 5 shows an Arrhenius plot of the conductivity for three different dopants at the 6 mole % level. (Unfortunately, Sc^{3+} could not be included since its solubility is close to 1%). It is interesting that the order of the σ values for these dopants is the same as for the 1% solid solutions. Further, from the low temperature slopes, values of h_σ, and of h_A (using $h_m = 0.61$eV), are obtained. The latter, tabulated in the last column of Table I, are in the same sequence as those for the 1% solutions but simply shifted upward by 0.04 - 0.05 eV. The simplicity of this variation in conductivity behavior is indeed surprising when we recall how drastic a difference in defect structure of 1% and 4% solutions is indicated by the internal friction results of Fig. 2.

If we wish to obtain information at even higher concentrations,

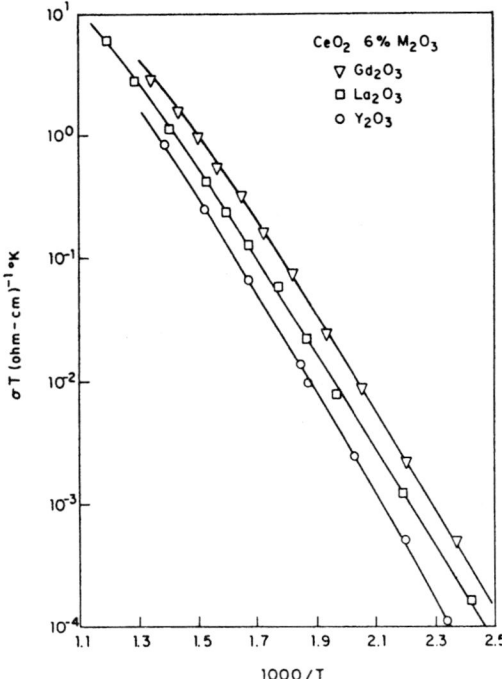

Fig. 5. Arrhenius Plots (log σT vs. T⁻¹) of the conductivity for three CeO₂: 6% M₂O₃ solid solutions.

it is worthwhile to reexamine the results of Kudo and Obayashi[15] on $Ce_{0.7}M_{0.3}O_{1.85}$, corresponding to 17.6 mole % M_2O_3, for a large number of different M^{3+} dopants. Since these measurements were only done at one frequency and complex-impedance analysis was not employed, their Arrhenius plots may be subject to appreciable errors. Instead we have chosen their σ values at one temperature (394 °C) in the mid-range, and compared them with our own results for the same temperature at 1% and 6% M_2O_3 in Fig. 6. The results show very similar behavior as a function of ionic radius, but gives many more points at the 17.6% composition because of the large number of solutes explored. (The reader should recognize that the increase in σ in going from 1% to 6% and its decrease again at high concentrations reflects the variation with concentration shown in Fig. 3.) In the case of Kudo's data the maximum as a function of ionic radius falls at Nd^{3+}, distinctly to the right of the Ce^{4+} radius, but the overall similarity to the data for the lower concentrations is, nevertheless, quite striking. It is therefore concluded that the same type of size effect found in the dilute range carries over into the concentrated range, although in the latter case we have no simple explanation to account for it.

To these results on ceria may be added those for zirconia solid solutions, which fall only in the concentrated range for the stabilized fluorite structure. The Zr^{4+} ion is much smaller than Ce^{4+}; accordingly most dopant ions are larger than Zr^{4+} so that the maximum such as that shown in Fig. 6 (or a minimum in h_σ) cannot be obtained. Nevertheless, a steady increase in h_σ with increasing ionic radius was demonstrated in early work of Strickler and Carlson[16].

118

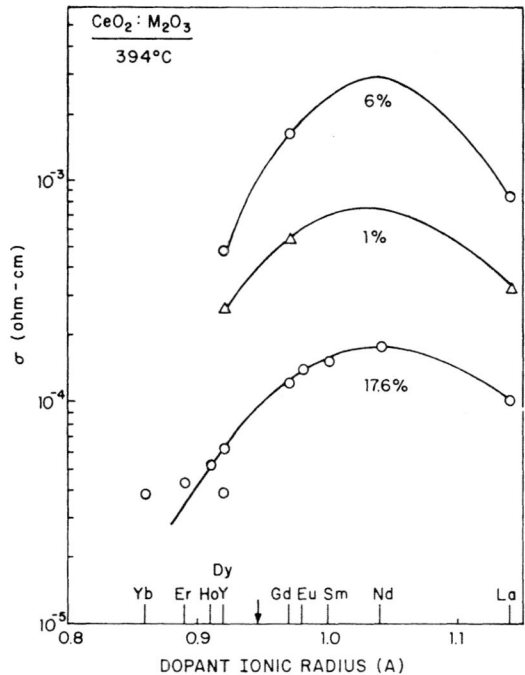

Fig. 6. Conductivity (at 394 °C) as a function of dopant ionic radius for $CeO_2:M_2O_3$ solid solutions containing three different mole % of M_2O_3. Arrow denotes the ionic radius of Ce.[4+] Data for 17.6% taken from ref. 15.

IV. FURTHER ANELASTIC MEASUREMENTS

In Fig. 2 we showed how useful internal friction measurements could be in showing, for Y_2O_3-doped ceria, when simple pairs were present and when more complex defects develop. More recently, we have studied anelastic relaxation in 1% solid solutions of other dopants. Figures 7 and 8 show the results for 1% Gd_2O_3 and 1% Sc_2O_3, respectively. (In contrast to Fig. 2, we have now plotted internal friction (Q^{-1}) on a <u>linear</u> scale versus 1/T.) In Fig. 7 the ideal Debye peak is drawn in. It is clear that, while the high-temperature side is close to a Debye peak, there is at least one additional peak on the low-temperature shoulder. Thus the Gd_2O_3 case is more complex than that of Y_2O_3 doping. Even greater complexity is shown for the case of 1% Sc_2O_3, in Fig. 8. Here the main peak is considerably broader (by $\sim 3\times$) than a Debye peak and, in addition, a low temperature peak is observed at 125K. (Such a peak is not present for any of the other dopants.) A careful study of variously double-doped samples has shown that the low-temperature peak is not due to $(ScV_O)^{\cdot}$ pairs but to isolated Sc'_{Ce} defects.[17] This means that the isolated Sc^{3+} cannot be located in a site of cubic symmetry, but rather that ionic readjustments to an off-center configuration occur. The interested reader may consult ref. 17; for

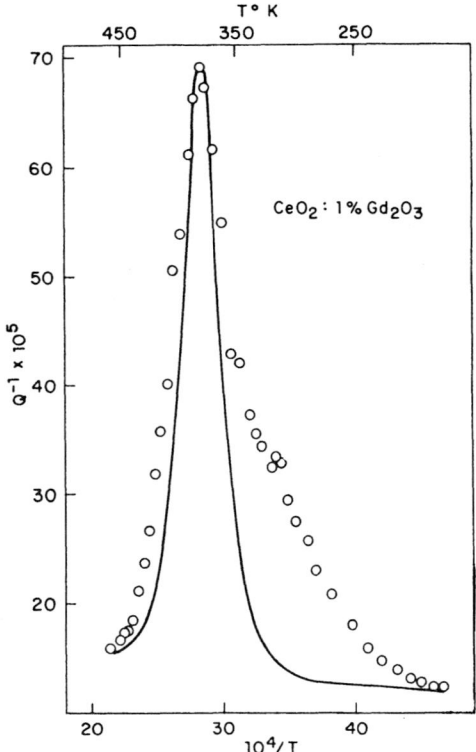

Fig. 7. Internal friction (Q^{-1}) at 8 kHz versus $1/T$ for CeO_2:1% Gd_2O_3.

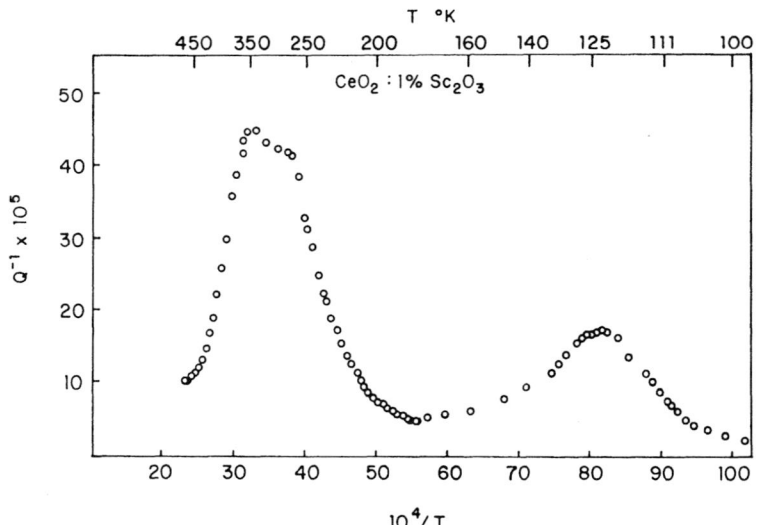

Fig. 8. Internal friction at 8 kHz versus $1/T$ for CeO_2:1% Sc_2O_3.

120

Table II.   Results of Anelastic Measurements on
            CeO_2:1% M_2O_3 Solid Solutions, Including
            Activation Enthalpy, h_r, and Peak Width
            Relative to Debye Peak (W/W_D).

Dopant	$r_{M^{3+}}/r_{Ce^{4+}}$	Peak Location $10^4/T$	h_r (eV)	W/W_D
La^{3+}	1.21	27.0	0.70	2.9
Gd^{3+}	1.03	28.3	0.65	1.3
Y^{3+}	0.98	28.6	0.64_5	1.3
Sc^{3+}	0.86	37.0	0.51	3.3

present purposes, however, we concentrate only on the higher
temperature peak. The results on Gd_2O_3, Sc_2O_3, Y_2O_3 and also La_2O_3-
doped ceria are summarized in Table II. This table compares the
peak locations, activation enthalpy for relaxation h_r (obtained
assuming $\nu'_o = 10^{14}$ sec^{-1}) and W/W_D, the peak width relative to that
of a perfect Debye peak. This table shows that for the two dopants
for which $r_{M^{3+}}/r_{Ce^{4+}}$ is close to unity, the peaks are only 30% wider
than Debye peaks. On the other hand, for La^{3+} and Sc^{3+} with large
size differences, peaks $\sim$ 3 $\times$ as broad as Debye peaks are observed.
In these latter cases, at least, the $(MV_o)^{\cdot}$ pair model, which has
served us so well for the case of Y_2O_3 doping, must be greatly
modified to represent the true picture. The resolution can probably
be traced to an M-M interaction energy (including nearby $V_o^{\cdot\cdot}$ defects)
in the case of dopants with large size differences. The existence
of such interaction can cause cation impurity pairing at the high
temperatures where cations are still mobile.

V. ELASTIC CONSTANTS

     Since the internal friction measurements are carried out on
flexurally vibrating resonant bars, the resonant frequency is
obtained in each case as a by-product of the measurement. This
resonant frequency is easily converted into Young's modulus of
elasticity, E. [3] Since the samples are polycrystalline, the measurement gives an average of E over all crystal orientations.

     The results show that doping increases E. This agrees with a
recent study[18] of single crystals of yttria-stabilized zirconia with
Y_2O_3 concentrations in the range from 11-18 mole %. This study
shows that the average modulus does increase with Y_2O_3 concentration,
primarily due to the increase in the elastic constant C_{44}. Our work
shows that the largest effect occurs for the smallest dopant ion

(Sc^{3+}), as shown in Fig. 9. The resemblance of this figure to Fig. 4 (showing h_A as derived from conductivity measurements) is quite striking.

The increase in elastic modulus with solute doping observed in these systems contrasts to a decrease generally found in metallic systems, and interpreted by Zener[19] as a strain energy effect. In the fluorite oxides, the increase in elastic modulus with doping has been attributed[18] to coulombic terms in the theory of Srinivasan.[20] Why the effect should be so much larger for a small cation dopant is not yet clear, however.

VI. CONCLUDING REMARKS

The behavior of M_2O_3-doped ceria in the dilute range (Section II) is relatively well understood. In this range, the dominant defects are the $(MV_O)^{\cdot}$ pair and isolated M'_{Ce} in nearly equal numbers. The size effects can be understood in terms of contributions of elastic interactions to the association enthalpy of the pair. The ionic conductivity at high concentrations preserves many of the features of the dilute systems, showing a minimum where the dopant and host cation radii are close. We do not have a detailed

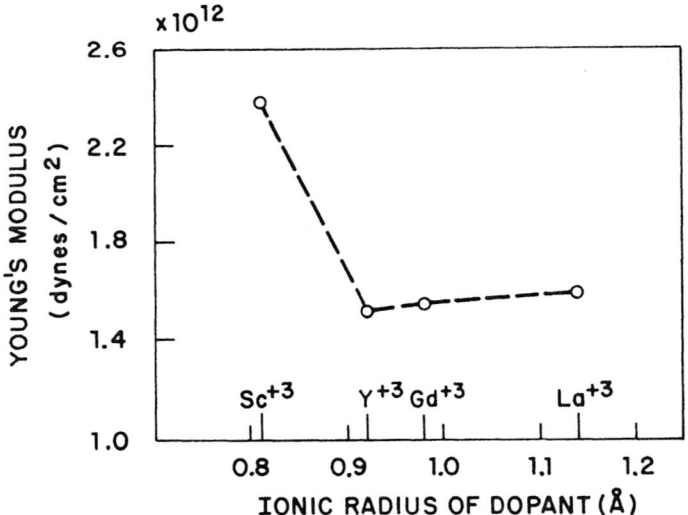

Fig. 9. Young's modulus as a function of dopant ionic radius for four 1% M_2O_3 solid solutions in ceria.

understanding of this behavior, however, because there is not a satisfactory theory for the high concentration region.

Internal friction studies on variously doped specimens have shown that even in 1 mole % solutions the defect structure is not simple for dopants that have a large size difference relative to the host. Multiple relaxation peaks suggest the presence of more complex clusters containing at least two M^{3+} ions frozen in from high temperatures. Finally, elastic constants are shown to increase with doping, particularly for the small dopant Sc^{3+} in ceria. A good interpretation is not yet available.

ACKNOWLEDGMENTS

This work was sponsored by the U.S. Department of Energy under contract DE-AC02-78 ER 04693.

References

1. A.S. Nowick, in "Diffusion in Crystalline Solids", G.E. Murch and A.S. Nowick, eds., Academic Press, New York (1984), Chapter 3.
2. P. Hagenmuller and W. van Gool, eds., "Solid Electrolytes", Academic Press, New York (1978), Chaps. 18, 25, 28, 29.
3. A.S. Nowick and B.S. Berry, "Anelastic Relaxation in Crystalline Solids", Academic Press, New York (1972).
4. A.S. Nowick, Adv. Phys. 16:1 (1967).
5. M.P. Anderson, M.S. Thesis, Columbia University (1979).
6. M.P. Anderson and A.S. Nowick, J. Phys. (Paris) 42:C5-823 (1981).
7. Da Yu Wang and A.S. Nowick, J. Phys. Chem. Solids 44:639 (1983).
8. Da Yu Wang and A.S. Nowick, J. Solid State Chem. 35:325 (1980).
9. Da Yu Wang, D.S. Park, J. Griffith and A.S. Nowick, Solid State Ionics 2:95 (1981).
10. R. Gerhardt-Anderson and A.S. Nowick, Solid State Ionics 5:547 (1981).
11. V. Butler, C.R.A. Catlow, B.E.F. Fender and J.H. Harding, Solid State Ionics 8:109 (1983).
12. W.C. Mackrodt and R.F. Stewart, J. Phys. C. 12:5015 (1979).
13. J.A. Kilner and R.J. Brook, Solid State Ionics 6:237 (1982).
14. E.A. Colbourn and W.C. Mackrodt, J. Nucl. Mat. 118:50 (1983).
15. T. Kudo and H. Obayashi, J. Electrochem. Soc. 122:142 (1975).
16. D.W. Strickler and W.G. Carlson, J. Amer. Cer. Soc. 48:286 (1965).

17. R. Gerhardt-Anderson, F. Zamani-Noor, A.S. Nowick, C.R.A. Catlow and A.N. Cormack, Solid State Ionics 9&10:931 (1983).
18. H.M. Kandil, J.D. Greiner and J.F. Smith, J. Amer. Cer. Soc. 67:341 (1984).
19. C. Zener, Acta Cryst. 2:163 (1949).
20. R. Srinivasan, Phys. Rev. 165:1041, 1054 (1968).

DIFFUSION PROPERTIES – PRIMARILY BULK

THE HAVEN RATIO IN TERMS OF THE SELF-DIFFUSION OF CHARGE

Alfred R. Cooper

Case Western Reserve
University
Cleveland, Ohio 44106

INTRODUCTION

The discrepancy between the measured electrical conductivity of ionic materials and that calculated from the Nernst-Einstein (N-E) equation has been known for a long time. In glasses the discrepancy was perhaps first documented by Fitzgerald[1] and has been confirmed by many subsequent works.[2] Le Claire suggested that the ratio of the ionic conductivity calculated from N-E and that actually determined be called the Haven Ratio, H_R. This terminology is widely accepted in the discussion of glassy as well as crystalline materials. The appropriateness of the choice relates to Haven's pioneering work in the understanding of the serial correlation of atomic jumps.[3] For crystalline materials with well-specified diffusion mechanisms, it is now routine to calculate the Haven Ratio based on the difference between the serial correlation of the jumps of lattice atoms and that of ionic defects.[4] Here the goal is to review the procedure and generalize it to apply to noncrystalline solids.

NERNST-EINSTEIN EQUATION WITH ONE MOBILE ION TYPE

The DC electrical conductivity of an ionic conductor with a single mobile ion type, u, is given by the Nernst-Einstein equation to be

$$\sigma_{NE} = \rho_u \frac{D_u}{kT} Z_u^2 e^2 \tag{1}$$

where ρ_u is the number density and $Z_u e$ is the charge of type u ions. However, it is often found that the actual conductivity, σ_a, is larger

127

than that predicted by the N-E equation, and this has been explained by asserting that with defect mechanisms for diffusion, the defects are the appropriate charge carriers, which modifies equation (1) to:

$$\sigma_d = \rho_d \frac{D_d}{kT} Z_d^2 e^2 \tag{2}$$

where ρ_d, D_d, and Z_d are the number density, diffusion coefficient, and virtual charge of the defect, respectively.

Comparing equation (2) with equation (1) gives the following expression for the Haven Ratio under the premise that defects are the appropriate charge carriers.

$$H_R^d = \frac{\rho_u D_u Z_u^2}{\rho_d D_d Z_d^2} \tag{3}$$

Another way to express the conductivity is:

$$\sigma_q = \rho_q \frac{Z D_q e^2}{kT} \tag{4}$$

where D_q is the self-diffusion coefficient for charge and ρ_q is the number density of the charge carriers, i.e., of the mobile ions. Comparing equation (1) and equation (4) allows the Haven Ratio to be written:

$$H_R^q = \frac{\rho_u D_u Z_u^2}{\rho_q D_q} = \frac{D_u Z_u^2}{D_q Z} \tag{5}$$

A purpose of this section is to examine the quantity H_R^q, describe the calculation of D_q, and show that for simple defect mechanisms, $H_R^q = H_R^d$.

SELF-DIFFUSION COEFFICIENTS

The self-diffusion coefficient of an atom or an ion is defined as the quotient of its mean square displacement after time, t, divided by 6t, i.e.,

$$D_u = \frac{\langle R_u^2 \rangle (t)}{6t} \tag{6}$$

where u refers to a particular species, e.g., to an ion type like

O^{--} or Na^+ and the averaging is over a sufficient time, t, or number of atoms so that the behavior is adequately sampled; i.e., continuing for longer time or using more ions would not significantly change D_u.

Here we review the calculation of R_u^2. The displacement after n jumps of x, an arbitrary type u ion is:

$$\vec{R}_\lambda(n) = \vec{\lambda}_{1,x} + \vec{\lambda}_{2,x} + \vec{\lambda}_{3,x} + \ldots \vec{\lambda}_{n,x} \tag{7}$$

where $\vec{\lambda}_{i,x}$ is the vector displacement of the ith jump of this ion. The mean square displacement

$$\langle R_x^2(n)\rangle = \sum_{i=1}^{n} \lambda_x^2 + 2\sum_{i=1}^{n-1} \vec{\lambda}_{i,x} \cdot \vec{\lambda}_{i+1,x} + 2\sum_{i=1}^{n-2} \vec{\lambda}_{i,x} \cdot \vec{\lambda}_{i+2,x}$$
$$+ \ldots \vec{\lambda}_{1x} \cdot \vec{\lambda}_{nx} \tag{8}$$

Since x was chosen arbitrarily:

$$\langle R_u^2(n)\rangle = n\langle\lambda_u^2\rangle + 2\left\{(n-1)\langle\vec{\lambda}_{i,x} \cdot \vec{\lambda}_{i+1,x}\rangle \cdot \cdot + (n-p)\langle\vec{\lambda}_{i,x}\right.$$
$$\left. \cdot \vec{\lambda}_{i+p,x}\rangle \cdot \cdot \cdot + \vec{\lambda}_{1x} \cdot \vec{\lambda}_{nx}\right\} \tag{9}$$

The quotient of $\vec{\lambda}_{i,x} \cdot \vec{\lambda}_{i+p,x}$ by λ^2 gives the correlation, α_{up}^{xx}, between the direction of the $(i+p)^{th}$ jump of an ion and the direction of the ith jump of the same ion. The double superscript, xx, indicates that correlation is between different jumps of the same ion. Clearly, $-1 \leq \alpha_{up}^{xx} \leq 1$, and typically $|\alpha_{up}^{xx}|$ decreases sharply as p increases.

Defining $\tau = t/n$, the mean waiting time between jumps, using equation (6) and the definition of α_{up}^{xx} and assuming that n is sufficiently large that $\alpha_{up}^{xx} \to 0$ at values of p where $(n-p)/n$ hardly differs from unity, gives:

$$D_u = \frac{\langle\lambda_u^2\rangle}{6\tau_u}\left[1 + 2\sum_{p=1}\alpha_{up}^{xx}\right] \equiv \frac{\langle\lambda_u^2\rangle}{6\tau_u} f_u^{xx} \tag{10}$$

The serial correlation coefficient for jumps of a particular type u ion, f_u^{xx}, is defined by equation (10).

Likewise, the self-diffusion coefficient for a defect, d, is written:

$$D_d = \frac{<\lambda_d^2>}{6\tau_d}[1 + 2\sum_{p=1}\alpha_{dp}^{xx}] = \frac{<\lambda_d^2>}{6\tau_d} f_d^{xx} \qquad (11)$$

where f_d^{xx} is the serial correlation coefficient for jumps of a particular defect.

The self-diffusion coefficient for charge is obtained in a similar way. The displacement of charge $R_q(n)$ per mobile ion is the sum after n successive jumps of the product of the charge of the jumping ion and its vector displacement divided by the number, N, of u ions. Following the procedure used to obtain equations (8) and (9), the mean square displacement of charge per carrier after n jumps, $R_q^2(n)$, is written:

$$N <R_q^2(n)> = n<\lambda^2 z^2> + 2\{(n-1)<z_i\vec{\lambda}_i \cdot z_{i+1}\vec{\lambda}_{i+1}> \cdots$$

$$+ (n-p)<z_i\vec{\lambda}_i \cdot z_{i+p}\vec{\lambda}_{i+p}> \cdots + z_1\vec{\lambda}_1 \cdot z_n\vec{\lambda}_n\} \qquad (12)$$

The absence of the subscript x indicates that the displacement of charge occurs from the motion of all mobile ions, i.e. the average number of jumps per ion is n/N.

Equation (12) can be written:

$$R_q^2(n) = (n/N)<\lambda_q^2 z_q^2> [1 + 2\sum_{p=1}\alpha_{up}^{xy}] \qquad (13)$$

where $\alpha_{qp}^{xy} = [(z\vec{\lambda})_i \cdot (z\vec{\lambda})_{i+p}]/z^2\lambda^2$ is the correlation between the direction of the jump of an ion and the pth subsequent jump (of any ion including itself). In fact, in the calculation of the cross terms, e.g., $[(z\vec{\lambda})_i \cdot (z\vec{\lambda})_{i+p}]$, only the ions in the vicinity of the ion making the ith jump are considered. Otherwise the correlation includes the effect from conservation of charge, a consideration which is irrelevant for the present purpose. Substituting $<R^2>(n)$ into equation (10) and noting that n/Nt is $1/\tau_u$ the average waiting time between jumps for a u ion, the self-diffusion coefficient for a charge, is given by

$$D_q = \frac{z_q^2<\lambda_q^2>}{6\tau_u}[1 + 2\sum_{p=1}\alpha_{up}^{xy}] \equiv \frac{z_q^2\lambda_q^2}{6\tau_u} f_u^{xy} \qquad (14)$$

130

where τ_q is the average time interval between jumps of any ion.
Thus, $\tau_q = \tau_u/N$. The serial correlation coefficient for charge,
f_u^{xy}, is defined by equation (14). The mixed superscript indicates
that correlation of the sequential jumps of neighboring ions is
included and the subscript u reminds us that only type u ions are
mobile.

Substituting from equations (10, (11), and (14) into equations
(3) and (5) gives, for the Haven Ratio:

$$H_R^d = \frac{\rho_u \, \lambda_u^2 \, \tau_d \, f_u^{xx} \, Z_u^2}{\rho_d \, \lambda_d^2 \, \tau_u \, f_d^{xx} \, Z_d^2} \tag{15a}$$

$$H_R^q = \frac{\rho_u \, \lambda_u^2 \, \tau_u \, f_u^{xx} \, Z_u^2}{\rho_q \, \lambda_q^2 \, \tau_u \, f_u^{xy} \, Z_q^2} \tag{15b}$$

The quantities λ_u, λ_q, are equal since both are the mean jump
distance of an ion. Likewise, ρ_u and ρ_q are both the density of u
ions. The values Z_u and Z_q are identical for a truly ionic solid
$Z_d = -Z_u$ in Kroger Vink notation. Thus $Z_u^2 = Z_q^2 = Z_d^2$ and equations 15
simplify to

$$H_R^d = \frac{f_u^{xx}}{f_d^{xx}} \, \frac{\rho_u \tau_d \lambda_u^2}{\tau_u \rho_d \lambda_d^2} \tag{16a}$$

$$H_R^q = \frac{f_u^{xx}}{f_u^{xy}} \tag{16b}$$

To illustrate and to compare the predictions of equation (16a)
with those from (16b) we consider schematically two simple cases--
vacancy diffusion and straight line interstitialcy diffusion.

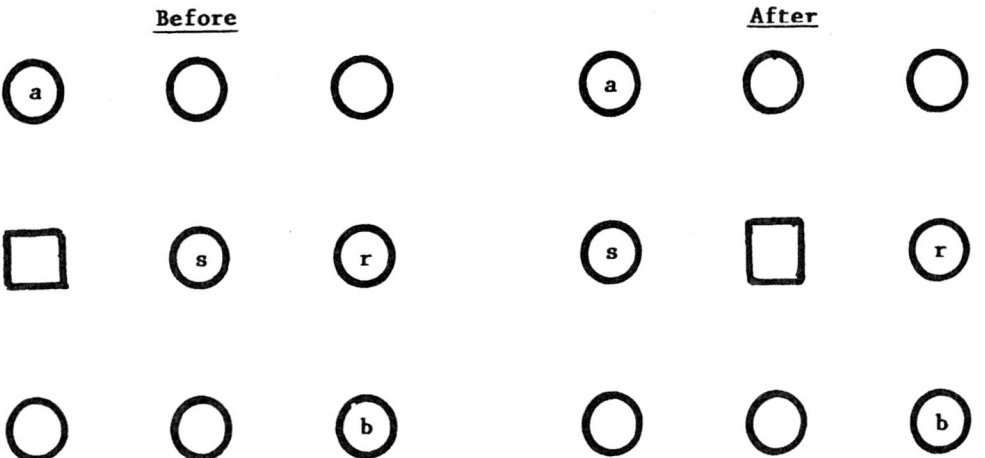

Figure 1. Schematic of a vacancy exchange with ion "s"

VACANCY MECHANISM (v)

Figure 1 shows a portion of a square 2d lattice of u ions in which a vacancy exchange with ion s has occurred. For clarity the counter ions are not shown. For a vacancy - ion exchange the number of ion jumps and vacancy jumps are identical and thus $\rho_u/\tau_u = \rho_d/\tau_d$ and the jump distances are equal so $\lambda_u^2 = \lambda_d^2$ which permits further simplification of 16a to:

$$H_R^d(v) = f_u^{xx}/f_v^{xx} \tag{17}$$

Because of its symmetrical position an isolated vacancy jumps randomly (i.e., without correlation) in a lattice without impurities. Hence, $f_v^{xx} = 1$, as is well known, and:

$$H_R^d(v) = f_u^{xx}(v) \tag{18}$$

Values for $f_u^{xx}(v)$, the correlation coefficient of ion jumps for a vacancy mechanism, have been calculated for a variety of crystal lattices. The enhanced tendency for a return jump ($\alpha_{u1}^{xx} < 0$) explains why $f_u^{xx}(v)$ and $H_R^d(v)$ are always less than unity.

The calculation of f_u^{xy} also takes advantage of the fact that the vacancy is at a local center of symmetry. The enhanced probability for the rth ion to jump in a return direction is exactly the same as the enhanced probability that the sth ion jumps in the forward direction. All ions can be paired (e.g., r and s or a and b) so that they are oppositely oriented to the vacancy. Thus, the correlations of their subsequent jumps with the jump of ion r are of equal magnitude

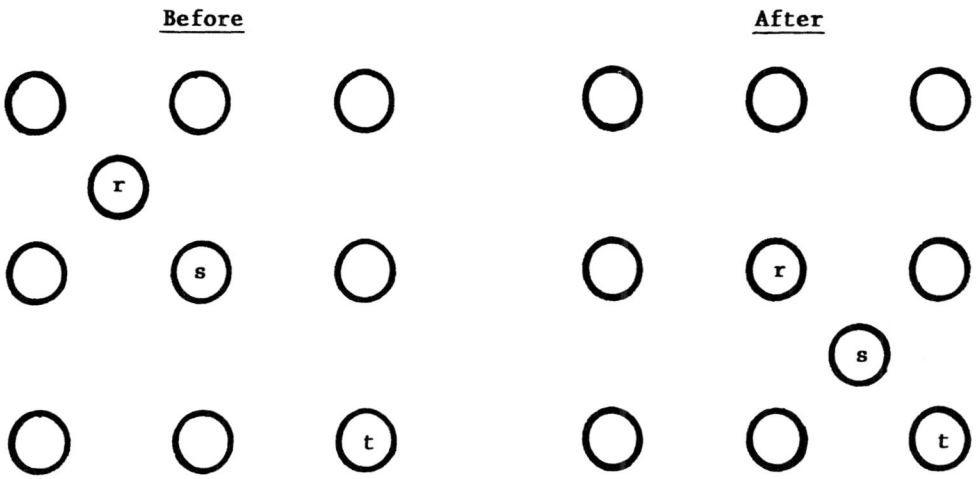

Before   **After**

Figure 2. Schematic of an interstitialcy jump

but opposite sign and hence, they are self-cancelling. $\Sigma\ \alpha_{up}^{xy} = 0$, $f_u^{xy} = 1$, and equations (16a) and (16b) give the same Haven Ratio

STRAIGHT LINE INTERSTITIALCY MECHANISM (I)

An interstitialcy mechanism is a two-ion jump in which an interstitial ion replaces a lattice ion which moves into another interstitial site (see illustration of an interstitialcy jump in a square lattice in Figure 2). "Straight line" refers to the fact that the jumps of the two ions are in the same direction. The "defect" in this case is initially ion r, on the interstitial site. The defect jumps twice as far as either ion r or ion s. Thus, $\lambda_I = 2\lambda_u$. Since two ions jump for each interstitial, interstitial jumps are half as frequent as ion jumps, and $\rho_I/\tau_I = \rho_u/2\tau_u$. Substituting into equation (16a)

$$H_R^d(I) = \frac{2}{4}\ f_u^{xx}(I) \tag{19}$$

The result, equation (19), is well known as are values for $f_u^{xx}(I)$ for various lattices. Each of the two jumping ions contribute equal weight. Thus

$$f_u^{xx}(I) = \left(f_u^{rr} + f_u^{ss} \right)/2 \tag{20}$$

Since the sth ion ends in an interstitial position, its jump will be in a random direction and $f_u^{ss} = 1$. However, the rth ion has enhanced probability of a return jump, making $f_u^{rr}(I) < 1$ and hence, $f_u^{xx}(I) < 1$, as is well known.

Calculation of f_u^{xy} likewise requires consideration of both the rth and the sth ion, i.e., $f_u^{xy} = (f_r^{xy} + f_s^{xy})/2$. While it is necessary to arbitrarily assign one to jump before the other, the result does not depend on the choice. Let r be considered the first to jump, then $\alpha_{r1}^{xy} = 1$. After the jump of r and s all ions are antisymmetrically paired with respect to the defect s and no further correlation exists. Thus, from equation (14), $f_r^{xy} = 3$, and $f_s^{xy} = 1$. Hence:

$$f_u^{xy} = \frac{4}{2} = 2 \tag{21}$$

Substituting equation (21) into equation (16b) gives the same result as equation (19), confirming again that it matters not whether the defects or the ions are considered as charge carriers. These two examples illustrate that when diffusion is by a defect mechanism, treating defects as carriers is equivalent to considering the jumps of charge, i.e., of all the ions.

In liquids and glasses where the definition of a defect is often obscure, the calculation of H_R may be most easily determined from equation (16b). It is evident from this equation that the Haven Ratio is the quotient of the serial correlation coefficient of the jump of a single ion, f_u^{xx}, by the total correlation coefficient of the jumps of all ions, f_u^{xy}. The examples show that these are independent quantities. Typically and perhaps invariably, for a system with simple defects and with a single mobile ion type, $f_u^{xx} \leq 1$ and $f_u^{xy} \geq 1$.

HAVEN RATIO WITH TWO MOBILE ION TYPES

On occasion there may be more than one type of ion in a system which is mobile. The behavior can be illustrated by restricting attention to a system with two types of mobile ions, say, u and w. For this case, the Nernst-Einstein equation is written:

$$\sigma_{NE} = (\rho_u D_u z_u^2 + \rho_w D_w z_w^2) e^2 / kT \tag{22}$$

The conductivity equation written in terms of the diffusion of charge, equation (14), needs no alteration for the case where there are more than one mobile ion, so the Haven Ratio is given by:

$$H_R^q = \frac{\rho_u D_u z_u^2 + \rho_w D_w z_w^2}{\rho_q D_q} = \frac{X_u D_u z_u^2 + X_w D_w z_w^2}{D_q} \tag{23}$$

where X_u is the fraction of the charge carriers which is type u ions.

The task remaining is to calculate D_q. The starting point is equation (12). Its first term can be decomposed into the separate contributions from u and w ions. Of the remaining terms, each is decomposed into four parts, one each for the combinations: [u,u]; [u,w]; [w,u]; [w,w], where the mean square displacement is then written:

$$NR^2(n) = n_u \langle (\lambda z)_u^2 \rangle + n_w \langle (\lambda z)_w^2 \rangle + \ldots \tag{24}$$

$$+ 2(n_u - p)[\langle (\vec{z\lambda})_{u,i} \cdot (\vec{z\lambda})_{u,i+p} \rangle + \langle (\vec{z\lambda})_{ui} \cdot (\vec{z\lambda})_{w,i+p} \rangle] +$$

$$2(n_w - p)[\langle (\vec{z\lambda})_{w,i} \cdot (\vec{z\lambda})_{u,i+p} \rangle + \langle \vec{z\lambda})_{w,i} \cdot (\vec{z\lambda})_{w,i+p} \rangle] + \ldots$$

The correlation of the jump of charge carried by an a ion and the pth subsequent jump of charge carried by a neighboring b ion (when a = b the subsequent jumps of the ion itself are included) is written:

$$\langle (\vec{\lambda z})_{ai} \cdot (\vec{\lambda z})_{b,i+p} \rangle / \langle (\lambda z)_a^2 \rangle \equiv \alpha_{abp}^{xy} \tag{25}$$

As usual, it is expected that correlation vanishes for p << n, and so a close approximation:

$$NR^2(n) = n_u \langle (\lambda z_u)^2 \rangle (1 + 2 \Sigma(\alpha_{uup}^{xy} + \alpha_{uwp}^{xy})) \tag{26}$$

$$+ n_w \langle (\lambda z_w)^2 \rangle (1 + 2 \Sigma(\alpha_{wup}^{xy} + \alpha_{wwp}^{xy}))$$

The terms in parentheses above are the corrections for the effect of correlation, i.e., they are the generalization of the correlation coefficients f_u^{xy} and f_w^{xy}. Inserting these coefficients into equation (26) and substituting into equation (6) gives:

$$D_q = \frac{n_u}{6Nt} \langle (\lambda z)_u^2 \rangle f_u^{xy} + \frac{n_w}{6Nt} \langle (\lambda z)_w^2 \rangle f_w^{xy} \tag{27}$$

Writing $N = N_u/X_u = N_w/X_w$, noting that n_u/N_u is the average number of jumps that a particular u ion makes in time, t, and recognizing that the quotient of the average number of jumps of a u ion in time, t, by t is the average waiting time, τ_u, for a type u ion, gives:

$$D_q = \frac{X_u}{6\tau_u} \langle (\lambda z)_u^2 \rangle f_u^{xy} + \frac{X_w}{6\tau_w} \langle (\lambda z)_w^2 \rangle f_w^{xy} \tag{28}$$

Thus, the Haven Ratio for a system with two mobile cations is:

$$H_R^q = \frac{f_u^{xx} X_u \langle (\lambda Z)_u^2 \rangle / \tau_u + f_w^{xx} X_w \langle (Z\lambda)_w^2 \rangle / \tau_w}{f_u^{xy} X_u \langle (\lambda Z)_u^2 \rangle / \tau_u + f_w^{xy} X_w \langle \lambda Z)_w^2 \rangle / \tau_w} \tag{29}$$

Equation (29) is a generalization of equation (16b). Its expansion to allow more than two mobile ions is straightforward. (This also requires expansion of f_u^{xy}, but that is likewise straightforward).

Although typically $f_u^{xy} \geq 1$ for a system with a single mobile ion, when there are mobile cations and anions then the above restriction need not apply.

Let us consider an example: Suppose Ca^{++} ions and O^{--} ions are the most mobile ions in a silicate glass. Typically, $\rho_{Ca} < \rho_O$. Because of the charge on the Ca^{++} ion, when it jumps it can produce an electrostatically favorable site for an O^{--} ion and leave behind O^{--} ions whose sites have become less favorable. Thus, there is a tendency for a neighboring O^{--} ion to jump in nearly the same direction as the Ca^{++} ion.

For simplicity, let us take as a premise an extreme case where every jump of a Ca^{++} ion is shortly followed by a jump of an O^{--} ion in the same direction and of the same length. Subsequent jumps of both ions are presumed to be uncorrelated. Thus, $\alpha_{Ca,O} = -1$ (minus because $Z_{Ca} = -Z_O$), $\alpha_{O,O} = \alpha_{Ca,Ca} = \alpha_{O,Ca} = 0$. Substituting these values into the definition of f_u^{xy} and f_w^{xy} gives $f_{Ca}^{xy} = -1$ and $f_O^{xy} = 1$. According to the premise, the number of Ca jumps is equal to the number of O jumps. Thus, $X_{Ca}/\tau_{Ca} = X_O/\tau_O$. Also, $\lambda_{Ca}^2 = \lambda_O^2$. Such a premise leads to a Haven Ratio of infinity. In reality, the correlation between Ca and O is never perfect; thus, finite values of H_R must be obtained. However, the possibility of values of H_R larger than 1, as observed by King and Koros[5] is consistent with positive correlation between Ca^{++} and O^{--} jumps.

CONCLUSIONS

The Haven Ratio for a system with a single mobile ion type is the ratio of the correlation coefficient for the jumps of a single ion to the correlation coefficient for the jumps of all ions. A generalization of this definition applies to systems with more than one mobile ion type.

ACKNOWLEDGEMENTS: The author appreciates support from MSF (DMR83-12301).

REFERENCES

1. J. V. Fitzgerald, *J. Chem. Phys.* 20:922 (1952).
2. A. D. LeClaire, Chapter 5, *in*: "Physical Chemistry," Vol. 10, H. Eyring, D. Henderson, and W. Jost, eds., Academic Press, New York (1970).
3. Y. Haven, Rept. Conf. Defects Crystalline Solids, London, p. 261 (1954).
4. (a) J. R. Manning, "Diffusion Kinetics for Atoms in Crystals," Van Nostrand, Princeton (1968).
   (b) A. D. LeClaire, Chapter 5, *in*: "Physical Chemistry," Vol. 10, H. Eyring, D. Henderson, and W. Jost, eds., Academic Press, New York (1970).
5. P. J. Koros and T. B. King, *Trans. AIME* 224:229 (1962).

USE AND MISUSE OF CHEMICAL DIFFUSION THEORY

D.S. Tannhauser

Department of Physics
Technion - Israel Institute of Technology
32000 Haifa, Israel

INTRODUCTION

The literature on chemical diffusion and interdiffusion contains many badly defined and/or wrongly used concepts. The reason they occur so frequently is that diffusion is based on thermodynamics where one can all too easily apply valid but very general formulae in a wrong manner. We would like here to clarify some of these concepts.

Chemical diffusion is defined as diffusion in a chemical concentration gradient. It therefore includes interdiffusion of two materials forming a solid solution as in Kirkendall experiments and in chemical doping of semiconductors as well as experiments in which the stoichiometry of a compound changes through changes in the defect concentration, such as oxidation/reduction of CoO_{1+x} to CoO_{1+y}. Here we shall use the term "chemical diffusion" in the restricted sense as applying only to experiments where the defect concentration changes and use the name "interdiffusion" for the other cases.

An equation derived by Darken[1] for the diffusion coefficient in interdiffusion experiments, has sometimes been applied to chemical diffusion processes. We shall show that this is not permitted. In order to do this we shall first review briefly the important equations of interdiffusion.

INTERDIFFUSION AND THE KIRKENDALL EFFECT

In a Kirkendall experiment one constructs a diffusion couple made of two pieces of an alloy AB with different compositions and measures as a function of time the distance between inert markers placed ·at the original boundary of the two pieces and the ends of the couple. A good discussion of the principles involved is found in Manning[2]. Here we shall not repeat the derivation of the equations, but would like to stress that an important and necessary assumption in the derivation is that the total number of atoms per unit volume is constant[1]. If the diffusion is by a vacancy mechanism, then this implies a concentration of vacancies independent of x and t.

Darken showed that the following equation gives the rate at which the composition gradient of the diffusion couple tends to smooth out

$$J'_A = -(c_A D^I_B + c_B D^I_A) \frac{\partial [A]}{\partial x} \qquad (1)$$

The effective diffusion constant, which we shall call D^{Dar}, is therefore

$$D^{Dar} = c_A D^I_B + c_B D^I_A \qquad (2)$$

This is known as Darken's equation.

In these equations J'_A is the flux of component A measured with respect to the ends of the couple (because of the above assumption about atoms per unit volume the length of the couple does not change with time), [A] is the concentration of component A (measured in units of particles/m^3), $c_A = [A]/([A] + [B])$, $c_B = [B]/([A] + [B])$ and D^I_A is the so called[2] "intrinsic diffusion" coefficient defined by

$$J_A = -D^I_A \frac{\partial [A]}{\partial x} \qquad (3)$$

with a similar definition for D^I_B. These D^I's which are called "diffusivities" by Darken, depend on the composition, i.e. on the ratio [A]/[B]. J, in contrast to J', is the flux density measured with respect to the lattice, i.e. operationally with respect to inert markers in the sample[1].

Manning, using a detailed microscopic theory, showed that Eq.(2) can be transformed as follows (his Eq. 5.87)

$$D^{Dar} = (c_A D^{Tr}_B + c_B D^{Tr}_A) \frac{d \ln a_A}{d \ln c_A} R \qquad (4)$$

with

140

$$R = 1 + \frac{(1 - f)c_A c_B (D_A^{Tr} - D_B^{Tr})^2}{f(c_A D_B^{Tr} + c_B D_A^{Tr})(c_A D_A^{Tr} + c_B D_B^{Tr})}$$

Here D_i^{Tr} is the radioactive tracer diffusion coefficient of component i, measured at the composition of the alloy, which is of course a function of x. a_A is the activity of component A, defined through the chemical potential μ_A by $\mu_A = kT \ln a_A$. f is the correlation factor for tracer diffusion in the lattice considered. It relates the mechanical mobility b_i (with dimensions $m \ sec^{-1} Nt^{-1}$) to D_i^{Tr} by $D_i^{Tr} = f_i kT b_i$ $(1 \geqslant f \geqslant 1/2)$. R, which is a measure of the "vacancy flow"[2], and usually differs from one by less than 10%. Eq. (4), without the factor R, had already been derived by Darken, but his treatment is only approximately correct. The expression $d \ln a_A/d \ln c_A$ is known as the thermodynamic factor. If the solution of A in B is ideal (this implies that the entropy is configurational only), then

$$\mu_A = kT \ln c_A + \text{const.} \tag{5}$$

and the thermodynamic factor is one.

From Eq. (4) we can calculate D^{Dar} for any composition provided D_A^{Tr}, D_B^{Tr}, f and the factor $d \ln a_A/d \ln c_A$ are all known for the composition of interest.

Two special cases will help to understand the concepts involved:
A. Measurement with radioactive tracers of the self diffusion in an element. We can regard this as interdiffusion of A^* and A, the radioactive and the nonradioactive isotope. In Eq. (4) the two diffusion coefficients are the same and the value of the thermodynamic factor is one, since the solution is ideal. Also $R = 1$, as can be verified from Eq. (4). We get therefore $D^{Dar} = D^{Tr}$. This must be so since there is no way of defining D^{Tr} operationally, except by measuring the interdiffusion of A^* and A.
B. Diffusion doping of a semiconductor, e.g. boron diffusing into silicon. Here we have interdiffusion of a minor constituent, (the dopant, A) and a major constituent (the semiconductor, B). Therefore $c_A \ll c_B$ and R can be shown to be one in this case. Eq. (4) gives now

$$D^{Dar} = D_A^{Tr} \frac{d \ln a_A}{d \ln c_A} \tag{6}$$

and because of the low concentration we would expect

d ln a_A = d ln c_A, i.e. $D^{Dar} = D_A^{Tr}$. Since, however, dopants in semiconductors diffuse usually not as atoms, but as ions and electrons, the relation is more complicated. The value of the factor multiplying D_A^{Tr} for this situation will be discussed later.

CAN DARKEN'S EQUATION BE APPLIED TO CHEMICAL DIFFUSION?

We discuss now a hypothetical binary compound MX in which we have a low concentration of vacancies V_M on the M sublattice. V_M is a neutral vacancy and causes diffusion of the M atoms. We assume that the X atoms stay in place so that the X sublattice is completely inert and can serve as reference lattice.

We now make an experiment which looks like a Kirkendall experiment (but is not, see below). We construct a diffusion couple with a step in the concentration of M-atoms (see Fig. 1). The initial concentrations are $[V_M]^1$ vacancies and $[M_M]^1$ atoms on the left and $[V_M]^2$ vacancies and $[M_M]^2$ atoms on the right, and we have

$[M_M] + [V_M]$ = const.

The couple will equilibrate with an effective diffusion coefficient $\tilde{D}$, the vacancies and the atoms will eventually be distributed uniformly. We see that this fits our definition of a chemical diffusion process so that $\tilde{D}$ is the chemical diffusion coefficient. $\tilde{D}$ can be measured as usual by comparing a measured concentration profile with a solution of the diffusion equation

$$\frac{\partial [M_M]}{\partial t} = \frac{\partial t}{\partial x} (\tilde{D} \frac{\partial [M_M]}{\partial x}) \qquad (7)$$

We note here that $\tilde{D}$ does not have an index. One usually measures relative rather than absolute concentrations so that there is no way of telling from the equilibration experiment whether the M atoms or the X atoms are mobile.

We now ask whether we can use Eq. (2), or the equivalent Eq. (4), to calculate $\tilde{D}$, as was done for closely related equilibration experiments by a number of authors[3,4,5].

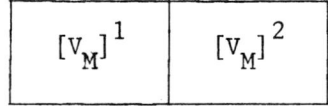

Fig. 1

If we do so, we get for $\tilde{D}$

$$\tilde{D} = c_M D_X^I + c_X D_M^I \tag{8}$$

and we have in our case $c_M \simeq c_X \simeq 1/2$. Since we assumed that $D_X^I = 0$, we get that $\tilde{D} = 1/2\, D_M^I$. However, Eq. (3) is for the M–atoms

$$J_M = -D_M^I \frac{\partial [M_M]}{\partial x} \tag{9}$$

Eq (7) implies that $J_M' = -\tilde{D}\, \dfrac{\partial [M_M]}{\partial x}$, but in our case J and J' are the same, since the X lattice, any plane of which can in principle be marked by radioactive X–atoms, does not move with respect to the ends of the couple. Therefore the effective diffusion coefficient characterizing the equilibration should be $\tilde{D} = D_M^I$, which obviously disagrees with Darken's equation.

We see that Darken's equation is not applicable here, as it leads to the wrong result. The basic reason is that Darken's equation applies to a system in which the total number of atoms per unit volume is constant, so that there is a concentration gradient of both M and X, rather than of M alone. Vacancies are then created on one side of the interface and annihilated on the other, whereas in our case vacancies are conserved, only redistributed.

We conclude that for our hypothetical experiment the correct expression is

$$\tilde{D} = D_M^I \tag{10}$$

Equation (10), with the explicit form of D^I to be derived in the next section, was implicitly used by Steele[4] for CdS, but it should be clear that it is not a limiting case of equation (2) as claimed by Steele. Wagner[3], who discussed FeO, calculated $\tilde{D}$ according to both equations (2) and (10), neither of which gave a very good fit to the experimental results.

RELATION BETWEEN CHEMICAL AND TRACER DIFFUSION

We shall now relate $\tilde{D}$, which is measurable for instance by a thermogravimetric absorption/desorption experiment (Heyne[6]) to D^{Tr} and the mobility of the mobile atoms. We shall do this for a compound MX in which as in the previous section the X sublattice is perfect and inert while the M sublattice contains normal M atoms, radioactive M* atoms and vacancies V. The concentrations are [M], [M*] and [V] and we have that

[M] + [M*] + [V] = [X] = constant. (From here on we leave out subscripts to M, V etc. when the meaning is clear) In view of the recent interest in highly non-stoichiometric compounds the discussion will not be limited to the case [V] << [X].

We have to assume here that the M-atoms move as ions, since only in such as system can we define operationally a mobility b_M by measuring the ionic current in an electric field. Since we assumed that the X sublattice is immobile, chemical diffusion will be carried by ions coupled to electrons; this is called ambipolar diffusion. We then have the following experimental complication. The magnitude of $\tilde{D}$ will only be defined by properties of the ions if the solid is predominantly an electronic conductor, i.e. if the transport number t_e of electrons is much larger than t_M of ions. Otherwise a measurement of $\tilde{D}$ will measure some combination of ionic and electronic properties. But if $t_e >> t_M$ then a measurement of b_M is not straightforward, we then have to either subtract the electronic current from the total or block the motion of electrons in our experiment.

If the relation $t_e >> t_M$ does not hold, then the properties of the ions determine only the ratio between b_M and D^{Tr}. This would apply for example to stabilized zirconia which is an ionic oxygen conductor.

The relation connecting b_M with D^{Tr} in a solid is known as the Einstein relation. However, the original relation was derived by Einstein to connect the mobility with the chemical diffusion constant for a system of noninteracting particles (see Murch[7] for details). The application to a solid is therefore not immediate.

TRANSPORT EQUATIONS

We start with the set of equations connecting ionic and electronic fluxes J and the corresponding forces X and include for generality cross-coefficients between the various types of particles.

$$J_M = L_{MM}(X_M - X_V) + L_{MM*}(X_{M*} - X_V) + L_{Me}X_e \qquad (11a)$$

$$J_{M*} = L_{M*M}(X_M - X_V) + L_{M*M*}(X_{M*} - X_V) + L_{M*e}X_e \qquad (11b)$$

$$J_e = L_{eM}(X_M - X_V) + L_{eM*}(X_{M*} - X_V) + L_{ee}X_e \qquad (11c)$$

In these equations X_M, X_{M*} and X_V are the negative gradients of the so-called virtual electrochemical potentials defined by $\tilde{\mu}_i = \mu_i + z_i q\phi$, with

$$\mu_i = (\frac{\partial G}{\partial n_i})_{n_i \neq j, T, P} \tag{12}$$

where G does not include the electric potential. $z_i q$ is the charge of a particle, with $z = 1$ for holes and $z = -1$ for electrons. These virtual potentials do not refer to a physical process since there is no way of adding an ion without changing the number of one of the other components. On the other hand, the combinations $(X_M - X_V)$ etc. are real forces in the sense that they are gradients of electrochemical potentials which refer to a real physical process, namely the insertion of an ion and elimination of a vacancy.

The Tracer Diffusion Coefficient

In a tracer diffusion experiment no field is applied and the electron concentration is constant. We have therefore $X_e = 0$. Since there is no field we are working only with chemical potentials. Also [V] is constant throughout the sample so that $X_V = 0$. Eq. (11b) gives therefore

$$J_{M*} = L_{M*M*}X_{M*} + L_{M*M}X_M = -L_{M*M*}\frac{\partial \mu_{M*}}{\partial x} - L_{M*M}\frac{\partial \mu_M}{\partial x} \tag{13}$$

We have $\mu_M = kT\ln \gamma [M]$, where the activity coefficient γ depends only on $([M] + [M*])$. With

$$\frac{\partial [M]}{\partial x} = -\frac{\partial [M*]}{\partial x} \tag{14}$$

and the definition of the tracer diffusion coefficient

$$J_{M*} = -D^{Tr}\frac{\partial [M*]}{\partial x} \tag{15}$$

we get

$$D^{Tr} = kT(\frac{L_{M*M*}}{[M*]} - \frac{L_{M*M}}{[M]}) \tag{16}$$

The Mobility

We now apply a field to the sample in which all the concentrations are constant. In order to do this the electrodes have to be nonblocking to ions and electrons. We get from Eq. (11a) that

$$J_M = z_M q(L_{MM} + L_{MM*})E + z_e qL_{Me}E \tag{17}$$

Since the mechanical mobility b_M (velocity/force) is defined through

$$J_M = b_M [M] z_M q E \tag{18}$$

we get that

$$b_M = \frac{1}{[M]} \left(L_{MM} + L_{MM*} + \frac{z_e}{z_M} L_{Me} \right) \tag{19}$$

We must have $b_M = b_{M*} = b_m$, where "m" refers to any ion. By using

$$[m] = [M] + [M*] \tag{20}$$

and defining

$$L_{mm} = L_{MM} + L_{MM*} + L_{M*M} + L_{M*M*}$$

$$L_{me} = L_{Me} + L_{M*e}$$

we get that

$$b_m = \frac{1}{[m]} \left(L_{mm} + \frac{z_e}{z_m} L_{me} \right) \tag{21}$$

This will be used later.

The ratio of b_M and D^{Tr} is from Eqs. (16) and (19)

$$\frac{b_M}{D^{Tr}} = \frac{1}{kT} \left[1 + \frac{1 + ([M]/[M*]) + (z_e L_{Me}/z_M L_{MM*})}{(L_{MM}/L_{MM*}) - ([M]/[M*])} \right] \tag{22}$$

Manning[2] showed (his Eqs. 4.86 to 4.90) that

$$\frac{b_M}{D^{Tr}} = \frac{1}{kTf} \frac{[X]}{[M] + [M*]} \tag{23}$$

For the case $L_{Me} = 0$ one can then express the ratio L_{MM*}/L_{MM} in the following form

$$\frac{L_{MM*}}{L_{MM}} = \frac{\frac{[M]}{[M*]} \frac{[X]}{[M] + [M*]} + f}{\frac{[X]}{[M] + [M*]} - f} \tag{24}$$

146

We note however, that the correlation factor f is usually calculated for low vacancy concentration where one can assume that the vacancies perform a random walk. Since f has to depend on the vacancy concentration (for $[V] \cong [X]$ we must have $f = 1$) we can not use the tabulated values of f (Manning[2], p.95) to calculate L_{MM*}/L_{MM} in the general case.

The Chemical Diffusion Coefficient

We have now a constant ratio of $[M]$ to $[M*]$ and a gradient of $[m]$. We can use the relation

$$X_M = X_{M*} \tag{25}$$

which follows from $\mu_M = kT\ln([M]\gamma([m]))$ and $[M]/[M*] = K$. We get from Eqs. (11a,b) and the definitions (20b,c) that

$$J_m = L_{mm}(X_m - X_V) + L_{me}X_e \tag{26}$$

and similarly from (11c) and the definition $L_{em} = L_{eM} + L_{eM*}$

$$J_e = L_{em}X_m + L_{ee}X_e \tag{27}$$

It can be shown that Eqs. (26) and (27) together with $z_m J_m + z_e J_e = 0$ lead to

$$J_m = \frac{L_{em}L_{me} - L_{mm}L_{ee}}{z_m^2 L_{mm} + z_m z_e (L_{me} + L_{em}) + z_e^2 L_{ee}} \, x$$

$$(z_e^2 \frac{\partial(\mu_m - \mu_V)}{\partial x} - z_m z_e \frac{\partial \mu_e}{\partial x}) \tag{28}$$

Since $\tilde{D}$ is defined by

$$J_m = -\tilde{D} \frac{\partial[atom]}{\partial x} \tag{29}$$

where [atom] means total concentration of "m" atoms, and since $(\mu_m - \mu_V) - (z_m/z_e)\mu_e = \mu_{atom}$, we get finally that

$$\tilde{D} = z_e^2 \frac{L_{mm}L_{ee} - L_{em}L_{me}}{z_m^2 L_{mm} + z_m z_e (L_{me} + L_{em}) + z_e^2 L_{ee}} \frac{\partial \mu_{atom}}{\partial[atom]} \tag{30}$$

With the help of the expression for the ionic conductivity (see Eq. (21)).

$$\sigma_m = b_m[m]z_m^2 q^2 = z_m^2 q^2 L_{mm} + z_m z_e q^2 L_{me} \tag{31}$$

147

and a similarly derived expression for the electronic conductivity

$$\sigma_e = b_e [e] z_e^2 q^2 = z_e^2 q^2 L_{ee} + z_e z_m q^2 L_{em} \tag{32}$$

we can transform Eq. (30) as follows

$$\tilde{D} = \frac{1}{z_m^2 q^2} (\sigma_m t_e - z_m z_e q^2 L_{em}) \frac{\partial \mu_{atom}}{\partial [atom]} \tag{33}$$

Here we have used $t_e = \sigma_e / (\sigma_e + \sigma_m)$ and also the Onsager relation $L_{me} = L_{em}$.

If $L_{me} = 0$, we get from Eq. (33) and $[atom] = [m]$

$$\tilde{D} = t_e b_m \frac{\partial \mu_{atom}}{\partial \ln [atom]} \tag{34}$$

which is a well-known result[6].

The full Eq. (33) shows the function of the cross-coupling term L_{me}. In principle we could measure L_{me} by the use of this equation, since all the other terms are accessible through independent measurements.

From Eqs. (23) and (34) we get

$$\tilde{D} = t_e \frac{D^{Tr}}{kTf} \frac{[X]}{[m]} \frac{\partial \mu_{atom}}{\partial \ln [atom]} \tag{35}$$

D^{Tr}, t_e and the thermodynamic factor are often measurable by independent experiments and for $[m] \simeq [X]$ f can be calculated. We can therefore test the validity of Eq. (35) for chemical diffusion.

Weppner and Huggins[8], who discussed this subject recently for the case $[m] \simeq [X]$ and $L_{me} = 0$, define an "enhancement factor," which they call W, in honor of Carl Wagner. It is

$$W = \frac{\tilde{D} f}{D^{Tr}} = \frac{t_e}{kT} \frac{\partial \mu_{atom}}{\partial \ln [atom]} \tag{36}$$

Eq. (36) with $t_e = 1$ should be applicable to the electronic conductors FeO, CoO, NiO, etc. This was done, as mentioned, by Wagner[3] as an alternative to using Darken's equation and also by Chu, Rickert and Weppner[9] who got better agreement.

SOME EXAMPLES OF THE ENHANCEMENT FACTOR

The enhancement factor is easiest to understand and calculate for a system with low defect concentration. We now evaluate Eq.(36) for a number of examples, including stabilized zirconia in which the defect concentration is high.

1. CoO

CoO contains up to ~1% holes and singly or doubly ionized cobalt vacancies, depending on the range of oxygen pressures[10]. Introduction of a neutral cobalt atom from rest at infinity corresponds to the following reaction

$$Co \ (\infty) \rightarrow Co_{Co}^{x} - V_{Co}^{z'} - zh^{\cdot} \qquad (z = 1 \quad or \quad 2) \qquad (37)$$

We use here the Kroeger-Vink notation in which ^{x}, ' and $^{\cdot}$ indicate effective charges (relative to the perfect lattice) of zero, $-q$ and $+q$ respectively. The chemical potential $\mu_{Co-atom}$ is therefore

$$\mu_{Co-atom} = const. - \mu_V - z\mu_h \qquad (38)$$

In our defect model $[h] = z[V_{Co}^{z'}]$ and because of the low concentration

$$\mu_V = kT \ \ell n[V_{Co}^{z'}] + const. \qquad (39)$$

and

$$\mu_h = kT \ \ell n[h] + const. \qquad (40)$$

We use $d[V_{Co}^{z'}] = -d[Co-atom]$ and get with some algebra

$$\frac{d\mu_{Co-atom}}{d \ \ell n[Co-atom]} = kT(z + 1) \frac{[Co-atom]}{[V_{Co}^{z'}]} \qquad (41)$$

Since in CoO $t_e = 1$, we get from Eq. (36) that

$$\tilde{D} = \frac{1}{f} D_{Co}^{Tr}(z + 1) \frac{[Co-atom]}{[V_{Co}^{z'}]} \qquad (42)$$

We see that the enhancement factor equals the reciprocal of the relative vacancy concentration times $(z + 1)$. If diffusion was dominated by neutral vacancies, we would have $(z + 1) = 1$.

The result of Eq.(42) is more intuitive if one remembers that $D^{Tr} = fD_V[V_{Co}^{z'}]/[Co-atom]$, where D_V is the "random walk" diffusion constant of vacancies. It reads then

$$\tilde{D} = (z + 1)D_V \qquad (43)$$

and means that $\tilde{D}$ equals D_V enhanced by a factor $(z + 1)$, depending on the ionization state of the vacancies. For neutral vacancies would have $\tilde{D} = D_V$.

2. UO_2

In UO_2 oxygen, which diffuses through vacancies, is the mobile ionic species and $t_e = 1$. The relation between oxygen self- and chemical diffusion in UO_{2+x} ($x \leqslant 0.16$) has recently been discussed by Breitung[5]. Breitung mistakenly uses Darken's equation and arrives at the result (his Eq.(17)):

$$\tilde{D} = D_0^{Tr} \frac{2 + x}{2RT} \frac{d\Delta G(O_2)}{dx} \qquad (44)$$

where $\Delta G(O_2)/2 = \mu_0$. If one uses Eq. (35) of the present paper one gets that

$$\tilde{D} = \frac{D_0^{Tr}}{f_0} \frac{(2 + x)(3 + x)}{2RT} \frac{d \Delta G(O_2)}{dx} \qquad (45)$$

and since $f_0 = 0.653$ for the simple cubic oxygen sublattice it follows that $\tilde{D}$ is larger by a factor ~ 4.5 than the values calculated by Breitung. The mistake originates of course in the use of $\tilde{D} = N_U D_0^I = \frac{1}{3} D_0^I$ by Breitung, instead of $\tilde{D} = D_0^I$, and his neglect of f_0. One sees from Fig. 5 of Breitung's paper that the bulk of the experimental results for $\tilde{D}$ lies indeed by a factor of 3 to 4 above the theoretical curve he calculated for $x \leqslant 0.01$. If one uses the correct theory, the agreement is obviously much better.

3. Stabilized Zirconia

This material, with a representative composition $Zr_{0.9}Y_{0.1}O_{1.95}$ is an oxygen ion conductor with a high concentration of oxygen vacancies ($2\frac{1}{2}\%$ in the above example) and $t_e \ll 1$[6]. Using $\sigma_0 = 4q^2 b_0 [O^=]$, $[O^=] = [O\text{-atom}]$ and Eq. (23) we can rewrite Eq. (36) for the present case:

$$W = \frac{1}{kT} \frac{\sigma_e}{\sigma_0} \frac{d\mu_{0\text{-atom}}}{d \ell n[O\text{-atom}]} = \frac{[O\text{-atom}]}{[O\text{-sites}]} \frac{\sigma_e}{4D_0^{Tr}} \frac{f_0}{q^2} \frac{d\mu_{0\text{-atom}}}{d[O\text{-atom}]} \qquad (46)$$

To evaluate this expression we note that σ_e is determined by holes near $P_{O_2} = 1$ atm., but by electrons at very low oxygen pressures. The incorporation reaction

150

$$0(\infty) \rightarrow 0_0^x - V_0^{\cdot\cdot} + 2h^{\cdot} \quad (\text{or } -2e') \tag{47}$$

then leads to

$$\mu_{0-atoms} = const. - \mu_V + 2\mu_h \quad (\text{or } - 2\mu_e) \tag{48}$$

and because of the high $[V_0^{\cdot\cdot}]$, μ_V is practically constant. There-fore near $P_{O_2} = 1$ atm:

$$\frac{d \mu_{0-atom}}{d[0-atom]} = 2 \frac{d \mu_h}{d[0-atom]} \tag{49}$$

Because of $[h] \ll 1$ we have $\mu_h = kT\ell n[h]$ and from Eq. (47) we get $d[h] = 2d[0-atom]$ so that

$$\frac{d\mu_{0-atom}}{d[0-atom]} = \frac{4kT}{[h]} \tag{50}$$

Eq. (36) and (46) give therefore

$$\tilde{D} = \frac{kT\sigma_e}{q^2[h]} = \frac{kTb_h q^2[h]}{q^2[h]} = kTb_h = D_h \tag{51}$$

and it can be shown similarly that at low pressures $\tilde{D} = D_e$. We neglected here the factor $[0-atom]/[0-sites]$ which is close to one.

We see that in stabilized zirconia (and similarly in any con-ductor with $t_e \ll 1$) $\tilde{D}$ is determined only by the diffusion coef-ficient of the electronic charge carriers and has nothing to do with the tracer diffusion coefficient of the ions. This result is very often overlooked, most recently for instance by Manasevit et al.[11]. In zirconia there are complications due to trapping, these are dis-cussed by Heyne.[6]

4. Diffusion Doping of Semiconductors

Boron and similar dopants diffuse into silicon substitutionally. This, as mentioned before, is a case of interdiffusion. Since boron is incorporated not as atoms, but as $B'_{Si} + h^{\cdot}$, we have ambipolar diffusion of the boron. As far as we know, the combination of ambi-polar and interdiffusion has not been treated in the literature, though the related problem of KCl-RbCl interdiffusion has been dis-cussed[12]. One usually neglects the interdiffusion aspect; the treatment of diffusion doping as a case of simple ambipolar diffusion then starts with the incorporation reaction

$$B(\infty) \rightarrow B'_{Si} + h^{\cdot} \tag{52}$$

which leads to

$$\mu_{B-atom} = \mu_{B'} + \mu_h \qquad (53)$$

Because of the low concentrations involved we have

$$\mu_{B'} = kT \, \ell n[B'], \qquad \mu_h = kT \, \ell n[h] \qquad (54)$$

We now have to differentiate between two cases:
a) If boron diffuses into intrinsic silicon then $[h] = [B']$.
Since $[B'] = [B-atom]$, we have in this case
$\mu_{B-atom} = 2 \, kT \, \ell n[B-atom]$, and get from Eq. (35) with $t_e = 1$ and $[m] = [X]$ that

$$\tilde{D} = 2 \, \frac{D_B^{Tr}}{f_B} . \qquad (55)$$

b) If boron diffuses into doped p-type silicon then $[h] = $ const.
and $\mu_{B-atom} = kT\ell n[B=atom] + $ const. In this case we get

$$\tilde{D} = \frac{D_B^{Tr}}{f_B} . \qquad (56)$$

The simple ambipolar diffusion approach gives therefore an enhancement factor of two or one, depending on the case. We note that, in contrast to the case of CoO, there is no vacancy concentration in the enhancement factor, the reason being of course that in interdiffusion we do not have a vacancy flow.

The factor f_B in Eqs. (55) and (56) is not really justified since, as mentioned, we have here an interdiffusion mechanism, which should be treated differently.

5. CdS

A variation of Eq.(35) can be applied to CdS. This was first mentioned by Steele[4], who realized that the kinetic argument used by Kumar and Kroeger[13] to explain their measurements is not really necessary. In the analysis of CdS there is the complication that it is mainly the sulfur sublattice which changes during the chemical diffusion process. The diffusing defects are highly mobile Cd interstitials, but the dominant defects are sulfur vacancies, i.e. $b_{Cd}[Cd_i^{\cdot}] \gg b_S[V_S^{\cdot\cdot}]$, but $[Cd_i^{\cdot}] \ll [V_S^{\cdot\cdot}]$. The two defect species are related through

$$Cd_{Cd} + V_S^{\cdot\cdot} + e' \longleftrightarrow Cd_i^{\cdot} \qquad (57)$$

It can be shown rigorously[14] that Eq. (35) converts in this case to

$$\tilde{D} = -t_e \frac{D_{Cd}^{Tr}}{f_{Cd}kT} \frac{d\mu_{Cd-atom}}{d\ln[S-atom]} \qquad (58)$$

Evaluation of this expression on the basis of the defect equilibria gives with $t_e = 1$ that

$$\tilde{D} = D_{Cd}^{Tr} \, 3 \, \frac{[S-atom]}{[V_S^{\cdot\cdot}]} \qquad (59)$$

which is the result derived by Kumar and Kroeger. The enhancement factor is therefore $3[S-atom]/[V_S^{\cdot\cdot}]$, it resembles the one for CoO. $D_{Cd}^{Tr}[S-atom]/[V_S^{\cdot\cdot}]$ is an effective diffusion coefficient for sulfur vacancies transporting Cd, and the factor "3" reflects again the fact that a neutral Cd atom is incorporated as three charged defects[14].

Acknowledgement

The author wishes to thank Prof. F.A. Kroeger and Prof. C. Kuper for advice and criticism. This research was supported by the Fund for the Promotion of Research at the Technion.

References

1. L. S. Darken, Trans. AIME 175:184 (1948).
2. J. R. Manning, "Diffusion Kinetics for Atoms in Crystals", van Nostrand (1968).
3. J. B. Wagner, Chemical Diffusion Coefficients for Some Non-stoichiometric Metal Oxides, in "Mass Transport in Oxides" NBS special publication No. 296, U.S. Govt. Printing Office, p. 65 (1968).
4. B. C. H. Steele, Chemical Diffusion, in "Fast Ion Transport in Solids" van Gool, Ed., North Holland, p. 103 (1973).
5. W. Breitung, J. Nucl. Mat. 74:10 (1978).
6. L. Heyne, Electrochemistry of Mixed Ionic Electronic Conductors, in "Solid Electrolytes", S. Geller, Ed., Springer, p. 169 (1977).
7. G. E. Murch, Phil. Mag. A 45:685 (1982).
8. W. Weppner and R. A. Huggins, J. Electrochem. Soc. 124:1569 (1977).
9. W. F. Chu, H. Rickert and W. Weppner, Electrochemical Investigation of Chemical Diffusion in Wustite and Silver-Supphide, Ref. 4, p. 181.
10. B. Fisher and D. S. Tannhauser, J. Chem. Phys. 44:1663 (1966).
11. H. M. Manasevit, I. Golecki, L. A. Moudy, J. J. Yang and J. E. Mee, J. Electrochem. Soc. 130:1752 (1983).

12. W. Muller and H. Schmalzried, Z. Phys. Chemie N.F. 54:203 (1968).
13. V. Kumar and F. A. Kroeger, J. S. S. Chem. 3:406 (1971).
14. D. S. Tannhauser, Clarification of Some Concepts in Chemical Diffusion or: Darken, Kirkendall and Other Sources of Difficulties and Confusion in Diffusion, J. S. S. Chem., to be published.

INFLUENCE OF HETEROVALENT IMPURITIES ON THE MASS

TRANSPORT IN p-TYPE SEMICONDUCTING OXIDES

G. Petot-Ervas*, F. Gesmundo**, and C. Petot*

*   CNRS, Laboratoire PMTM, avenue J.B. Clément
    93430 Villetaneuse (France)
** CNR Istituto di Chimica Fisica Applicata dei
    Materiali, Lundobisagno Istria
    34-16141 Genova (Italy)

ABSTRACT

In this study, the effect of the presence of heterovalent impurities on the chemical diffusion coefficient of an oxide AO having a p-type semiconductivity is considered. The defect structure of the oxide has been represented assuming the presence of metal vacancies with an average effective charge α. This approximation can include the presence of interactions among defects and the possibility of formation of associates between defects or between defects and impurity atoms.

Using an expression for the fluxes of the metal ions which takes into account their dependence on the gradients of oxygen activity and of impurity concentration and the definition of $\widetilde{D}$, the chemical diffusion coefficient in the doped oxide is obtained in the form :

$$\widetilde{D} = D_{V^{\alpha'}}\left[(\alpha + 1) + t\alpha/[h^{\cdot}]\right]$$

for a trivalent impurity with a mole fraction t, and :

$$\widetilde{D} = D_{V^{\alpha}}\left[(\alpha + 1) - m\alpha/[h^{\cdot}]\right]$$

for a monovalent impurity with a mole fraction m, where $D_{V^{\alpha}}$ is the diffusion coefficient of the vacancies and $[h^{\cdot}]$ the concentration of the electron holes. These expressions predict that the addition of a trivalent impurity to an oxide AO should produce an increase of the chemical diffusion coefficient while the addition of a monovalent impurity should produce a decrease of this coefficient. These changes are in part related to changes in α induced by the impurities but depend more strongly on the correction terms containing the ratios between the concentration of the impurity and that of the electron holes.

The above predictions are in agreement with the experimental results for $\tilde{D}$ obtained using single crystals of cobaltous oxide doped with chromium (mole fractions : $1.04 \; 10^{-3}$ and $3.83 \; 10^{-3}$) and lithium (mole fractions : $5.13 \; 10^{-3}$ and $11.8 \; 10^{-3}$) and deduced from electrical conductivity measurements performed in the transient period in relaxation-type experiments.

INTRODUCTION

When a chemical potential gradient appears in a compound AX due to the sudden change in the activity of one of the components, a mass transport is observed and the rate of equilibration of the material is controlled by the chemical diffusion coefficient. Some experimental information is now available concerning the chemical diffusion coefficient values in pure compounds /1-6/. In previous papers the calculation of these diffusion coefficients has been considered in pure and doped oxides in relation to the prevailing point defects /7-10/ and to their diffusion coefficients /5,6/. However, the influence of impurities on these mass transport processes is not well known and little experimental information is available so far in the literature /6/. From a practical point of view the knowledge of these diffusion coefficients and of the impurity effects is important, for example, for the improvement of the kinetics of reduction of oxides or of the protecting effects of the oxide layers formed at the surface of metals during their high temperature oxidation.

In the following paper we will consider only the p-type semi-conducting oxides AO whose prevailing defects are vacancies on the cationic sublattice. The present analysis does not include a detailed model for the defect structure of the oxide which should be taken into account in a more refined theory. Nevertheless, in many cases the relevant information is not available both for the pure and doped oxides and the defect structure is represented by assuming the presence of metal vacancies with an average effective charge α. This approximation can be considered to include the presence of interactions among defects and the possibility of formation of associates between defects or between defects and impurity atoms. The present analysis will be made in the case of the AO oxides but the treatment can also be applied to AX p-type semi-conducting compounds where A is the metal and X the non-metal. The relationships obtained have allowed us to evaluate the chemical diffusion coefficient in cobaltous oxide doped with lithium and chromium and to compare these results with the experimental values.

I. GENERAL EQUATIONS

The flux of matter which appears in a crystal under the influence of a chemical potential gradient in the x direction may be written according to the first Fick law as :

$$j = - \widetilde{D} \frac{\partial \widetilde{c}}{\partial x} \tag{1}$$

where $\widetilde{D}$ is the chemical diffusion coefficient and $\widetilde{c}$ the absolute value of the deficiency of the metal or non metal in the compound in moles per cm^{-3}. In the case where $\widetilde{c}$ is equal to the concentration of the vacancies in the cation sublattice one can write :

$$\widetilde{c} = \left[V^{\alpha'} \right] c_M \tag{2}$$

with $c_M = d/M$
where d is the density, M the molar mass of the oxide and $\left[V^{\alpha'} \right]$ the mole fraction of vacancies in the cation sublattice.

According to Eq. 2, to the electroneutrality condition and to the coupled current condition ($j_{h^{\cdot}} = \alpha . j_{V^{\alpha}}$) Eq. 1 can also be written in both the following forms :

$$j_{V^{\alpha}} = - c_M \widetilde{D} \frac{\partial \left[V^{\alpha'} \right]}{\partial x} \tag{3}$$

$$j_{h^{\cdot}} = - c_M \widetilde{D} \frac{\partial \left[h^{\cdot} \right]}{\partial x} \tag{4}$$

a) Chemical diffusion in compounds containing monovalent impurities

The fluxes of the ions A and B which appear in the crystal due to coupled effects of gradients of oxygen potential and oxide compositions can be written in the case of compounds involving prevailing electronic conduction, as shown previously by Wagner /11/, as :

157

$$J_A = - c_M D_A (1-m) \frac{\partial \ln a_A}{\partial x} \tag{5}$$

$$J_B = - c_M D_B \, m \, \frac{\partial \ln a_B}{\partial x} \tag{6}$$

where m is the local concentration of the monovalent impurity in the oxide (in mole fraction), D_A and D_B the diffusion coefficients of cations A and B.

The activities a_A and a_B of the metals A and B may be expressed as functions of the oxygen activity. Indeed, if one considers the local chemical equilibrium between the different neutral species in each element of volume of the oxide ($1/2 \, O_2 + A \rightleftarrows AO$), the corresponding equation of equilibrium in terms of chemical potential, can be written as follows :

$$\mu_A^o + RT \ln a_A + z_A/z_O(\mu_O^o + RT \ln a_O) = \mu_{A(0)}^o + RT \ln a_{A(0)} \tag{7}$$

$$\mu_B^o + RT \ln a_B + z_B/z_O(\mu_O^o + RT \ln a_O) = \mu_{B(0)}^o + RT \ln a_{B(0)} \tag{8}$$

where μ_i^o is the chemical potential of constituent i in its standard state and z_i its valency ; $a_{A(0)}$ and $a_{B(0)}$ are the activities of the constituents of the quasi binary system A(0) - B(0) ; a_0 is the oxygen activity related to the oxygen partial pressure in equilibrium with the oxide by the relation :

$$a_0 = P_{O_2}^{1/2} \tag{9}$$

where P_{O_2} is expressed in atmospheres.

The sum of the fluxes of matter which appear in the crystal is equal to the fluxes of the prevailing lattice point defects but of opposite sign. In this case one can write :

$$J_{V^\alpha} = - J_A - J_{B'} \tag{10}$$

Furthermore, according to the coupled current condition, we can write in the case of a p-type semiconducting oxide doped with a monovalent impurity :

$$J_{o_h} = J_{B'} + \alpha J_{V^\alpha} \tag{11}$$

158

This relation combined with Eq. (10) yields :

$$J_{o_h} = (\alpha - 1)J_{V^\alpha} - J_A \tag{12}$$

Using Eq. 4 and Eq. 12 and the fact that the hole concentration is a function of the impurity concentration and of the oxygen partial pressure we have shown /10/ that the chemical diffusion coefficient may then be written as :

$$\tilde{D} = - \frac{(\alpha - 1)J_{V^\alpha} - J_A}{c_M\left[\left(\dfrac{\partial [h^\circ]}{\partial m}\right)_{a_0} \dfrac{\partial m}{\partial x} + \left(\dfrac{\partial [h^\circ]}{\partial \ln a_0}\right)_m \dfrac{\partial \ln a_0}{\partial x}\right]} \tag{13}$$

It is possible to express the flux of the ions A as a function of the different gradients which appear in the crystal /10/. Indeed, in the case of impurity concentrations low enough so that the Raoult law may be applied to the quasibinary system AO-BO, one can show /12/ that :

$$\left(\frac{\partial \ln a_{(AO)}}{\partial m}\right)_{a_0} \simeq - \frac{1}{1 - m} \tag{14}$$

By differentiating Eq. 7 it follows from Eq. 5 and Eq. 14, that :

$$J_A = c_M \left[D_A \frac{\partial m}{\partial x} + D_A(1-m) \frac{\partial \ln a_0}{\partial x}\right] \tag{15}$$

If the concentration of impurities is low enough to neglect their contribution to the overall vacancy flux ($m \ll 1$), $J_{V^\alpha'}$ may be expressed, as we have shown previously in the case of a pure oxide /5,7,13/, by the relation :

$$J_{V^\alpha} = - c_M D_{V^\alpha'}[V^{\alpha'}] \frac{\partial \ln a_0}{\partial x} \tag{16}$$

Substituting Eq. 15 and Eq. 16 into Eq. 13 and assuming that $D_A \simeq D_{V^\alpha'}[V^{\alpha'}]$, one obtains the following general expression for the chemical diffusion coefficient :

159

$$\tilde{D} = \frac{\alpha\left[V^{\alpha'}\right] D_{V^{\alpha'}} \dfrac{\partial \ln a_0}{\partial x} + D_A \dfrac{\partial m}{\partial x}}{\left(\dfrac{\partial \left[h^\circ\right]}{\partial m}\right)_{a_0} \dfrac{\partial m}{\partial x} + \left(\dfrac{\partial \left[h^\circ\right]}{\partial \ln a_0}\right)_m \dfrac{\partial \ln a_0}{\partial x}} \qquad (17)$$

If the monovalent impurity is assumed to be uniformly distributed in the crystal this expression may be simplified, yielding :

$$\tilde{D} = \frac{\alpha\left[V^{\alpha'}\right]}{\left[h^\circ\right]} D_{V^{\alpha'}} \left(\frac{\partial \ln a_0}{\partial \ln \left[h^\circ\right]}\right)_m \qquad (18)$$

According to the mass action law applied to the equation of formation of the defects :

$$1/2\ O_2 \underset{\leftarrow}{\rightarrow} V^{\alpha'} + \alpha h^\circ + O_O^x \ , \qquad (19)$$

it follows that :

$$\alpha\ d\ \ln \left[h^\circ\right] + d\ \ln \left(\left[h^\circ\right] - m\right) = d\ \ln a_0 \qquad (20)$$

with

$$\alpha\left[V^{\alpha'}\right] = \left[h^\circ\right] - m \qquad (21)$$

Combining Eq. 21 and Eq. 9, yields :

$$\left(\frac{\partial \ln a_0}{\partial \ln \left[h^\circ\right]}\right)_m = \alpha + \frac{\left[h^\circ\right]}{\left[h^\circ\right] - m} \qquad (22)$$

By then inserting Eq. 22 into Eq. 18, it follows that :

$$\tilde{D} = (\alpha + 1)\ D_{V^{\alpha'}} - \frac{m\alpha}{\left[h^\circ\right]}\ D_{V^{\alpha'}} \qquad (23)$$

In the extrinsic range, the hole concentration is controlled by the amount of the monovalent impurity ($\left[h^\bullet\right] \simeq m$). Consequently, Eq. 23 allows us to predict that the chemical diffusion coefficient values must decrease when the concentration of the monovalent impurity increases.

b) Chemical diffusion in compounds doped with trivalent impurities

According to the coupled current condition, the flux of the electronic defects may be written as :

$$J_{h^\circ} = (\alpha + 1)\, J_{V^{\alpha\prime}} + J_A \tag{24}$$

The expressions giving the fluxes of the ions A (Eq. 15) and of the cationic vacancies (Eq. 16) are unchanged in this case. Consequently, if t is the concentration of the trivalent impurity, one obtains the general expression :

$$\widetilde{D} = \frac{\alpha\left[V^{\alpha\prime}\right] D_{V^{\alpha\prime}} \dfrac{\partial \ln a_0}{\partial x} - \left[V^{\alpha\prime}\right] D_{V^{\alpha\prime}} \dfrac{\partial t}{\partial x}}{\left(\dfrac{\partial\left[h^\circ\right]}{\partial t}\right)_{a_0} \dfrac{\partial t}{\partial x} + \left(\dfrac{\partial\left[h^\circ\right]}{\partial \ln a_0}\right)_t \dfrac{\partial \ln a_0}{\partial x}} \tag{25}$$

In the case of a uniform distribution of the impurity in the crystal this expression may be simplified, yielding :

$$\widetilde{D} = \frac{\alpha\left[V^{\alpha\prime}\right]}{\left[h^\circ\right]} D_{V^{\alpha\prime}} \left(\frac{\partial \ln a_0}{\partial \ln\left[h^\circ\right]}\right)_t \tag{26}$$

The mass action law applied to the equation of formation of the defects (Eq. 19) allows one to calculate :

$$\left(\frac{\partial \ln a_0}{\partial \ln\left[h^\circ\right]}\right)_t = \frac{\alpha^2\left[V^{\alpha\prime}\right] + \left[h^\circ\right]}{\alpha\left[V^{\alpha\prime}\right]} \tag{27}$$

with $\alpha\left[V^{\alpha\prime}\right] = \left[h^\circ\right] + t$ \hfill (28)

Substituting Eq. 27 into Eq. 26 yields the following expression for the chemical diffusion coefficient :

$$\widetilde{D} = (\alpha + 1)\, D_{V^{\alpha\prime}} + \frac{\alpha t}{\left[h^\circ\right]} D_{V^{\alpha\prime}} \tag{29}$$

In the extrinsic range $t > [h^{\bullet}]$, consequently, Eq. 29 allows us to predict an increase of the chemical diffusion coefficient when the trivalent impurity concentration increases. Obviously, this conclusion may be generalized to impurities of higher valencies.

II. CHEMICAL DIFFUSION COEFFICIENT IN DOPED COBALTOUS OXIDE

The chemical diffusion coefficient values of pure and doped cobaltous oxide have been deduced from electrical conductivity measurements performed during the equilibration of the gas-oxide system /6,14,15/. In this work possible complications concerning the evaluation of $\widetilde{D}$ from electrical conductivity measurements are neglected /9/. The results obtained under air with the chromium and lithium doped samples are compared in figure 1 with those obtained with the pure oxide in the same conditions.

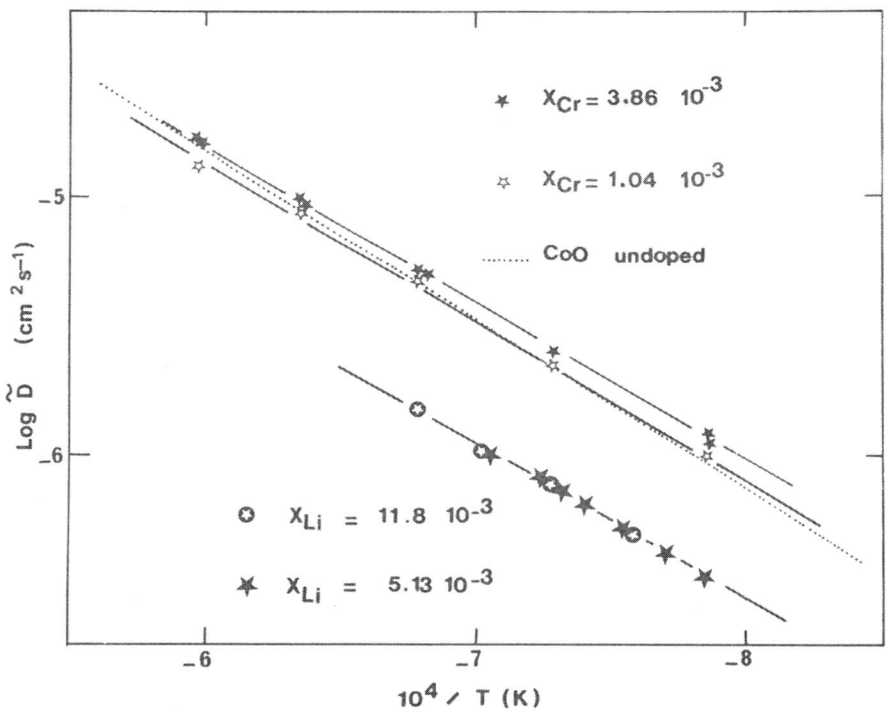

Figure 1

Chemical diffusion coefficients in pure and doped CoO

In figure 2 we have reported the chemical diffusion coefficient values of the pure oxide as a function of the oxygen partial pressure. These last results have allowed us /6,15/ to obtain in-

162

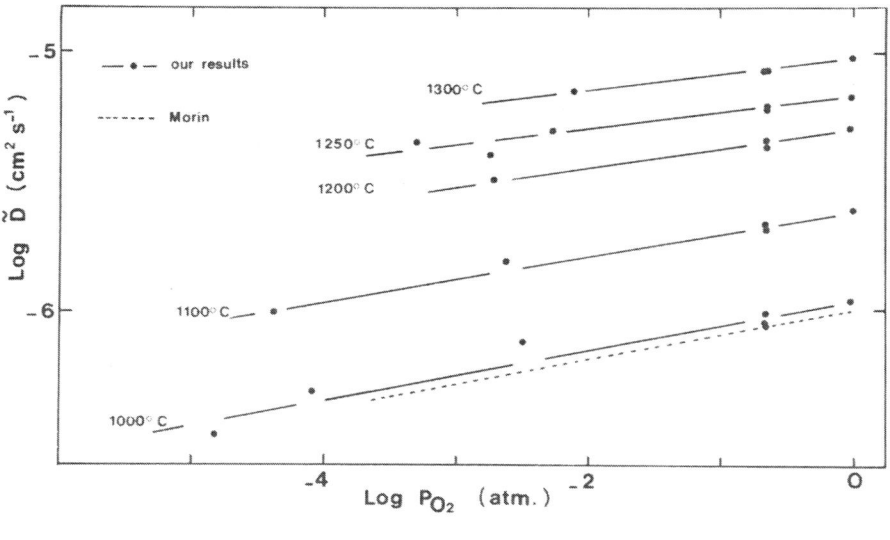

Figure 2

Chemical diffusion coefficients in pure cobaltous oxide
as a function of the oxygen partial pressure

formation on the diffusion coefficient values of the different
types of point defects in CoO. Indeed, if one assumes that the
prevailing point defects in pure CoO are merely cationic vacancies
α times ionized, the chemical diffusion coefficient is given by the
well known relation :

$$\tilde{D} = (1 + \alpha) D_{V^\alpha}, \qquad (30)$$

The mean charge of the point defects (α) has been determined
/16,18/ as a function of the oxygen partial pressure from electri-
cal conductivity measurements $(1/(1+\alpha) = (\partial \ln \sigma / \partial \ln P_{O_2})_T)$. As

shown in table 1, α decreases when P_{O_2} increases. Consequently, the

P_{O_2} dependence of the chemical diffusion coefficient (figure 2)

suggests that the mobility of the vacancies increases with the
oxygen partial pressure, that is when the concentration of the
defects and their interactions increase. As we shall see later,
such a variation of the mobility of the defects seems to be confirmed by the results of $\tilde{D}$ determined with the doped samples.

Table 1. Values of the parameter α in pure cobaltous oxide /16, 18/.

$T°C$ \\ PO_2 atm.	0.99	0.21	10^{-1}	10^{-2}	10^{-3}	10^{-4}
1000	0.67	0.84	0.87	0.99	1.12	1.27
1100	0.62	0.81	0.85	0.98	1.13	1.30
1200	0.79	0.87	0.90	1.04	1.19	1.37
1300	0.74	0.90	0.94	1.07	1.21	1.38

In order to calculate the chemical diffusion coefficient in the doped samples according to Eq. 23 and Eq. 29 it is necessary to know both the hole concentration and the values of α in the doped samples as well as the diffusion coefficient of the vacancies. The values of $D_V \alpha'$ have been calculated from the chemical diffusion coefficient data obtained with the pure samples exposed to air /6,15/ and from the knowledge of the mean charge of the point defects for P_{O_2} = 0.21 atm /16,18/ (table 1). The values of $D_V \alpha'$

Table 2. Diffusion coefficients in pure Co O.

	PO_2 atm. \\ $T°C$	1000	1100	1200	1300
$\tilde{D}$ $cm^2 s^{-1}$	0.21	$8.5\ 10^{-7}$	$20.8\ 10^{-7}$	$44.6\ 10^{-7}$	$86.6\ 10^{-7}$
	$10^{-2.5}$	$6.4\ 10^{-7}$	$14.3\ 10^{-7}$	$32.0\ 10^{-7}$	$47.4\ 10^{-7}$
$D_V \alpha'$	0.21	$4.6\ 10^{-7}$	$11.5\ 10^{-7}$	$23.8\ 10^{-7}$	$45.6\ 10^{-7}$
	$10^{-2.5}$	$3.2\ 10^{-7}$	$7.1\ 10^{-7}$	$16.0\ 10^{-7}$	$23.7\ 10^{-7}$

calculated for P_{O_2} = 0.21 atm and P_{O_2} = $10^{-2.5}$ atm are listed in table 2. The hole concentration in the doped samples under air (table 3) has been calculated from the electrical conductivity data /14-16,18/ using the following relation /16-18/ for the hole mobility :

$$\mu = 0.52 \exp(- 1600 \text{ cal mol}^{-1}/RT) \text{ cm}^2 \text{ V}^{-1} \text{ s}^{-1} \qquad (31)$$

The values of the chemical diffusion coefficient of the doped samples calculated according to Eq. 23 and Eq. 29 are listed in table 3 where they are compared with the experimental values. The agreement between these two sets of data is good in the case of the chromium doped samples but the chemical diffusion coefficient values calculated for the lithium doped samples are higher than those determined experimentally.

Table 3. Chemical diffusion coefficients (in cm² s⁻¹) in doped cobaltous oxide. The values calculated have been obtained using the data of $D_{v\alpha}$, obtained with the pure sample exposed to air.

T°C		1000	1100	1200	1300
$D_{v\alpha}$, cm² s⁻¹		$4.6\ 10^{-7}$	$11.5\ 10^{-7}$	$23.8\ 10^{-7}$	$45.6\ 10^{-7}$
x_{Li}=5.13 10⁻³	α	1	1		
	[h°]	$10.8\ 10^{-3}$	$12.7\ 10^{-3}$		
	$\tilde{D}$ experimental	$3.7\ 10^{-7}$	$8.0\ 10^{-7}$		
	$\tilde{D}$ calculated	$7.1\ 10^{-7}$	$18.4\ 10^{-7}$		
x_{Li}=11.8 10⁻³	α	1	1		
	[h°]	$17.3\ 10^{-3}$	$19.3\ 10^{-3}$		
	$\tilde{D}$ experimental	$3.7\ 10^{-7}$	$8.2\ 10^{-7}$		
	$\tilde{D}$ calculated	$6.1\ 10^{-7}$	$15.9\ 10^{-7}$		
x_{Cr}=1.04 10⁻³	α	0.82	0.86	0.87	0.90
	[h°]	$5.2\ 10^{-3}$	$6.8\ 10^{-3}$	$8.6\ 10^{-3}$	$11.0\ 10^{-3}$
	$\tilde{D}$ experimental	$10.5\ 10^{-7}$	$22.4\ 10^{-7}$	$46.6\ 10^{-7}$	$86.6\ 10^{-7}$
	$\tilde{D}$ calculated	$9.3\ 10^{-7}$	$22.1\ 10^{-7}$	$47.0\ 10^{-7}$	$90.5\ 10^{-7}$
x_{Cr}=3.86 10⁻³	α	0.82	0.86	0.87	0.90
	[h°]	$6.5\ 10^{-3}$	$8.9\ 10^{-3}$	$12.0\ 10^{-3}$	$14.7\ 10^{-3}$
	$\tilde{D}$ experimental	$11.3\ 10^{-7}$	$24.4\ 10^{-7}$	$51.6\ 10^{-7}$	$99.6\ 10^{-7}$
	$\tilde{D}$ calculated	$10.6\ 10^{-7}$	$25.7\ 10^{-7}$	$51.2\ 10^{-7}$	$97.4\ 10^{-7}$

In the following we have tried to correlate the chemical diffusion results with those concerning the structure of the defects. In the case of pure and chromium doped cobaltous oxide, the electrical conductivity data plotted as a function of the oxygen partial pressure /14/ (figure 3) show that the mean charge

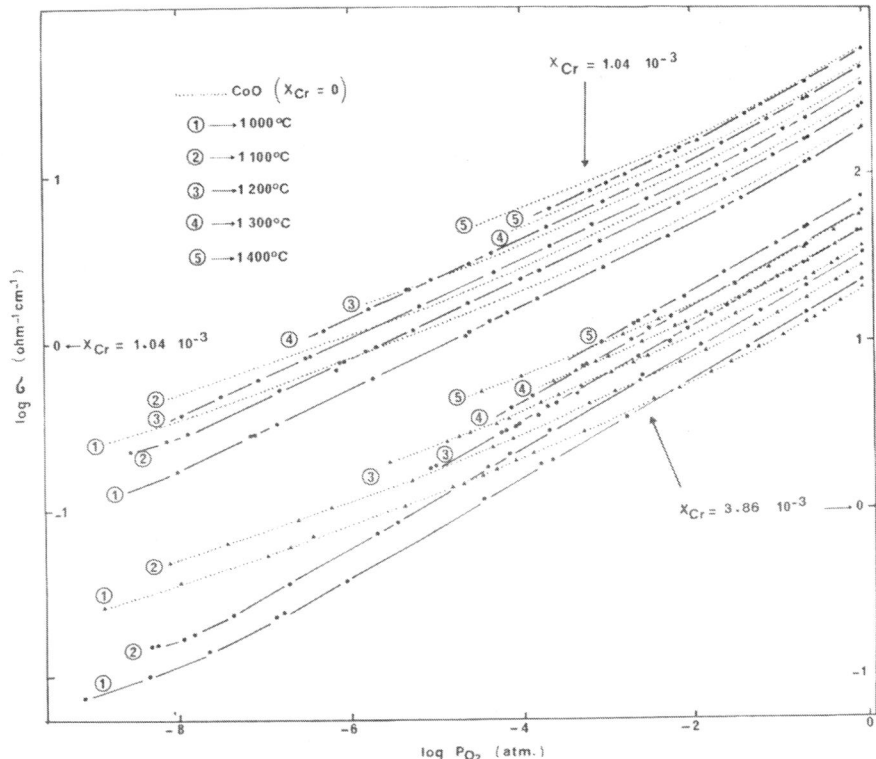

Figure 3

Electrical conductivity in pure and chromium doped cobaltous oxide
/14,16/

of the defects $(1/(1+\alpha) = (\partial \ln \sigma / \partial \ln P_{O_2})_T)$ is the same for these different samples under air. Consequently, this result suggests that the structure /16/ of the defects is similar for these oxides. On the contrary, in the case of lithium doped cobaltous oxide we have shown /16,18/ that the mean charge of the point defects is close to one in the presence of air. For the pure oxides, values of α close to one have been obtained /16/ in the oxygen partial pressure range

Table 4 Chemical diffusion coefficients (in cm^2 s^{-1}) in lithium doped cobaltous oxide. The values calculated have been obtained using the data of $D_{V\alpha'}$ obtained with the pure sample for P_{O_2} = $10^{-2.5}$ atm.

T°C	$D_{V\alpha'}(cm^2 s^{-1})$	x_{Li} = 5.13 10^{-3}		x_{Li} = 11.8 10^{-3}	
		$\tilde{D}$ experimental	$\tilde{D}$ calculated	$\tilde{D}$ experimental	$\tilde{D}$ calculated
1000	3.2 10^{-7}	3.7 10^{-7}	4.2 10^{-7}	3.7 10^{-7}	4.8 10^{-7}
1100	7.1 10^{-7}	8.0 10^{-7}	11.3 10^{-7}	8.2 10^{-7}	9.8 10^{-7}

10^{-2} - 10^{-3} atm (table 1) which seems to indicate that the interactions among defects are lowered in presence of lithium. In order to take these interactions into account we have recalculated the data of $\tilde{D}$ for the lithium doped samples using the values of the diffusion coefficient of the cationic vacancies having a mean charge close to one, that is the data of $D_{V}\alpha'$ determined for the pure oxide in the range 10^{-3} - 10^{-2} atm (table 2). The better agreement then observed (table 4) between the experimental and calculated results seems therefore to confirm the influence of the interactions among defects on their mobility already observed in previous studies /5,6/ as well as the good agreement between our different experimental results.

III. CONCLUSION

An ambipolar treatment of the diffusion of the defects in a p-type semiconducting oxide has allowed us to show the influence of the heterovalent impurities on the chemical diffusion coefficient values. In the case of pure as well as chromium and lithium doped cobaltous oxide the differences observed between the experimental values of $\tilde{D}$ are in relatively good agreement with those predicted theoretically. Furthermore, these results seem to confirm the influence of the interactions among defects on their mobility already observed in the case of pure samples.

REFERENCES

1. J.B. Wagner, Mass Transport in Oxides, Ed. J.B. Wachtman and A.D. Franklin, Nat. Bur. Stand. 296:65 (1968).

2. P.E. Childs and J.B. Wagner, Heterogeneous Kinetics at Elevated Temperatures, Ed. G.R. Belton and W.L. Worrel, Plenum Press, 269, 1970.

3. P.E. Childs, L.W. Laub and J.B. Wagner, Proc. Brit. Cer. Soc. 19:29 (1971).

4. J. Maluenda, R. Farhi and G. Petot-Ervas, J. Phys. Chem. Solids, 42:697 (1981).

5. P. Ochin, G. Petot-Ervas and C. Petot, to be published.

6. G. Petot-Ervas, O. Radji, B. Sossa, P. Ochin, Rad. Effects 75:301 (1983).

7. F. Gesmundo, F. Viani, J. Phys. Chem. Solids 42:777 (1981).

8. F. Gesmundo, Solid State Ionics 12:79 (1984) and Transport in Nonstoichiometric Compounds, Ed. G. Petot-Ervas, Hj. Matzke and C. Monty, North Holland, Publ. Comp. 79, 1984.

9. F. Gesmundo, Solid State Ionics, to be published.

10. F. Gesmundo, J. Phys. Chem. Solids, to be published.

11. C. Wagner, Corrosion Science 9:91 (1969).

12. F.A. Kröger, The Chemistry of Imperfect Crystals, North Holland Pub. Comp. 1974.

13. P. Ochin, thesis, Paris 1983.

14. C. Clauss, thesis, Paris 1984.

15. G. Petot-Ervas, Basic Properties of Oxides, Ed. A. Dominguez-Rodriguez, J. Castaing, University of Sevilla, 131, 1984.

16. G. Petot-Ervas, P. Ochin, B. Sossa, Solid State Ionics 12:277 (1984) and Transport in Nonstoichiometric Compounds Ed. G. Petot-Ervas, Hj. Matzke and C. Monty, North Holland Pub. Comp., 277, 1984.

17. G. Petot-Ervas, P. Ochin, T.O. Mason, this publication.

18. B. Sossa, thesis, Paris 1982.

CONCENTRATION DISTRIBUTION FOR DIFFUSION IN A TEMPERATURE GRADIENT

Han-Ill Yoo and B. J. Wuensch

Department of Materials Science and Engineering
Massachusetts Institute of Technology
Cambridge, Massachusetts 02139

INTRODUCTION

Many refractory materials see service in high-temperature environments in which they may be subject to large temperature gradients. Such temperature gradients serve as a driving force for mass transport in addition to the normal diffusion produced by a gradient in concentration or chemical potential. The cross effect which couples mass flux, J, to a temperature gradient, $\partial T/\partial z$, has been variously called the Sorét effect, thermal diffusion, thermotransport or thermomigration.

The phenomenological equations governing thermomigration have been established under the principles of irreversible thermodynamics (de Groot, 1951; de Groot and Mazur, 1962; Haase, 1969). As a specific illustration, we will first consider the case of the diffusion of an interstitial solute in an isotropic elemental metal. It may be assumed that a steady-state temperature distribution (i.e., $\partial T/\partial t = 0$) will be established within times which are negligibly small in comparison with those required to produce measurable changes in concentration as heat conduction in solids is usually much more rapid than mass transport. Under these conditions the mass flux of the interstitial species is given by (Howard and Lidiard, 1964):

$$J_i = -D_i \left(\frac{\partial C_i}{\partial z}\right) - D_i \, C_i \, \frac{q_i^*}{kT^2} \left(\frac{dT}{dz}\right) \tag{1}$$

where D_i is the normal diffusion coefficient for the interstitial solute, C_i is concentration and k is the Boltzmann constant. The terms preceding the temperature gradient in Eq. (1) may be defined (de Groot, 1951) as the thermal diffusion coefficient, D_i',

$$D_i' \equiv D_i \frac{q_i^*}{kT^2} \tag{2}$$

and the ratio of the thermal diffusion coefficient to the normal diffusion coefficient is defined as the Sorét coefficient, S_T

$$S_T \equiv \frac{D_i'}{D_i} = \frac{q_i^*}{kT^2}. \tag{3}$$

The influence of the temperature gradient on solute redistribution is determined by q_i^*, a parameter with the dimensions of energy which is termed the reduced heat of transport of the interstitial. Defined relative to a reference frame fixed to the host lattice, the parameter may be viewed as the heat flux per unit flux of the solute species in the absence of a temperature gradient and, accordingly, provides a measure of the energy which must be added or removed from a volume element to maintain it at constant temperature upon the entrance of diffusing atoms. Equation (1) shows that diffusion species will migrate up the temperature gradient toward the hot end of a specimen if q_i^* is negative and will segregate to the cold end if q_i^* is positive. Study of thermomigration is thus essentially a concern with theoretical formulation or experimental measurement of q^*.

Although thermomigration in liquids was observed more than a century ago by Ludwig (1856) and by Sorét (1879), understanding of the process in solids remains in a rudimentary state. A review of atomistic interpretations of the heat of transport has been recently provided by Gillan (1983) along with the development of a statistical mechanical treatment. Few reliable measurements of q^* are available, on the order of only twenty measurements having been performed, for example, for even simple systems such as an interstitial solute in an elemental metal (Allnatt and Chadwick, 1967; Oriani, 1969). The paucity of data stems in part from the difficulty of the experiments. It is necessary to maintain a sample in a controlled, steep temperature gradient for extended periods of time in order to produce measurable effects. Most measurements of the heat of transport are based on the steady-state Sorét effect--that is, measurement of the stationary concentration distribution after unmixing from an initially uniform distribution of solute (e.g., Shewmon, 1960). The flux, J_i, in such circumstances is zero throughout the specimen and Eq. (1) reduces to

$$\frac{d \ln C_i}{d(1/T)} = \frac{q_i^*}{k} \; .$$

(4)

A plot of the logarithm of measured concentration as a function of the reciprocal of temperature along the sample should thus be linear, with a slope which provides a direct measure of q_i^*. It may be noted that the analysis does not involve a value for D_i and, as the slope of $\ln C_i$ is employed, concentration need not be measured on an absolute scale. Silver linings, however, are frequently enveloped in a cloud. The time required to attain a stationary state for the concentration distribution may be unreasonably long. De Groot (1942) has shown that the stationary state is approached in proportion to $\exp(-t/\Theta)$ where the time constant for a given system is

$$\Theta = L^2/\pi^2 D$$

(5)

and L is the dimension of the sample in the direction along which the temperature gradient is applied. The system will have achieved 99.3% of the stationary state when $t = 5\Theta$. For a sample .3 cm in thickness and D on the order of $10^{-10} cm^2/sec$, for example, the time 5Θ which is required to achieve the distribution is about 14 years.

To our knowledge only little progress has been made toward a general time-dependent solution to the equations controlling the redistribution of solute in a temperature gradient. The present analysis was undertaken with the intuitive expectation that a diffusion specimen prepared under usual boundary conditions would, if annealed in a temperature gradient, display a concentration gradient which might bear perturbations or a displacement of its centroid which would contain information on the heat of transport. The directed walk driven by the temperature gradient superposed on the random walks of normal diffusion were thus envisioned as the thermal analogue of the classic Chemla (1956) experiment in which a diffusion sample is annealed in an electric potential gradient. It has been possible in the present work to use perturbation methods to obtain approximate but explicit and highly accurate solutions to the problem for the boundary and initial conditions employed in practice (semi-infinite, thin film and thick film initial sources). The form of the solutions for thermomigration of an interstitial solute suggests several types of experiments through which a value for the heat of transport might be obtained. Such experiments promise to be more than an order of magnitude shorter in duration than measurements based upon the stationary-state Sorét effect. In the case of diffusion by means of Schottky vacancies, however, it is shown that the problem reduces to a form similar to that for diffusion by a vacancy mechanism in an elemental metal (Mock, 1969).

All information on the heat of transport thus vanishes from a concentration distribution measured relative to a fixed laboratory coordinate system. Perturbations induced by thermomigration are exactly compensated by local motion of the lattice. Measurement of the velocity of the local lattice through marker experiments provides a means for obtaining a value of the heat of transport, provided that diffusion rates are not so sluggish as to require unreasonably long annealing times.

TRANSPORT EQUATION FOR DIFFUSION OF AN INTERSTITIAL SOLUTE

The diffusion of an interstitial impurity in an elemental metal will be specifically considered (although the analysis obviously applies to any system for which a flux equation of the form of Eq. (1) is valid). It will be assumed that the host lattice forms an immobile cage for the solute atoms, and that the migration of the interstitials is not coupled to motion of the host atoms. The host lattice provides an appropriate frame of reference with respect to which the flux of interstitials may be measured. This corresponds to the Hittorf (Haase, 1969) or the solvent-fixed (Kirkwood et al., 1960) frame of reference. To the extent to which the volume change upon incorporation of an interstitial is negligible, this frame of reference will be approximately coincident with the Fick (Haase, 1969) or volume-fixed (Kirkwood et al., 1960) frame. The flux specified by Eq. (1) is thus experimentally accessible.

Upon substitution of the flux equation (1) into the equation of continuity

$$\frac{\partial C_i}{\partial t} = -\text{div } J_i \tag{6}$$

one obtains

$$\frac{\partial C_i}{\partial t} = \frac{\partial}{\partial z}\left(D_i \frac{\partial C_i}{\partial z}\right) - T_i^* \frac{\partial}{\partial z}\left[D_i C_i \frac{\partial}{\partial z}\left(\frac{1}{T}\right)\right] \tag{7}$$

where, for brevity, we have defined a temperature of transport (assumed to be independent of temperature) as $T_i^* \equiv q_i^*/k$. Analogously, an activation temperature will be defined by $T_m \equiv Q/k$, where Q is the activation enthalpy for diffusion.

It will be assumed that the local diffusion coefficient along the temperature gradient, $D_i(z)$, will be the isothermal diffusion coefficient at the local temperature, $T(z)$. [The assumption of local thermodynamic equilibrium is a postulate of irreversible

172

thermodynamics which is inherent in the formulation of the initial phenomenological equations for thermomigration.] Thus

$$D_i(z) = D_o \exp \left[-Q/kT(z)\right] = D_o \exp \left[-T_m/T(z)\right] \qquad (8)$$

which, upon substitution into the equation of continuity (7), provides

$$\frac{\partial c_i}{\partial t} = D_i \left\{ \frac{\partial^2 c_i}{\partial z^2} + \frac{T_m + T_i^*}{T^2} \frac{\partial c_i}{\partial z} \frac{dT}{dz} \right.$$

$$+ \frac{T_i^* (T_m - 2T)}{T^4} c_i \left(\frac{dT}{dz}\right)^2 \qquad (9)$$

$$\left. + \frac{T_i^*}{T^2} c_i \frac{d^2 T}{dz^2} \right\} .$$

No exact solution to Eq. (9) is possible, but an approximate and highly accurate explicit solution may be obtained by perturbation methods.

Let the origin of the coordinate system z = 0 be taken at the interface at which two semi-infinite slabs containing different solute concentrations are joined or, alternatively, the position of an initial layer of solute sandwiched between two semi-infinite slabs of host material. If the temperature gradient is a, the temperature distribution in the sample is

$$T = T_o + az = T_o(1 + \frac{az}{T_o}) \qquad (10)$$

where T_o is the temperature at the interface and a linear station-ary-state temperature distribution has been assumed. Consider a short-time diffusion annealing in this temperature gradient which produces a concentration gradient over a very shallow region δ in the neighborhood of z = 0, such that in Eq. (10),

$$\frac{a\delta}{T_o} \sim \varepsilon^2; \ \varepsilon^2 \ll 1. \qquad (11)$$

For example, if $\delta = 10 \ \mu$, $T_o = 10^3$ K and $a = 10^3$ K/cm = 2 T_o/L, then $\varepsilon^2 = 10^{-3}$. The diffusion coefficient may then be expanded in this region in a Taylor series about z = 0 (T = T_o) and will be well represented within an accuracy of ε^2 by

$$D_i \approx \Delta_o + \Delta_1 z \qquad (12)$$

with

$$\Delta_0 = D_i(T_0)$$

$$\Delta_1 = \frac{\partial D_i}{\partial T}\frac{dT}{dz} = \left(\frac{a}{T_0}\right)\left(\frac{T_m}{T_0}\right)D_i(T_0). \tag{13}$$

Similarly, the higher order terms, $T(z)^n$, which appear in (9) may be approximated to within ε^4 by

$$T(z)^n \approx T_0^{~n}\left[1 + n\left(\frac{az}{T_0}\right)\right]. \tag{14}$$

The approximations (12) and (14), when introduced into Eq. (9), provide

$$\frac{\partial C_i}{\partial t} = \Delta_0\frac{\partial^2 C_i}{\partial z^2} + \Delta_1 z\frac{\partial^2 C_i}{\partial z^2} + \Delta_0\frac{\alpha}{T_0^2}\frac{\partial C_i}{\partial z} + \Delta_1\frac{\alpha z}{T_0^2}\frac{\partial C_i}{\partial z} - 2\Delta_0\frac{a\alpha z}{T_0^3}\frac{\partial C_i}{\partial z}$$

$$- 2\Delta_1\frac{a\alpha z}{T_0^3}\frac{\partial C_i}{\partial z} - \Delta_0\frac{\beta}{T_0^3}C_i - \Delta_1\frac{\beta z}{T_0^3}C_i + 3\Delta_0\frac{a\beta z}{T_0^4}C_i$$

$$\tag{15}$$

$$+ 3\Delta_1\frac{a\beta z^2}{T_0^4}C_i + \Delta_0\frac{\gamma}{T_0^4}C_i + \Delta_1\frac{\gamma z}{T_0^4}C_i - 4\Delta_0\frac{a\gamma z}{T_0^5}C_i$$

$$- 4\Delta_1\frac{a\gamma z^2}{T_0^5}C_i$$

where

$$\begin{aligned}\alpha &\equiv a(T_m + T_i^*)\\ \beta &\equiv 2a^2 T_i^*\\ \gamma &\equiv a^2 T_m T_i^*.\end{aligned} \tag{16}$$

In assessing the relative magnitude of the terms in Eq. (15), we note that typical values for the activation energies and reduced heats of transport for interstitials are of order of magnitude 1 eV or less. Thus $T_i^* \lesssim T_m \sim 10^4$ K. One may thus establish

$$\frac{T_m}{T_0} \approx \frac{T_i^*}{T_0} \approx 10 \sim \frac{1}{\varepsilon} \tag{17}$$

under typical experimental conditions. The terms in Eq. (15) are presented in order of decreasing magnitude on this basis. The

174

ratio of the second and third terms in Eq. (15) to the leading term are $(a\delta/T_o)(T_m/T_o)$ and $(a\delta/T_o)[(T_m + T_i^*)/T_o]$, respectively. Both ratios, accordingly are of order of magnitude ε. The ratios of the fourth through fourteenth terms in Eq. (15) range, relative to the leading term, from order ε^2 to ε^5. An approximate differential equation, accurate to ε^2, may thus be obtained by retaining only the first three terms on the right of Eq. (15):

$$\frac{\partial C_i}{\partial t} = \Delta_o \left\{ \frac{\partial^2 C_i}{\partial z^2} + \left(\frac{a}{T_o}\right)\left(\frac{T_m}{T_o}\right) z \frac{\partial^2 C_i}{\partial z^2} + \frac{a(T_m + T_i^*)}{T_o^2} \frac{\partial C_i}{\partial z} \right\} . \tag{18}$$

This equation may be solved by Fourier Transform methods. The solutions so obtained are presented in the following sections for the boundary and initial conditions which are commonly used in diffusion experiments.

SOLUTIONS FOR INTERSTITIAL SOLUTE DISTRIBUTIONS

Semi-infinite Source Boundary Conditions

For the diffusion experiment in which a semi-infinite slab containing solute concentration C_s is joined to a semi-infinite host crystal containing no interstitial atoms, the initial and boundary conditions are

$$\left. \begin{array}{ll} C_i \ (z < 0, \ t = 0) = C_s & C_i \ (z = -\infty, \ t > 0) = C_s \\ C_i \ (z > 0, \ t = 0) = 0 & C_i \ (z = \infty, \ t > 0) = 0. \end{array} \right\} \tag{19}$$

The time-dependent solution is

$$\frac{C_i(z,t)}{C_s} = \tfrac{1}{2}\mathrm{erfc}\ [z(4\Delta_o t)^{-\frac{1}{2}}]$$

$$- \frac{1}{4\sqrt{\pi}} \frac{a(T_m + 2T_i^*)}{T_o^2} (\Delta_o t)^{\frac{1}{2}} \exp\left(- \frac{z^2}{4\Delta_o t}\right) \tag{20}$$

$$+ \frac{1}{8\sqrt{\pi}} \frac{a\,T_m}{T_o^2} \frac{z^2}{(\Delta_o t)^{\frac{1}{2}}} \exp\left(- \frac{z^2}{4\Delta_o t}\right)$$

where erfc is the complementary Gaussian error function. This result may be condensed by introducing the reduced variables

$$\xi \equiv (4\Delta_o t)^{\frac{1}{2}} \ \text{and} \ \eta = z/\xi. \tag{21}$$

Upon collecting terms in T_m and T_i^* one then obtains

$$\frac{C_i(\eta,\xi)}{C_s} = \frac{1}{2} \, \text{erfc} \, (\eta)$$

$$- \frac{1}{4\sqrt{\pi}} \frac{aT_i^*}{T_o^2} \, \xi \, \exp \, (-\eta^2) \tag{22}$$

$$+ \frac{1}{8\sqrt{\pi}} \frac{aT_m}{T_o^2} \, \xi \, (2\eta^2 - 1) \, \exp \, (-\eta^2).$$

Thin Initial Film Boundary Conditions

Diffusion experiments which employ a tracer isotope as solute usually sandwich a thin initial layer of tracer between semi-infinite slabs of host crystal. The boundary and initial conditions are

$$\left. \begin{array}{l} C_i \, (z = 0, \, t = 0) = \infty \\ C_i \, (|z| > 0, \, t = 0) = 0 \\ C_i \, (|z| = \infty, \, t > 0) = 0. \end{array} \right\} \tag{23}$$

The total amount of solute in the initial layer, M, is given by

$$\int_{-\infty}^{\infty} C_i \, (z,t) \, dz = M. \tag{24}$$

The time-dependent solution is

$$\frac{\sqrt{\pi}}{M} \, C_i \, (z,t) = (4\Delta_o t)^{-\frac{1}{2}} \exp \, (- \frac{z^2}{4\Delta_o t})$$

$$- \frac{1}{8} \frac{a(T_m + 2T_i^*)}{T_o^2} \frac{z}{(\Delta_o t)^{\frac{1}{2}}} \exp \, (\frac{-z^2}{4\Delta_o t}) \tag{25}$$

$$+ \frac{1}{16} \frac{aT_m}{T_o^2} \frac{z^3}{(\Delta_o t)^{3/2}} \exp \, (- \frac{z^2}{4\Delta_o t}).$$

Upon substitution of the reduced variables ξ and η, Eq. (21), and collection of terms in T_i^* and T_m one obtains

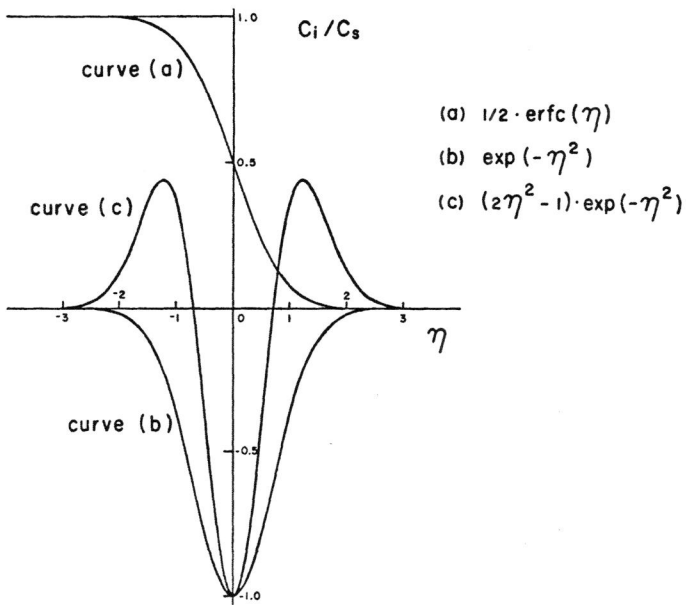

Fig. 1. Variation with reduced penetration, $\eta = z(4\Delta_0 t)^{-\frac{1}{2}}$, of the terms whose sum provides the expected redistribution of interstitial solute from a semi-infinite source, Eq. (22). Curve (a) is the isothermal distribution which would be produced at $T(\eta = 0)$, curve (b) is proportional to the reduced heat of transport, and curve (c) is proportional to the activation enthalpy for migration.

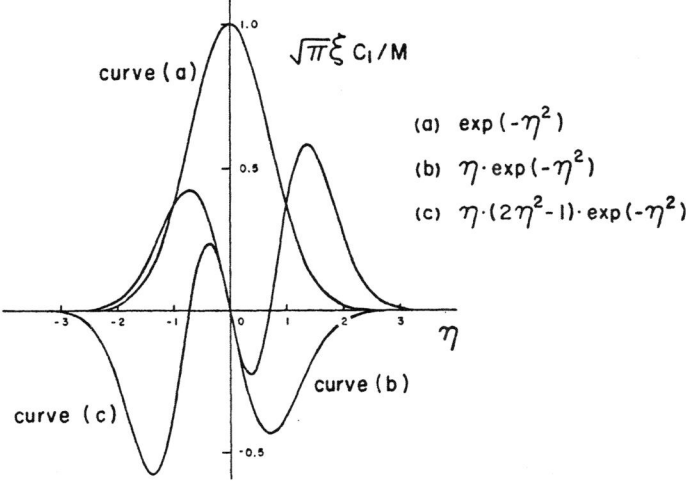

Fig. 2. Variation of normalized concentration with reduced penetration, $\eta = z(4\Delta_0 t)^{-\frac{1}{2}}$, of the terms whose sum provides the expected redistribution of solute supplied from a thin-film initial source, Eq. (26). Curve (b) is proportional to the reduced heat of transport and curve (c) is proportional to the activation enthalpy for migration.

$$\frac{\sqrt{\pi}}{M} C_i \ (\eta, \xi) = \frac{1}{\xi} \exp \ (-\eta^2)$$

$$- \frac{1}{2} \frac{aT_i^*}{T_o^2} \eta \exp \ (-\eta^2) \tag{26}$$

$$+ \frac{1}{4} \frac{aT_m}{T_o^2} \eta \ (2\eta^2 - 1) \exp \ (-\eta^2).$$

Thick Initial Film Boundary Conditions

On occasion one encounters a diffusion experiment in which the thickness of the initial layer of solute is comparable to the extent of the diffusion zone. Although less frequently employed than the thin-film conditions, these initial conditions are encountered when it proves necessary to employ a large total amount of solute, M (e.g., diffusion of a stable isotope in the presence of a large background natural abundance), or when slow diffusion rates or short diffusion annealings are involved. The latter situation is likely to be encountered in experiments for which the present solutions apply as the analysis is valid only for short diffusion gradients. The boundary and initial conditions for a finite source of thickness 2h are

$$\left. \begin{array}{l} C_i \ (|z| < h, \ t = 0) = C_s \\ C_i \ (|z| > h, \ t = 0) = 0 \\ C_i \ (|z| = \infty, \ t \geqslant 0) = 0. \end{array} \right\} \tag{27}$$

The solution is

$$\frac{C_i(z,t)}{C_s} =$$

$$\frac{1}{2} \left\{ \text{erf} \left[\frac{h + z}{(4\Delta_o t)^{\frac{1}{2}}} \right] + \text{erf} \left[\frac{h - z}{(4\Delta_o t)^{\frac{1}{2}}} \right] \right\}$$

$$+ \frac{1}{4\sqrt{\pi}} \frac{a}{T_o} \frac{(T_m + 2T_i^*)}{T_o} (\Delta_o t)^{\frac{1}{2}} \left\{ \exp\left[-\frac{(h + z)^2}{4\Delta_o t} \right] - \exp \left[-\frac{(h - z)^2}{4\Delta_o t} \right] \right\}$$

$$+ \frac{1}{8\sqrt{\pi}} \frac{a}{T_o} \frac{T_m}{T_o} \frac{(h^2 - z^2)}{(\Delta_o t)^{\frac{1}{2}}} \left\{ \exp\left[-\frac{(h + z)^2}{4\Delta_o t} \right] - \exp \left[-\frac{(h - z)^2}{4\Delta_o t} \right] \right\}. \tag{28}$$

Upon substitution of the reduced variables

$$\rho \equiv z/h \text{ and } \zeta \equiv (4\Delta_0 t)^{\frac{1}{2}}/h \qquad (29)$$

one obtains

$$\frac{C_i(\rho,\zeta)}{C_s} = \frac{1}{2}\left[\text{erf}\left(\frac{1+\rho}{\zeta}\right) + \text{erf}\left(\frac{1-\rho}{\zeta}\right)\right.$$

$$+ \frac{1}{4\sqrt{\pi}}\frac{a}{T_0}\frac{T_i^*}{T_0}h\zeta \left\{ \exp\left[-\left(\frac{1+\rho}{\zeta}\right)^2\right] - \exp\left[-\left(\frac{1-\rho}{\zeta}\right)^2\right]\right\}$$

$$+ \frac{1}{4\sqrt{\pi}}\frac{a}{T_0}\frac{T_m}{T_0}h\zeta \left[\frac{1}{2} + \left(\frac{1+\rho}{\zeta}\right)\left(\frac{1-\rho}{\zeta}\right)\right] \cdot$$

$$\left\{\exp\left[-\left(\frac{1+\rho}{\zeta}\right)^2\right] - \exp\left[-\left(\frac{1-\rho}{\zeta}\right)^2\right]\right\}. \qquad (30)$$

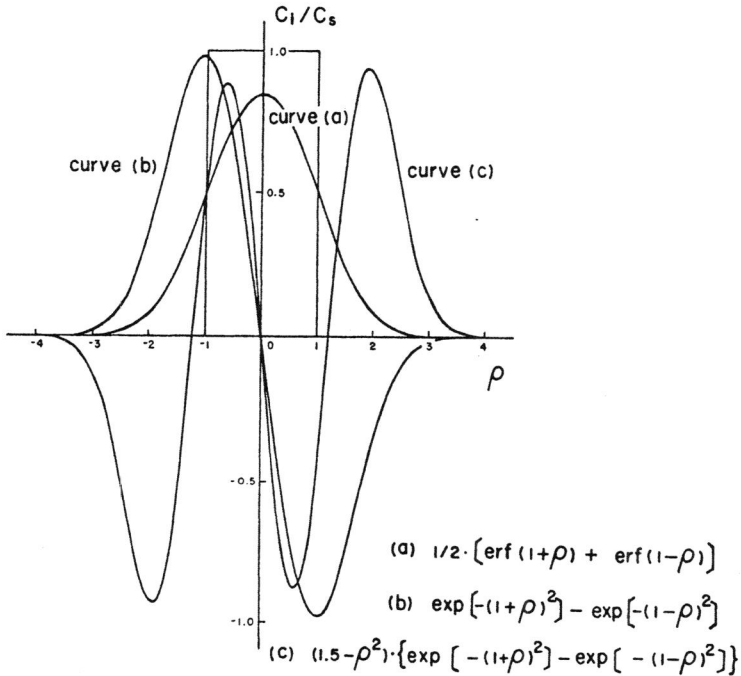

(a) $1/2 \cdot \left[\text{erf}(1+\rho) + \text{erf}(1-\rho)\right]$

(b) $\exp\left[-(1+\rho)^2\right] - \exp\left[-(1-\rho)^2\right]$

(c) $(1.5-\rho^2)\cdot\left\{\exp\left[-(1+\rho)^2\right] - \exp\left[-(1-\rho)^2\right]\right\}$

Fig. 3. Variation of normalized concentration with reduced penetration, $\rho = z/h$, of the terms whose sum provides the
expected redistribution of solute supplied from an initial
layer of thickness $z = 2h$, Eq. (30). The value of $\zeta = (4\Delta_0 t)^{\frac{1}{2}}/h$ has been selected as unity. Curves (b) and (c)
are proportional to the reduced heat of transport and
the activation enthalpy for diffusion, respectively.

CHARACTERISTICS OF THE TIME-DEPENDENT SOLUTIONS FOR REDISTRIBUTION OF INTERSTITIAL SOLUTE

The concentration profiles $C_i(\eta,\xi)$ of Eqs. (22) and (26) and $C_i(\rho,\zeta)$ of Eq. (30) which are expected for semi-infinite source, thin film and thick film initial conditions, respectively, are plotted in Figs. 1 through 3. The three solutions have certain features in common. The leading term is the normal isothermal distribution of solute which would be produced by annealing at T_0, the temperature at the interface. Two perturbations, comparable in magnitude to one another if $T_i^* \approx T_m$, modify this distribution. The first term is directly proportional to the temperature of transport, T_i^*, and contains the information on thermomigration. The second perturbation is directly proportional to the activation temperature, T_m, and arises from the temperature and thus positional dependence of the normal diffusion coefficient. It may be seen that both perturbations are directly proportional to the temperature gradient, a, so that the magnitude of the perturbations is enhanced by steep temperature gradients.

It is of interest to note that, in the two distributions anticipated from finite sources of solute, Figs. 2 and 3, the perturbations are odd functions of penetration whose integral $\int_{-\infty}^{\infty} C_i \, dz$ will vanish. As the integral of the leading isothermal term in the distribution satisfies the requirement of mass conservation, Eq. (24), the full solution will fulfill this condition as well. Both perturbations for semi-infinite source boundary conditions, Fig. 1, are even functions so that $\int_0^{C_S} z \, dC_i = 0$. The solute gained on one side of the original interface is thus equal to the depletion of solute on the other and the condition of mass conservation again is satisfied. It may also be noted that the perturbation which arises from the temperature dependence of D_i^* is zero at $z \approx \pm(2\Delta_0 t)^{\frac{1}{2}}$ or $\eta \approx \pm 0.7$ for the distributions anticipated under both semi-infinite source and thin film initial conditions. A similar condition occurs for the thick-film initial conditions, Fig. 3, at $z \approx \pm(h^2 + 2\Delta_0 t)^{\frac{1}{2}}$ or $\rho \approx \pm(1 + \frac{1}{2}\zeta^2)^{\frac{1}{2}}$. Any perturbation and departure from the symmetry of the isothermal distribution must accordingly arise entirely from the influence of the heat of transport in this region of penetration.

TRANSPORT EQUATION FOR DIFFUSION VIA SCHOTTKY DEFECTS IN AN IONIC COMPOUND

We briefly consider next the contrasting behavior of self-thermomigration via Schottky defects in a binary ionic solid. The case of $Mg(^{18}O_x \; ^{16}O_{1-x})$ will be specifically examined for illustration. Five fluxes must be specified for a pure and perfectly stoichiometric material: (1) $^{18}O_O^{2-}$, (2) $^{16}O_O^{2-}$ and (3) $V_O^{\cdot\cdot}$ on the anion sublattice, plus (4) Mg_{Mg}^{2+} and (5) V_{Mg}'' on the cation sublattice.

Let the fluxes and concentrations of these species be distinguished through subscripts ranging 1 through 5. The fluxes are not independent. The total concentration of species on the cation sublattice, $C_+ = C_4 + C_5$, must be equal to $C_- = C_1 + C_2 + C_3$. The sum of fluxes of cationic species and anionic species must be zero for this condition to be fulfilled

$$_LJ_1 + {}_LJ_2 + {}_LJ_3 = 0 \tag{31}$$

and

$$_LJ_4 + {}_LJ_5 = 0 \tag{32}$$

where the prefix L emphasizes that the fluxes are defined in a frame of reference relative to the local crystalline lattice. The linear dependency allows two fluxes to be eliminated. After development of expressions for the various phenomenological coefficients and the introduction of some approximations, (details are contained in Yoo, 1984 and will be published elsewhere) one may show:

$$\begin{cases} _LJ_1 = -D_-^* \dfrac{\partial C_1}{\partial z} - C_1 \dfrac{D_- D_+}{D_- + D_+} \dfrac{q_-^* + q_+^* - h_s}{kT^2} \dfrac{dT}{dz} \\[2ex] _LJ_2 = -D_-^* \dfrac{\partial C_2}{\partial z} - C_2 \dfrac{D_- D_+}{D_- + D_+} \dfrac{q_-^* + q_+^* - h_s}{kT^2} \dfrac{dT}{dz} \\[2ex] _LJ_4 = \qquad\qquad - C_+ \dfrac{D_- D_+}{D_- + D_+} \dfrac{q_-^* + q_+^* - h_s}{kT^2} \dfrac{dT}{dz} \end{cases} \tag{33}$$

where the gradient of cation concentration $\partial C_4/\partial z$, has been assumed to be negligibly small; D_-^*, D_- and D_+ are anion tracer diffusion, anion self-diffusion, and cation self-diffusion coefficients, respectively; q_+^* and q_-^* are the reduced heats of transport for cation and anion, and h_s is the enthalpy for Schottky pair formation.

Unlike the flux equations for diffusion of an interstitial solute, the relations of Eq. (33) describe net (and equal) fluxes of anion and cation vacancies which arise solely from the applied temperature gradient:

$$_LJ_5 = {}_LJ_3 = C_- \frac{D_- D_+}{D_- + D_+} \frac{(q_-^* + q_+^* - h_s)}{kT} \frac{dT}{dz} . \tag{34}$$

This renders the local lattice mobile, with velocity

$$u_L = \frac{1}{C_-} {}_LJ_3 = \frac{D_- D_+}{D_- + D_+} \frac{(q_-^* + q_+^* - h_s)}{kT} \frac{dT}{dz} \tag{35}$$

181

relative to the Fick or volume-fixed frame of reference. The latter frame of reference will coincide with the laboratory frame of reference as no volume change will occur during the isotope intermixing under consideration in the present problem. Fluxes relative to a moving frame of reference are not accessible to an observer in the laboratory and Fick frame of reference. The fluxes of (33) must accordingly be transformed to the Fick frame of reference according to

$$_F J_1 = _L J_1 + C_1 u_L \qquad (36)$$

where the prefix F signifies Fickian, volume-fixed reference axes. Substitution of (35) and (33) into (36) provides

$$
\left.
\begin{aligned}
_F J_1 &= -D_-^* \ \partial C_1/\partial z \\
_F J_2 &= -D_-^* \ \partial C_2/\partial z \\
_F J_4 &= 0 \ .
\end{aligned}
\right\} \qquad (37)
$$

Equations (37) show that, in the Fick and laboratory-fixed frame of reference the coupling of the temperature gradient with mass flux has completely vanished. Application of the equation of continuity to (37) and solution for the time-dependent distribution of tracer isotope as performed above in Eqs. (7)-(30) will provide solutions identical to those of Eqs. (22), (26) and (30), but with T_i^* set equal to zero. No information about the reduced heat of transport can be extracted from measurement of the isotope intermixing profile produced by self-diffusion in a temperature gradient.

EXPERIMENTS FOR MEASUREMENT OF THE HEAT OF TRANSPORT

The form of the anticipated distributions of solute, Eqs. (22), (26) and (30) suggests several types of experiments through which the heat of transport for interstitial solutes may be accurately and quickly measured. A first obvious method is to fit, by least-squares methods, the full solution to a measured concentration distribution treating T_i^* (and perhaps also Δ_0, if not precisely known) as an adjustable variable. In the solutions for thin film and thick film boundary conditions it may be noted that the leading isothermal term is a symmetric, even function. The difference in concentration at $C(z)$ and $C(-z)$ thus will provide the sum of the two perturbations alone.

Alternatively, it is possible to determine T_i^* from the characteristics of the profile itself. The form of the perturbations added to the isothermal distribution, Figs. 1-3, are such that the position where $C_i/C_s = \frac{1}{2}$ under semi-infinite source initial conditions, or the position of maximum concentration under thin or thick film initial conditions will move with time. For similar reasons, the normalized concentration at $z = 0$, (which would remain fixed at $\frac{1}{2}$ under semi-infinite source initial conditions in an isothermal annealing) will change with time. Measurement of either the change in position of a fixed concentration or the rate of decrease of concentration at fixed position provides alternate means for the determination of T_i^*. The profile for semi-infinite source boundary conditions will be specifically examined in this respect.

Displacement of Fixed Concentration of Interstitial with Time

Let the expression for the solute distribution for semi-infinite source boundary conditions, Eq. (22), be written in terms of series expansions of the complementary error function and exponential functions which appear in the first and second plus third terms, respectively:

$$C_i/C_s = \frac{1}{2}(1 - \frac{2}{\sqrt{\pi}} \eta + \ldots)$$

$$- \frac{1}{4\sqrt{\pi}} \frac{aT_i^*}{T_o^2} \xi \, (1 - \eta^2 + \frac{1}{2}\eta^4 \ldots) \tag{38}$$

$$- \frac{1}{8\sqrt{\pi}} \frac{aT_m}{T_o^2} \xi \, (-1 + 3\eta^2 - \frac{5}{2}\eta^4 + \ldots)$$

Let the position at which $C_i/C_s = \frac{1}{2}$ ($z = \eta = 0$ under isothermal conditions) be defined as $\eta_o = z_o (4\Delta_o t)^{-\frac{1}{2}}$. Substituting η_o in Eq. (38), setting $C_i/C_s = \frac{1}{2}$ on the left, and neglecting terms of second and higher order in η, one obtains as an expression valid for small η_o

$$\eta_o = - \frac{1}{8} \frac{a(T_m + 2T_i^*)}{T_o^2} \xi \tag{39}$$

or

$$z_o = - \frac{1}{2} \frac{a(T_m + 2T_i^*)}{T_o^2} \Delta_o t \, . \tag{40}$$

Measurement of the velocity z_o/t of the location at which $C_i/C_s = \frac{1}{2}$ thus provides a measure of T_i^* if Δ_o, T_o, T_m and a are accurately known. The relation is strictly valid only as η_o and z_o approach zero as motion of the concentration to a location of different temperature and diffusion coefficient will cause the velocity to be dependent upon position. Displacement Δz_0 after an elapsed time Δt will thus provide an average velocity and one may wish, in practice, to make several such measurements and extrapolate the results to $t = 0$. One may show, through a similar analysis, that the velocity of the location of maximum concentration in the distribution expected under thin-film initial conditions is given by exactly the same expression as Eq. (40).

Rate of Decrease of Interstitial Concentration at z = 0

At $z = 0$ the concentration in the semi-infinite source distribution, Eq. (20), will change with time according to

$$\frac{C_i(0,t)}{C_s} = \tfrac{1}{2} - \alpha \sqrt{t} \tag{41}$$

with

$$\alpha = \frac{1}{4\sqrt{\pi}} \; \frac{a}{T_0^2} \; (T_m + 2T_i^*) \; \Delta_0^{\tfrac{1}{2}} \tag{42}$$

The time-dependent term arises from the first perturbation term in (20), the expression which contains information on the heat of transport. The normalized concentration at the original interface, $z = 0$, may thus be expected to vary linearly with $t^{\tfrac{1}{2}}$ and the proportionality constant α provides a direct measure of $(T_m + 2T_i^*)$ if the temperature at the interface, T_0, the diffusion coefficient at this temperature Δ_0, and the temperature gradient, a, are known.

The value of α contains a term $\Delta_0^{\tfrac{1}{2}} = D_0^{\tfrac{1}{2}} \exp(-T_m/2T_0)$ in the numerator and a term T_0^2 in the denominator, Eq. (42). Setting the differential of this ratio with respect to T_0 equal to zero shows that an experiment of maximum sensitivity will result when the interface temperature is selected to be $T_0 = T_m/4$. Under such conditions α attains a maximum value

$$\alpha_{max} = \frac{8}{\sqrt{\pi} \, e^2} \; D_0^{\tfrac{1}{2}} \, \frac{a}{T_m} \; (\tfrac{1}{2} + \frac{T_i^*}{T_m}) \tag{43}$$

Measurement of Combined Heats of Transport for Diffusion by a Schottky Vacancy Mechanism

As shown by (37), it is not possible to obtain a value for the reduced heat of transport of either anion or cation through measurement of the tracer self-diffusion profile which is produced by annealing in a temperature gradient. The reason for this result is the strong coupling of any fluxes on the anion and cation sublattices, a coupling which is necessary if electroneutrality is to be preserved. The net flux of vacancies causes movement of the lattice which exactly cancels the directed walk imposed by the temperature gradient. In other words, while the center-of-mass of a distribution of interstitial solute will shift relative to the host lattice, the constraint that stoichiometry be preserved prohibits a relative shift of the anion and cation center-of-masses with respect to the lattice.

One can use markers, however, to measure the velocity of the local lattice. This rate, provided above by Eq. (35),[*] is proportional to dT/dz and can provide the value of $(q_-^* + q_+^* - h_s)$. In common with the steady-state Sorét effect, however, such an experiment may, in practice, involve very long diffusion annealings under harsh conditions if measurable effects are to be produced. For example, in the oxygen thermomigration experiment with $Mg(^{18}O, {}^{16}O)$ which was originally envisioned, one may estimate $q_-^* \leqslant h_{m-} = 2.38$ eV, $q_+^* \leqslant h_{m+} = 2.16$ eV, $h_s = 7.5$ eV (Mackrodt and Stewart, 1979), $T = 2100$ K, $D_- D_+ (D_- + D_+)^{-1} \approx D_- = 3 \ 10^{-14}$ cm^2/sec (Yoo et al., 1984), and $dT/dz = 2100$ K/cm. Under these conditions a time on the order of 4 years will be required to produce marker movement of 1 μm. Such times could be considerably reduced if it would be possible to develop procedures for depositing markers (e.g., thin films or implanted layers) which would permit measurement of marker movement on the 50-200 Å scale which is available for measurement through, for example, SIMS analysis. Measurements of this sort for an ionic material such as MgO would be of great interest in conjunction with a measurement of the thermoelectric power (Wagner, 1972). The latter measurement, for a material in which cation transport is dominant, would provide q_+^* while the marker-movement experiments, Eq. (35), yield $q_-^* + q_+^*$. Separation of heats of transport for anion and cation would thus be possible if data from the two types of measurements were combined.

RANGE OF VALIDITY OF THE SOLUTIONS AND EFFICIENCY OF THE PROCEDURE

The assumptions made in neglecting higher-order terms in the perturbation analysis were given as equations (17) and (11) which restated, are

---

[*]The quasi-molecular nature of the strongly-coupled cation and anion vacancy fluxes is reflected in the form of Eq. (35). Thermomigration by a vacancy mechanism in an elemental metal results in an analogous lattice velocity $D(q^* - h_v)/kT \cdot dT/dz$, and it has been previously shown as well that information on the heat of transport is lost from the concentration profile when it is measured in the laboratory frame of reference (Mock, 1969).

Eq. (17): $\begin{cases} T_m/T_o \sim 1/\varepsilon \\ T_i^*/T_o \sim 1/\varepsilon \end{cases}$　　　Eq. (11): $\dfrac{a\delta}{T_o} \sim \varepsilon^2$

On the basis of theoretical models and all experimental data accumulated to date, the magnitude of the reduced heat of transport seems to be at most $|q_i^*| \leqslant H_m$, the enthalpy of motion. One may thus combine the first two relations into the single condition $T_m/T_o \sim 1/\varepsilon$, or $T_o \sim \varepsilon T_m$. Substitution of this expression for T_o into Eq. (11) leads to a single requirement

$$\left|\frac{a\delta}{T_m}\right| \sim \varepsilon^3 \tag{44}$$

in order that the solutions be valid within an accuracy of ε^2. One can demonstrate, through direct substitution of the approximate solutions (20), (25) and (28) in original differential equation (9), that this condition is indeed sufficient to insure a solution which is accurate within $(a\xi/T_o)^2 \, (T_m/T_o)^2 \approx \varepsilon^2$.

The parameters available in the design of an experiment are the temperature gradient, a, the interface temperature, T_o, the width of the diffusion gradient, δ (or, in other terms, the duration of the diffusion annealing) and the desired accuracy of the solution, ε^2. Upon selecting a value for ε, the activation energy, through Eq. (17), may be viewed as fixing the necessary interface temperature, T_o. Two experimental parameters (e.g., the time of the diffusion annealing or δ, and the magnitude of the temperature gradient, a) may thus be selected to insure that the experimental conditions conform to Eq. (11). The necessary conditions should thus be readily realizable.

The approximations made in the perturbation analysis cause the solutions to be valid only for short diffusion times and very shallow concentration gradients. This, however, is precisely the alternative to the time-consuming measurements of the stationary-state Sorét effect which one desires. The ratio of the time scales of the two sorts of experiments may be demonstrated by comparing, for example, the time necessary to produce a measurable motion of the location z_o, Eq. (40), under semi-infinite source boundary conditions, to the time scale Θ, Eq. (5), of the stationary-state Sorét effect:

$$\frac{t}{\Theta} = \pi^2 \left(\frac{T_o}{T_m + 2T_i^*}\right) z_o/L \tag{45}$$

where the value $2T_o/L$ has been used for the temperature gradient, a. Using the relations $T_m/T_o \approx T_i^*/T_o \approx 1/\varepsilon$, this result may be expressed in terms of ε, the square root of the accuracy of the present solutions

$$t/\Theta = \pi^2/3 \; \varepsilon \; z_o/L \tag{46}$$

Alternatively, for typical values which might be encountered in practice, $T_o \approx 10^3$ K, $T_m + 2T_i^* \approx 10^4$ K, one would expect $t/\Theta \approx z_o/L$. The time scales of the experiments are thus on the order of the ratio of the migration distance to the total dimension of the specimens, an economy in annealing time of easily one or more orders of magnitude.

ACKNOWLEDGEMENT

This work was supported under contract DE-AC02-76ER02923 with the Office of Basic Energy Sciences, Division of Materials Sciences, of the U.S. Department of Energy.

REFERENCES

Allnatt,A. R., and Chadwick, A. V., 1967, Chem. Rev., 67:681.

Chemla, M., 1956, Ann. Phys., Ser. 13, 1:959.

de Groot, S. R., 1942, Physica, 9:699.

de Groot, S. R., 1951, "Thermodynamics of Irreversible Processes", North Holland Publishing Co., Amsterdam.

de Groot, S. R., and Mazur, P., 1962, "Non-Equilibrium Thermodynamics", North Holland Publishing Co., Amsterdam.

Gillan, M. J., 1983, Diffusion in a Temperature Gradient, in: "Mass Transport in Solids", F. Bénière and C. R. A. Catlow, Eds., Plenum Press, N.Y., 227-250.

Haase, R., 1969, "Thermodynamics of Irreversible Processes", Addison-Wesley Publishing Co., Reading, MA.

Howard, R. S. and Lidiard, A. B., 1964, Rep. Prog. Phys., 27:161.

Kirkwood, J. G. Baldwin, R. L., Dunlop, P. J., Gosting, L. J., and Kegeles, G., 1960, J. Chem. Phys., 33:1505.

Lugwig, C., 1856, Sitzber. Akad. Wiss. Wien, Math.-Naturw. Kl, 20:539.

Mackrodt, W. C., and Stewart, R. F., 1979, J. Phys. C., 12:5015.

Mock, W., Jr., 1969, Phys. Rev., 179:663.

Oriani, R. A., 1969, J. Phys. Chem. Solids, 30:339.

Sorét, C., Arch. Sci. Phys. Nat., 3:48

Shewmon, P. G., 1960, Acta Metall., 8:605.

Yoo, H.-I., 1984 "Isothermal and Non-isothermal Diffusion of Oxygen in Single Crystal MgO", Ph.D. Thesis, Department of Materials Science and Engineering, Massachusetts Institute of Technology, Cambridge, MA.

Yoo, H.-I., Wuensch, B. J., and Petuskey, W. T., 1984, Adv. Ceramics, 12.

Wagner, C., 1972, Progr. Sol. State Chem., 7:1.

OXYGEN SELF-DIFFUSION IN CUBIC ZrO$_2$ SOLID SOLUTIONS

Yasumichi Oishi* and Ken Ando

Department of Nuclear Engineering
Faculty of Engineering
Kyushu University
Fukuoka 812, Japan

INTRODUCTION

Because of the interest in their application to solid electrolyte, electrical conductivity has been extensively determined for various stabilized ZrO$_2$ by many investigators. In contrast, direct determination of the oxygen tracer diffusion coefficient has been made only for CaO-stabilized ZrO$_2$ by Kingery *et al*. (1959) and later by Simpson and Carter (1966). Agreement of the oxygen self-diffusion coefficient with the electrical conductivity determined by Kingery *et al*. proved that CaO-stabilized ZrO$_2$ is an oxygen ionic conductor due to the temperature-independent oxygen vacancies resultant from substitution of divalent Ca ions for Zr ions in the fluorite structure. The determined activation energy, 127 kJ/mol, was interpreted to be the migration energy for the oxygen ion.

In interpreting those results, however, there is a problem with respect to phase equilibrium. When those determinations were made, the temperature ranges <900°C (Kingery *et al*.) and <1100°C (Simpson and Carter) were in a cubic single-phase region on the CaO-ZrO$_2$ phase diagram constructed by Duwez *et al*. (1952). According to the recent phase diagram by Hellmann and Stubican (1983), however, those temperature ranges are now in two-phase regions of a tetragonal solid solution plus CaZr$_4$O$_9$ or of a monoclinic solid solution plus CaZr$_4$O$_9$. In other words, oxygen self-diffusion coefficients for CaO-stabilized ZrO$_2$ by Kingery *et al*. and Simpson and Carter are interpreted to have been determined in a meta-stable state. Then a question arises

* Present address: R & D Laboratories, NGK Insulators, Ltd.
Suda-cho, Mizuho-ku, Nagoya 467, Japan

whether the diffusion mechanism is the same or different in the meta-stable state and in the true cubic phase.

This question is a problem directly related to the electrical conductivity of stabilized ZrO_2. Hoffmann and Fisher (1962) pointed out that the temperature dependence of the conductivity of CaO-stabilized ZrO_2 broke at 1300°C and also Hohnke (1979) that the activation energy for the conductivity decreased in the high-temperature range. These observations suggest the mechanism of the oxygen diffusion varying from the low-temperature range to the high-temperature range.

Table 1 shows activation energies of the oxygen migration determined by tracer technique and theoretically calculated for various fluorite-cubic oxides, where activation energies for CaO-stabilized ZrO_2 are higher in comparison with those for other fluorite-cubic oxides. This suggests that those high activation energies for CaO-stabilized $ZrO2$ in a meta-stable state are not simply the migration energy for the oxygen vacancy but involve some other effect and that

Table 1. Activation energies for oxygen diffusion in fluorite-type oxides.

	Substance	Q (kJ/mol)	Author(s)
Experimental	UO_{2-x}	49	Kim & Olander
	PuO_{2-x} (chemical diffusion)	46	Bayoglu & Lorenzelli
	ThO_2 (extrinsic)	74	Ando *et al.*
	$CeO_2 (P_{O_2} = 2.1 \times 10^4 \text{Pa})$	104	Floyd
	$CeO_2 (P_{O_2} = 2.0 \times 10^3 \text{Pa})$	73	Floyd
	$CeO_{1.92}$	50	Floyd
	$CeO_{1.80}$	15	Floyd
	$Ce_{0.9}Y_{0.1}O_{1.95}$	80	Floyd
	$Ce_{0.8}Y_{0.2}O_{1.90}$	77	Floyd
	$Ce_{0.6}Y_{0.4}O_{1.8}$	89	Floyd
	$Zr_{0.85}Ca_{0.15}O_{1.85}$	127	Kingery *et al.*
	$Zr_{0.858}Ca_{0.142}O_{1.858}$	131	Simpson & Carter
Theoretical	ThO_2	75	Colbourn & Mackrodt
	UO_2	24	Catlow & Lidiard

another diffusion mechanism with a lower activation energy than those values may occur at more elevated temperatures where only the fluorite-cubic phase is stable.

Such change of diffusion mechanism can be caused not only by phase transition but also by defect association or formation of micro-domains. Structure change from the oxygen-vacancy type to the cation-interstitial type in the solid solution also may lead to occurrence of the second diffusion mechanism. To clarify these questions, determination of the oxygen self-diffusion coefficient was extended, in the present work, from two-phase regions toward high temperatures to the single-cubic-phase region as indicated on Hellmann and Stubican's $CaO-ZrO_2$ phase diagram. Since the problem is not only of CaO-stabilized ZrO_2 but related to other ZrO_2 solid solutions, similar studies were carried out also for MgO- and Y_2O_3-stabilized ZrO_2. Oxygen self-diffusion coefficients were determined by isotope exchange technique (Ando *et al.*, 1976). Specimens are mostly polycrystalline unless otherwise specified.

SYSTEM $MgO-ZrO_2$

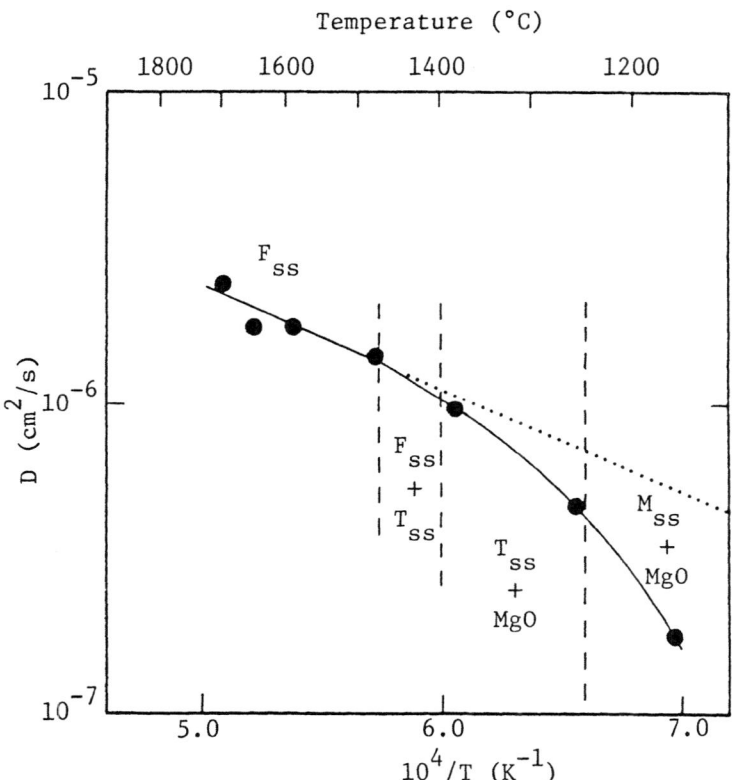

Fig. 1. Oxygen self-diffusion coefficients for $12MgO \cdot 88ZrO_2$.

Oxygen self-diffusion coefficients determined for $12MgO \cdot 88ZrO_2$ are shown in Fig. 1 as a function of temperature where vertical dashed lines indicate phase boundary temperatures read from the $MgO-ZrO_2$ phase diagram constructed by Grain (1967). The whole diffusion coefficients can not be properly represented by a single Arrhenius equation, but the cubic single-phase region in the high temperature range can be represented by Eq. (1).

$$D = 1.02 \times 10^{-4} \exp[-62.8(kJ/mol)/RT] \ cm^2/s \qquad (1)$$

As compared with the extrapolation of those results shown by the dotted line, the determined oxygen diffusion coefficients tend to deviate from Eq. (1) with increasing apparent activation energy, as the temperature decreases through two-phase regions, indicating change in diffusion mechanism.

That change of diffusion mechanism is not due to the structural transition in the cubic ZrO_2 solid solution as confirmed by the density measurement (Ando *et al.*, 1984). As shown in Fig. 2, the pycnometric densities measured for four compositions quenched from 1800°C are close to the values theoretically calculated on the assumption of the anion-vacancy type solid solution. The anion-vacancy type is understood to be the stable structure in the low temperature range of interest by analogy with system $CaO-ZrO_2$, where occurrence of the cation-interstitial type is limited only in the high temperature side (Diness and Roy, 1965). Consequently, formation of the lattice defect in $MgO-ZrO_2$ solid solution is primarily described by Eq. (2) in the Kröger-Vink notation.

$$MgO \xrightarrow{ZrO_2} Mg''_{Zr} + V_O^{\cdot\cdot} + O_O^x \qquad (2)$$

Samples were equilibrated at 1580°C prior to the diffusion experiment, so that no appreciable phase transition is expected to proceed from the single-cubic-phase state during the diffusion experiment, since the cation interdiffusion

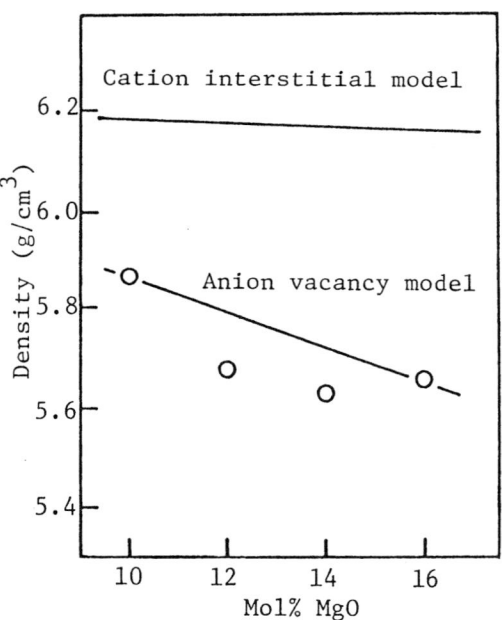

Fig. 2. Pycnometric and theoretically calculated densities for system $MgO-ZrO_2$.

required for the phase transition is slow (Sakka *et al.*, 1982). Nonetheless, as was shown in Fig. 1, the deviation from the linear temperature dependence of the oxygen self-diffusion in the low-temperature range appears to be in accordance with the phase transitions indicated on the phase diagram. The deviation becomes noticeable below 1400°C where the cubic phase disappears leaving a tetragonal phase of reduced MgO content and the MgO phase. If this transition takes place, it should result in a substantial decrease of the oxygen vacancies and consequently, in reduced oxygen diffusivity.

In the higher temperature range, 1470°C-1400°C, where the tetragonal phase occurs in the host cubic phase, the cubic solid solution is the major, continuous phase which determines the overall oxygen diffusivity of the system. Thus, even if the phase transition takes place, the expected decrease in the oxygen diffusivity is not necessarily pronounced in the temperature range, as shown in Fig. 1.

The pronounced decrease in the oxygen diffusivity from the tetragonal phase plus MgO region to the monoclinic phase plus MgO region is interpreted to be due to the oxygen diffusivity decreasing from the tetragonal to the monoclinic solid solution which has lower MgO solubility than the tetragonal phase, as indicated on the phase diagram (Grain, 1967).

As shown in Fig. 3, the oxygen self-diffusion coefficient of $16MgO \cdot 84ZrO_2$ exhibits a temperature dependence similar to that of $12MgO \cdot 88ZrO_2$ in the single-cubic-phase region through two-phase regions except that the decrease occurring in two-phase regions becomes more pronounced with increasing MgO content.

The activation energy for the cubic single-phase region of $12MgO \cdot 88ZrO_2$, 62.8 kJ/mol as given by Eq. (1), is approximately one half of that reported for the oxygen self-diffusion in CaO-

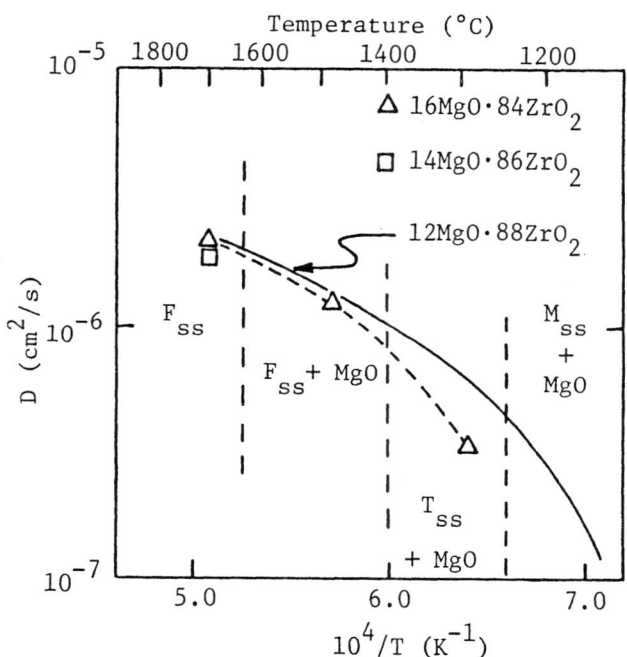

Fig. 3. Oxygen self-diffusion coefficients for $14MgO \cdot 86ZrO_2$ and $16MgO \cdot 84ZrO_2$ in comparison with $12MgO \cdot 88ZrO_2$.

stabilized ZrO_2. This magnitude is interpreted to be the migration energy of the oxygen ion typical of the fluorite-cubic $MgO-ZrO_2$ solid solution.

In contrast, the high activation energies reported for CaO-stabilized ZrO_2, 127 kJ/mol (Kingery *et al.*, 1959) and 131 kJ/mol (Simpson and Carter, 1966), are not only the migration energy of the oxygen ion but involve other energy, as will be discussed later. The higher temperature dependences determined in the two-phase regions for $12MgO \cdot 88ZrO_2$ (Fig. 1) and $16MgO \cdot 84ZrO_2$ (Fig. 3) will vary depending on thermal history of the samples and hence have no simple physical meaning for the oxygen self-diffusion.

SYSTEM $CaO-ZrO_2$

Oxygen self-diffusion coefficients of system $CaO-ZrO_2$ were determined for four compositions of 13, 15, 17, and 19 mol% CaO in the temperature ranges over the cubic single-phase region to two-phase regions, as plotted in Fig. 4 on the Hellmann and Stubican phase diagram. Figure 5 shows the results for $15CaO \cdot 85ZrO_2$ determined by using three different specimens, two of them repeatedly, along with the results of Kingery *et al.* and Simpson and Carter. One specimen, indicated by closed circles, showed a break of the temperature dependence at 1400°C in a similar manner to the case of system $MgO-ZrO_2$ in Figs. 1 and 3. The low-temperature regime with the higher temperature dependence agrees with the results of Kingery *et al.* and Simpson and Carter.

The break point, 1400°C, is 100°C higher than the phase boundary temperature between the single-phase region and the two-phase region; that is, the break occurred within the single phase region. The temperature of 1400°C is the temperature where Tien and Subbarao (1963) observed disappearance of the ordered structure in CaO-stabilized ZrO_2 by X-ray diffraction technique. After the break at 1400°C, no significant breaks seem to have occurred at phase boundary temperatures. The temperature dependence of the oxygen diffusivity in CaO-stabilized $ZrO2$ does not appear to be immediately related to the phase transitions indicated on the

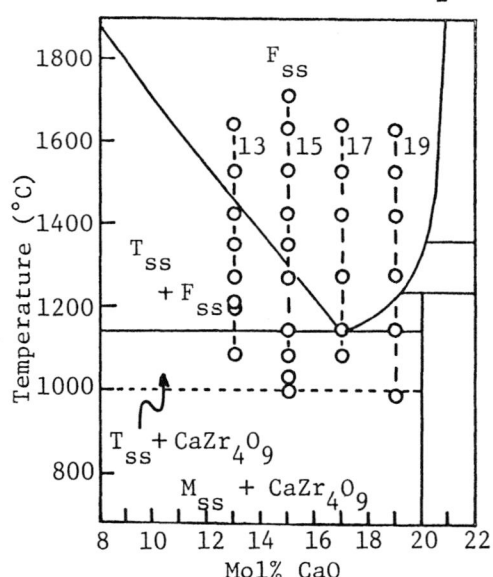

Fig. 4. Compositions and temperatures for determination of oxygen self-diffusion coefficients for system $CaO-ZrO_2$ plotted on Hellmann and Stubican's phase diagram.

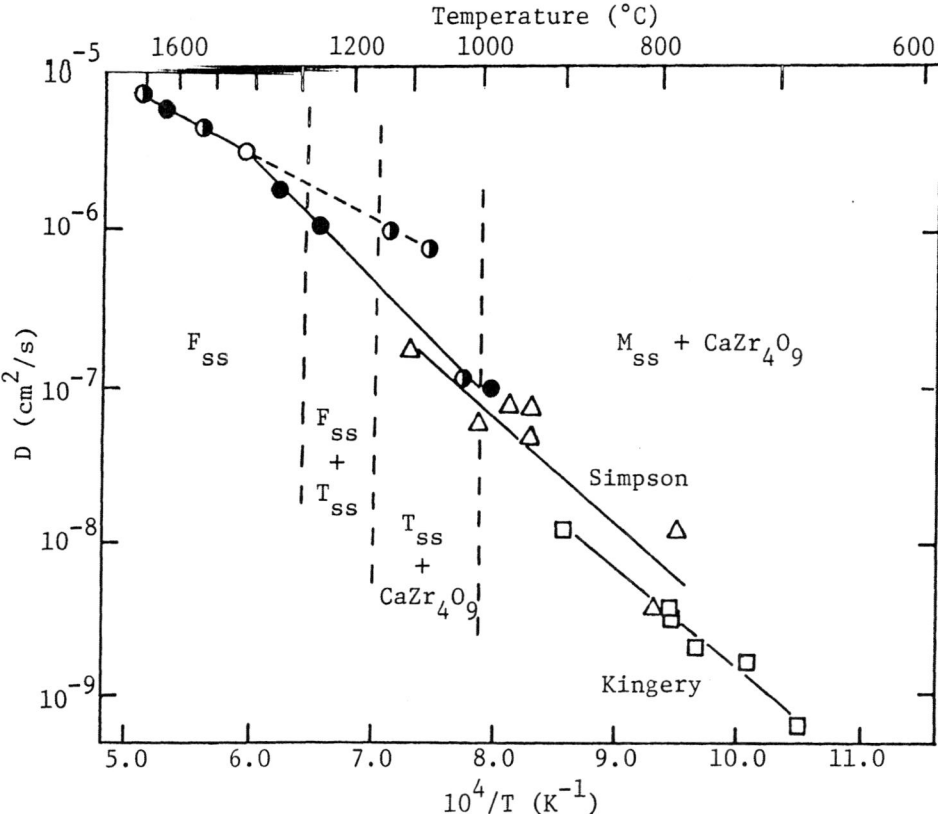

Fig. 5. Oxygen self-diffusion coefficients for 15CaO·85ZrO$_2$.

phase diagram, in contrast to system MgO-ZrO$_2$.

However, the agreement of three measurements by the different investigators suggests that the oxygen diffusion in the low-temperature range is controlled by the same single mechanism, being independent of the phase transition on the phase diagram. The results are interpreted that the high-temperature regime with the lower activation energy is for the disordered structure and the low-temperature regime with the higher activation energy for some ordered structure in a meta-stable state. The oxygen diffusivity and consequently, the ionic conductivity in this low-temperature range should be affected more or less by thermal history of the sample.

The second specimen indicated by half-closed circles in Fig. 5, did not exhibit the off-breaking at 1400°C but did at 1050°C, falling onto the low-temperature regime of the first specimen. The relatively high self-diffusion coefficients in the neighborhood of 1100°C lie on the extrapolation of the high-temperature regime. This high diffusisity is understood to be due to the frozen state of the high-temperature disordered structure.

The high-temperature regime for the three specimens is represented by Eq. (3).

$$D = 1.0 \times 10^{-3} \exp[- 81.2(kJ/mol)/RT] \; cm^2/s \qquad (3)$$

The activation energy, 81.2 kJ/mol, is interpreted to be the migration energy for a single oxygen vacancy in the disordered fluorite-cubic structure. The magnitude is comparable with that for the cubic single-phase range of system MgO-ZrO$_2$ in Eq. (1).

The low-temperature regime for the first specimen is described by Eq. (4).

$$D = 0.10 \exp[- 146(kJ/mol)/RT] \; cm^2/s \qquad (4)$$

The activation energy, 146 kJ/mol, which is 65 kJ/mol higher than that for the high-temperature regime, is understood to involve some energy other than the migration energy of the oxygen vacancy. Since this effect starts already within the cubic single-phase region, it is not due to the phase transition but presumably due to defect association.

The oxygen self-diffusion coefficients determined for 13CaO·87ZrO$_2$ exhibited a break similar to that for 15CaO·85ZrO$_2$ at 1430°C, which is close to the phase boundary temperature between the cubic single-phase and the cubic-tetragonal two-phase regions. The high temperature regime is described by Eq. (5).

$$D = 2.1 \times 10^{-4} \exp[- 66.7(kJ/mol)/RT] \; cm^2/s \qquad (5)$$

In the case of 17 CaO·83ZrO$_2$ and 19CaO·81ZrO$_2$, the high-temperature regime with low activation energy was not observed but only

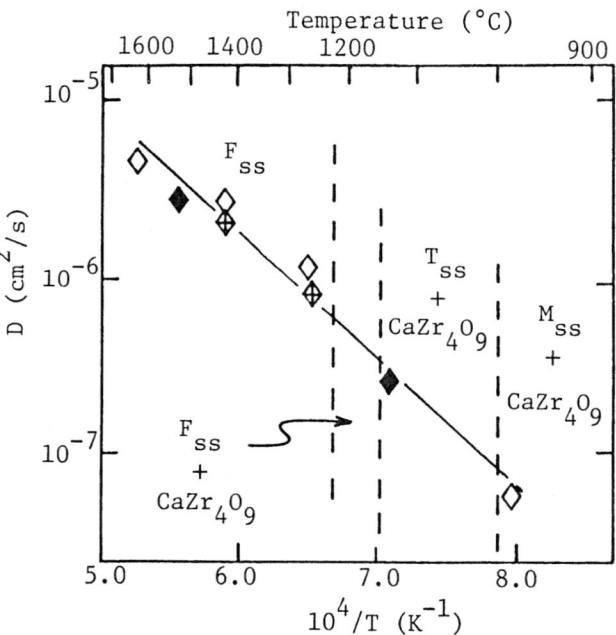

Fig. 6. Oxygen self-diffusion coefficients for 19CaO·81ZrO$_2$.

the low-temperature regime was observed. The results for 19CaO·81ZrO$_2$ are shown in Fig. 6 and described by Eq. (6),

$$D = 2.1 \times 10^{-2} \exp[- 132(kJ/mol)/RT]\ cm^2/s \qquad (6)$$

where the low temperature mechanism dominates the oxygen diffusion in the cubic single-phase regions; the break point from the low-temperature to high-temperature regime has been shifted toward the higher-temperature side by the increased CaO content.

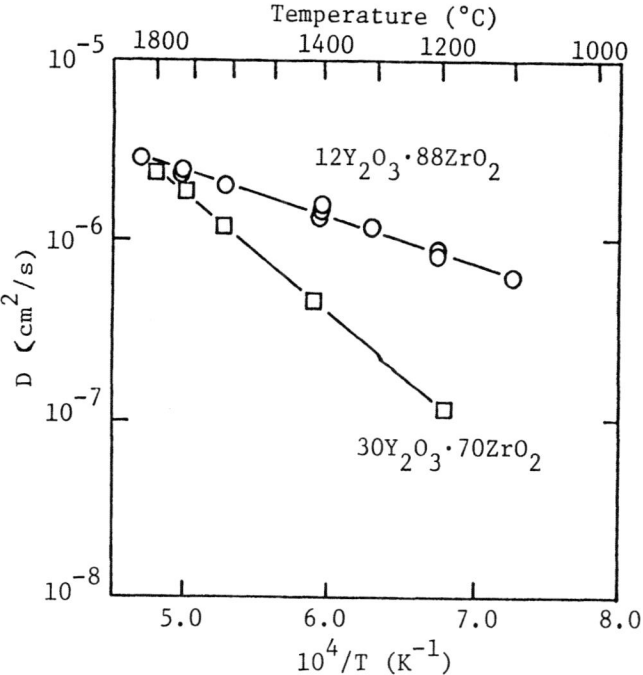

Temperature (°C)

Fig. 7. Oxygen self-diffusion coefficients for 12Y$_2$O$_3$·88ZrO$_2$ and 30Y$_2$O$_3$·70ZrO$_2$.

SYSTEM Y$_2$O$_3$-ZrO$_2$

The oxygen diffusion mechanism varying within the fluorite-cubic phase region was not only observed in system CaO-ZrO$_2$ but also confirmed in system Y$_2$O$_3$-ZrO$_2$. Figure 7 shows oxygen self-diffusion coefficients determined within the fluorite-cubic single-phase region for 12Y$_2$O$_3$·88ZrO$_2$ and 30Y$_2$O$_3$·70ZrO$_2$ which were prepared by sintering at 2200°C. The results for 12Y$_2$O$_3$·88ZrO$_2$ are described by Eq. (7).

$$D = 4.07 \times 10^{-5} \exp[- 47.7(kJ/mol)/RT]\ cm^2/s \qquad (7)$$

The results for 30Y$_2$O$_3$·70ZrO$_2$ are described by Eq. (8).

$$D = 3.67 \times 10^{-3} \exp[- 127(kJ/mol)/RT]\ cm^2/s \qquad (8)$$

It is understood that the former corresponds to the high-temperature type and the latter with the higher activation energy to the low-temperature type diffusion mechanism of system CaO-ZrO$_2$. The possibility of the oxygen diffusivity of 30Y$_2$O$_3$·70ZrO$_2$ being higher than that of 12Y$_2$O$_3$·88ZrO$_2$ is expected at only temperatures higher than 1800°C in Fig. 7.

The results for 12Y$_2$O$_3$·88ZrO$_2$ shown in Fig. 7 were determined within the fluorite-cubic single-phase region but their lower temperature side was proved to be not in the equilibrium state; as shown in Fig. 8, the 1700°C-treated sample exhibits oxygen diffusivities

lower than those of the
2200°C-treated samples of
the same composition.
This difference is not due
to sample preparation as
confirmed by two different
samples, one prepared by
coprecipitation technique
and the other by mixing
the two component oxides.
The reproducible results
for two samples of dif-
ferent preparation suggest
that a meta-stable struc-
ture is easily attained
by the 1700°C-treatment
and farther transition
from the meta-stable state
is slow. Reheating the
1700°C-treated sample at
2200°C was confirmed to
give the 2200°C-treated
type oxygen diffusivities
shown in Fig. 8. These
variations of the diffu-
sivity are understood to
be related to some defect

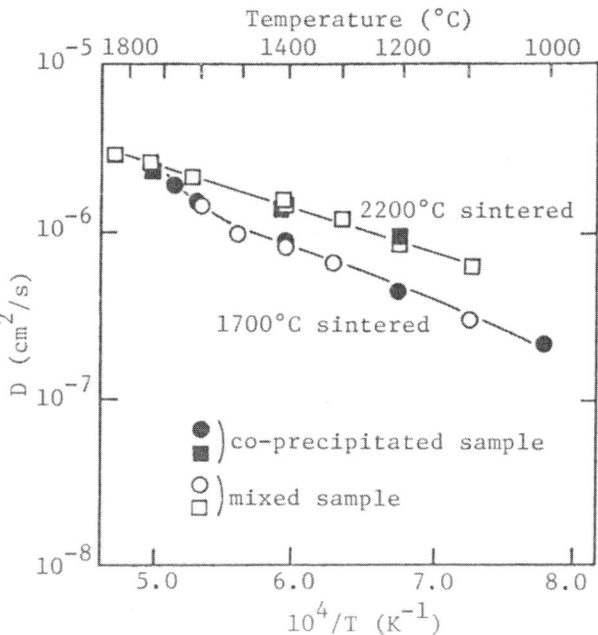

Fig. 8. Oxygen self-diffusion coefficients
for $12Y_2O_3 \cdot 88ZrO_2$ sintered at 2200°
C and 1700°C.

equilibrium occurring within the cubic single-phase region, but the
true equilibrium has not been attained due to insufficient treatment
time.

Figure 9 shows oxygen self-diffusion coefficients determined at
1400°C for 2200°C-pretreated and 1700°C-pretreated polycrystals and
a 2200°C-pretreated single crystal, both of $12Y_2O_3 \cdot 88ZrO_2$, as a function of the 1400°C-annealing time. The 1700°C-pretreated sample does
not exhibit significant change but the 2200°C-pretreated samples,
both polycrystal and single crystal, gradually decrease the oxygen
diffusivities over 200h toward the 1700°C type. The slow variation
suggests that the kinetics is controlled by long-distance diffusion
of the cations which have low diffusivity. Equilibrium electrical
conductivities attained by slow rates have been observed for Y_2O_3-
stabilized ZrO_2 cooled from the high-temperature side and heated from
the low-temperature side within the cubic phase region (Vlasov, 1983;
Suzuki, 1983).

Similar results for the polycrystal and the single crystal
suggest that the structural change is not occurring preferentially
along grain boundaries but occurring in the crystal bulk and that
the effect of grain boundaries on the oxygen diffusivity is not
significant at this high temperature.

DISCUSSION

The oxygen diffusion mechanism varying from high temperature range to low temperature range with increased activation energy was observed in common for MgO-, CaO-, and Y_2O_3-stabilized ZrO_2. The lower activation energies determined for the high-temperature regimes, 48-81 kJ/mol, are interpreted to be the migration energy of the oxygen ion in the disordered fluorite-cubic structure. The activation energies determined for the low-temperature regime, 127-146 kJ/mol, involve the effect of structural change or defect association.

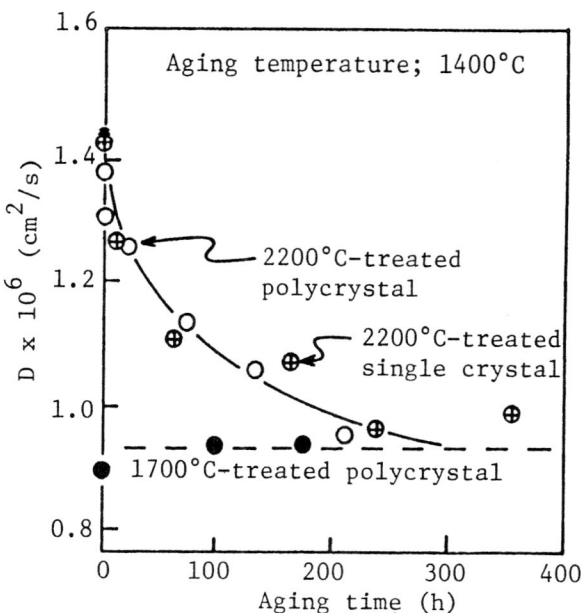

Fig. 9. Oxygen self-diffusion coefficients for polycrystalline and single-crystal $12Y_2O_3 \cdot 88ZrO_2$ as a function of annealing time at 1400°C.

The low-temperature regime of system $MgO-ZrO_2$ is controlled by phase transition. In contrast, that of system $CaO-ZrO_2$ is independent of phase transition and starts to occur in the cubic single-phase region. The difference of system $MgO-ZrO_2$ from system $CaO-ZrO_2$ comes from the difference in stability of the fluorite-cubic structure. The relatively high cationic interdiffusion coefficient of system $MgO-ZrO_2$ (Oishi $et\ al.$, 1983) leads to the faster phase transition and consequently, to decreased oxygen diffusivity, while the low cationic interdiffusion coefficient tends to stabilize the fluorite-cubic structure with high oxygen diffusivity in system $CaO-ZrO_2$ as well as expected in system $Y_2O_3-ZrO_2$.

The low-temperature type oxygen diffusion in system $CaO-ZrO_2$ appears to be controlled by the same mechanism from the cubic single-phase through two-phase regions. In a theoretical calculation for CaO-doped ZrO_2, formation energies for the pair of Ca-substitutional and oxygen vacancy and for microdomains are in a close balance (Butler $et\ al.$, 1983). However, the defect pair can be formed by migration of only the oxygen vacancy which has a high diffusivity. This is inconsistent with the freezing of the high-temperature structure observed for $15CaO \cdot 85ZrO_2$. The slow kinetics is not inconsistent with growth of microdomains.

The slow decrease of the oxygen diffusivity determined at 1400°C

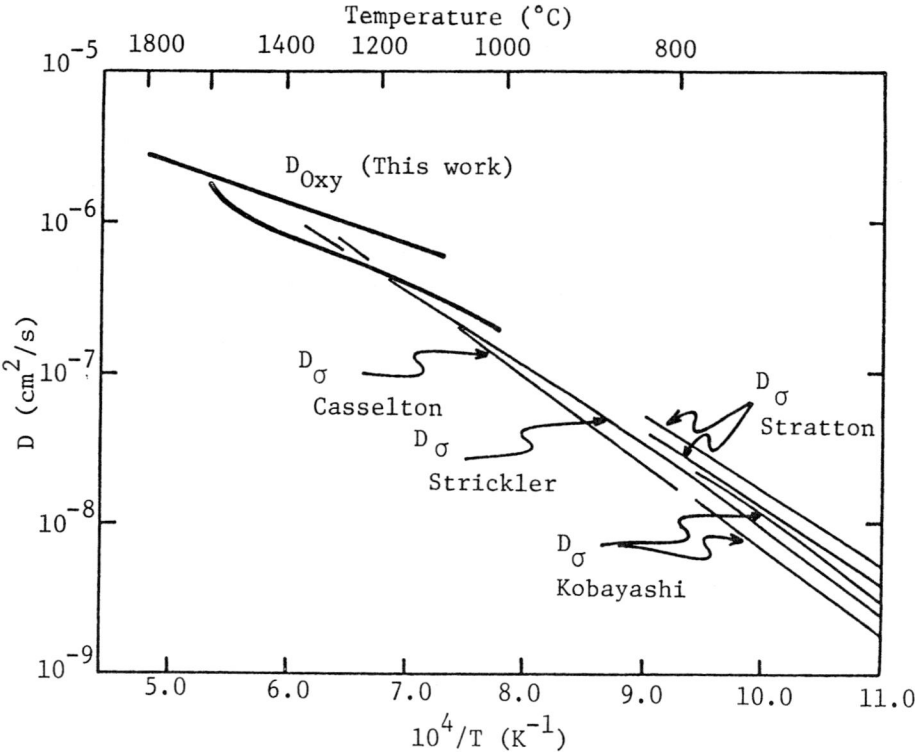

Fig. 10. Comparison of oxygen self-diffusion coefficients determined
by tracer technique and calculated from electrical conductivities of $12Y_2O_3 \cdot 88ZrO_2$.

for the 2200°C-pretreated $12Y_2O_3 \cdot 88ZrO_2$ (in Fig. 9) suggests that the
kinetics is controlled by long-distance diffusion of the cations which
have low diffusivity. In theoretical calculation of the binding
energies for various clusters of the dopant substitutionals and oxygen
vacancies in Mg-, Ca-, and Y-doped ZrO_2, the binding energy for the
cluster of one oxygen vacancy and two cation substitutionals is relatively high (Butler *et al.*, 1981). However, formation of such isolated defect associations does not explain the slow kinetics, since
the migration distance of the cations required for their formation
should be relatively short.

The diffusion coefficients calculated using the Nernst-Einstein
relation for the electrical conductivities of Y_2O_3-stabilized ZrO_2 in
the literature are shown in Fig. 10, where the calculated diffusion
coefficients are understood to correspond to the low-temperature type
oxygen diffusivity. Similar correspondence is found also between the
electrical conductivity and the low-temperature type oxygen diffusivity of CaO-stabilized ZrO_2. The discrepancy of the electrical
conductivities for Y_2O_3-stabilized ZrO_2 reported by different investigators suggests the effect of differing thermal histories of the
samples. However, the discrepancy is found to be smaller than that

for CaO-stabilized ZrO_2. This
is consistent with the smaller
difference between the high-
temperature and low-temperature
type diffusivities for Y_2O_3-
stabilized ZrO_2 in Fig. 8 than
for the CaO-stabilized ZrO_2 in
Fig. 5.

High concentrations of
dopants tend to enhance the occurrence of the low-temperature
mechanism in all three systems.
Resultant dopant concentration
dependence of ionic conduction
is schematically demonstrated
in Fig. 11, where increasing
ionic conductivity with increasing dopant concentration
is expected at only elevated
temperatures.

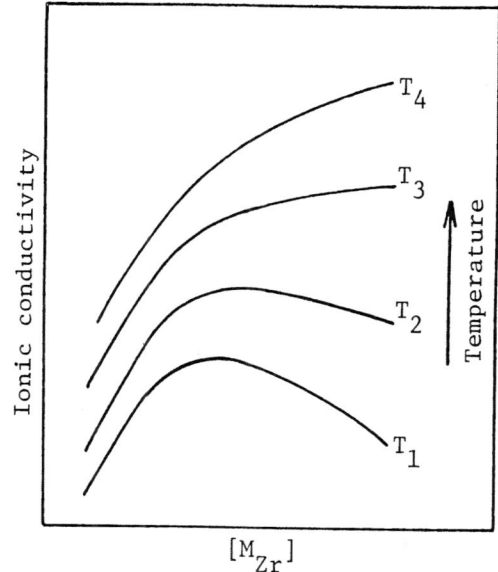

Fig. 11. Schematic composition dependence of ionic conductivity
of ZrO_2 solid solution.

REFERENCES

K. Ando, Y. Oishi and Y. Hidaka, J. Chem. Phys. 65:2751 (1976).

K. Ando, Y. Oishi, H. Koizumi and Y. Sakka, J. Mat. Sci. Let.,
to be published (1984).

A. S. Bayoglu and R. Lorenzelli, J. Nucl. Mat. 82:403 (1979).

V. Butler, C. R. A. Catlow and B. E. F. Fender, Solid State Ionics
5:539 (1981).

V. Butler, C. R. A. Catlow and B. E. F. Fender, Radiation Effect
73:273 (1983).

R. E. W. Casselton, Phys. Stat. Sol. (a)22:571 (1970).

C. R. A. Catlow and A. B. Lidiard, Theoretical Studies of Point-
Defect Properties of Uranium Dioxide, IAEA Symposium on the
Thermodynamics of Nuclear Materials, Vienna, Oct. 190/13:27
(1974).

E. A. Colbourn and W. C. Mackrodt, J. Nucl.Mat. 118:50 (1983).

A. M. Diness and R. Roy, Solid State Comm. 3:123 (1965).

P. Duwez, F. Odell, and F. H. Brown, Jr., J. Am. Ceram. Soc.
35:107 (1952).

J. M. Floyd, Indian J. Technol. 11:589 (1973).

C. F. Grain, J. Am. Ceram. Soc. 50:288 (1967).

J. R. Hellmann and V. S. Stubican, ibid., 66:260 (1983).

A. Hoffmann and W. A. Fisher, Z. Phys. Chem. 35:95 (1962).

D. K. Hohnke, Ionic Conduction in Doped Zirconia, in Fast Ion Transport in Solids, Eds., Vashishta, P., J. N. Mundy and G. K.
Shenoy, North-Holland, pp. 669.

K. C. Kim and D. R. Olander, J. Nucl. Matl. 102:192 (1981).

W. D. Kingery, J. Pappis, M. E. Doty and D. C. Hill, J. Am. Ceram. Soc. 42:393 (1959).

S. Kobayashi, Yogyo-Kyokai-Shi 89:14.

Y. Oishi, K. Ando and Y. Sakka, Lattice and Grain-Boundary Diffusion Coefficients of Cations in Stabilized Zirconias, Advances in Ceramics, Vol. 7, Additives and Interfaces in Electronic Ceramics, Eds., M. F. Yan and A. H. Heuer, pp. 208 (1983).

Y. Sakka, Y. Oishi and K. Ando, Bull. Chem. Soc. Jpn. 55:420 (1982).

L. A. Simpson and R. E. Carter, J. Am. Ceram. Soc. 49:139 (1966).

T. G. Stratton, D. Reed and H. L. Tuller, Study of Boundary Effects in Stabilized Zirconia Electrolytes, Advances in Ceramics, Vol. 1, Grain Boundary Phenomena in Electronic Ceramics, Ed., L. M. Levinson, pp. 114 (1981).

D. W. Strickler and W. G. Carlson, J. Am. Ceram. Soc., 48:286 (1965).

Y. Suzuki, private communication (1983).

T. Y. Tien and E. C. Subbarao, J. Chem. Phys., 39:1041 (1963).

A. N. Vlasov, Elektrokhimiya 19:1624 (1983).

OXYGEN SELF-DIFFUSION IN Y_2O_3 AND Y_2O_3-ZrO_2 SOLID SOLUTION

Ken Ando and Yasumichi Oishi

Department of Nuclear Engineering
Faculty of Engineering
Kyushu University
Fukuoka 812
Japan

INTRODUCTION

Yttrium oxide (Y_2O_3) has the C-type cubic rare-earth-oxide crystal structure. This structure is described as a modified fluorite-type cubic structure where one fourth of the anion sites are vacant and regularly arranged. Some studies concerning the point defect disorder (Tallan and Vest, 1966; Odier et al., 1971; Rifflet et al., 1975) and self-diffusion (Wirkus et al., 1967; Berard et al., 1968; Berard and Wilder, 1969; Gaboriaud, 1980) have been done for this oxide.

Oxygen diffusion experiments were carried out for single crystal using weight-gain technique by oxidizing reduced materials by Wirkus et al. (1967) and Berard et al. (1968). In these studies, large oxygen diffusion coefficients comparable with those in CaO-stabilized ZrO_2 (Oishi and Ando, 1984), UO_2 (Marin and Contamin, 1969), and ThO_2 (Ando et al., 1976), and relatively low activation energies were obtained. Although they interpreted those values as the oxygen self-diffusion coefficients, those are not self-diffusion coefficients but should be interpreted as chemical diffusion coefficients determined under a concentration gradient of the defect.

Yttria is a fairly stable compound but a slight deviation from the stoichiometric composition is to be expected. To determine the defect type, Tallan and Vest (1966) measured the oxygen partial pressure dependence of the electrical conductivity. They found that the p-type conduction was operative in the high oxygen pressure range at high temperatures, and interpreted that this conduction mechanism depended upon the presence of fully ionized yttrium vacancies, V_Y''', and compensating holes. The results mean that the defect type of Y_2O_3 at high oxygen partial pressures can be expressed as $Y_{2-x}O_3$. In contrast, Odier et al. (1971) explained the p-type conduction range as being due to introduction of interstitial oxygen ions. This explanation means that the defect type is expressed as Y_2O_{3+x}.

The above question with respect to the defect type can not be clarified by only the electrical conductivity measurement but requires information on self-diffusion of constituent ions. For this purpose, in this work, oxygen self-diffusion coefficients in single crystal Y_2O_3 and in single crystal and polycrystalline ZrO_2-doped Y_2O_3 were determined as a function of temperature and oxygen partial pressure.

EXPERIMENTAL

Sample Preparation

Single crystal. Spherical and disk-shaped specimens were prepared from single crystals grown by a floating zone method (Kitazawa et al., 1977) for undoped, 0.30 and 5.0 mol% ZrO_2-doped Y_2O_3. The sphere was ground using SiC paper followed by diamond paste. Approximately 100 μm of the surface was then removed by chemical polishing with hot $n-H_3PO_4$. The diameter of the sphere was ≃1 mm.

To prepare the disk-shaped sample, a disk ≃1 mm thick was cut from a single crystal rod. After the specimen was ground using diamond paste, ≃10 μm of the surface was removed by an Ar-ion bombardment at an accelerating voltage of 6.5 kV. To determine the effect of surface finish, one of the samples received only diamond-paste finish.

Polycrystal. Polycrystals were prepared for compositions of 5.0 and 10.0 mol% ZrO_2-doped Y_2O_3. An oxide mixture was prepared by coprecipitation method from Y_2O_3 oxide and $ZrOCl_2 \cdot 8H_2O$ (99.6 % pure). Polycrystalline pellets were sintered in air at 1800°C for 25h. The density was 98 % of the theoretical. The diffusion coefficient was determined using disk-shaped specimen of which the surface was finished with diamond paste.

Determination of the Oxygen Self-Diffusion Coefficient

The oxygen self-diffusion coefficient was determined by a gas-solid isotope exchange technique using ^{18}O as a tracer (Ando et al., 1983). Oxygen partial pressure was controlled using O_2, CO_2, or H_2/CO_2 mixture gases. The sample was preannealed under a designed oxygen partial pressure consisting of natural oxygen at the same temperature as used in the diffusion anneal, for 5 to 7 h (2 to 3 times the subsequent diffusion anneal time).

RESULTS

A) Oxygen Self-Diffusion Coefficient In Single Crystal Y_2O_3

Self-diffusion coefficients for single-crystal Y_2O_3 obtained
under a constant oxygen pressure of about 150 Torr are shown in
Fig. 1 (Ando et al., 1983). There were no significant differences

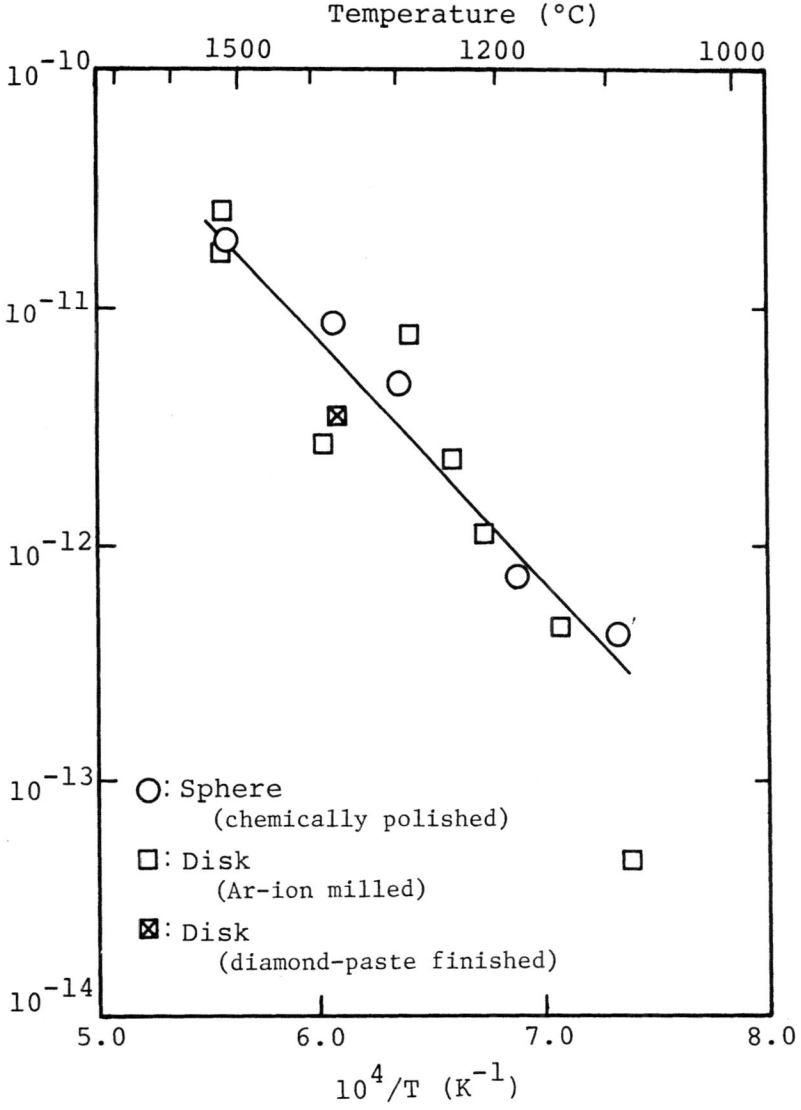

Fig. 1. Oxygen self-diffusion coefficients in Y_2O_3 single crystal.

between the values obtained from different lots or samples with different shapes and surface finishings. The temperature dependence of the self-diffusion coefficient for the temperature range 1100° to 1500°C is expressed as

$$D = 7.3 \times 10^{-6} \exp [- 191 \ (kJ/mol)/RT] \ cm^2/s. \tag{1}$$

B) Oxygen Partial Pressure Dependence of The Self-Diffusion Coefficient

Oxygen pressure dependences of oxygen self-diffusion coefficients in undoped Y_2O_3 and 0.3 mol% ZrO_2-doped Y_2O_3 are shown in Fig. 2. The oxygen self-diffusion coefficient in 0.3 mol% ZrO_2-doped Y_2O_3 was independent of the oxygen pressure. On the other hand, the oxygen self-diffusion coefficients in undoped Y_2O_3 increased with increasing oxygen pressure. The self-diffusion coefficients for undoped Y_2O_3 were approximately proportional to $p_{O_2}^{1/6}$.

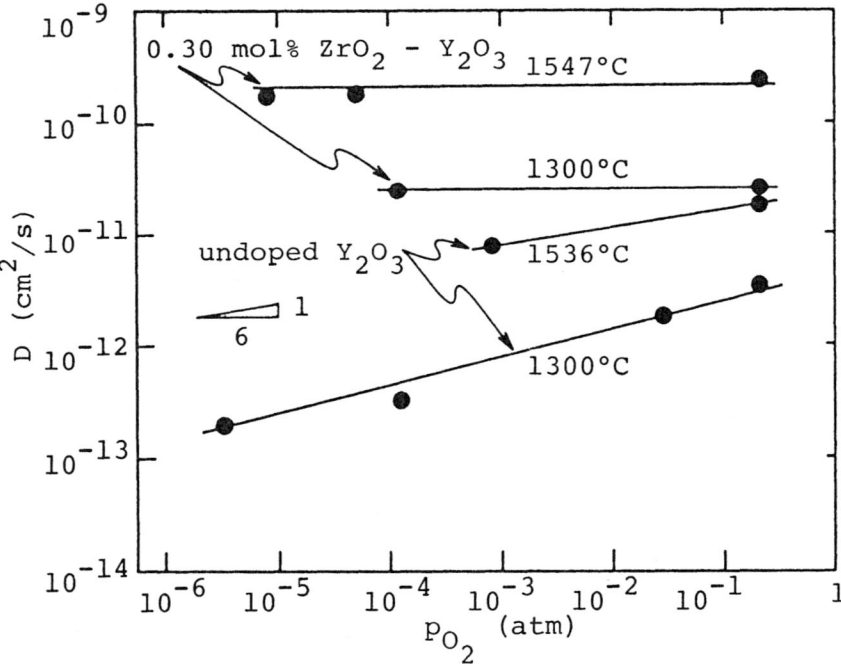

Fig. 2. Oxygen pressure dependences of oxygen self-diffusion coefficients in undoped Y_2O_3 and 0.3 mol% ZrO_2-doped Y_2O_3.

C) Oxygen Self-Diffusion Coefficients in ZrO_2-doped Y_2O_3

Oxygen self-diffusion coefficients in ZrO_2-doped Y_2O_3 under an oxygen pressure of 150 Torr are shown in Fig. 3 along with the self-

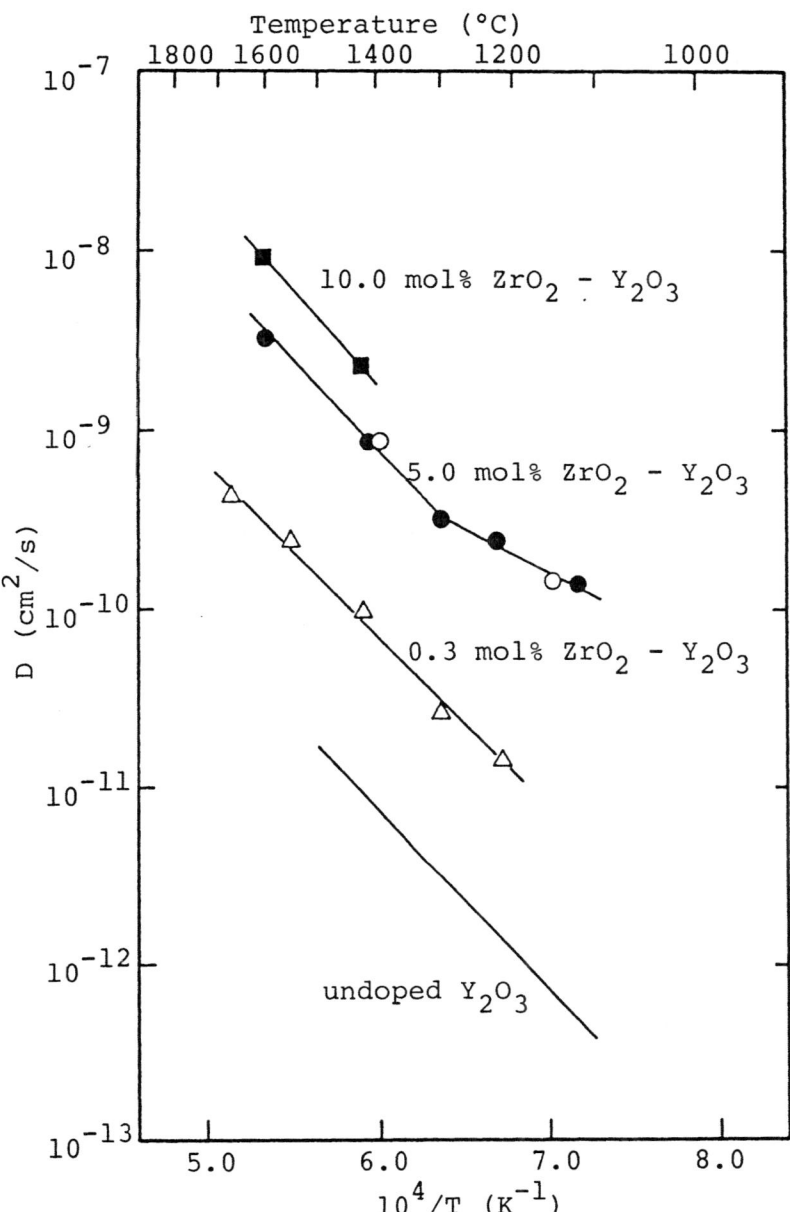

Fig. 3. Self-diffusion coefficients of oxygen ion in ZrO_2-doped Y_2O_3 and undoped Y_2O_3.

diffusion coefficient in undoped Y_2O_3. The temperature dependence of the self-diffusion coefficient in 0.3 mol% ZrO_2-doped Y_2O_3 single crystal in the temperature range 1215° to 1670°C is given by

$$D = 7.5 \times 10^{-5} \exp [- 192 \; (kJ/mol)/RT] \; cm^2/s. \qquad (2)$$

For the 5.0 mol% ZrO_2-doped Y_2O_3, closed circles are the results determined for polycrystalline samples and open circles are those for the single crystal, showing no significant difference. A break was observed at 1300°C in the temperature dependence of the self-diffusion coefficient. The break was proved to be due to dendritic precipitation observed in the diffusion sample after annealing at 1200 °C under an optical microscope; the low temperature region with a smaller activation energy corresponds to the heterogeneous microstructure introduced by some phase separation. The temperature dependence of the self-diffusion coefficient in the high temperature range 1300° to 1600°C is expressed as

$$D = 6.3 \times 10^{-4} \exp [- 189 \; (kJ/mol)/RT] \; cm^2/s. \qquad (3)$$

For the 10.0 mol% ZrO_2-doped Y_2O_3, only two measurements were done using polycrystalline samples at 1422° and 1603°C. The temperature dependence calculated from these two is expressed as

$$D = 6.0 \times 10^{-3} \exp [- 209 \; (kJ/mol)/RT] \; cm^2/s. \qquad (4)$$

DISCUSSION

A) Comparison of Oxygen Diffusivities in Y_2O_3 and Fluorite-Type Oxides

The self-diffusion coefficient of the oxygen ion in single-crystal Y_2O_3 obtained in this work is approximately four orders of magnitude smaller than that obtained by Wirkus et al. (1967) and Berard et al. (1968). This discrepancy is interpreted as the difference of the self-diffusion and chemical diffusion coefficients, as mentioned earlier.

The oxygen self-diffusion coefficient for single-crystal Y_2O_3 obtained in the present work is compared in Fig. 4 with those for fluorite-type oxides. Although the melting points of ThO_2 (3300°C) and UO_2 (2800°C) are higher than that of Y_2O_3 (2450°C), the oxygen self-diffusion coefficient in Y_2O_3 is about three orders of magnitude lower than those in UO_2 (Marin and Contamin, 1969) and ThO_2 in the intrinsic region (Ando et al., 1976). A difference of more than six orders of magnitude of the oxygen self-diffusion coefficient is observed between CaO-stabilized zirconia (Oishi and Ando, 1984) and Y_2O_3.

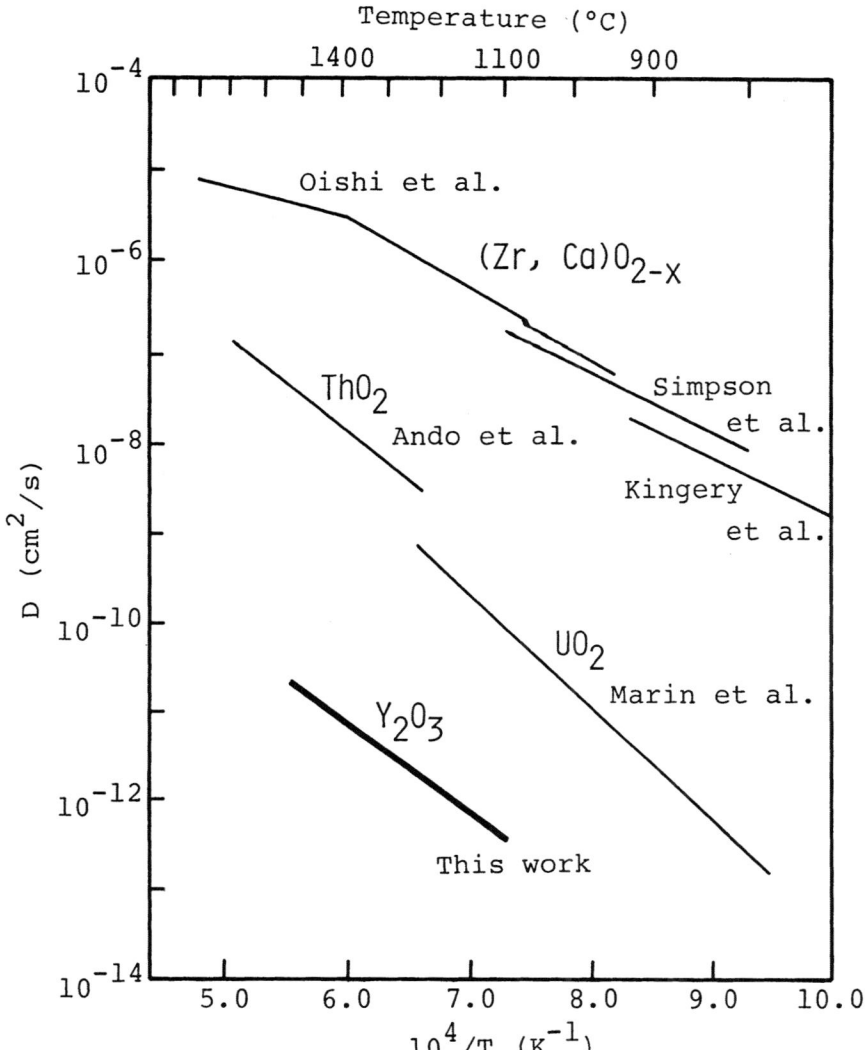

Fig. 4. Comparison of self-diffusion coefficients of oxygen ions in Y_2O_3 and fluorite-type oxides.

The activation energy of oxygen diffusion for Y_2O_3 obtained in this work, 191 kJ/mol, is far greater than those for the stabilized zirconia, 62.8 kJ/mol for $MgO-ZrO_2$, 81.2 kJ/mol for $CaO-ZrO_2$ in the high temperature region (Oishi and Ando, 1984), and 130 kJ/mol for $CaO-ZrO_2$ in the low temperature region (Kingery et al., 1959; Simpson and Carter, 1966), and close to those for In_2O_3, which is isostructural with Y_2O_3, 166 kJ/mol for the interstitial mechanism and 190 kJ/mol for the vacancy mechanism (Wirtz and Takiar, 1981).

Although the crystal structures of the C-type rare-earth oxides and the fluorite-type oxides are similar, diffusion characteristics of the oxygen ions in those structures are different; the structural vacant sites which are in an ordered arrangement do not contribute effectively to the oxygen diffusion in the C-type structure (Ando et al., 1983).

B) Comparison of Self-Diffusion Coefficients of The Oxygen and Yttrium Ions in Single-Crystal Y_2O_3

The yttrium self-diffusion coefficient in Y_2O_3 obtained for the single crystal (Gaboriaud, 1980) is several orders of magnitude smaller than those for polycrystals (Berard and Wilder, 1969). Gaboriaud discussed the discrepancy but reached no decisive conclusion. Such a large difference of the self-diffusion coefficients in single crystal and polycrystal will be due to the enhancement of the cation diffusion at grain boundaries, as is the case for Y_2O_3-doped ZrO_2 (Sakka et al., 1982).

In contrast to the cation diffusion, as shown in Fig. 3, the oxygen self-diffusion in 5.0 mol% ZrO_2-doped Y_2O_3 was not enhanced by grain boundaries; the presence of grain boundaries in ionic crystals enhances diffusion of slower species but not the diffusion of the faster species.

The self-diffusion coefficient of the oxygen ion is several orders of magnitude larger than that of the cation in single-crystal Y_2O_3 as shown in Fig. 5. The activation energy for the oxygen diffusion, 191 kJ/mol, is smaller than that for the yttrium ion, 301 kJ/mol. This relativity is similar to that for fluorite-type oxides, suggesting that the predominant defect in Y_2O_3 is on the oxygen sublattice. This will be discussed later in terms of the oxygen pressure dependence of the oxygen self-diffusion coefficient.

C) Diffusion Mechanisms for Undoped Y_2O_3 and ZrO_2-doped Y_2O_3

The oxygen self-diffusion coefficient for the 0.3 mol% ZrO_2-doped Y_2O_3 independent of the oxygen pressure means that the concentration of defects which contribute to the diffusion does not change in the present oxygen pressure range. On the other hand, the oxygen self-diffusion coefficient in undoped Y_2O_3 increases with increasing oxygen pressure. This means that the defect concentration increases with increasing oxygen pressure. The oxygen self-diffusion coefficient proportional to the oxygen pressure with the exponent of approximately 1/6 implies that the defect type is not $Y_{2-x}O_3$ but Y_2O_{3+x}. This oxygen pressure dependence is consistent with the assumption of anti-Frenkel type defect in pure Y_2O_3.

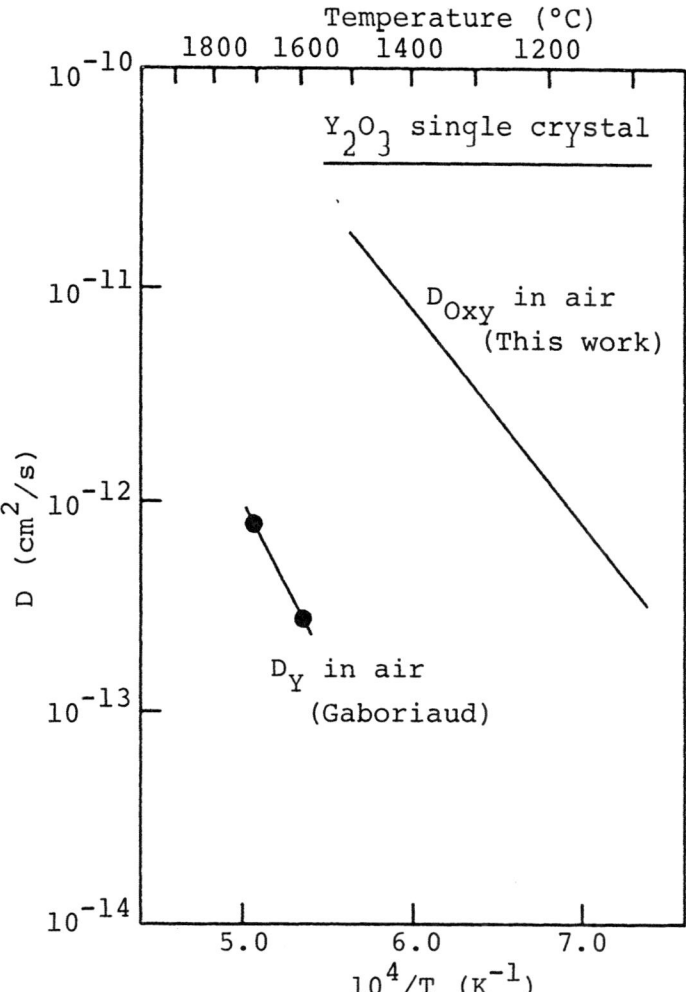

Fig. 5.   Comparison of self-diffusion coefficients of the oxygen
and yttrium ions in Y_2O_3 single crystal.

It is well known that Y_2O_3 is a p-type conductor in the high
oxygen pressure range like in air (Tallan and Vest, 1966; Odier et
al., 1971). As mentioned earlier, the oxygen diffusivity increases
with increasing oxygen pressure. Consequently, under the condition
of a high oxygen pressure, since the concentration of electron holes
equals to the concentration of interstitial oxygen ions, $[O_i'']$, the
$[O_i'']$ can be expressed by Eq. (5),

$$[O_i''] = (1/4)^{1/3} K_{O_i}^{1/3} K_d^{1/3} p_{O_2}^{1/6} \qquad (5)$$

211

where K_{Oi} and K_d are equilibrium constants for reactions (6) and (7), respectively.

$$\frac{1}{2} O_2(g) = O_i^x(s) \tag{6}$$

$$O_i^x(s) = O_i'' + 2h^\cdot \tag{7}$$

Therefore, the oxygen self-diffusion coefficient can be expressed by Eq. (8).

$$D = (1/4)^{1/3} \alpha\, a_o^2\, \nu\, p_{O_2}^{1/6} \exp\left[\frac{\frac{1}{3}(\Delta S_{Oi} + \Delta S_d) + \Delta S_m}{R}\right]$$

$$\exp\left[\frac{\frac{1}{3}(\Delta H_{Oi} + \Delta H_d) + \Delta H_m}{RT}\right] \tag{8}$$

where α is a constant determined by crystal structure and diffusion mechanism, a_o the lattice constant, ν the vibrational frequency, ΔS_{Oi} and ΔS_d entropy changes for reactions (6) and (7), respectively, ΔH_{Oi} and ΔH_d the enthalpy changes for reactions (6) and (7), respectively, and ΔH_m the migration energy for oxygen self-diffusion.

The exponent of 1/6 is consistent with the present experimental results under the constant temperature. Under the condition of a constant oxygen pressure, the activation energy for the diffusion is expressed by Eq. (9).

$$Q_D = \frac{1}{3}(\Delta H_{Oi} + \Delta H_d) + \Delta H_m \tag{9}$$

Odier et al. obtained the value of $(\Delta H_{Oi} + \Delta H_d)$ as 550 kJ/mol by thermoelectronic emission technique. By substituting the diffusion activation energy obtained in this work, $Q_D = 191$ kJ/mol, and $(\Delta H_{Oi} + \Delta H_d) = 550$ kJ/mol into Eq. (9), we obtain the value of ΔH_m as 7 kJ/mol. This calculated migration energy is too small as to the migration energy of the interstitial oxygen ion in Y_2O_3, as compared to $\simeq 90$ kJ/mol reported for UO_2 (Contamin et al., 1972) and 166 kJ/mol reported for In_2O_3 (Wirtz and Takiar, 1981).

It was clarified by Bratton (1969) by measuring lattice constants and densities that the defect structure for system ZrO_2-Y_2O_3 is interstitial oxygen type. Formation of this type defect can be expressed by Eq. (10).

$$2ZrO_2 \xrightarrow{\;Y_2O_3\;} 2Zr_Y^\cdot + 3O_O^x + O_i'' \tag{10}$$

If the interstitial ions introduced by doped ZrO_2 contribute to the oxygen diffusion, the self-diffusion coefficient should increase as the dopant concentration increases (and consequently, the activation energy should be equal to the migration energy of the interstitial oxygen ion). This agrees with the experimentally determined oxygen self-diffusion coefficients for compositions of 0.3, 5.0 and 10 mol% ZrO_2-doped Y_2O_3 as shown in Fig. 3. This is consistent also with the fact that the oxygen diffusivity in ZrO_2-doped Y_2O_3 is independent of the oxygen partial pressure.

We must explain, however, the experimental results that the activation energies for the solid solutions are not smaller as

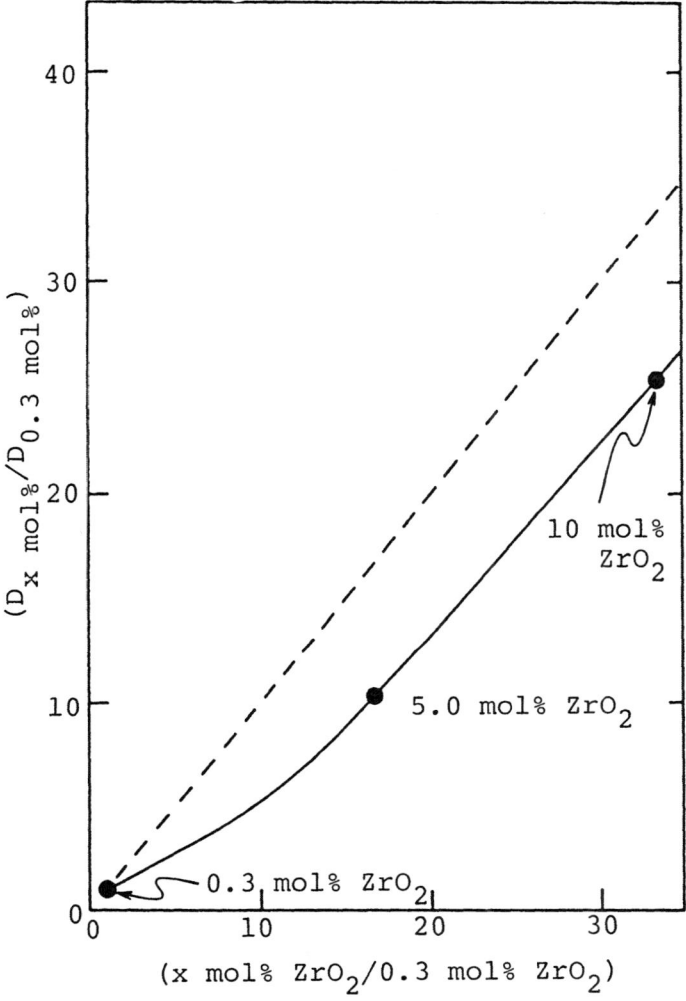

Fig. 6.   Relative oxygen self-diffusion coefficients as a function of relative ZrO_2 concentration.

expected but similar to that in undoped Y_2O_3. Moreover, these values ($\simeq$200 kJ/mol) are much higher compared with those obtained for stabilized zirconia, 60 $\sim$ 80 kJ/mol at high temperatures (Oishi and Ando, 1984) and $\simeq$130 kJ/mol at low temperatures (Kingery et al., 1959; Simpson and Carter, 1966) and that for hyper-stoichiometric uranium dioxide, $\simeq$90 kJ/mol (Contamin et al., 1972), and rather close to the interstitial oxygen migration energy for In_2O_3, 166 kJ/mol (Wirtz and Takiar, 1981).

To clarify the above question, the relative oxygen self-diffusion coefficients for 5.0 and 10 mol% ZrO_2-Y_2O_3 against the 0.3 mol% ZrO_2-doped Y_2O_3 are plotted as a function of the relative ZrO_2 concentration as shown in Fig. 6. The closed circles are the results calculated using the self-diffusion coefficients determined at 1500°C. If all the introduced interstitial ions are free, the relative diffusivity should go on along the broken line. The curved experimental results, however, suggest that some of the interstitial oxygen ions are associated with doped zirconium ions and do not contribute to the diffusion; the diffusion activation energies for the ZrO_2-Y_2O_3 solid solutions involve energy for some defect association, ΔH_a, which is not necessarily involved in the case of pure Y_2O_3.

$$Q_D = \Delta H_m + \Delta H_a \tag{11}$$

The similar values of the diffusion activation energies found for undoped and ZrO_2-doped Y_2O_3 are understood to be a coincidence.

ACKNOWLEDGMENT

The authors wish to thank K. Kitazawa for his preparation of single crystals. The authors thank H. Hase and K. Sogabe for the help in experiment.

REFERENCES

K. Ando, Y. Oishi and Y. Hidaka, J. Chem. Phys. 65:2751 (1976).
K. Ando, Y. Oishi, H. Hase and K. Kitazawa, J. Am. Ceram. Soc. 66:C-222 (1983).
M. F. Berard, C. D. Wirkus and D. R. Wilder, J. Am. Ceram. Soc. 51:643 (1968).
M. F. Berard and D. R. Wilder, J. Am. Ceram. Soc. 52:85 (1969).
R. J. Bratton, J. Am. Ceram. Soc. 52:213 (1969).
P. Contamin, J. J. Bacmann and J. F. Marin, J. Nucl. Mat. 42:54 (1972).
R. J. Gaboriaud, J. Solid State Chem. 35:252 (1980).
W. D. Kingery, J. Pappis, M. E. Doty and D. C. Hill, J. Am. Ceram. Soc., 42:393 (1959).

K. Kitazawa, K. Nagashima, T. Mizutani, K. Fueki and T. Mukaibo, J. Cryst. Growth 39:211 (1977).

J. F. Marin and P. Contamin, J. Nucl. Mat., 30:16 (1969).

P. Odier, J. P. Loup and A. M. Anthony, Rev. Int. Hautes Temp. Refract. 8:243 (1971).

Y. Oishi and K. Ando, in the present proceedings (1984).

J. C. Rifflet, P. Odier and A. M. Anthony and J. P. Loup, J. Am. Ceram. Soc., 58:493 (1975).

Y. Sakka, Y. Oishi and K. Ando, J. Mat. Sci. 17:3101 (1982).

L. A. Simpson and R. E. Carter, J. Am. Ceram. Soc. 49:139 (1966).

N. M. Tallan and R. W. Vest, J. Am. Ceram. Soc. 49:401 (1966).

C. D. Wirkus, M. F. Berard and D. R. Wilder, J. Am. Ceram. Soc. 50:113 (1967).

G. P. Wirtz and H. P. Takiar, J. Am. Ceram. Soc. 64:748 (1981).

DETERMINATION OF OXYGEN DIFFUSIVITIES IN β AND β" ALUMINA BY $^{18}O/^{16}O$ EXCHANGE

A.E. McHale, J.A. Kilner and B.C.H. Steele

Wolfson Unit for Solid State Ionics
Imperial College
London, SW7, 2BP

ABSTRACT

Mode II failure or slow degradation of Na β-Alumina and Na β"-Alumina requires the motion of both oxygen point defects and electronic carriers if it is to lead to the ultimate failure of the electrolyte. The oxygen diffusivity in these materials has been measured using the technique of oxygen isotopic $^{18}O/^{16}O$ exchange followed by examination using negative ion Secondary Mass Spectroscopy (SIMS) under equilibrium conditions over the temperature range 350°C-500°C within the conduction plane, perpendicular to the c-axis. The results are discussed with reference to the ionic defect structures that have been proposed for non-stoichiometric Na β-Alumina and Na β"-Alumina.

INTRODUCTION

Na β-Alumina and Na β"-Alumina are structurally related compounds which have been developed in recent years as cation superionic conductors for use as solid electrolytes. Na β-Alumina, with stoichiometric formula $Na_2O \cdot 11Al_2O_3$, has a hexagonal crystal structure characterized by an alumina "spinel-block" layers separated by low occupancy Na-planes perpendicular to the c-axis. These planes consist of hexagonal array of chemically equivalent Na-sites of which only half are normally occupied. To enhance superionic conduction, non-stoichiometric material containing an excess of sodium is used. Incorporation of excess Na_2O in Na β-Alumina can <u>be described</u> via the reaction[*]

---

[*]Kroger-Vink notation has been used, with fully ionized defects indicated throughout. The actual ionization states of ionic defects in these materials have not been determined.

217

$$[1] \quad Na_2O \rightarrow 2\,Na^+_{i(in\ plane)} + O_i$$

Excess sodium can be accommodated in the pre-existing Na-type sites in the conduction plane; the excess oxygen as interstitials is also accommodated in the conduction plane at the "mid-oxygen" site, forming a stable defect complex with off-site aluminium ions in the adjacent spinel blocks (1,2). The structure of the conduction plane in Na β-Alumina is shown schematically in Figure 1a.

The related Na β"-Alumina structure is rhombohedral and differs from Na β-Alumina in that the spinel blocks contain "stabilizing" Mg^{2+} in addition to Al^{3+}. The conduction plane is still a hexagonal-type array of Na^+ sites but these are now fully occupied at the stoichiometric composition $Na_2O \cdot MgO \cdot 5Al_2O_3$. The structure of the conduction plane is shown in Figure 1b. Deviation from stoichiometry to sodium deficiency is necessary to maximize superionic Na^+ conduction, resulting in a general formula $(1-x)Na_2O \cdot (1-2x)MgO \cdot (5+x)Al_2O_3$ ($x \leq 0.17$), where defect compensation is generally assumed to be via a reaction of the type

$$[2] \quad Al_2O_3 \rightarrow 2Al^\circ_{Mg} + 2V'_{Na} + 3O^x_O$$

However, recent structural work has suggested that the defect compensation mechanism may be more complex, involving substitutional disorder on the Al-sublattice and the possibility of oxygen vacancy formation (3). Electron microscopy has also indicated that complex planar defects may form to accommodate high degrees of nonstoichio-

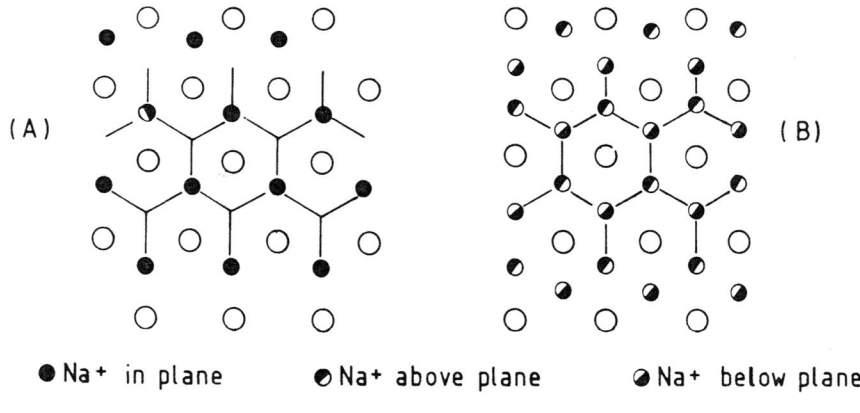

● Na⁺ in plane   ◐ Na⁺ above plane   ◐ Na⁺ below plane

○ Oxygen

Figure 1. (a) Schematic projection of the conduction plane in Na β-Alumina, (b) as above, for Na β"-Alumina.

metry leading to a local collapse of the conduction plane structure (4).

The failure mechanisms observed in Na β-Alumina and Na β"-Alumina have been discussed by many investigators and are broadly divided into two basic modes; Mode I is a catastrophic failure resulting from the combined effects of mechanical surface flaws, surface impurities and high (local) current loading during cell operation. Mode II "slow degradation" is a progressive reduction of the electrolyte accompanied by increases in the electronic conductivity (5). The stages of Mode II degradation in Na β-Alumina have been described in terms of electrical and chemical modifications as detected by ESR (6) and correlated to physical observation (7). Similar systematic studies of the process of Mode II failure in Na β"-Alumina has not been reported in the literature. Degradation proceeds in two related stages; an initial coloration due to electrons "trapped" at impurity cation defects in the spinel blocks (noted in contact with high Na activity environments) progressing to the formation of colloidal Na with the evolution of $O_{2(gas)}$ if an electric field is applied. Coloration of the electrolyte proceeds isotropically through the crystal from the surface in contact with the high Na activity and is reversible, but this "bleaching" process is observed to proceed only in the direction of the conduction planes and only in the presence of oxygen (8).

Speculation on the role of oxygen transport in slow degradation, as shown by the "bleaching" studies on Na β-Alumina, has led to the proposal that oxygen transport is predominantly within the conduction plane via mobile oxygen interstitials. Extension of these speculations to the much less studied case of Na β"-Alumina has led to the expectation of a relatively higher resistance to reduction and degradation in the latter material as oxygen defects are not generally thought to be created for charge compensation of nonstoichiometry.

To obtain a direct measure of the possible role of oxygen transport in the slow degradation process in Na β-Alumina and Na β"-Alumina, the oxygen diffusivities within the conduction plane have been measured in single crystals of both materials.[*]

---

[*]The contribution of oxgyen transport parallel to the c-axis (perpendicular to the conduction plane or across the spinel blocks) was discounted in this study as a result of preliminary measurements which indicated no detectable oxygen transport in this direction.

EXPERIMENTAL

A. Oxygen Diffusivity Determination via Isotopic $^{18}O/^{16}O$ Exchange.

The technique of oxygen isotopic exchange followed by analysis using negative ion Secondary Mass Spectroscopy (SIMS) used in this study has been described previously by Kilner et al. (9). Isotopic exchange is performed via an equilibrium anneal of the specimen in an enriched atmosphere of the stable isotope ^{18}O. The resulting isotopic enrichment profile, $^{18}O/(^{18}O + ^{16}O)$, is then determined by SIMS profiling as a function of sputter depth at a sampling rate of ∿ 100 data points per micron total sputter depth.

The appropriate solution to the diffusion equation corresponding to the experimental conditions is that for a semi-infinite medium where the boundary conditions must be related to the rate of transfer of the diffusing species at the surface of the specimen. The assumption is made that the rate of exchange is proportional to the difference between the concentration in the gas and the concentration in the surface at any time, resulting in the boundary condition,

$$[3] \quad -D(dC/dx)_{x=o} = K(C_g - C_s)$$

The solution to the diffusion equation then follows as (10),

$$[4] \quad C_x = (C - C_{bg}) / (C_g - C_{bg})$$
$$= erfc[x/2(Dt)^{\frac{1}{2}}] -$$
$$exp(hx + h^2Dt)erfc[x/2(Dt)^{\frac{1}{2}} + h/(Dt)^{\frac{1}{2}}]$$

where   C = isotopic concentration at depth x.
        C_{bg} = isotopic background concentration.
            (C_{bg} = 0.2% at mean sea level)

and   h = K/D where K is the surface exchange coefficient.

B. Exchange Specimen Preparation

Small crystals of Na β-Alumina* and Na β"-Alumina** were prepared for this study with flat (hkO) faces (i.e. perpendicular to the conduction plane) to enable direct depth profiling in the direction of the conduction plane. As this is also a strong cleavage plane in both materials, special care is required in the preparation of these faces without the development of cleavage cracks which would interfere with the isotopic profile measurement. To avoid such cracking and facilitate handling of the small crystals

---

*Na β-Alumina crystal obtained from AERE, Harwell (U.K.).
**Mg-stabilized Na β"-Alumina crystal supplied by J. Bates, Oak Ridge National Laboratory, (U.S.A.).

used, a mounting scheme was devised in which the crystals were
encapsulated in a "solder glass" (Corning 7578) having a softening
temperature $\sim$ 550°C. The general specimen dimensions and configuration are shown in Figure 2. The thermal expansion coefficient of
the glass ($\sim 8 \times 10^{-6}$) was selected as slightly greater than the
approximate average thermal expansion coefficient of Na β-Alumina
to ensure that the crystals would be held in a slightly compressive
stress state when cooled to room temperature for slicing and
polishing. The low softening temperature of the glass was such as
to ensure no sodium loss from the crystals during mounting. Several millimeter-thick wafers were cut from each prepared crystal.
The wafers were then polished on one face only to a 1/4μm diamond
finish, cleaned and annealed at $\sim$ 525°C for several hours in air
prior to the exchange experiments. No chemical polishes or
etchants were employed in specimen preparation and anhydrous solvents and lubricants were used throughout diamond polishing.

Isotopic $^{18}O/^{16}O$ exchange was performed two specimens at a
time, one Na β-Alumina and one Na β"-Alumina being given identical
treatment. The exchange apparatus has been described previously
by Kilner et al.(9), with modification prior to these experiments
to enable continuous monitoring of the gas phase composition using
a quadripole mass spectrometer*. Prior to ^{16}O equilibration and
^{18}O exchange, the specimens were vacuum baked at $P_{total} \sim 10^{-3}$ torr
and $\sim$ 495°C overnight to remove residual adsorbed H_2O and trace
volatile contaminants from cleaning solvents. Sample temperature

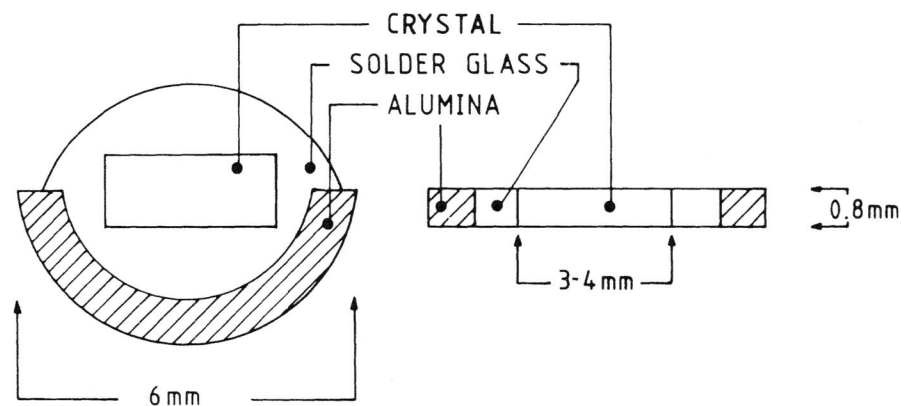

Figure 2. Mounting configuration plan and cross-section of diffusion specimens. All dimensions are approximate.

---

*Spectralab SX200, VG Gas Analysis, Cheshire.

was then adjusted to the desired experimental temperature and ^{16}O (P_{total} = 200 torr, ^{18}O concentration = C_{bg}) introduced. After the equilibration period ($t_{equilibration} > t_{exchange}$), the samples were cooled quickly to $\sim$ 50°C and the system re-evacuated. A pressure of ^{18}O (P_{total} = 200 torr, 2-3% ^{16}O) was then introduced into the cooled system, and the samples were rapidly reheated to the experimental temperature. After the desired exchange period, the samples were again cooled quickly to room temperature, the ^{18}O collected and the samples removed from the exchange apparatus to be stored in a vacuum desiccator prior to SIMS analysis. The experimental conditions used in these exchange experiments have been summarized in Table 1.

Isotopic profiles were obtained by using the negative ion SIMS technique previously described. A rastered Ar^+ primary beam is employed together with a gated area sampling technique to avoid crater edge effects. Low energy e^- flooding of the specimen surface was used to avoid sample charging. The data were analysed using a standard multiple regression technique to obtain a fit to Eq. [4]. Best fit was determined by a minimization of the unweighted sum of the residuals squared. No weighting of the input data was used in this fitting procedure.

RESULTS AND DISCUSSION

The diffusion profiles obtained in this work correspond most usually to the case of very slow surface exchange kinetics coupled with relatively rapid oxygen diffusivity. A typical example diffusion profile is given in Figure 3. This type of exchange behaviour

Table 1. Summary of experimental conditions during
          isotopic ^{18}O exchange anneals.

T (°C)	t (min)*	P ^{18}O (torr)
350±5	960	205±5
426±2	180	200±2
498±5	60	202±2

* Time given is uncorrected for heat-up and cool-down periods and represents the actual time at T $\geq$ (T_{exp} -20)°C.

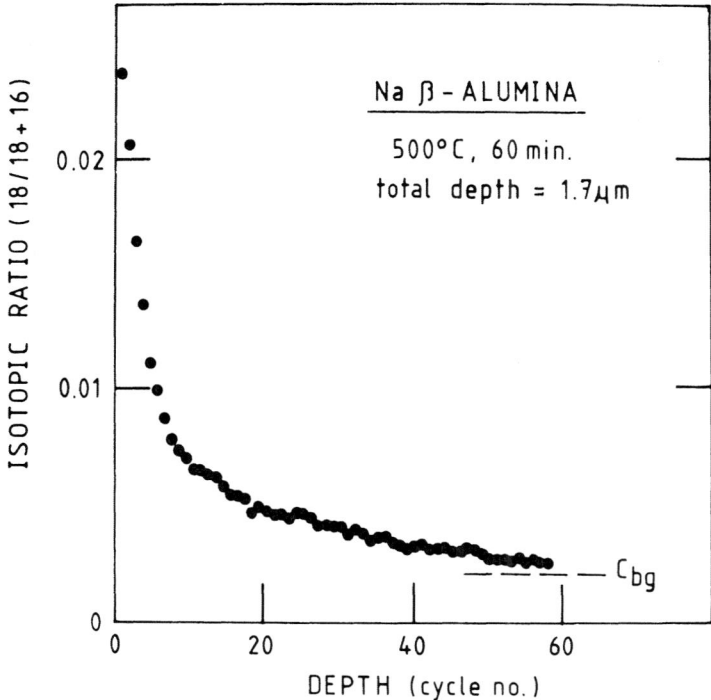

Figure 3. Example of diffusion profile obtained. (Na β-Alumina; 500°C, 60 min).

is well treated using the solution to the diffusion equation given previously as Eq. [4].

The oxygen diffusion coefficients calculated using Eq. [4] from the data for Na β-Alumina and Na β"-Alumina are shown in Figures 4 and 5, respectively. The error bars indicated correspond to the maximum experimental uncertainties expected in anneal time, temperature and sputtered crater depth during SIMS analysis. The large error bar for the Na β-Alumina sample annealed at 425°C results from not only these systematic errors but also from the effect of an additive second component in the diffusion profile due probably to residual surface damage or surface microcracking. While such a two stage process is not adequately modelled by Eq. [4], this preliminary study yielded insufficient data to enable development of a more satisfactory mathematical form for this case of simultaneous diffusion processes.

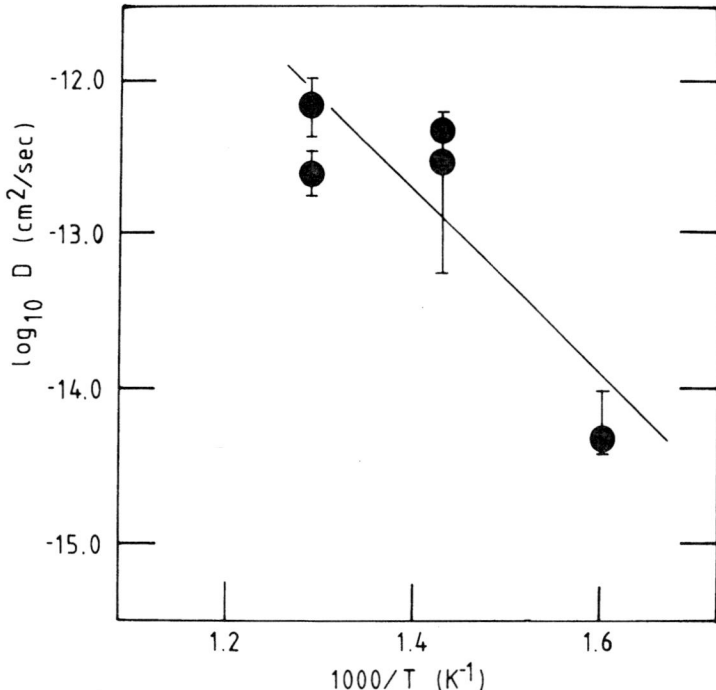

Figure 4. Oxygen diffusivity as a function of 1/T for Na β-Alumina

The apparent self diffusion coefficients for oxygen are calculated as the best linear fit to the Arrhenius plots of the experimental data. This leads to the result for Na β-Alumina of D = 1.2 x 10^{-5} exp (-1.1 ev/kT) cm^2/sec and for Na β"-Alumina of D = 2.2 x 10^6 exp (2.6 ev/kT) cm^2/sec*. The magnitude of the experimental oxygen diffusivity in Na β-Alumina agrees well with an estimate prepared by DeJonghe et al.(11) as an upper bound for interstitial diffusion in this material. The experimental activation energy is also appropriate for this mechanism.

In contrast, the observed apparent activation energy for oxygen diffusion in Na β"-Alumina is rather higher than would be expected for an extrinsic interstitial mechanism. Consideration of the relatively close packed conduction plane structure in Na β"-Alumina also weighs against the importance of an interstitial mechanism. Considering the relatively high magnitude of the

---

*It should be noted that the apparent self diffusion coefficient for Na β-Alumina is calculated as 3.2 x 10^{-5} exp(1.2ev/kT) cm^2/sec if the 425° anneal data is excluded.

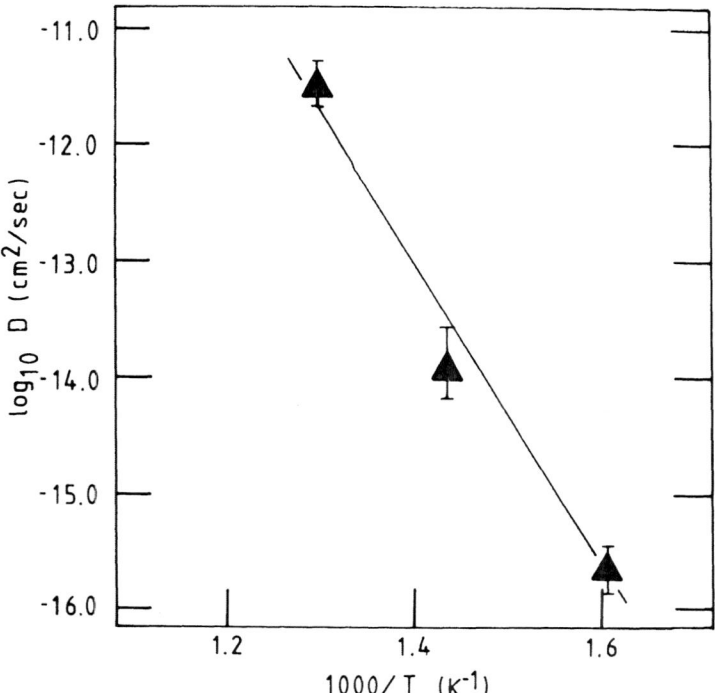

Figure 5. Oxygen diffusivity as a function of 1/T for Na β"-Alumina

apparent oxygen diffusivity, the relatively large activation
energy observed for oxygen transport may well include a defect
formation enthalpy as would be expected in the case of a vacancy
diffusion mechanism.

At present, insufficient study of oxygen defect structures
and transport in Na β"-Alumina hinder the further interpretation
of our results on this material. Further systematic study of
oxygen transport in Na β"-Alumina is planned to attempt to elucidate the operative defect mechanisms in this material.

REFERENCES

1. J.M. Newsam and B.C. Tofield, Solid State Ionics, 5:59 (1981).
2. W.A. England, A.J. Jacobson and B.C. Tofield, Solid State
      Ionics, 6:21 (1982).
3. K.G. Frase, J.O. Thomas and G.C. Farrington, Solid State
      Ionics '83, 307 (1983)
4. L.C. DeJonghe, Acta Cryst., A36:831 (1980).

5. D.S. Parker, R.W. Powers and M.W. Breiter, Solid State Ionics, 5:271 (1981).
6. J.P. Barret, D. Gourier and D. Vivien, Solid State Ionics, (in press).
7. L.C. DeJonghe, L. Feldman and A. Beuchele, J. Mat. Sci., 16:780 (1981).
8. L.C. DeJonghe and A. Beuchele, J. Mat. Sci., 17:885 (1982).
9. J.A. Kilner, B.C.H. Steele and L. Ilkov, Solid State Ionics, 12:89 (1984).
10. J. Crank in The Mathematics of Diffusion, Clarendon Press, Oxford (1956).
11. L.C. DeJonghe, A. Beuchele and M. Armand, Solid State Ionics '83, Part 1, 165 (1983).

CHEMICAL DIFFUSION IN CoO

Janusz Nowotny

Institute of Catalysis
and Surface Chemistry
Polish Academy of Sciences
ul. Niezapominajek
30-239 Krakow, Poland

Andrzej Sadowski

Institute of Metallurgy
Academy of Mining and
  Metallurgy
al. Mickiewicza 30
30-059 Krakow, Poland

INTRODUCTION

The available experimental data for the chemical diffusion in CoO[1-12] can be grouped according to the reported dependence of $\tilde{D}$ on P_{O_2} and the slope of log $\tilde{D}$ vs. $1/T$ (Table I). The character of the dependence of $\tilde{D}$ vs. P_{O_2} is important for understanding the nature of interactions between diffusing defects. When defects form an ideal solution in the lattice with neglegible interactions, i.e. undoped NiO[13], then $\tilde{D}$ should be essentially independent of defect concentration and P_{O_2}. This type of behavior has been reported for CoO by Mrowec et al.[4], Wimmer et al.[5], Chadzhieva et al.[8] and Chowdhry and Coble[9]. However, Morin and Dieckmann[6,10] and Petot-Ervas et al.[1,12] have observed changes in $\tilde{D}$ with P_{O_2}, which are consistent with strong interactions between defects in CoO[7,15-17].

Conflicting data have also been reported for the $\tilde{D}$ vs $1/T$ dependence. One group reports values for the activation energy between 95-100 kJ mol^{-1} [1-5,9]. Smaller values are observed with polycrystalline samples, but in this case a contribution from a near-surface transport may be involved[9], as it has been shown that the transport of defects within near-surface layers of CoO is much faster than that in the crystalline bulk[18]. Recently Chadzhieva et al.[8] and Petot-Ervas et al.[11] have reported significantly higher values of the activation energy ranging between 125 and 132 kJ mol^{-1}. Furthermore, data for $\tilde{D}$ calculated from the tracer diffusion of Co in CoO give an activation energy of 136 kJ mol^{-1} [7]. It is difficult to explain such marked differences in the diffusion data.

227

TABLE I

Literature data of chemical diffusion coefficient and diffusion coefficient of defects in CoO

No.	Authors	$\tilde{D}$ vs. 1/T; [$\tilde{D}$ in cm^2s^{-1}]; [E_a in $J\ mol^{-1}$]	Temperature Range [deg. K]
1	Price and Wagner[1]	$\tilde{D} = 4.33 \times 10^{-3}$ exp $(-100.480/RT)$	1073-1373
2	Meurer[2]	$\tilde{D} = 4.4 \times 10^{-3}$ exp $(-100.480/RT)$	
3	Koel, Gellings[3]	$D_{V'_{Co}} = 8.3 \times 10^{-3}$ exp $(-100.480/RT)$	1223-1523
4	Koel, Gellings[3]	$D'_{V_{Co}} = 2.1 \times 10^{-2}$ exp $(-108.860/RT)$	1223-1523
5	Fryt, Mrowec, Walec[4]	$\tilde{D} = 6.6 \pm 0.6 \times 10^{-3}$ exp $(-102.160/RT)$	1273-1524
6	Wimmer, Blumenthal Bransky[5]	$\tilde{D} = 4.8 \times 10^{-3}$ exp $(-94.200/RT)$	1173-1573
7	Morin[6]	$\tilde{D} = 5.5 \times 10^{-7}$ $10^2 Pa - 1.2 \times 10^{-6}$ $1 \times 10^5 Pa$	1273
8	Dieckmann[7]	$\tilde{D} = 0.113$ exp $(-136.000/RT)$	1273-1673
9	Chadzijeva[8]	$\tilde{D} = 14.28 \times 10^{-2}$ exp $(-133.140/RT)$	1188-1472
10	Chadzijeva[8]	$\tilde{D} = 5.04 \times 10^{-2}$ exp $(-123.510/RT)$	
11	Chowdhry, Coble[9]	$\tilde{D} = 9.31 \times 10^{-3}$ exp $(-98.950/RT)$	1073-1313
12	Chowdhry, Coble[9]	$\tilde{D} = 4.07 \times 10^{-3}$ exp $(-80.960/RT)$	
13	Morin, Dieckmann[10]	$\tilde{D} = 5.2 \times 10^{-6}$ $3 \times 10^{-2} Pa$ $-8.6 \times 10^{-7}\ 8 \times 10^4 Pa$	1273
14	Petot-Ervas[11,12]	$\tilde{D} = 0.13$ exp $(-125.560/RT)$	1273-1673
15	Petot-Ervas[11,12]	$\tilde{D} = 0.17$ exp $(-131.880/RT)$	
16	Petot-Ervas[11,12]	$\tilde{D} = 1.4 \times 10^{-2}$ exp $(-113.040/RT)$	

Table I. (continued)

No.	P_{O_2} Range [Pa]	Method	$\tilde{D}$ vs. P_{O_2}	Remarks
1	$3 \times 10^{-1} - 0.5 \times 10^5$	electrical conductivity		
2		tensivolu-metric		
3		thermogravi-metry		
4		electrical conductivity		
5	$3 \times 10^2 - 10^5$	Rosenburg	No dependence	
6	$1 - 10^5$	thermogravi-metry	No dependence	
7	$10^2 - 10^5$	electrical conductivity	$\tilde{D}$ increases vs. P_{O_2}	
8		tracer diffusion and electrical conductivity		
9	$10^2 - 10^5$	electrical conductivity	No dependence	Reduction
10				Oxidation
11	$10 - 10^5$	electrical conductivity thermogravimetry	No dependence	Single crystal
12	$10 - 10^5$	electrical conductivity thermogravimetry		Polycrystal-grain size 13.5 μm
13	$10^2 - 10^5$	electrical conductivity	$\tilde{D}$ increases vs. P_{O_2}	
14	$60 - 2.1 \times 10^4$	electrical conductivity	$\tilde{D}$ increases vs. P_{O_2}	
15	$60 - 10^5$	electrical conductivity		oxidation
16	$60 - 2.1 \times 10^4$	electrical conductivity	Li-doped CoO	Li-doped CoO

This short survey of the literature data for $\tilde{D}$ in CoO indicates
that the transport mechanism of defects in this material is still
unresolved. The above discrepancies stimulated our re-investiga-
tions of the temperature and oxygen pressure dependence of $\tilde{D}$ in
undoped CoO.

CoO is a metal deficient oxide exhibiting p-type conductivity
in which cation vacancies are the major defects. Predominant
diffusion of cobalt ions in the CoO lattice occurs by a vacancy
mechanism. Interstitial cations were also reported for CoO[12] but
their existence was concluded from speculations based on a comparison of different kinds of diffusion data. According to theoretical
considerations, their concentration should be appreciable only at
very low values of P_{O_2}[7,19].

The chemical diffusion coefficient is usually determined by
relaxation methods which have been described in detail in several
publications[1-3,5,8-11,14-16,20]. This method consists in an isothermal change of P_{O_2} over an initially equilibrated sample which
establishes a new defect concentration at the surface. The rate of
propagation of the defect concentration into the crystal is monitored by a physical property which is proportional to defect concentration. Both thermogravimetry and electrical conductivity
measurements have been used to monitor the equilibration kinetics.
The work-function method has also been applied in studies of near-
surface transport[18]. $\tilde{D}$ may be calculated using two approximate
solutions of Fick's second law:

1. Parabolic equation:

$$\left(\frac{\Delta w_t}{\Delta w_\infty}\right)^2 = \frac{4}{\Pi} t \left(\frac{q}{\delta}\right)^2 \tilde{D} \tag{1}$$

where:

Δw_t = weight change in time t,

Δw_∞ = total weight change between two equilibrium states,

q = surface area,

δ = volume.

This equation is valid for shorter re-equilibration times when

$\Delta w_t / \Delta w_\infty < 0.7$

2. Logarithmic equation:

$$1 - \frac{\Delta w_t}{\Delta w_\infty} = \frac{512}{\Pi^6} \exp \left[\left(-\frac{\tilde{D}\, t\, \Pi^2}{4} \right) \left(\frac{1}{a^2} + \frac{1}{b^2} + \frac{1}{c^2} \right) \right] \qquad (2)$$

where:

2a, 2b, 2c are crystal dimensions.

This equation is valid for longer times when $\Delta w_t / \Delta w_\infty > 0.5$.

The most direct method is thermogravimetry from which one can directly calculate changes in the crystal nonstoichiometry. The change in defect concentration can also be related to the conductivity change if the resulting electron carrier mobility is independent of P_{O_2}[1]. Both thermogravimetry and electrical conductivity have been verified in diffusion studies of CoO leading to identical results[5,9].

It is also important that the experimental change of P_{O_2} should remain within the stability range of a given ionization degree of the predominant defects. In the case of CoO, we should therefore avoid oxygen pressures corresponding to the transition region between singly and doubly ionized cation vacancies. At higher oxygen pressures, the cation vacancies in CoO are singly charged, while appreciable concentrations of doubly ionized cation vacancies are present below 0.1 Pa[21]. Fig. 1 illustrates defect concentrations in CoO vs. log P_{O_2} at 1300°K, calculated from Seebeck effect and electrical conductivity data[17]. Singly ionized cation vacancies are predominant above 0.1 Pa, in agreement with data of Fisher and Tannhauser[21].

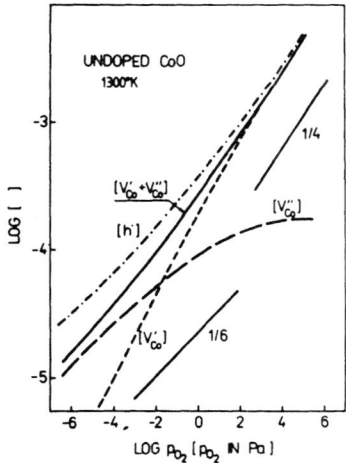

Fig. 1. Concentration of defects in undoped CoO at 1300°K [17].

It is generally assumed that the reequilibration in CoO crystals is rate controlled by the diffusional transport of cobalt vacancies from the surface into the bulk (oxidation) or from the bulk to the surface (reduction). A purely bulk controlled diffusion mechanism requires that the oxidation kinetics be identical to the reduction and that the near-surface diffusive resistance be negligible[22-23]. Moreover, the diffusion data should be self-consistent, i.e. Eq. 1 and Eq. 2 should provide identical values of $\tilde{D}$. Such agreement has been claimed to occur[5,9] but the diffusion data have not been reported. Thus the purpose of this work is to verify the chemical diffusion data for CoO.

EXPERIMENTAL

A rectangular sample (2.0 x 1.5 x 9.0 mm) was cut from a single crystal boule of CoO grown by the Vernuille method[24]. Results of the mass spectrographic analysis were already reported[24].

The four-probe dc conductivity method was applied. Details of the experimental set-up has been already described[14]. Temperature was controlled with an accuracy ± 2°K. Air or Ar-O_2 gas mixtures were used to establish the required oxygen activity in the reaction chamber. The CoO crystal was put into an appropriate holder[17] in a vertical alundum tube furnace. The reaction gas mixture was passed over the sample with a velocity of 0.9 cm s^{-1}. The electrical conductivity was monitored after each change of P_{O_2} until a new equilibrium was reached. The measurements were performed at P_{O_2} between 3 and 10^5 Pa and temperatures between 1213-1423°K.

The crystal was initially equilibrated at a given T and P_{O_2}. Then the P_{O_2} was changed isothermally by admitting a new gas mixture which reached the crystal in 1.5 min. The equilibration kinetics were measured for oxidation and reduction, i.e. when P_{O_2} was increased and decreased, respectively. At least three oxidation and three reduction experiments were performed at each T.

RESULTS AND DISCUSSION

Fig. 2 illustrates the dependence of log σ upon log P_{O_2}. The power of P_{O_2} dependence $1/n = 1/4$ indicates singly ionized cobalt vacancies as the predominant defects in the experimental P_{O_2} range. The electrical conductivity data are identical for both reduction and oxidation, which indicates that the sample was well equilibrated.

As theoretically predicted, the re-equilibration data follow Eq. 1 at shorter times and Eq. 2 at longer times (Fig. 3 and 4). The dotted lines in Fig. 3 and 4 indicate the ranges in which experi-

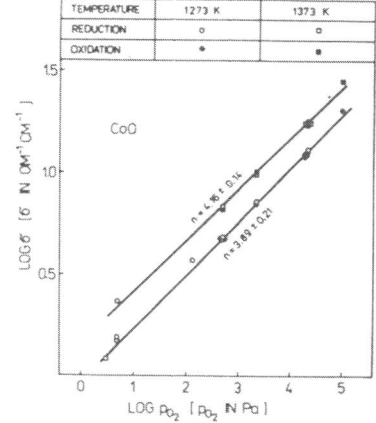

Fig. 2. Log σ vs. log P_{O_2}

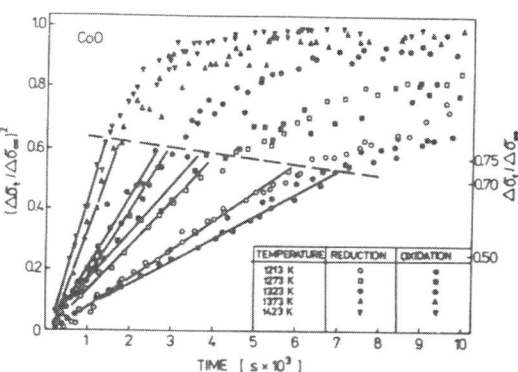

Fig. 3. Parabolic plot of electrical conductivity changes vs. time

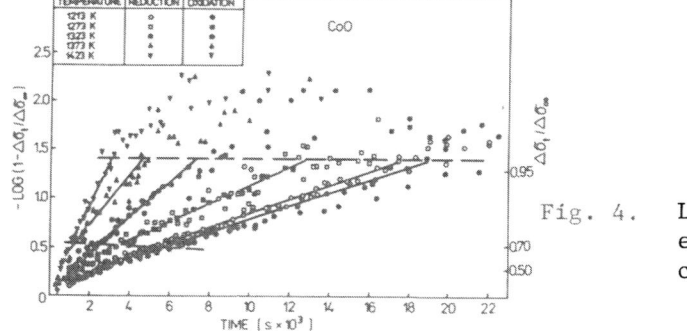

Fig. 4. Logarithmic plot of electrical conductivity changes vs. time

mental data fit the kinetic equations. At 1213°K the kinetic data for oxidation and reduction runs are slightly different, but any observed differences at higher temperatures are within the experimental error.

The determined values of $\tilde{D}$ at 1213°K are slightly higher for reduction than those for oxidation. The average data of $\tilde{D}$ have been calculated independently for oxidation and reduction. The least squares analysis leads to the following expressions:

logarithmic equations:

1. $\tilde{D}_{red} = 5.25 \ 10^{-2} \ \exp \left(- \dfrac{116,030 \pm 6,600 \ [J \ mol^{-1}]}{RT} \right)$ (3)

2. $\tilde{D}_{oxid} = 0.1891 \ \exp \left(- \dfrac{129,790 \pm 5,930 \ [J \ mol^{-1}]}{RT} \right)$ (4)

3. $\tilde{D}_{oxid} = 9.93 \ 10^{-2} \exp\left(- \dfrac{127,650 \pm 7,230 \ [\text{J mol}^{-1}]}{RT}\right)$ (5)

parabolic equations:

4. $\tilde{D}_{red} = 2.42 \ 10^{-2} \exp\left(- \dfrac{117,250 \pm 8,480 \ [\text{J mol}^{-1}]}{RT}\right)$ (6)

5. $\tilde{D}_{oxid} = 5.60 \ 10^{-2} \exp\left(- \dfrac{126,610 \pm 8,770 \ [\text{J mol}^{-1}]}{RT}\right)$ (7)

The results of log $\tilde{D}$ vs 1/T are plotted in Fig. 5. As seen Eq. 1 gives slightly higher values of $\tilde{D}$ than Eq. 2, with almost the same activation energy in the two cases. Moreover the oxidation kinetics exhibit a slightly higher activation energy than the reduction data. It is interesting to note that the difference is displayed for diffusion data calculated according to both kinetic equations (1) and (2). The observed difference is, however, within the experimental error. Fig. 6 summarizes some of the reported values of $\tilde{D}$ vs. 1/T including results obtained in this work.

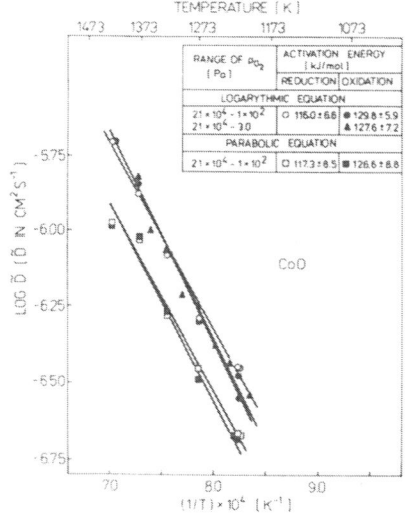
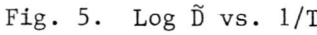

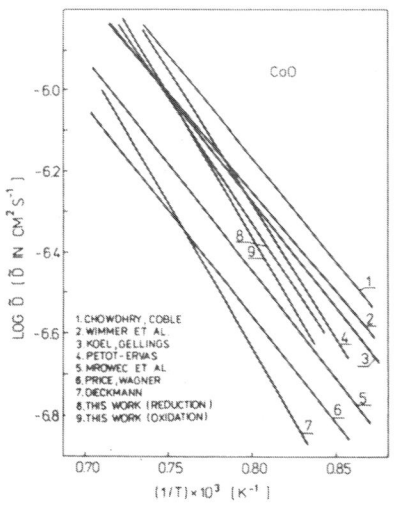

Fig. 5. Log $\tilde{D}$ vs. 1/T

Fig. 6. Comparison of present results with literature data of log $\tilde{D}$ vs. 1/T

At two temperatures, $\tilde{D}$ was determined vs. P_{O_2}. Fig. 7 illustrates $\tilde{D}$ measured at various P_{O_2}. The observed change of $\tilde{D}$ with oxide composition is in a qualitative agreement with some earlier results[6,10,11]. For comparison the present results at 1273°K are plotted in Fig. 8 along with those of Morin and Dieckmann[6,10], Petot-Ervas et al.[11], Fryt et al.[4], Koel and Gellings[3] and Price and Wagner[1].

234

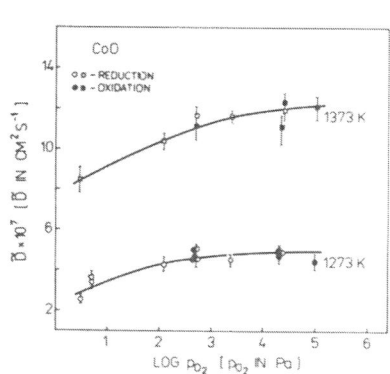

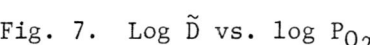

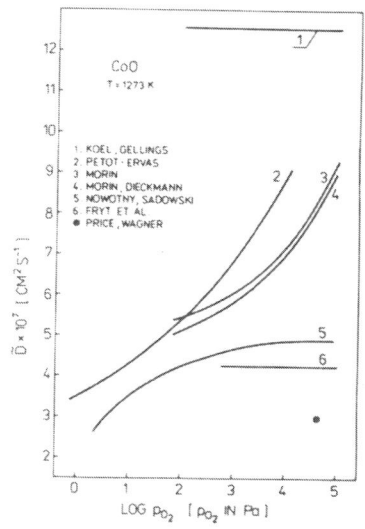

Fig. 7. Log $\tilde{D}$ vs. log P_{O_2} Fig. 8. Comparison of present
results with literature
data of log $\tilde{D}$ vs. 1/T

As seen the results are similar at lower P_{O_2}, but at higher P_{O_2} the data of Morin and Dieckmann[6,10] and Petot-Ervas et al.[11] exhibit a sharp increase with composition while our results are essentially constant. The reported dependence of $\tilde{D}$ vs. P_{O_2} is also similar to that reported for FeO by Rickert and Weppner[25].

The observed changes of $\tilde{D}$ with oxide composition may comprise several effects. The first effect may consist in strong inter-actions between defects which have been already postulated[10,15,17]. At higher nonstoichiometry vacancies may migrate via complexes like $(Co_iV_{Co})'$ or four vacancy clusters which according to Catlow et al.[26] are faster than isolated cation vacancies. According to Petot-Ervas et al.[11,12] the variation of $\tilde{D}$ with P_{O_2} corresponds to changes of defect mobility as a result of change in its ioniza-tion degree. This explanation is based on the assumption that the mobility of singly ionized cation vacancy (higher P_{O_2}) is higher than of the doubly ionized defect. This assumption, how-ever, seems to be unrealistic.

The absolute value of $\tilde{D}$ calculated according to Eq. 1 or Eq. 2 should be also considered in terms of its physical meaning. It has been shown that the correct description of the transport of defects influenced by an electric field should involve simultaneous solution of Fick's second law and Poisson's equation which describes the distribution of the electric field in the boundary layer[23]. It has been also reported that segregation of defects in oxides leads to the formation of the electrical potential barrier across the near-surface layer involving the chemical heterogenity. Accordingly, in the case of the retarding electric field, the calculated values of $\tilde{D}$ may involve the near-surface diffusive resistance which depends on the value of the near-surface potential barrier as well as on the thickness of the boundary layer δ_0. Fig. 9 illustrates the effect of the surface potential ψ (at constant δ_0) on the near-surface diffusive resistance. It has been assumed that the line for $\psi=0$ represents bulk diffusion. As seen the effect of ψ is considerable and cannot be neglected. In case of CoO the segregation may involve both intrinsic defects (cation vacancies and interstitials) and impurities. It has been reported that the surface of CoO is stabilized by a thin surface layer of Co_3O_4 [27]. The surface layer may also involve defects which are not displayed in the bulk, e.g. anion vacancies[28]. Variations in P_{O_2} may lead to changes in surface concentration profiles and thus to changes in ψ.

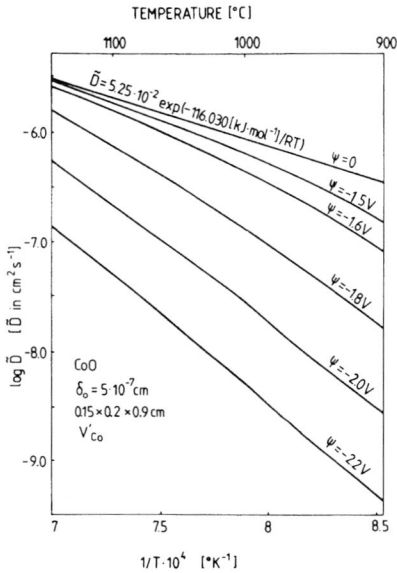

Fig. 9. Plots of log $\tilde{D}$ vs. 1/T involving the near-surface diffusive resistance corresponding to the retarding surface potential barrier ψ (calculations are performed for the following parameters: thickness of the boundary layer $\delta_0 = 5\ 10^{-7}$ cm, crystal dimensions 0.15 x 0.02 x 0.9 cm, singly ionized cation vacancies as a major kind of defect).

From the theory of transport in nonstoichiometric compounds[1,22] a simple relationship is predicted between the self diffusion coefficient of metal ions (D_s), the diffusion coefficient of defects (D_V) and the concentration of defects $[V_{Co}^z]$:

$$D_s = D_V\ [V_{Co}^z] \tag{8}$$

where z is ionization degree of the defects. On the other hand D_V and D are interrelated by:

$$\tilde{D} = (1 + z)D_V \qquad (9)$$

Accordingly:

$$[V_{Co}^z] = \frac{D_s(1 + z)}{\tilde{D}} = y \qquad (10)$$

Mrowec and Przybylski[13,29] interpret $[V_{Co}^z]$ as a total concentration of defects in the cation sublattice of $Co_{1-y}O$ in contrast to the concentration term obtained in thermogravimetric measurements N_d which refers to the apparent deviation from stoichiometry. According to Mrowec and Przybylski there is a strong difference between $[V_{Co}^z]$ and N_d as a result of the interstitial Co ions formed by the Frenkel mechanism. For calculations of $[V_{Co}^z]$ from Eq. 10 they assume data of $\tilde{D}$ after Koel and Gellings[3] (Fig. 10).

Fig. 10.
Temperature dependence of defect concentration in CoO after Mrowec and Przybylski[26].

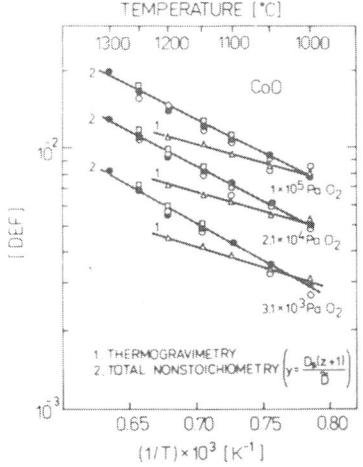

Replacing D_s by tracer diffusion coefficient Eq. 10 assumes the form:

$$[V_{Co}^z] = \frac{D^T (1 + z)}{\tilde{D} f} \qquad (11)$$

where f is correlation coefficient (f=0.78). Assuming z = 1 we receive:

$$[V_{Co}'] = 2.564 \frac{D^T}{\tilde{D}} \qquad (12)$$

For calculations of $[V_{Co}']$ we assume values of $\tilde{D}$ obtained in this work and tracer diffusion data reported by Chen and Peterson[30]:

$$D^T = 5.0 \pm 0.4 \; 10^{-3} \exp (-160{,}730 \; [J \; mol^{-1}]/RT) \qquad (13)$$

Substituting expressions (3) – (7) into (12) we receive respective relations:

$$[V'_{Co}] = 2.448 \ 10^{-1} \exp \ (-44{,}700 \ [\text{J mol}^{-1}]/RT) \tag{14}$$

$$[V'_{Co}] = 6.779 \ 10^{-2} \exp \ (-30{,}940 \ [\text{J mol}^{-1}]/RT) \tag{15}$$

$$[V'_{Co}] = 1.294 \ 10^{-1} \exp \ (-33{,}080 \ [\text{J mol}^{-1}]/RT) \tag{16}$$

$$[V'_{Co}] = 5.296 \ 10^{-2} \exp \ (-43{,}480 \ [\text{J mol}^{-1}]/RT) \tag{17}$$

$$[V'_{Co}] = 2.299 \ 10^{-1} \exp \ (-34{,}120 \ [\text{J mol}^{-1}]/RT) \tag{18}$$

The expressions (14) – (18) can be compared with data of the deviation from stoichiometry reported in literature:

$$[V'_{Co}] = 9.95 \ 10^{-2} \exp \ (-32{,}600 \ [\text{J mol}^{-1}]/RT) \tag{19}$$

after Koel and Gellings[3],

$$[V'_{Co}] = 6.36 \ 10^{-2} \exp \ (-27{,}210 \ [\text{J mol}^{-1}]/RT) \tag{20}$$

after Fisher and Tannhauser[21],

$$[V'_{Co}] = 9.0 \ 10^{-2} \exp \ (-30{,}140 \ [\text{J mol}^{-1}]/RT) \tag{21}$$

after Eror and Wagner[31], and

$$[V'_{Co}] = 3.1 \ 10^{-1} \exp \ (-40{,}310 \ [\text{J mol}^{-1}]/RT) \tag{22}$$

after Bransky and Wimmer[32].

As seen both the preexponential factor and the activation energy of the two sets of data are within the same range. Thus, the agreement between the literature data and those calculated from Eq. 12 is satisfactory within the reported reproducibility of $[V'_{Co}]$. Accordingly a conclusion about interstitials in CoO can not be drawn from the available data as far as the crystalline bulk is considered. On the other hand it has been postulated that considerable concentration of cobalt interstitials is displayed in the near-surface layer[18,33]. Interstitially located cobalt ions lead to lattice expansion and consequently to the decrease of the activation energy of chemical diffusion in the surface layer[18]. This is in agreement with data of chemical diffusion for both single- and polycrystalline material[9]. It has been shown that a decrease in the crystal size leads to a decrease of the activation energy of chemical diffusion. This effect may be related to the increasing ratio of the surface and grain boundaries. Consequently a considerable concentration of interstitials comes into the picture when polycrystalline material is

238

considered. Then the ratio of different kinds of defects essentially depends on crystal dimensions. A quantitative description of this problem requires further studies of the surface properties of oxide materials in equilibrium with the gas phase.

Fig. 11 shows the plot of log σ vs. 1/T. The activation energy of electrical conductivity can be used to calculate the enthalpy of defect formation ΔH_f using the relation[31]:

$$E_\sigma = \frac{2}{n} \Delta H_f + \Delta H_m^{h\cdot} \tag{23}$$

where $\Delta H_m^{h\cdot}$ is the activation enthalpy of motion of electron holes. Table II shows values of ΔH_f which are calculated assuming $\Delta H_m^{h\cdot} = 20,890 \text{ J mol}^{-1}$ [21]. Good agreement with the value reported by Chowdhry and Coble[9] $\Delta H_f = 31.08 \text{ kJ mol}^{-1}$ is obtained.

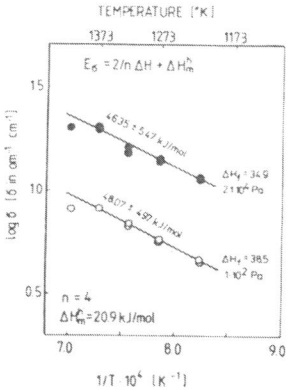

Fig. 11.   Log σ vs. 1/T.

CONCLUSIONS

Both parabolic and logarithmic kinetics equations yield consistent data for the chemical diffusion coefficient in undoped CoO with almost identical values for the activation energy and slight differences for the preexponential factor. The present results of $\tilde{D}$ vs. 1/T exhibit an activation energy between 116 and 129 kJ mol^{-1} in contrast to previous data exhibiting an activation energy of 100.5 kJ mol^{-1}. There is good agreement between the deviation from stoichiometry in CoO calculated from the present diffusion data and the results obtained by thermogravimetry. Cation vacancies are major type of defects as far as moncrystalline material is considered. However, due to segregation, the defect structure becomes more complicated within grain boundaries and surface layers.

TABLE II

The activation energy of electrical conductivity E_σ and the activation enthalphy of formation of cation vacancies ΔH_f.

Oxygen Pressure [Pa]	$\dfrac{E_\sigma}{kJ\ mol^{-1}}$	ΔH_f
2.1×10^4	46.09 ± 5.47	34.86
1×10^2	48.07 ± 4.97	38.45

Changes of $\tilde{D}$ vs. P_{O_2} may be considered in terms of inter-actions between defects, if a purely bulk controlled mechanism is taken into account. When segregation of defects leads to formation of a retarding potential barrier across the near-surface (or the grain boundary layer) the calculated value of $\tilde{D}$ may involve the near-surface diffusive resistance. Determinations of the sign and the value of the barrier as well as of the surface layer thickness are required for evaluation of the magnitude of the resistance.

ACKNOWLEDGEMENTS

The authors are grateful to Professor Wayne L. Worrell for helpful comments.

REFERENCES

1. J. B. Price and J. B. Wagner, Jr., Z. Physik. Chem., Neue Folge, 49:257 (1966).
2. H. Meurer, Thesis, Clausal University, (1970).
3. G. J. Koel and P. J. Gellings, Oxid. Metals, 5:185 (1972).
4. E. Fryt, S. Mrowec and T. Walec, Oxid. Metals, 7:117 (1973).
5. J. M. Wimmer, R. N. Blumenthal and I. Bransky, J. Phys. Chem. Solids., 36:269 (1975).
6. F. Morin, Canad. Metallurg. Quart., 14:105 (1975).
7. R. Dieckmann, Z. Physik. Chem., Neue Folge, 107:189 (1977).
8. M. Chadzhieva, Ph.D. Thesis, Academy of Mining and Metallurgy, Krakow (1980).
9. U. Chowdhry and R. L. Coble, J. Am. Ceram. Soc., 65:336 (1982).
10. F. Morin and R. Dieckmann, Z. Physik. Chem., Neue Folge, 129:219 (1982).
11. G. Petot-Ervas, O. Radji, B. Sossa and P. Ochin, Radiation Effects, 75:301 (1983).
12. G. Petot-Ervas, Lecture presented on January, 1983 at the Institute of Metallurgy, Academy of Mining and Metallurgy, Krakow.
13. S. Mrowec, Ceram. Intern., 4:47 (1978).
14. J. Nowotny and A. Sadowski, J. Am. Ceram. Soc., 62:24 (1979).
15. H.-C. Chen and T. O. Mason, J. Am. Ceram. Soc., 64:C-130 (1981).
16. F. Gesmundo and F. Viani, J. Electrochem. Soc., 128:460 (1981).
17. J. Nowotny, M. Rekas and I. Sikora, J. Electrochem. Soc., 131:94 (1984).
18. J. Nowotny, I. Sikora and J. B. Wagner, Jr., Proc. 9th Intern. Symp. Reactivity of Solids, Elsevier, Amsterdam, (1982) p. 225.
19. B. Fisher and J. B. Wagner, Jr., Physics Letters, 21:606 (1966).
20. A. B. Newman, Trans. Am. Inst. Chem. Eng., 27:203 (1931).
21. B. Fisher and D. S. Tannhauser, J. Chem. Phys., 44:1663 (1966).
22. J. Nowotny, J. B. Wagner, Jr., Oxid. Metals, 15:169 (1981).
23. Z. Adamczyk, J. Nowotny, submitted to J. Electrochem. Soc.

24. J. Nowotny, I. Sikota and J. B. Wagner, Jr., J. Am. Ceram. Soc. 65:192 (1982).
25. H. Rickert and W. Weppner, Z. Naturforsch., 29A: 1849 (1974).
26. C. R. A. Catlow, B. Fender and D. G. Muxworthy, J. Phys. (Paris) C7:67 (1977).
27. J. Haber, J. Stoch and L. Ungier, J. Electron Spectroscopy, 9:459 (1976).
28. J. Nowotny, J. Mater. Sci., 12:1143 (1977).
29. S. Mrowec and K. Przybylski, Rev. Int. Htes. Temp. et Refract., 14:285 (1977).
30. W. K. Chen, N. L. Peterson and W. T. Reeves, Phys. Rev., 186:887 (1969).
31. N. G. Eror and J. B. Wagner, Jr., J. Phys. Chem. Solids, 29:1597 (1968).
32. I. Bransky and J. M. Wimmer, J. Phys. Chem. Solids, 33:801 (1972).
33. W. Komatsu, Y. Chida and T. Maruyama, Proc. 9th Intern. Symp. Reactivity of Solids, Elsevier, Amsterdam (1982) p. 430.

ALTERATION OF CoO WAFERS IN AN OXYGEN CHEMICAL

POTENTIAL GRADIENT

M.J.Shingler, K.M.Vedula and J.W.Halloran

Department of Metallurgy and Materials Science
Case Western Reserve University
Cleveland, Ohio 44106

INTRODUCTION

The Wagner Oxidation Theory, and its later developments, allow the parabolic rate constant to be predicted from fundamental transport coefficients and point defect thermodynamics, provided that the oxidation reaction is controlled by well-understood lattice diffusion mechanisms. In many practical cases, however, diffusion along fast paths such as grain boundaries is dominant and the actual oxidation rate is strongly influenced by the microstructure of the oxide scale. Moreover, the oxide microstructure is not static, but rather continuously evolves as new oxide grows and as the existing oxide re-structures. The growth mechanism depends upon the oxide microstructure, while the oxide microstructure is determined by the details of the growth mechanism. This creates what is essentially a feedback loop system in which the scale microstructure both influences and is influenced by the growth mechanism. A description of this loop system seems to be necessary to develop a richer understanding of oxidation.

In this paper we are concerned with one aspect of this system - the re-structuring of polycrystalline and single crystalline transition metal oxides by a cation flux induced by an oxygen chemical potential gradient. This cation flux causes one interface of the wafer to erode, while oxide growth occurs on the opposite interface. We believe that the growing interface closely models the actual processes which occur at the oxide/gas surface of an outwardly-growing CoO scale. The receding interface, in particular the pattern of erosion, is also of great interest and may be quite relevent to detachment, creep, and pore formation in actual oxide scales.

243

Research in this area was pioneered by Schmalzried and his colleagues[1,2]. We have adopted the same method used by Yurek and Schmalzried in their elegant 1975 paper[1]. In this paper we describe some effects similar to those discussed by them and present novel observations on the structure of the growing surface. This paper represents a first report on an on-going investigation of micro-structural alterations in CoO[3].

The fluxes of cations, cation vacancies and electron holes in our experiment are illustrated in Figure 1. A wafer of initial thickness X is exposed to a low oxygen activity on the left face and a higher oxygen activity on the right face. The advancing interface grows by deposition of new oxide through the reaction:

$$Co_{Co} + \frac{1}{2} O_2 \rightarrow CoO + V'_{Co} + h \cdot \tag{1}$$

with this surface acting as a source of cation vacancies and holes. The receding interface is the sink for these vacancies and holes, and erodes by the reverse of reaction 1. After a brief transient, the steady state cobalt flux develops, given by:

$$J_{Co} = \frac{\tilde{D} \, C_{Co}}{RT} \nabla \mu_{Co} \tag{2}$$

where μ_{Co} is the cobalt chemical potential in J/mole
C_{Co} is the cobalt concentration in moles/cm³

Here $\tilde{D}$ is the chemical diffusivity which is equal to twice the cobalt vacancy diffusivity[4] for the case of singly charged cobalt vacancies.

The cobalt chemical potential gradient can be related to the applied oxygen chemical potential gradient by the Gibbs-Duhem relation $\nabla \mu_{Co} = \nabla \mu_O$, so that Equation 2 can be re-written as:

$$J_{Co} = + \frac{\tilde{D} \, C_{Co}}{RT} \nabla \mu_O \tag{3}$$

As this flux impinges on the right side of the wafer, it causes that side to advance by deposition of new oxide. The velocity of the advancing interface (with respect to a fixed oxygen sublattice) is given by:

$$V = \Omega \, J_{Co} \tag{4}$$

But since the molar volume Ω is simply the reciprical of C_{Co}, Equation (4) becomes:

244

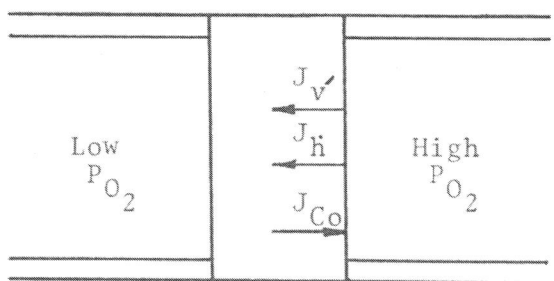

Fig. 1. Schematic of fluxes in a CoO wafer exposed to an oxygen
chemical potential gradient.

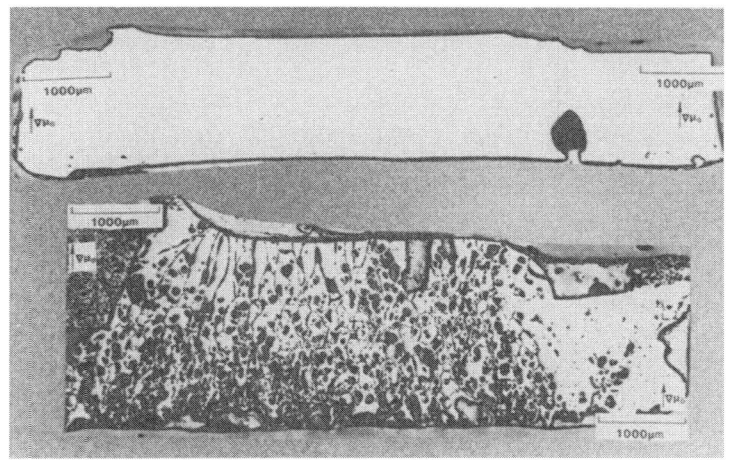

Fig. 2. Cross-sections of specimens after exposure.

$$V = \frac{\tilde{D}}{RT} \nabla \tilde{\mu}_O \approx \frac{\tilde{D}}{RT} \frac{\Delta \tilde{\mu}_O}{X} \qquad (5)$$

For comparison with experimental data it is convenient to express $\nabla \mu_O$ in terms of the mole fraction of cation vacancies, which are known as a function of oxygen activity, so that Equation 5 is expressed as:

$$V = \tilde{D} \frac{\Delta [V_{Co}]}{X} \qquad (6)$$

where X = Thickness of the wafer

$\Delta [V_{Co}]$ = Difference in mole fraction of cation vacancies at the two surfaces.

Experimental Procedure

Wafers of single crystal* and polycrystalline** CoO were exposed to oxygen chemical potential gradients of about 50-100 kJ/mole-mm at 1200-1300°C. The wafer was secured between high purity alumina tubes. Rings of platinum foil separated the wafers from the alumina tubes and served as gas-tight gaskets when pressed together with gentle spring tension.

One face of the wafer was exposed to air, while the opposite face experienced the low P_{O_2} of either Ar-O_2 or CO-CO_2 mixtures. Oxygen activity in the low P_{O_2} side was monitored with a zirconia cell downstream from the specimen. The surfaces of the wafers were prepared by polishing with 1 micron diamond paste. Single crystal surfaces were (100) planes. After exposure, the advancing and eroding surfaces were examined with Scanning Electron Microscopy. Cross-sections, impregnated with a lead borosilicate frit to preserve delicate features, were prepared for standard optical metallography.

We will present the results for three specimens. Crystal A was exposed for 24 hours at 1200°C with oxygen partial pressures of 0.21 and 1.9×10^{-5} atmospheres on the opposite faces to create an average gradient of 91.1 kJ/mole-mm in oxygen chemical potential. Crystal B was subjected to a gradient of 49.7 kJ/mole-mm by exposing the surfaces to 0.21 and 6.7×10^{-4} (Ar-O_2) atmospheres of oxygen at 1300°C. The polycrystalline specimen was exposed 24 hours at 1300°C to 0.21 and 5.8×10^{-5} atm (CO-CO_2) with an average gradient of 50 kJ/mole-mm.

---

* Supplied by N.L.Peterson and described in (5).
** Hot pressed CoO prepared by M. Notis and described in (6). Purity is 99.9%, 50 micron grain size, 99% dense.

Results and Discussion

A. Extent of Growth

The cross-sections of single crystal B and the polycrystalline specimen are shown in Figure 2. Note the displacement of the advancing interface and the erosion of the receding interface. The observed displacement of the advancing interface should be comparable to the product of the exposure time and the steady state velocity given by Equation 6. This velocity was calculated using Diekmann's data for the vacancy concentration[7]. Chemical diffusivities of 9.0×10^{-6} cm^2/sec at 1300°C and 4.5×10^{-6} cm^2/sec at 1200°C were obtained from[8]. Table I displays the calculated velocities and surface displacements, compared with the observed displacements.

TABLE I. Calculated and Observed Displacement
Of The Advancing Interface

Specimen	Calculated V	Calculated	Observed
Crystal A 1200°C	7.8 μ/h.r	187 μ	70 ± 20 μ
Crystal B 1300°C	13.3 μ/hr.	319 μ	230 ± 50 μ
Polycrystal 1300°C	15 μ/hr.	360 μ	Approx. 600 μ

Notice that the observed displacements for the two single crystals are substantially smaller than the calculated values. This is puzzling since chemical diffusivities and vacancy concentrations are fairly well established in CoO, leading one to expect near-quantitative agreement. One factor neglected by Equation 6 is the severe erosion of the receding interface. However, the pitting of receding interface should make the gradient, and hence the observed velocity, larger than predicted by Equation 6. However, it is possible that the actual gradient in the wafer was lower than had been assumed. Perhaps this is due to an experimental problem such as small leaks in the seals. The actual gradient might also be diminished if a significant part of the $\Delta\mu_o$ is dissipated by driving the interfacial reaction 1, rather than being entirely used to force the diffusive flux. This issue is still being investigated.

The polycrystalline specimen displayed a surface displacement much larger than predicted by bulk diffusion, indicating that even at 1300°C grain boundary diffusion contributes significantly to the total flux.

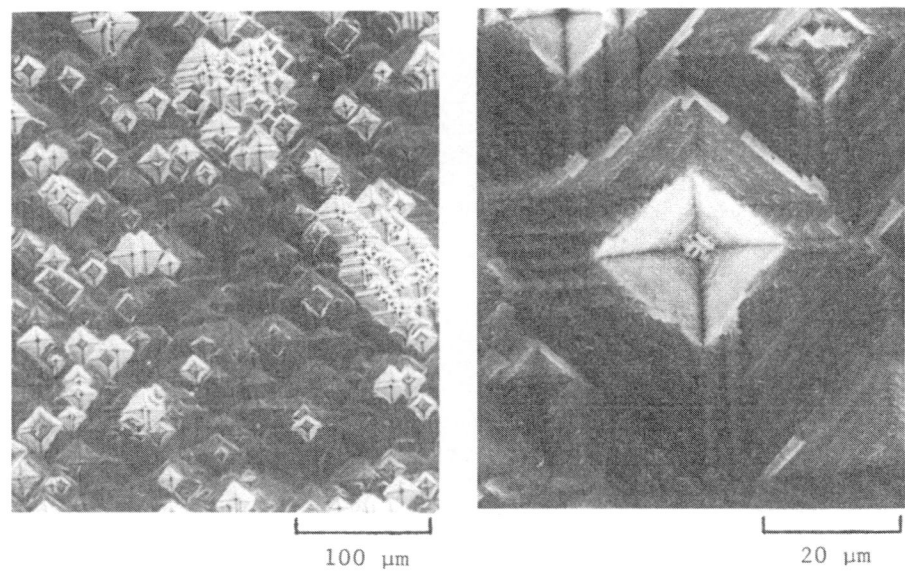

100 μm

20 μm

Fig. 3. Advancing face of crystal
         A (200X)

Fig. 4. Detail of facets on
         advancing face of
         crystal A (1000X)

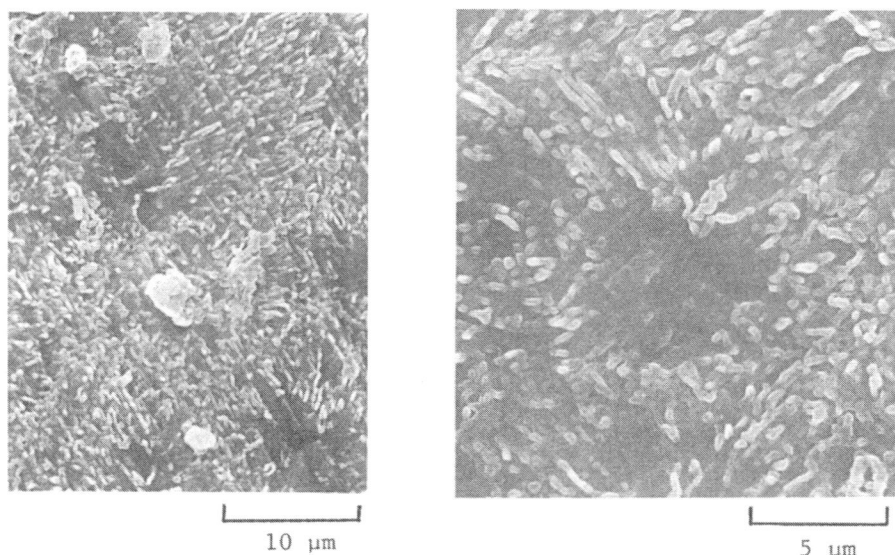

10 μm

5 μm

Fig. 5. Advancing face of crystal
         B.

Fig. 6. Whisker-like features
         on advancing face of
         crystal B

B. Structure of the Advancing Interface

The advancing face of the single crystal A exposed at 1200°C was planar on a macroscopic scale, but faceted microscopically, as shown in Figure 3. The primary features decorating the advancing surfaces were four-sided pyramids about 10-30 microns along a side. The pyramids were distributed non-uniformly on the advancing surface, being relatively sparse at the center of the wafer, where they numbered about $200/mm^2$, and most numerous at the edges near the Pt gasket where the density was as high as $3000/mm^2$. Distinct rows or clusters of pyramids were common. Although their appearance is reminiscent of dislocation etch-pits, the pyramid density is far too small to be consistent with typical dislocation densities.

On a finer scale the advancing surface appears entirely faceted, consisting of a herringbone pattern of ledges with 1-2 μ step size. These ledges run in ⟨110⟩ directions and consist of facets of {100} planes. From Figure 4 it is apparent that the pyramids are in fact "step pyramids" constructed of these {100} facets. The growth mechanism of the crystal under these conditions is thus the formation of {100} ledges running along ⟨110⟩ direction. The origin of the distinct pyramidal features has not yet been fully identified. Certain pyramids are crowned by smooth-sided faces which show no faceting at the level of resolution of the SEM. These appear in brighter contrast in Figure 3. Trace analysis suggests that these faces are {110} planes. Some of these smooth-sided pyramids are sharply pointed at their apex. Others are either truncated by smooth {100} planes, or have faceted features decorating their apex.

A distinctly different set of features was observed on the advancing interface in the single crystal B exposed at 1300°C. On a macroscopic scale the advancing surface of crystal B was also planar. Under optical examination a small number of widely spaced facets were apparent running in ⟨110⟩ directions, separated by 50-100 μ. However, a fine scale structure was apparent upon SEM examination.

Figure 5 shows the growing surface of crystal B at a location near the center of the wafer. A pattern of fine striations, 2μ wide, run along ⟨110⟩ directions with no intersections. Small whisker-like features, about 0.3 μ wide and 2 μ long, appear to lie on the surface, generally pointing along ⟨110⟩ normal to the direction of the striations. The size of the whiskers is remarkably uniform over the surface, with a density on the order of $0.5-1.0 \times 10^6/mm^2$. In other regions the whiskers appear to radiate from distinct centers, as illustrated by Figure 6, which shows whiskers oriented along several ⟨110⟩ type directions about a central faceted feature. In spite of this whisker-like

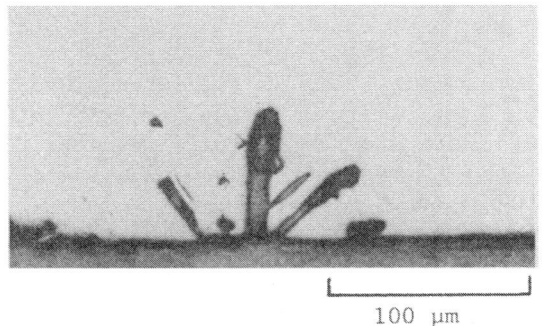

100 μm

Fig. 7. Cross-section of crystal A, near receding interface.

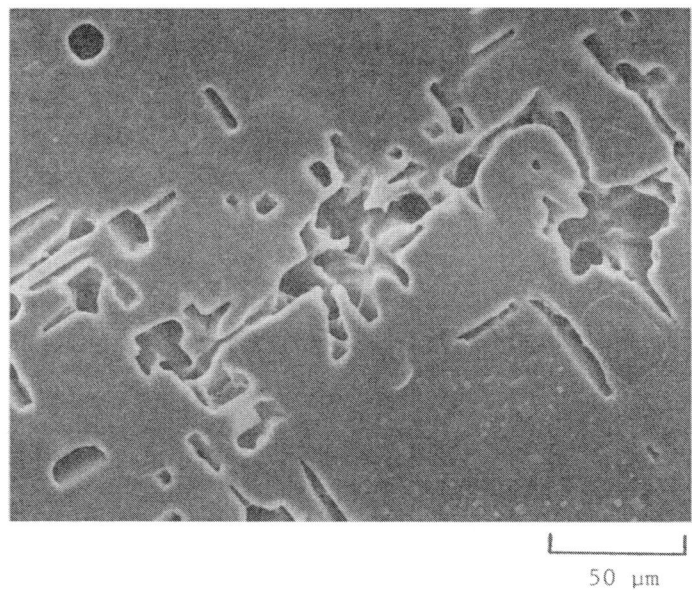

50 μm

Fig. 8. Receding surface of crystal A.

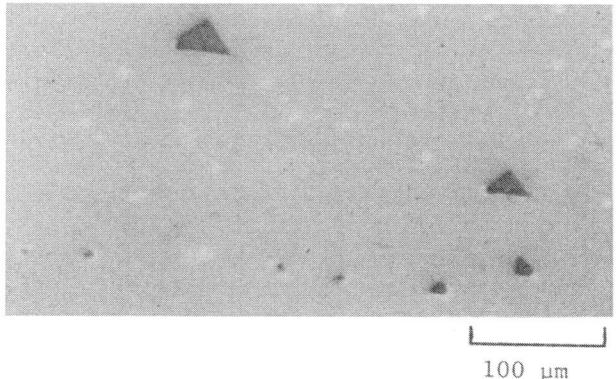

100 μm

Fig. 9. Porosity in crystal B shown in cross-section.

morphology, the new oxide growth is single crystalline and has no optically-resolvable porosity. Note that in this case the interface advanced 230 μ, which is about 500 times the size of the individual whiskers.

We do not understand why the growth morphologies of crystal A and B were so different. Aside from the difference in temperature, the growth velocity of B was three times as fast as crystal A so the amount of growth in crystal B (230 μ) was three times the amount of growth in crystal A (70 μ). Experiments are now underway to ascertain whether the difference in growth mechanism is a simple temperature effect, an artifact of crystal growth rate, or due to some uncontrolled variable (such as ambient humidity, which is known to affect whisker formation on iron oxides[9]).

C. Structure of the Receding Interface

In all three specimens the receding interface was unstable, causing the surface to become extensively pitted. It is apparent, however, that a substantial portion of the vacancy flux impinged uniformly along the low P_{O_2} surface, as there was both uniform recession and pitting. However, the pattern of instability was distinctly different in each case.

Figure 7 shows a cross-section of the crystal A, exposed in CO/CO_2 at 10^{-5} atm O_2. Faceted protrusions, with a dendritic appearance, penetrate typically 30 μ. The largest observed in a single section was 100 μ deep. These protrusions intersect the (100) surface of the crystal at angles of 45°. Pore-like features in Figure 7 are probably the intersection of dendritic protrusions with the plane of polish. The nature of these protrusions is more apparent in SEM micrographs of the receding surface. Figure 8 illustrates the three types of features observed on the receding surface of crystal A. The surface is decorated with many shallow rectangular bumps, 2-10 μ along a side, aligned along ⟨110⟩ directions of the crystals. The larger bumps assume more rounded contours. They can be seen in faint contrast in the lower right of Figure 8. The major features are deep protrusions into the surface. A few protrusions are round in cross-section (an example is in the upper left of Figure 8), but most are sharply faceted and slot-like. These are the dendritic features in Figure 7. The protrusions tend to be clustered, but the clusters are uniformly distributed across the surface with a density of about 200-350 protrusions/mm². The smallest are 3 μ across, with large protrusions reaching 50 μm in cross-section. The slot-like features are bounded by {110} planes and intersect the surface along ⟨110⟩ directions.

Crystal B, exposed on its low P_{O_2} face to Ar with 10^{-5} atm O_2 at 1300°C, shows an entirely different pattern of erosion. The cross-section of the entire specimen, shown in Figure 2 shows that there has been at least 100 µ of uniform recession of the low P_{O_2} surface, probably representing the majority of mass transport. However, the surface was also pock-marked with very large protrusions, some easily visible with the unaided eye. Most were about 250 µ in cross-section, being either round or bounded by $\{110\}$ planes. Approximate density of protrusion was $5/mm^2$. Some of the protrusions were shallow depressions, but others extended hundreds of microns into the crystal. The large protrusion shown in Figure 2 is about 650 µ deep, extending halfway across the crystal. Usually the protrusions had a rounded shape with suggestion of $\{110\}$ faceting.

D. Structure of the Polycrystalline Specimen

The polycrystalline specimen, initially equiaxed and dense, underwent extensive re-structuring so as to develop the pronounced columnar texture obvious in Figure 2. The rapid growth of this specimen transferred almost 40% of its mass to the advancing surface. It appears that the surface advanced by the columnar growth of existing grains since the origin of the columnar grains is near the original position of the air surface. The erosion of the low P_{O_2} interface is particularly non-uniform. Remnants of eroded grains mark the original position of the interface, while cavities extend deep into the specimen. Both the old oxide and the new oxide have become quite porous. Many of the pores seem to be inside grains, rather than on grain boundaries. It is not known whether the pores were generated in the interior of the wafer, or whether they migrated in from the low P_{O_2} interface, as observed by Yurek and Schmalzried[1].

Oxide scales commonly display quite similar porosity[10]. The fact that these pores were created by application of a rather mild gradient suggests that similar effects must occur under the more intense gradients characteristic of actual oxidation. These effects and other examples of diffusion-induced recrystallization, are discussed in detail in references (3).

E. Porosity in Single Crystalline Specimens

In contrast with Yurek and Schmalzried's observations[1], we find very little pore formation in our single crystals. The pore-like features in crystal A are probably not porosity but rather portions of the dendritic protrusions. Crystal B was pore-free except for a distinct line of small pores 750 microns from the original position of the receding interface. Figure 9 shows that the pores are pyramids which point directly in the growth direction.

252

The sides of the pyramidal pores are {110} planes, with {100} planes as a base.

SUMMARY AND CONCLUSIONS

Cobalt oxide wafers are extensively re-structured by exposure to an oxygen chemical potential gradient. The displacement of the advancing interface is smaller than that predicted for lattice diffusion for single crystals, but larger than the predicted displacement for polycrystalline CoO.

The growth mechanism for the advancing interface involves the formation of {100} ledges running in <110> directions at 1200°C, and fine whiskers lying in <110> directions at 1300°C. The receding interface becomes severely eroded, forming small {110} faceted protrusions at 1200°C and large protrusion at 1300°C. Polycrystalline CoO becomes extensively pitted. Initially dense equiaxed polycrystalline CoO becomes grossly porous and develops a columnar grain structure.

References

1. G.J.Yurek and H.Schmalzried, Ber. Bunsenges, Phys., Chem., 79:255 (1975).
2. H.Schmalzried, W.Laqua, and P.L.Lin, Z.Naturforsch. 34a:192 (1979); H.Schmalzried and L.Laqua, Oxidation of Metals, 15:339 (1981).
3. M.Shingler, M.S. Thesis, Case Western Reserve University, (1984).
4. H.Schmalzried, Solid State Reactions, Verlag Chemie Gmbh. Weinheim Academic Press, N.Y. (1974).
5. W.K.Chen and N.L.Peterson, J. Phys. Solids, 34:1093 (1973).
6. P.A.Urick and M.R.Notis, J. American Ceramic Soc., 56:570 (1973).
7. R.Diekmann, Z. Physikalische Chemie N.F., 107:189 (1977).
8. M.Schnehage, R.Diekmann, and H.Schmalzried, Ber. Bunsenges. Phys. Chem., 86:1061 (1982).
9. D.H.Voss, E.P.Butler and T.E.Mitchell, Metallurgical Transactions, 13A:929 (1982).
10. P.A.Labun, J.Covington, K.Kuroda, and G.Welsch, Metallurgical Transactions, 13A:2103 (1982).

SELF-DIFFUSION AND CREEP STUDIES OF THE OXYGEN SUBLATTICE POINT DEFECTS IN CoO

C. Clauss, R.-J. Tarento, C. Monty, A. Dominguez-Rodriguez*, J. Castaing, and J. Philibert

Laboratoire de Physique des Matériaux, C.N.R.S. Bellevue - 92195 Meudon Cedex (France)

INTRODUCTION

Point defects in nonstoichiometric transition metal oxides NiO, CoO, MnO, FeO, have been extensively studied for several years [1], [2]. The defects responsible for the departure from stoichiometry are relatively well known as several methods are able to give information about them : cationic self diffusion, electrical conductivity, thermogravimetry are commonly used. In pure CoO for example, the majority defects are cationic vacancies $V_{Co}^{\alpha'}$ with a negative charge $- \alpha |e|$ ($\alpha = 0 - 2$) and holes h· ; this has been found in the temperature and oxygen partial pressure ranges where CoO is stable and can be easily investigated (T = 1000 - 1800°C and $P_{O_2} \geqslant 10^{-6}$ atm). Nevertheless, the possibilities of complex defects occuring at high departures from stoichiometry [3] or cobalt interstitials near the limit Co/CoO [4] are still controversial.

In constrast, little is known about the dominating point defects of the oxygen sublattice, which are minority point defects in these oxides. The main reason is their very low concentration level and the small number of specific ways to observe them. Two methods can be used : oxygen self diffusion and high temperature creep tests. The data collected on minority point defects can then be compared with deductions from computed energy values [5].

A diffusion coefficient is proportional to the concentration of point defects responsible for the diffusion mechanism and to the diffusion coefficient of the point defects themselves. The

---

*Permanent address: Departamento de Optica, Facultad de Ciencias Fisicas, Universidad de Sevilla, España.

oxygen diffusion coefficient can be written in the general form [2] :

$$D_0 = A \, P_{O_2}{}^m \, \exp - \frac{\Delta H_0^d}{kT} \qquad (1)$$

where the P_{O_2} exponent m characterizes the type and the charge of point defects ; ΔH_0^d is the sum $\Delta H_{\nabla}^f + \Delta H_{\nabla}^m$ of the "apparent formation enthalpy" (Temperature dependence of the concentration) of a defect ∇, and of ΔH_{∇}^m, its migration enthalpy ; entropy terms are included in A.

If the point defects are charged, the m and $\Delta \tilde{H}_{\nabla}^f$ values for oxygen self diffusion depend on corresponding parameters for the majority point defect. The knowledge of the hole concentration dependence on oxygen partial pressure and on temperature is sufficient to describe the oxygen diffusion behaviour. A detailed knowledge of the majority point defect population is not necessary. Knowledge of the electrical conductivity is therefore sufficient [8].

It is generally accepted that the steady state creep rate is controlled by the climb velocity of dislocations. Most of the models predict that the creep rate is proportional to a self diffusion coefficient which is that of the slowest species in a compound, namely the oxygen in CoO.

The stationary creep rate can be written [6] [7] :

$$\dot{\varepsilon}_s \cong B \, \sigma^n \, D_0 = B' \, \sigma^n \, P_{O_2}{}^m \, \exp - \frac{\Delta Q}{kT} \qquad (2)$$

σ = applied stress ; B and B' are constant.

The oxygen pressure dependence comes from the diffusion coefficient only, while the activation energy Q could in some cases be related not only to the oxygen self diffusion but to other temperature dependent quantities such as the jog formation energy, the stacking fault or the elastic constants ...[6], [7], [8]. Creep tests can provide in non stoichiometric oxides such as CoO, information on the nature and the charge state of the point defects involved in oxygen self diffusion. When the only contribution to the temperature dependence is diffusion, the activation energy for creep gives the sum $\Delta \tilde{H}_{\nabla}^f + \Delta H_{\nabla}^m$ of the formation and of the migration enthalpies of these point defects.

Creep studies have already been performed on pure CoO [9], [10]. Assuming that the steady state creep was controlled by oxygen diffusion, oxygen vacancies and oxygen interstitials both appeared to be responsible for the observed behaviour, i.e. vacancies at low P_{O_2}, interstitials at high P_{O_2}. Constant strain rate mechanical behaviour studies [11] are in agreement with this model.

The purpose of this paper is to present a recent investigation of oxygen self diffusion, especially the oxygen partial pressure dependence, and to describe some new creep results obtained on Cr doped CoO which all agree with the proposed model and give new insight on the charge of the defects.

1°/ POINT DEFECT CHARACTERIZATION

Considering only simple point defects such as β times positively charged oxygen vacancies, $V_O^{\beta\bullet}$, and γ times negatively charged interstitials, $O_i^{\gamma'}$, we establish below the relations between the oxygen partial pressure and the defect concentrations according to the charge of the defects. We shall determine also the relations between the measured activation energy and the point defect formation process.

Vacancies and interstitials are created by oxygen exchange between the oxide and the surrounding atmosphere, we can describe these processes as reactions of oxidation or reduction [12].

$$\frac{1}{2} O_2(g) \rightarrow O_i^{\gamma'} + \gamma h^\bullet \tag{3}$$

$$O_O \rightarrow V_O^{\beta\bullet} + \beta e' + \frac{1}{2} O_2(g) \tag{4}$$

Applying the mass action law to such reactions we obtain :

$$[O_i^{\gamma'}] [h^\bullet]^\gamma = K_{i\gamma}^f \; P_{O_2}^{1/2} \tag{5}$$

$$[V_O^{\beta\bullet}] [e']^\beta = K_V^f \; P_{O_2}^{-1/2} \tag{6}$$

electrons and holes are in thermal equilibrium :

$$[e'] [h^\bullet] = K_G \tag{7}$$

The reaction constants are thermally activated. The point defect concentrations are given by :

$$[O_i^{\gamma'}] = A_1 [h^\bullet]^{-\gamma} P_{O_2}^{1/2} \exp - \frac{\Delta H_{i\gamma}^f}{kT} \tag{8}$$

$$[V_O^{\beta\bullet}] = A_2 [h^\bullet]^\beta \; P_{O_2}^{1/2} \exp - \frac{(\Delta H_{V\beta}^f -, \beta \Delta H_G)}{kT} \tag{9}$$

where $\Delta H_{i\gamma}^f$ relates to $K_{i\gamma}^f$, $\Delta H_{V\beta}^f$ to $K_{V\beta}^f$ and ΔH_G to K_G. A_1 and A_2 include the preexponential terms.

From (8) and (9), we see that the behaviour of minority defects is related to [h˙] which is known from electrical conductity σ_{el} given in a p-type semiconductor by the relation :

$$\sigma_{el} = [h˙] \, e \, \mu_h \tag{10}$$

where [h˙] is the number of holes by volume unit and μh is the mobility of holes. If μ_h does not depend on P_{O_2} and is thermally activated with a small activation energy ΔH, σ_{el} and [h˙] have the same P_{O_2} dependence. We can write :

$$\sigma_{el} = \sigma_0 \, P_{O_2}^{\,1} \exp - \frac{(\Delta H_h + \Delta H_\mu)}{kT} \tag{11}$$

l and ΔH_h being generally dependent on P_{O_2} and T.

Using the approximation [h˙] α σ_{el} , we can write (8) and (9) on the following form :

$$[O_i^{\gamma'}] = A_1' \, P_{O_2}^{\,-1\gamma + \frac{1}{2}} \exp - \frac{\Delta H_{i\gamma}^f - \gamma \Delta H_h}{kT} \tag{12}$$

$$[V_0^{\beta˙}] = A_2' \, P_{O_2}^{\,1\beta - \frac{1}{2}} \exp \frac{(\Delta H_{V\beta}^f - \beta \Delta H_G + \beta \Delta H_h)}{kT} \tag{13}$$

where A_1' and A_2' are new constants ; l and ΔH_h can be obtained directly from electrical conductivity according to (11).

From this we can deduce the P_{O_2} exponents from minority defects given by m = - lγ + ½ or lβ - ½.

We have calculated m for $O_i^{\gamma'}$ and $V_0^{\beta˙}$ in a MO oxide, using l between 0. and 0.5 ; the results are displayed in fig. 1. In special conditions (l $\cong$ 0.25) the sign of the exponent changes from positive to negative for two minority defects (see O_i'' and $V_0^{··}$).

Such relations also hold in the case of extrinsic behaviour when h˙ becomes a minority defect, as long as the relation (10) is always describing the conductivity. That is an important simplification in order to characterize the oxygen sublattice point defects in doped samples. One can compare the measured values of m with those of fig. 1 and deduce the type of minority defects, provided l is known. Even if cationic impurities dominate giving an extrinsic conductivity with l = 0, the defect concentration depends on P_{O_2} (m ≠ 0) (fig. 1).

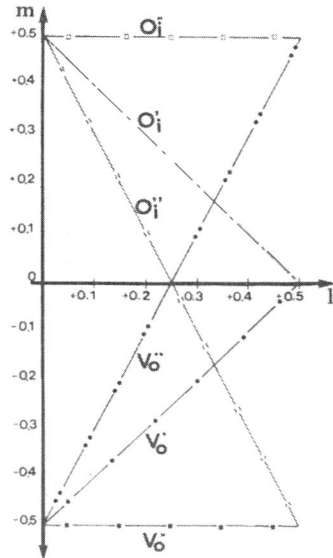

Fig. 1 : The concentration of indicated point defects is proportional to $P_{O_2}^m$; m, for a given defect, depends on the concentration of holes which is proportional to $P_{O_2}^l$. The figure shows the relation between m and l calculated according to eq. (12) and (13).

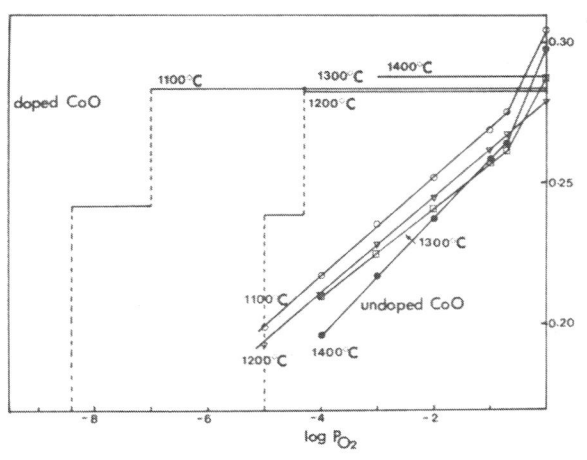

Fig. 2 : l versus log P_{O_2} deduced from electrical conductivity measurements [3], [15] ($\sigma_{el} \propto P_{O_2}^l$).For undoped CoO, we have used the values of ref [3]. "Doped CoO" has been doped with 3860 at ppm. Cr ; we have decomposed the log σ_{el} - log P_{O_2} curves in straight lines, giving constant l values.

Figure 2 shows the variation of l vs P_{O_2}, l is the exponent deduced from electrical conductivity of pure CoO [3] and of Cr doped CoO (3860 ppm) [15]. For P_{O_2} above 10^{-2} atm, l values are very similar which is expected in that case since the departure from stoichiometry of CoO is larger than the chromium concentration [4]. For P_{O_2} below 10^{-2} atm, chromium influences the electrical conductivity - P_{O_2} relation.

2°/ EXPERIMENTAL TECHNIQUES

a) Specimen preparation and characterization

As for previous studies [9] [10], single crystals have been grown by the zone melting technique using an arc image furnace. For Cr doped CoO, the polycrystalline rods which feed the melt were made by sintering a mixture of Co_3O_4 and Cr_2O_3 in air at 1300°C. Crystals were annealed at 1300°C in air. Chromium concentration was checked by chemical analysis and its uniformity was assessed using an electron microprobe. Creep tests were performed on specimens containing 3.86×10^{-3} at.Cr. Electrical conductivity, creep and diffusion on pure and Cr doped CoO have been performed on crystals prepared in the set-up of L.P.M.T.M. (Université Paris-Nord) [3].

b) Creep tests

Constant load compression creep tests were performed on parallelopipeds (size 2 x 2 x 5 mm) with (100) faces, using a machine already used in previous work [9]. Temperatures were between 1100°C and 1400°C and stresses σ between 13 MPa and 44 MPa. The oxygen activity was controlled by setting P_{O_2} in a flow

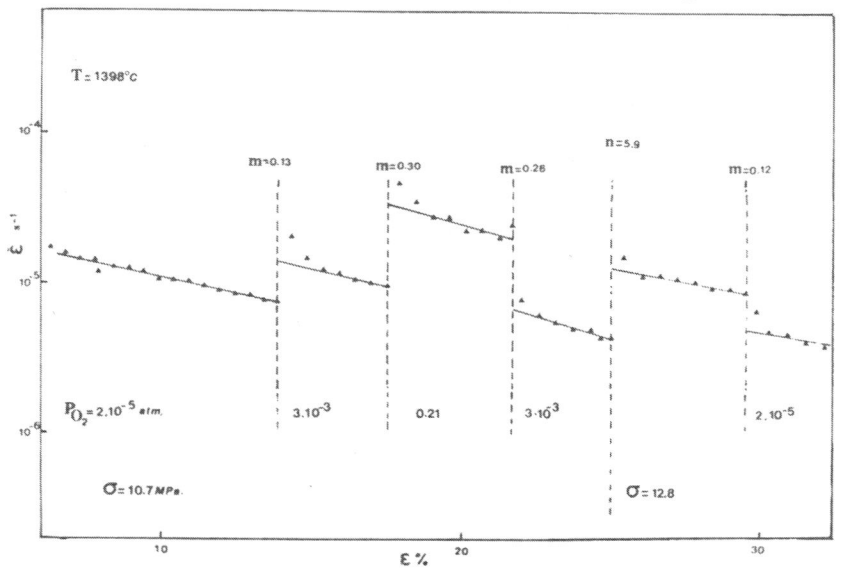

Fig. 3 : typical creep curves with one stress change and four P_{O_2} changes.

of air, of argon/oxygen or CO/CO2 mixtures ; P_{O_2} was measured by passing the gas through a zirconia cell placed before or after the creep machine. Care was taken to make sure that specimens have reached equilibrium after setting new thermodynamic parameters T and P_{O_2}.

The exponents in the creep equation (2) are determined by changing σ , T or P_{O_2} ; this is shown on fig. 3 where we show several P_{O_2} changes and one σ change. This allows calculation of m and n (eq. (2)).

c) <u>Diffusion experiments</u>

A thin layer of ^{18}O enriched cobalt oxide was deposited on a CoO single crystal by radio frequency sputtering. Before that, the CoO specimen had been mechanically polished and annealed in the conditions (T, P_{O_2}) of the diffusion measurements. An argon/oxygen mixture was used to control P_{O_2}. It was measured by a zirconia cell located in the specimen chamber.

At the end of the diffusion annealing, the specimen was quickly moved to a cold part of the furnace to avoid the formation of Co_3O_4. The ^{18}O concentration vs penetrations were measured in an ion probe analyser (SIMS). The curves show a maximum which is not located at the initial surface (fig. 4). That is due to ^{18}O diffusion through the surface, in the furnace atmosphere. The curves can be fitted by adjusting the parameters : D self-diffusion coefficient : P transparency coefficient of the surface; A, R_p, ΔR_p being respectively the initial number of deposited atoms, the position of the maximum and the spread of the initial distribution of the tracer, which has been assumed to be a gaussian. The Fick's law solution in this case can be written [16].

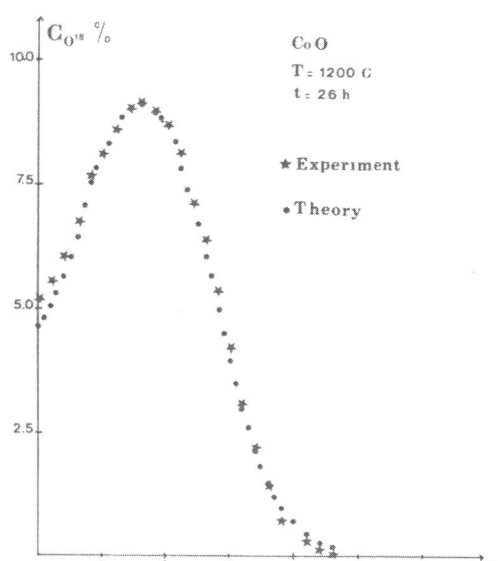

Fig. 4 : ^{18}O concentration vs penetration profile. The experimental points are well fitted to the analytical form (14).

$$c(x,t) = \frac{A}{2\sqrt{2\pi}\,\delta} \left\{ \exp - \frac{(R_p - x)^2}{2\,\delta^2} \left[1 + \text{erf} \frac{(a^2 R_p + x)}{a\sqrt{2\delta}} \right. \right.$$

$$\left. \left. ...- P \exp\left[- \frac{(R_p + x)^2}{2\,\delta^2}\right] \left[1 + \text{erf} \frac{(a^2\ R_p - x)}{a\sqrt{2\delta}} \right] \right\} \right. \qquad (14)$$

with $\delta = (R_p^2 + 2Dt)^{1/2}$ and $a = \frac{(2Dt)^{1/2}}{\Delta R_p}$

The tracer has been introduced as a thin film. After a small time, it can be described by a gaussian curve. The solution (14) describes its evolution taking into account a loss of tracer through the surface.

3°/ RESULTS

a) Creep experiments in pure CoO

Creep experiments in pure CoO have been discussed in previous papers [8], [9] and we know how the stationary creep rate $\dot{\varepsilon}_s$ depends on T, P_{O_2}, the stress σ and on the microstructure [6]. One consequence is that, at constant T and P_{O_2}, $\dot{\varepsilon}_s$ is not constant (fig. 3) because the microstructure and σ change with the strain In these conditions it is difficult to obtain an absolute curve of $\dot{\varepsilon}_s$ versus P_{O_2} for example.

Therefore, we have measured the influence of P_{O_2} by changing its value, keeping all other parameters constant (fig. 3). In order to make the comparison between diffusion and creep easier, we have displayed (fig. 5) the m values by representing $\log \dot{\varepsilon}_s$ vs $\log P_{O_2}$, using arbitrary units for $\dot{\varepsilon}_s$. The slopes of the straight lines are equal to experimental m values. They represent the chords instead of the tangents to the curve. This method tends to smooth the non-linear variations. For CoO, it can be seen on fig. 5 that the slope changes from a negative value to a positive one for P_{O_2} of $10^{-5} - 10^{-4}$ atm at lower temperatures. This has already been interpreted as evidence of a diffusion process controlled by a mixture of $V_O^{\bullet\bullet}$ and O_i^* point defects [10].

b) Oxygen diffusion measurements in pure CoO

Oxygen self diffusion coefficients have been measured at 1200°C with oxygen partial pressures ranging between 10^{-6} and 0.2 atm. The results have been plotted as $\log D$ vs $\log P_{O_2}$ on figure 6.

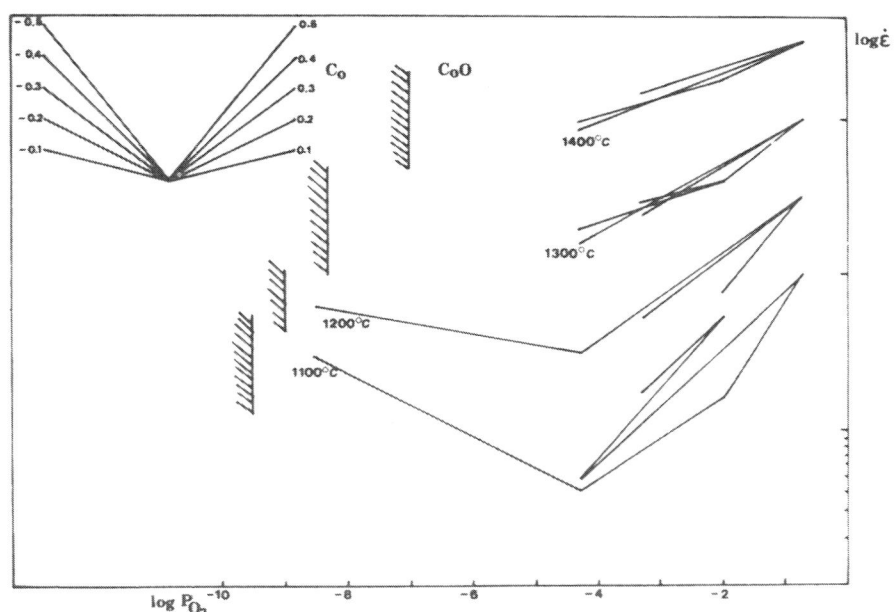

Fig. 5 : $\dot{\varepsilon}_S$ - P_{O_2} relation in pure CoO calculated using average m values ($\dot{\varepsilon}_S \propto P_{O_2}{}^m$) determined from experiments. The origins for $\dot{\varepsilon}_S$ are arbitrary. Reference slopes and corresponding m values have been indicated in the upper left corner.

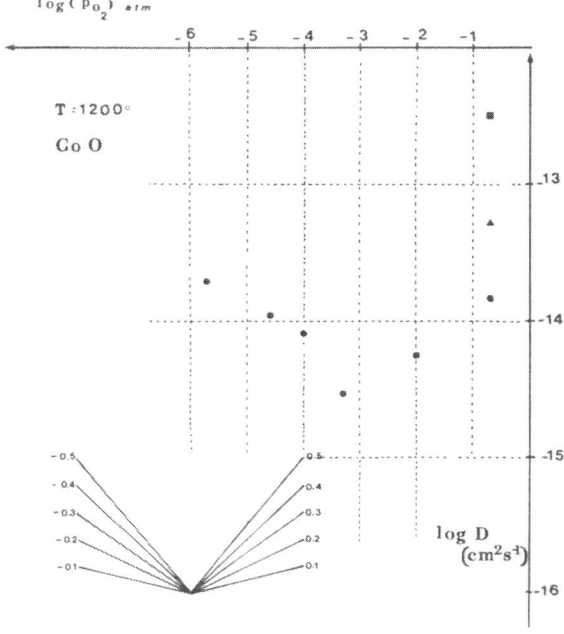

Fig. 6 : oxygen diffusion coefficient versus P_{O_2} at T = 1200°C in pure CoO. (points : this study, triangle : [18], square : [20]) Typical slopes have been indicated in the lower left corner.

Two different behaviours clearly appear : below about 10^{-3} atm the oxygen self diffusion coefficient decreases with increasing P_{O_2}. This result agrees well with a model which considers that at least two different types of defects are important in the oxygen sublattice.

There have been three studies of oxygen diffusion in CoO [18] [19] [20]. We can compare our results to those obtained by other authors [18], [20]. Practically the values can be compared at only one P_{O_2} (0.21 atm). It appears that the literature data are higher than ours. The tracer has been introduced in these studies by the isotopic exchange method ; in one case [18] the authors did not establish the tracer profiles and could therefore not detect any anomaly; they assumed an analytical profile to deduce D. The effect of doping led the authors to the conclusion that the main defects were oxygen vacancies [18]. The effect of oxygen partial pressure on diffusion was apparently the same as in our work and in that of ref. [20] which gives partial data at 1050°C.

c) Creep experiments in doped CoO

Measurements have been carried out on Cr doped CoO (fig. 7) in the same temperature (1130 - 1400°C) and P_{O_2} (10^{-9} - 0.2 atm) ranges as used for pure CoO. The results have been plotted as log $\dot{\varepsilon}_s$ vs log P_{O_2} on fig. 7 according to the same procedure used for pure CoO. The behaviours appear to be similar for pure and doped CoO at large P_{O_2} but doping substantially affects the low P_{O_2} regime.

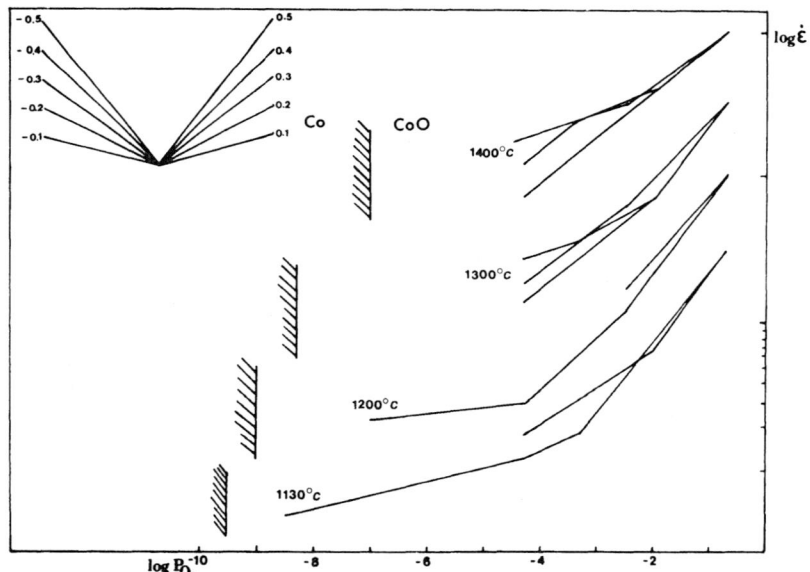

Fig. 7 : Same as fig. 5 for CoO doped with 3860 at. ppm Cr.

4°/ DISCUSSION

We have measured the P_{O_2} dependence of oxygen self diffusion in pure CoO and of creep rate in pure and Cr doped CoO ; it appears that at least two types of point defect play a role in the oxygen sublattice. Assuming simple point defects, with an oxygen vacancy at low P_{O_2} and an oxygen interstitial at high P_{O_2} we can explain our results. Nevertheless to assess this model and to obtain more information on the charge of the defects we have to look at the results in more detail.

The experimental slope of log $\dot{\varepsilon}_s$-log P_{O_2} or log D_0-log P_{O_2} in a given P_{O_2} range gives an m exponent which indicates the main defect responsible for the oxygen diffusion provided we take into account the behaviour of hole concentration depicted by l (fig. 2). By reporting in fig. 1 the m and l values corresponding to the same P_{O_2} and T, we can deduce the anionic defect controlling oxygen diffusion. The characterization depends on the accuracy of l and m measurements.

l values have been deduced from electrical conductivity measurements on the same single crystals as those used in creep and diffusion experiments [3], [15] ; this avoids many problems. The accuracy depends on the P_{O_2} range but is probably of the order of 0.02. The values higher than 0.25 obtained at highest P_{O_2} are significant. They can be interpreted as evidence for complex defects [17] or they could be due to a slight dependence on P_{O_2} of the hole mobility ($P_{O_2}^{1/10}$ to $P_{O_2}^{1/20}$). In the first case our analysis holds as l gives the true P_{O_2} dependence of the hole concentration, but it introduces an error in the second case.

The accuracy on measured m exponents is worse than on l. In creep tests, as it is not the true m but the slope of a chord in place of the tangent to the real curve, there is generaly a tendency to reduce m. The error may be particularly important when the curvature varies quickly, e.g. around a minimum (fig. 5). An error of 0.1 or more is possible, the scatter of experimental data being of the order of $\pm$ 0.05 [6]. In the diffusion experiments, the accuracy is in principle better ; but in this case, the number of points is small ; the uncertainty in D_0 measurements can reach 20 % ; a total uncertainty of 0.1 on m is possible. If the slope varies in a small P_{O_2} range the uncertainties can reach an even higher value.

Turning now to the results, it appears that in pure CoO creep tests indicate slopes between 0.2 and 0.5 when P_{O_2} ranges between 10^{-4} and 0.21 atm ; that seems a clear indication that neutral interstitials O_i^* play a role. The slope decreases when T increases (fig. 5) ; this indicates that another defect is significant. In that P_{O_2} range, l varies between 0.23 and 0.3.

Fig. 1 shows that it may be due to the ionization of O_i^* or to the influence of V_O, probably ionized.

At P_{O_2} below 10^{-4} atm, the slope changes and has an average value of -0.1, -0.2. In the same range l is between 0.15 and 0.21 $V_{\ddot{O}}$ is clearly compatible with these observations.

Oxygen diffusion in pure CoO at 1200°C shows a negative slope below 10^{-3} atm and a positive one above with $0.2 < |m| < 0.3$. Fig. 1 and 2 shows that $V_{\ddot{O}}$ alone can almost explain the results in the whole field but the appearance, $O_i^.$ at high P_{O_2} and $V_{\ddot{O}}$ or $V_{\dot{O}}$ at low P_{O_2} gives a better account of the diffusion mechanism.

Creep tests on Cr doped CoO behaves as on pure CoO above 10^{-4} atm (fig. 5 and 7). That is in complete agreement with the fact that l is about the same in this range (fig 2), which means that the oxide is insensitive to the doping impurity (doping level less than departure from stoichiometry). But, at low P_{O_2}, the slopes are reversed from those in pure CoO ; in that P_{O_2} range, l is higher in doped than in pure CoO ($\cong 0.2$) the results can be explained if $V_{\ddot{O}}$ are present, the mixture with interstitials being also possible.

It is difficult to draw a definite conclusion regarding the ionization behaviour of minority point defects. However, taking into account the uncertainties, the most probable defects are a mixing of O_i^* and $V_{\ddot{O}}$, the interstitial dominating at high P_{O_2} and the vacancy at low P_{O_2}. Defects such as $O_i^{\dot{}}$ and $V_{\dot{O}}$ could influence the results particularly in the intermediate P_{O_2}.

CONCLUSION

The most striking result of this work is probably the similar behaviour observed between oxygen self diffusion and stationary creep rate dependences on oxygen partial pressure. This is expected from creep models [6] [7] [8] and has been observed also for Cu_2O [6] [21]. Uncertainties in data and measurements lead to a semi quantitative agreement only, but such a result is of great importance regarding the description of high temperature mechanical properties of compounds.

As already pointed out [8] the importance of a detailed knowledge of electrical conductivity behaviour of the studied samples is essential for interpreting the diffusion and creep results.

Diffusion studies are in progress, they should bring more information concerning the details of point defect populations on the oxygen sublattice, which are so rarely explored.

REFERENCES

[1]   P. KOFSTADT "Non stoichiometry, Diffusion and electrical
      conductivity in binary metal oxides" (Wiley-Interscience,
      New York, 1972)
[2]   C. MONTY, Radiation Effects 74:29 (1983).
[3]   G. PETOT-ERVAS, P. OCHIN, B. SOSSA, Solid State Ionics
      12:277 (1984).
[4]   R. DIECKMANN, Zeit. Physik. Chem. Neue Folge 107:189
      (1977).
[5]   C.R.A. CATLOW in "Non stoichiometric oxides" O. Toft
      Sorensen (ed.) (Academic Press, 1981, p. 61-98)
[6]   A. DOMINGUEZ-RODRIGUEZ, J. CASTAING, Rad. Effects
      75:309 (1983).
[7]   J. PHILIBERT, Solid State Ionics 12:321 (1984).
[8]   J. CASTAING, A. DOMINGUEZ-RODRIGUEZ, C. MONTY, Int. Symp. on
      "Plastic deformation of Ceramic materials". The Pennsyl-
      vania State University, July 20-22 (1983) Plenum Press
      (to be published)
[9]   A. DOMINGUEZ-RODRIGUEZ, M. SANCHEZ, R. MARQUEZ, J. CASTAING,
      C. MONTY, J. Phys., Paris C.3 42:67 (1981).
[10] A. DOMINGUEZ-RODRIGUEZ, M. SANCHEZ, R. MARQUEZ, J. CASTAING,
      C. MONTY, J. PHILIBERT, Phil Mag A 46A:411 (1982).
[11] J.L. ROUTBORT, Acta Met. 30, 663-671 (1982)
[12] C. MONTY, Chap. XII in "Défauts ponctuels dans les Solides -
      Confolant 1977" Editions de Physique, 91402 Orsay,
      France (1978)
[13] R. GOMRI, H. BOUSSETTA, C. BAHEZRE, C. MONTY, Solid State
      Ionics 12:227 (1984).
[14] C. DUBOIS, C. MONTY, J. PHILIBERT, Phil. Mag.A
      46A:419 (1982).
[15] C. CLAUSS, Thèse de 3è cycle - Paris 1984 (to be published)
[16] M. MEYER, B. BARBEZAT, C. EL HOUCH, R. TALON, J. de Physique
      C6, sup. 7, 41:327 (1980).
[17] G. PETOT-ERVAS "Colloquium on basic properties of binary
      oxides" Sevilla 12-16 Sept. 1983 - Ed. A. DOMINGUEZ-
      RODRIGUEZ, J. CASTAING, R. MARQUEZ, Sevilla University
      Press (1984)
[18] W.K. CHEN, R.A. JACKSON, J. Phys. Chem. Sol. 30:1309 (1969).
[19] J.B. HOLT, Proc. Brit. Cer. Soc. 11:157 (1967).
[20] S. YAMAGUCHI, M. SOMENO, Trans. Japan Inst. Metals
      23:259 (1982).
[21] T. BRETHEAU, B. PELLISSIER, B. SIEBER, Acta Metall.
      29:1617 (1981).

MECHANISMS OF IMPURITY DIFFUSION IN RUTILE[*]

N. L. Peterson and J. Sasaki[**]

Materials Science and Technology Division
Argonne National Laboratory, Argonne, IL 60439

ABSTRACT

Tracer diffusion of ^{46}Sc, ^{51}Cr, ^{54}Mn, ^{59}Fe, ^{60}Co, ^{63}Ni, and ^{95}Zr, was measured as functions of crystal orientation, temperature, and oxygen partial pressure in rutile single crystals using the radioactive tracer sectioning technique. Compared to cation self-diffusion, divalent impurities (e.g., Co and Ni) diffuse extremely rapidly in TiO_2 and exhibit a large anisotropy in the diffusion behavior; divalent-impurity diffusion parallel to the c-axis is much larger than it is perpendicular to the c-axis. The diffusion of trivalent impurity ions (Sc and Cr) and tetravalent impurity ions (Zr) is similar to cation self-diffusion, as a function of temperature and of oxygen partial pressure. The divalent impurity ions Co and Ni apparently diffuse as interstitial ions along open channels parallel to the c-axis. The results suggest that Sc, Cr, and Zr ions diffuse by an interstitialcy mechanism involving the simultaneous and cooperative migration of tetravalent interstitial titanium ions and the tracer-impurity ions. Iron ions diffused both as divalent and as trivalent ions.

[*]Work supported by the U.S. Department of Energy.
[**]Present address: Mitsubishi Chemical, Tokyo, Japan.

1. INTRODUCTION

The crystal structure and the atomic-defect structure significantly influence the mass-transport process in solids. The crystal structure of rutile is noncubic as may be seen in Fig. 1. The sublattice of the Ti^{4+} ions is body-centered tetragonal. Each Ti^{4+} ion is surrounded by a slightly distorted octahedron of six O^{2-} ions. These TiO_6 octahedra share edges and corners in such a way that each oxygen ion belongs to three neighboring octahedra. The titanium ions lie in rows directly along the c-axis. When the structure is viewed along the c-axis (Fig. 1(b)), it is seen that the rutile lattice has "open channels" parallel to the c-axis; open channels perpendicular to the c-axis are not apparent. The open channels may cause anisotropy in the diffusion process and may allow rapid diffusion of smaller ions parallel to the c-axis.

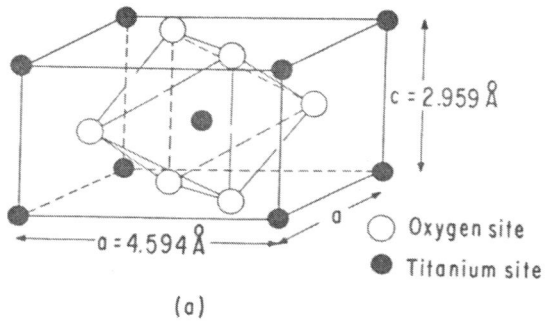

(a)

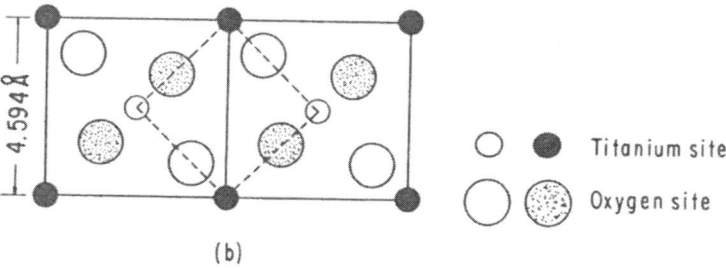

(b)

Fig. 1. (a) A unit cell of the rutile (TiO_2) structure. (b) An end view of the atomic arrangement for the [001] axial direction. The open channel along the c-axis is indicated by the dashed lines.

Rutile is a metal-excess (or oxygen deficient) nonstoichiometric oxide. The extent of the nonstoichiometry and the major defects in rutile is still controversial. Recently Akse and Whitehurst[1] have measured cation self-diffusion over a limited range of oxygen partial pressure p_{O_2}. They conclude that tetravalent interstitial titanium ions $Ti_i^{\bullet\bullet\bullet\bullet}$ are the major atomic defects in rutile. However, measurements of nonstoichiometry and electrical conductivity suggest that oxygen vacancies and possibly trivalent interstitial titanium ions are also important defects in rutile.[2] In addition, the usual presence of small concentrations of Al and other impurities in rutile crystals further complicates the defect equilibrium near the stoichiometric composition.

In the present study, we present measurements of impurity tracer diffusion in rutile as functions of crystal orientation, temperature, charge state of the impurity ion, oxygen partial pressure, and concentration of Al impurity. These results provide insight into the importance of open channels for diffusion in rutile and help identify the most probable impurity-diffusion mechanisms. Since the defect structure of rutile is of importance in the interpretation of the diffusion data, the defect structure as a function of p_{O_2}, impurity content, and temperature is discussed in the next section before the diffusion measurements are presented.

2. DEFECT STRUCTURE IN RUTILE

Extensive studies of defects in rutile have been conducted as a function of oxygen partial pressure by means of thermogravimetry,[3-7] electrochemical titration,[8-10] electrical conductivity,[7,11-14] tracer diffusion,[1] and ion-beam channeling[15] measurements. Many of the results were interpreted in terms of only one defect, usually tetravalent interstitial titanium ions $Ti_i^{\bullet\bullet\bullet\bullet}$ or doubly charged oxygen vacancies $V_O^{\bullet\bullet}$. Kofstad[2] was one of the first to propose the coexistence of several defects in rutile, namely trivalent interstitial titanium ions $Ti_i^{\bullet\bullet\bullet}$, $Ti_i^{\bullet\bullet\bullet\bullet}$, and $V_O^{\bullet\bullet}$. Although Kofstad's model accurately describes the nonstoichiometry data, it is less consistent with the electrical-conductivity data. Recently Marucco et al.,[7] using thermogravimetric and conductivity measurements, carried out an extensive study of unusually pure rutile. They obtained a satisfactory fit to the data using a defect model involving $Ti_i^{\bullet\bullet\bullet\bullet}$, $V_O^{\bullet\bullet}$, and $V_O^{\bullet}$. The existence of $V_O^{\bullet}$ is still questionable and is considered to play a significant role in the defect equilibrium only at compositions near stoichiometry. For the purpose of interpreting the impurity-diffusion data presented in Section 3, it is more reasonable to develop a defect model for rutile based on the simultaneous existence of $Ti_i^{\cdots\cdot}$ and $V_0^{\cdots}$.

The formation of the principle atomic defects $Ti_i^{\cdots\cdot}$ and $V_0^{\cdots}$ in rutile can be described by the reactions

$$Ti_{Ti}^x + 2O_0^x \rightleftarrows Ti_i^{\cdots\cdot} + 4e' + O_2(g) \tag{1}$$

$$O_0^x \rightleftarrows V_0^{\cdots} + 2e' + \frac{1}{2} O_2(g), \tag{2}$$

where Ti_{Ti}^x, O_0^x, and e' denote a neutral cation on a cation site, a neutral anion on an anion site, and an electron, respectively. Application of the law of mass action to these reactions yields

$$K_1 = [Ti_i^{\cdots\cdot}][e']^4 \, p_{O_2} \tag{3}$$

and

$$K_2 = [V_0^{\cdots}][e']^2 \, p_{O_2}^{1/2}, \tag{4}$$

where the square brackets denote concentrations of defects per mole of TiO_2, and K_1 and K_2 denote equilibrium constants for reaction 1 and 2, respectively. The deviation from stoichiometry x in TiO_{2-x} is given by

$$x = 2[Ti_i^{\cdots\cdot}] + [V_0^{\cdots}]. \tag{5}$$

The effect of impurities may become important to the study of those properties of rutile which are sensitive to the defect concentration. The major impurity in the rutile crystals used in the present study (and a common impurity in most of the previous studies) is Al. In Al-doped rutile, the value of [e'] will be lower than in undoped rutile. Hence the reaction

$$null \rightleftarrows e' + h^{\cdot} \tag{6}$$

must be considered at high p_{O_2} and low T. Application of the law of mass action to eqn (6) yields

$$K_i = [e'][h^{\cdot}]. \tag{7}$$

In addition to the reactions expressed in eqns (1), (2), and (6), we must consider the following two reactions for Al-doped rutile

$$TiO_2 + 2Al_2O_3 \rightleftarrows 4Al_{Ti}' + 8O_0^x + Ti_i^{\cdots\cdot} \tag{8}$$

272

$$Al_2O_3 \xrightleftharpoons{} 2Al'_{Ti} + 3O^x_O + V^{\bullet\bullet}_O. \tag{9}$$

If $V^{\bullet\bullet}_O$, $Ti^{\bullet\bullet\bullet\bullet}_i$, Al'_{Ti}, e', and $h^\bullet$ are the primary defects in Al-doped rutile, electroneutrality requires

$$[Al'_{Ti}] + [e'] = 4[Ti^{\bullet\bullet\bullet\bullet}_i] + 2[V^{\bullet\bullet}_O] + [h^\bullet]. \tag{10}$$

If Al dissolves substitutionally in $(Ti_{1-y}Al_y)O_{2-x}$, the deviation from stoichiometry is given by eqn (5). Using the definition

$$g \equiv [e']/[Al'_{Ti}] \tag{11}$$

and eqns (3), (4), (7), and (10), one can write

$$[V^{\bullet\bullet}_O] = \frac{K^2_2}{4K_1} \left[-1 + \sqrt{1 + \frac{4K_1}{K^2_2}[Al'_{Ti}] \left(1 + g - \frac{K_1}{g[Al'_{Ti}]^2}\right)} \right] \tag{12a}$$

$$[Ti^{\bullet\bullet\bullet\bullet}_i] = \frac{K_1}{K^2_2}[V^{\bullet\bullet}_O]^2 \tag{12b}$$

and

$$p_{O_2} = \frac{K_1}{[Ti^{\bullet\bullet\bullet\bullet}_i] \, g^4 [Al'_{Ti}]^4}. \tag{12c}$$

Using values of K_1 and K_2 estimated from Marucco et al.,[7]

$$K_1 = 3.3 \times 10^{11} \ \exp\left(-233.3 \ \frac{kcal}{mole} \ / \ RT\right)$$

$$K_2 = 3.0 \times 10^2 \ \exp\left(-105.4 \ \frac{kcal}{mole} \ / \ RT\right)$$

and eqn (5), one can assume values of g and calculate $[Ti^{\bullet\bullet\bullet\bullet}_i]$, $[V^{\bullet\bullet}_O]$, and x as functions of p_{O_2}, T, and $[Al_{Ti}]$. Figure 2 shows the p_{O_2} dependence of $[Ti^{\bullet\bullet\bullet\bullet}_i]$, $[V^{\bullet\bullet}_O]$, and x at 1200°C for $[Al'_{Ti}] = 4.74 \times 10^{-4}$ (160 ppm by wt. Al in TiO_2). The contribution to the defect concentrations from eqn (6) is small and has been omitted from the calculations.[16] The asymptotic slopes 1/5 and 1/10 are shown in appropriate locations on Fig. 2. The Magneli

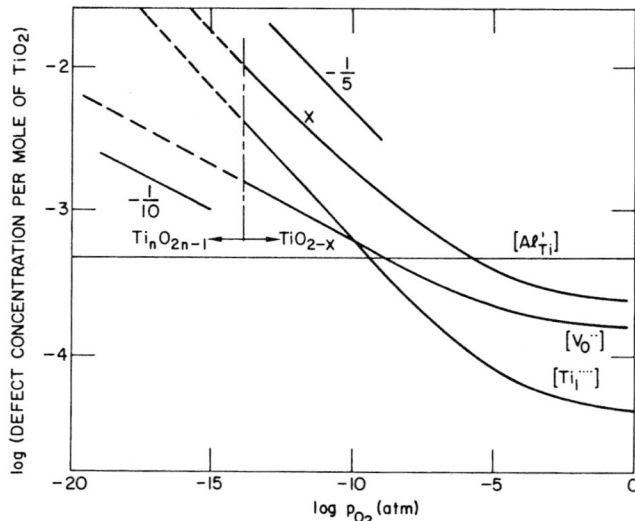

Fig. 2.
Oxygen-partial
pressure de-
Oxygen-partial
pressure depence
of defect concen-
trations in 160
ppm by wt. Al-
doped TiO_{2-x} at
$1200^\circ C$.

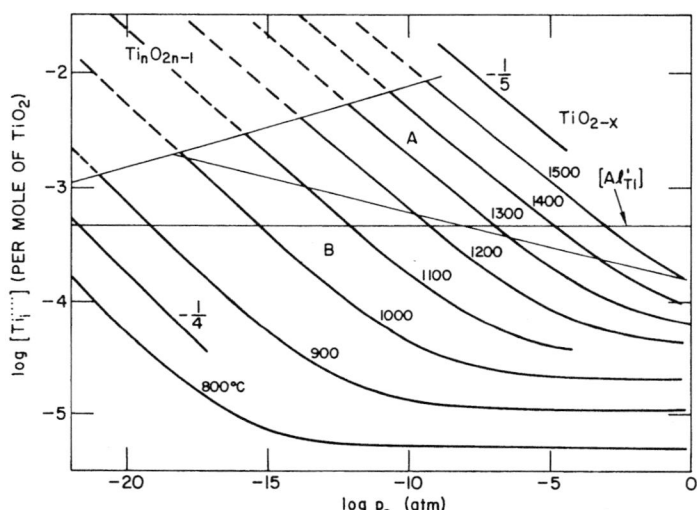

Fig. 3. Oxygen-partial-pressure dependence of $[Ti_i^{\bullet\bullet\bullet\bullet}]$ in 160 ppm by wt. Al-doped TiO_{2-x}. $[Ti_i^{\bullet\bullet\bullet\bullet}] > [V_0^{\bullet\bullet}]$ in region A, and $[V_0^{\bullet\bullet}] > [Ti_i^{\bullet\bullet\bullet\bullet}]$ in region B.

phases Ti_nO_{2n-1} are the stable forms of TiO_{2-x} at low values of p_{O_2}. In this region, the simple point-defect theory is inappropriate, and the defect concentrations are shown as dashed lines. The location of the boundary between Ti_nO_{2n-1} and the

274

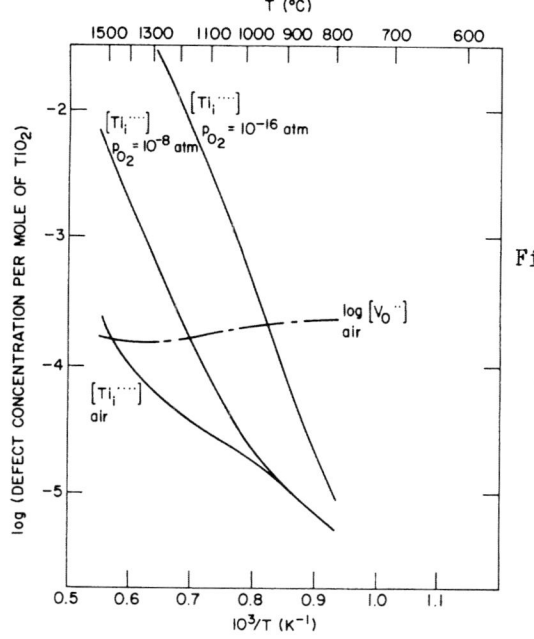

Fig. 4. Temperature dependence of the defect concentrations in 160 ppm by wt. Al-doped TiO_{2-x}.

TiO_{2-x} phase is approximate.

Figure 3 shows the p_{O_2} dependence of $[Ti_i^{\cdots\cdot}]$ at several temperatures. The value of $[Ti_i^{\cdots\cdot}]$ varies with temperature at constant p_{O_2} even in the extrinsic region (Fig. 3); whereas $[V_O^{\cdot\cdot}]$ is virtually independent of temperature in the same p_{O_2} region. This behavior is due to the fact that $Ti_i^{\cdots\cdot}$ is not the major defect in this p_{O_2} regime. The temperature dependencies of $[Ti_i^{\cdots\cdot}]$ and $[V_O^{\cdot\cdot}]$ are shown in Fig. 4 in the form of Arrhenius plots.

Impurities may dissolve interstitially and diffuse as a freely migrating interstitial ion. The diffusion coefficient for impurity diffusion by this mechanism may be expected to be highly anisotropic with rapid transport parallel to the c-axis and may be independent of p_{O_2}.

Impurities may prefer to dissolve substitutionally and diffuse by the interstitialcy mechanism. In this mechanism, an interstitial titanium ion may jump to a substitutional site displacing the substitutional impurity ion into a neighboring interstitial site. The interstitial impurity ion may then jump to a neighboring substitutional site displacing the lattice titanium ion into a neighboring interstitial site. The diffusion coefficient for impurities diffusing by the interstitialcy mechanism should vary with p_{O_2} and $[Al'_{Ti}]$ in the same manner as does $[Ti_i^{\cdots\cdot}]$. Hence, the variation of the impurity diffusion coefficient with varying p_{O_2}, $[Al'_{Ti}]$, temperature, and crystallographic direction may identify the mechanism of diffusion.

3. EXPERIMENTAL PROCEDURE

The tracer-sectioning technique has been used to determine the tracer-diffusion coefficients of ^{46}Sc, ^{51}Cr, ^{54}Mn, ^{59}Fe, ^{60}Co, ^{63}Ni, and ^{95}Zr as functions of p_{O_2}, temperature, and crystallographic orientation in single crystals of rutile. Several single-crystal rutile boules were purchased from Commercial Crystal Laboratories, Inc. The major impurity Al was specifically analyzed for, and found to be, 160 ppm by wt. in one boule and

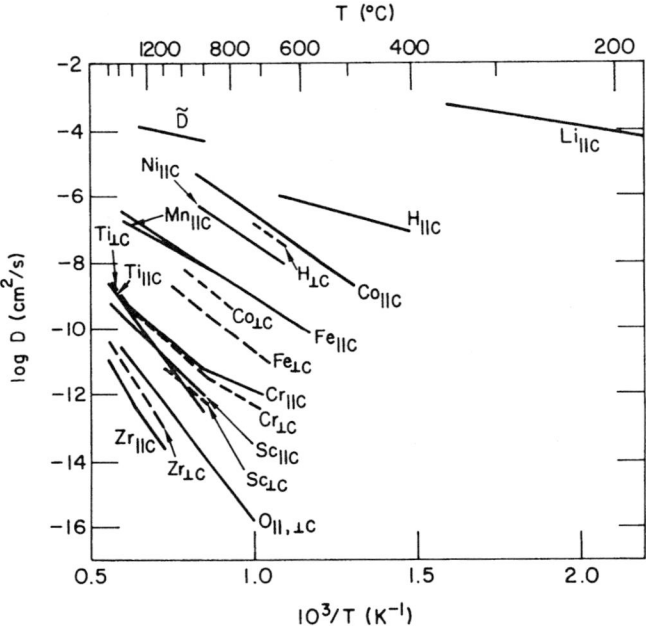

Fig. 5. Temperature dependencies of impurity diffusion coefficients in rutile in air. Cation[17] and anion[18,19] self-diffusion in air, chemical diffusion $\tilde{D}$ in CO/CO_2 gas mixtures,[20] lithium chemical diffusion in vacuum,[21] and hydrogen chemical diffusion in H_2/H_2O gas mixtures[22] are also shown. Diffusion $\perp c$ is denoted by dashed lines.

50 ppm by wt. in a second boule. The details of the experimental procedures may be found in Ref. 16.

4. RESULTS AND DISCUSSION

4.1 Impurity Diffusion in Rutile in Air

The values of D^* for the diffusion of ^{60}Co, ^{63}Ni, ^{59}Fe, ^{54}Mn, ^{51}Cr, ^{46}Sc, and ^{95}Zr in rutile in air are plotted in Fig. 5 in the form of Arrhenius plots. Diffusion coefficients for other impurities reported by previous investigators are also shown in Fig. 5.

The magnitude of the diffusion coefficient is very sensitive to the charge states of the impurity. Generally, the tracer diffusion coefficient for monovalent impurities (Li,H) is larger than that for divalent impurities (Co,Ni), which is larger than that for trivalent impurities (Cr,Sc), which is larger than that for tetravalent impurities (Zr). Also, the larger the diffusion coefficient, the greater the anisotropy in the diffusion coefficient. Although the data for Li diffusion in pure rutile are not well established (the published values involve considerable concentration gradients), the value of $D_{//c}(Li)$ is very large, and Johnson[1] suggests that $D_{//c}(Li) \approx 10^8 D_{\perp c}(Li)$. For the diffusion of divalent impurities like Co in rutile, $D_{//c}^*(Co) > 10^3 D_{\perp c}^*(Co)$. Rather little anisotropy is observed for the diffusion of the trivalent impurities Sc and Cr. Impurities that commonly coexist in both the divalent and trivalent states (Fe,Mn) diffuse at rates and with anisotropies that lie between the values observed for divalent impurities (Co,Ni) and trivalent impurities (Cr,Sc). The anisotropy in the diffusion of the tetravalent impurity Zr is opposite to that observed for the divalent impurities; $D_{\perp c}^*(Zr)$ is nearly 5 times larger than $D_{//c}^*(Zr)$. The mechanisms of diffusion responsible for this diverse impurity-diffusion behavior in rutile are considered in the next section where the p_{O_2} dependence of the impurity diffusion coefficients is discussed.

4.2. p_{O_2} Dependence of Impurity Diffusion in Rutile

The impurity tracer-diffusion coefficients in rutile have been measured as a function of p_{O_2} for the diffusion of ^{60}Co, ^{46}Sc, ^{59}Fe, ^{51}Cr, and ^{95}Zr. The results are particularly useful in identifying the mechanism of diffusion and are discussed in detail for ^{60}Co and ^{46}Sc. Due to space limitations, the results for ^{59}Fe, ^{51}Cr, and ^{95}Zr are only mentioned.

4.2.1. ^{60}Co diffusion in rutile. Our first measurements of the p_{O_2} dependence of ^{60}Co diffusion were limited to 800 and 1000°C because of the very large values of $D^*_{\parallel c}(Co)$ at high temperatures. These values of D^*_{Co} were independent of p_{O_2} because they were predominantly in the extrinsic region. Recent measurements of $D^*_{\parallel c}(Co)$ at 1100 and 1200°C in the more pure rutile (50 ppm by wt. Al) by Hoshino and Peterson[23] are shown in Fig. 6. The concentration of $Ti_i^{\cdots\cdots}$, calculated from eqn (12) for rutile containing 50 ppm by wt. trivalent ions, is also shown in the figure.

There are two features of the data in Fig. 6 that must be considered. First, the p_{O_2} dependencies of $D^*_{\parallel c}(Co)$ and $[Ti_i^{\cdots\cdots}]$ are identical. The large anisotropy in the diffusion coefficients, $D^*_{\parallel c}(Co) \ggg D^*_{\perp c}(Co)$ (Fig. 5), results from rapid interstitial migration along the open channels parallel to the c-axis in the rutile structure. The p_{O_2} dependencies of $D^*(Co)$ and $[Ti_i^{\cdots\cdots}]$ suggest that cobalt prefers to dissolve substitutionally in TiO_2; the cobalt ion is pushed into the open channels by an

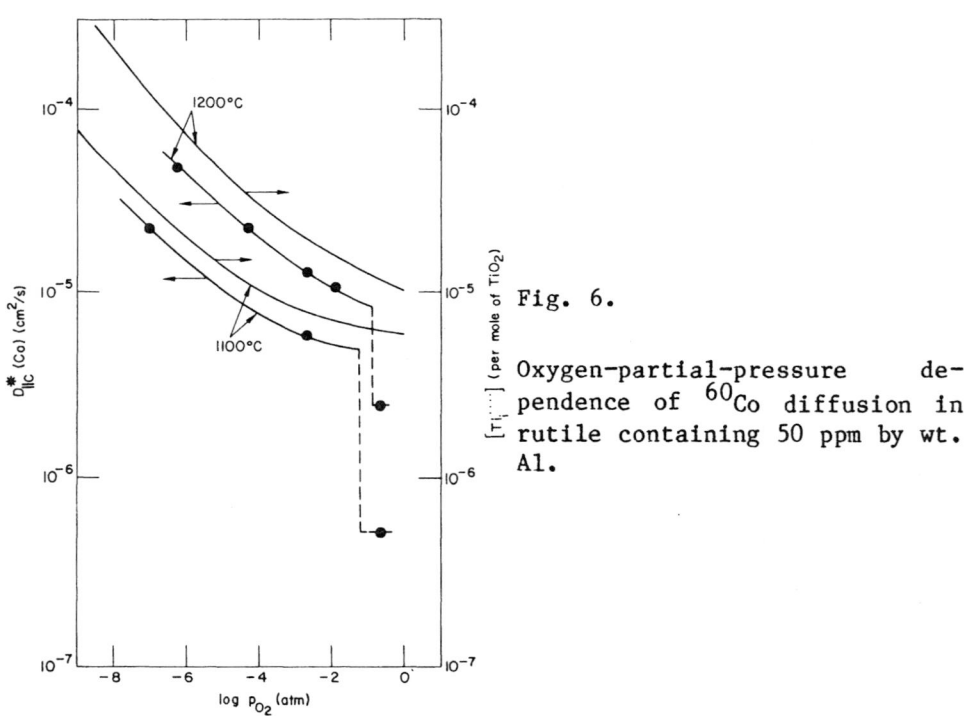

Fig. 6.

Oxygen-partial-pressure dependence of ^{60}Co diffusion in rutile containing 50 ppm by wt. Al.

interstitialcy mechanism involving the simultaneous motion of $Ti_i^{\bullet\bullet\bullet\bullet}$ defects and Co ions.

Second, both the earlier data at 800°C and the data at 1100 and 1200°C in Fig. 6 show a considerable change in $D_{||c}^*(Co)$ near $p_{O_2} = 10^{-2}$ atm. This same behavior is noted in Sec. 4.2.3 for Fe diffusion in rutile. A possible explanation for this behavior is as follows. As the p_{O_2} increases, the value of [e'] decreases and the point of intrinsic electronic equilibrium (eqn (6)) is reached in trivalent-impurity doped TiO_2. According to Yahia[24] and Rudolph,[25] trivalent-impurity doped rutile shows a change of electronic conductivity from n type to p type in the range $10^{-4} < p_{O_2} < 10^{-2}$ at 800°C, precisely the region where the diffusion coefficient changes rapidly with varying p_{O_2}. This change in conduction mechanism may produce an increase in the $[Co^{3+}]/[Co^{2+}]$ ratio. Since Co^{3+} may be expected to diffuse by the interstitialcy mechanism (with little anisotropy in D^*) and Co^{2+} may diffuse by an interstitial mechanism along the open channels at a much more rapid rate, a small increase in the $[Co^{3+}]/[Co^{2+}]$ ratio may produce a significant decrease in $D_{||c}^*(Co)$. The fact that the effective activation energy for $D^*(Co)$ is at least 10 kcal/mole smaller at $p_{O_2} = 10^{-4}$ atm than at p_{O_2} for air is consistent with this explanation. No change in D^* is observed in this p_{O_2} regime for impurities whose charge state is expected to be independent of the conduction mechanism.

The limited data for the diffusion of ^{63}Ni in TiO_2 (see Fig. 5) would suggest that other divalent transition-metal impurities may diffuse rapidly by the same diffusion mechanism responsible for ^{60}Co migration in rutile, with considerable anisotropy in D^*.

4.2.2. ^{46}Sc diffusion in rutile. The p_{O_2} dependence of the diffusion of ^{46}Sc in rutile is shown in Fig. 7; the results by Akse and Whitehurst[1] for cation self-diffusion in TiO_2 are also shown. Values of $D_{||c}^*(Sc)$ are proportional to $p_{O_2}^{-1/5}$ in the intrinsic region, and the anisotropy and p_{O_2} dependence of $D^*(Sc)$ are quite similar to those for cation self-diffusion in TiO_2. The p_{O_2} dependence of $[Ti_i^{\bullet\bullet\bullet\bullet}]$ in 160 ppm by wt. Al-doped rutile is represented by the solid lines in Fig. 7; the values of $D_{||c}^*(Sc)$ are in good agreement with these lines. At 1100°C,

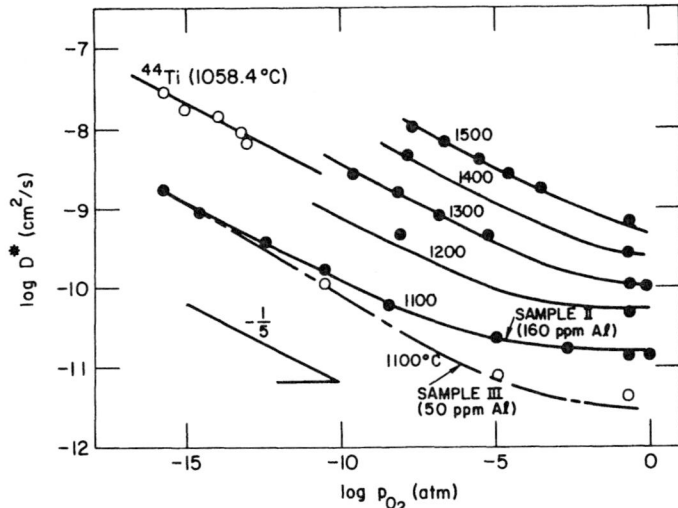

Fig. 7. Oxygen-partial-pressure dependence of ^{46}Sc diffusion parallel to the c-axis in rutile. All data are for rutile containing 160 ppm by wt. Al except for three points at 1100°C for rutile containing 50 ppm by wt. Al. The data for cation self-diffusion at 1058.4°C are also shown.[1]

rutile samples containing two different amounts of Al impurity were used. The dot-dash line in Fig. 7 shows the theoretical curve for $D_{||c}^*(Sc)$ estimated from the theory described in Section 2 (see eqn (12)) for a sample containing 50 ppm by wt. Al. Clearly, $D_{||c}^*(Sc)$ is strongly coupled to the $[Ti_i^{\bullet\bullet\bullet\bullet}]$ which suggests that Sc ions move by the interstitialcy mechanism involving the simultaneous motion of Ti defects and Sc ions. This also suggests that cation self-diffusion in rutile occurs by the interstitialcy mechanism.

The temperature dependence of $D^*(Sc)$ involves the temperature dependence of $[Ti_i^{\bullet\bullet\bullet\bullet}]$. Figure 8 shows Arrhenius plots for $D^*(Sc)$ in rutile in air and in $p_{O_2} = 10^{-8}$ atm. Using the temperature dependencies of $[Ti_i^{\bullet\bullet\bullet\bullet}]$ for air and for $p_{O_2} = 10^{-8}$ atm shown in Fig. 4, one may obtain values of $D_{||c}^*(Sc)/[Ti_i^{\bullet\bullet\bullet\bullet}]$ as a function of temperature (see Figure 8). The slope of the Arrhenius plot for $D_{||c}^*(Sc)/[Ti_i^{\bullet\bullet\bullet\bullet}]$, 16.7 kcal/mole, gives the migration energy for Sc^{3+} ions via the interstitialcy mechanism plus the impurity-defect binding energy minus the temperature dependence of the correlation factor for impurity diffusion; the latter two terms tend to cancel one another and are generally small compared to the

280

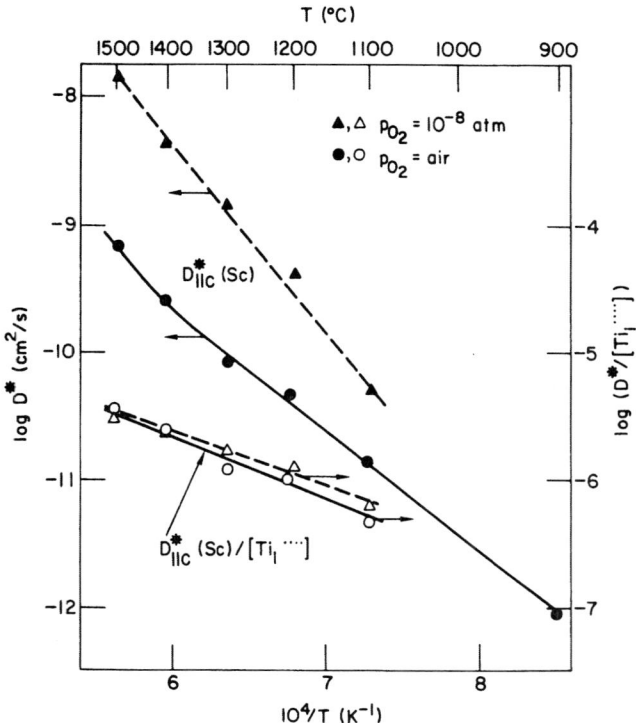

Fig. 8. Temperature dependence of ^{46}Sc diffusion parallel to the c-axis in rutile containing 160 ppm by wt. Al. The values of $D^*(Sc)/[Ti_i^{····}]$ are also shown.

migration energy. The near equality of $D^*_{//c}(Sc)/[Ti_i^{····}]$ at $p_{O_2} = 10^{-8}$ atm and at p_{O_2} for air suggests that the mobility of Sc^{3+} ions is independent of p_{O_2} and that $Ti_i^{····}$ is the principle cation defect at both values of p_{O_2}.

Scandium and other trivalent ions such as Cr (Sec. 4.2.4) diffuse through the rutile lattice by an interstitialcy mechanism and avoid the open channels parallel to the c-axis even though Cr^{3+} has a smaller ionic radius than Co^{2+} [26]. Hence the charge state of the impurity ion plays a more major role than does the size of the ion in determining the site occupancy for impurity ions in rutile.

4.2.3. ^{59}Fe <u>diffusion in rutile</u>. As observed for ^{60}Co tracer diffusion in rutile, the p_{O_2} dependence of ^{59}Fe diffusion shows a large change near $p_{O_2} = 10^{-2}$ atm. The same explanation offered to account for this phenomenon for $D^*(Co)$ should also be appropriate for $D^*(Fe)$ in rutile.

The rather large anisotropy in the diffusion coefficients, $D^*_{\parallel c}(Fe) \gg D^*_{\perp c}(Fe)$ (Fig. 5), suggests that at least a fraction of the iron ions diffuses as Fe^{2+} along the open channels parallel to the c-axis as observed for other divalent ions (Co,Ni). However, the p_{O_2} dependence of $D^*(Fe)$ suggests that a significant fraction of the iron ions dissolves substitutionally as Fe^{3+} and diffuses by the interstitialcy mechanism, as observed for other trivalent ions (Sc,Cr). The magnitude of $D^*(Fe)$ is between $D^*(Co)$ and $D^*(Sc)$, which adds further support to the idea that iron dissolves both as Fe^{2+} and as Fe^{3+}. For a detailed analysis of Fe diffusion in rutile, see Ref. 16. The limited data available for ^{54}Mn diffusion in rutile in air (Fig. 5) are very similar to those for $D^*(Fe)$, thus suggesting that both Mn^{2+} and Mn^{3+} ions contribute to diffusion.

4.2.4. ^{51}Cr and ^{95}Zr diffusion in rutile. The p_{O_2} dependencies of ^{51}Cr and ^{95}Zr diffusion were determined both parallel to and perpendicular to the c-axis. $D^*(Cr)$ and $D^*(Zr)$ are proportional to $p_{O_2}^{-1/5}$, in good agreement with the p_{O_2} dependence of $[Ti_i^{\cdots\cdots}]$. Hence, like scandium, chromium and zirconium ions move by the interstitialcy mechanism involving the simultaneous motion of a titanium ion and an impurity ion. The slopes of the Arrhenius plots for $D^*/[Ti_i^{\cdots\cdots}]$ suggest that the activation energy for impurity migration by the interstitialcy mechanism exhibits very little anisotropy.

5. CONCLUSIONS

The experimental results for diffusion of ^{46}Sc, ^{51}Cr, ^{54}Mn, ^{59}Fe, ^{60}Co, ^{63}Ni, and ^{95}Zr as functions of crystal orientation, oxygen partial pressure, and temperature in rutile suggest the following conclusions. (1) Compared to cation self-diffusion, divalent impurity ions (Co and Ni) diffuse extremely rapidly in rutile and exhibit a large anisotropy in the diffusion coefficients. The divalent impurity ions diffuse rapidly along open channels parallel to the c-axis and are knocked into the channels by an interstitialcy mechanism involving interstitial Ti ions. (2) Trivalent (Sc,Cr) and tetravalent (Zr) impurity ions dissolve substitutionally in rutile and diffuse by an interstitialcy mechanism involving the simultaneous and cooperative motion of titanium ions and impurity ions. (3) Mixed-valence impurity ions (Fe,Mn) diffuse both as divalent ions by the interstitial mechanism along the channels and as trivalent ions by the interstitialcy mechanism. (4) The charge state of the impurity ion seems to play a more important role than does the size of the impurity ion in determining whether an impurity ion enters the open channels parallel to the c-axis in the rutile structure. Small trivalent impurity ions avoid the open channels; whereas larger divalent ions diffuse rapidly along these channels.

ACKNOWLEDGEMENTS

The authors wish to thank C. L. Wiley, G. J. Talaber, and L. Nowicki for their assistance with the experimental studies.

REFERENCES

1. J. R. Akse and H. B. Whitehurst, J. Phys. Chem. Solids 39:457 (1978).
2. P. Kofstad, J. Less-Common Metals 13:635 (1967).
3. P. Kofstad, J. Phys. Chem. Solids 23:1579 (1962).
4. K. S. Førland, Acta Chem. Scand. 18:1267 (1964).
5. J. B. Moser, R. N. Blumenthal, and D. H. Whitmore, J. Am. Ceram. Soc. 48:384 (1965).
6. R. T. Dirstine and C. J. Rosa, Z. Metallkde 70:322 (1979).
7. J.-F. Marucco, J. Gantron, and P. Lemasson, J. Phys. Chem. Solids 42:363 (1981).
8. R. N. Blumenthal and D. H. Whitmore, J. Electrochem. Soc. 110:92 (1963).
9. C. B. Alcock, S. Zador, and B. C. H. Steele, "Electromotive Force Measurements in High Temperature Systems" (Institute of Mining and Metallurgy, London, 1968), pp. 231-45.
10. A. Koztowska-Rog, S. Kozinski, and G. Rog., Rocz. Chem. 50:2153 (1976).
11. D. S. Tannhauser, Solid State Commun. 1:223 (1963).
12. R. N. Blumenthal, J. Coburn, J. Bankus, and W. M. Hirthe, J. Phys. Chem. Solids 27:643 (1966).
13. J. F. Baumard, D. Panis, and A. M. Anthony, J. Solid State Chem. 20:43 (1977).
14. G. Levin and C. J. Rosa, Z. Metallkde. 70:646 (1979).
15. E. Yagi, A. Koyama, H. Sakairi, and R. R. Hasiguti, J. Phys. Soc. Japan 42:939 (1977).
16 J. Sasaki and N. L. Peterson, to be published.
17. T. S. Lundy and W. A. Coghlan, J. Phys. (Paris), Colloq. 24:C9-299 (1972).
18. R. Haul and G. Dümbgen, J. Phys. Chem. Solids 26:1 (1965).
19. M. Arita, M. Hosoya, M. Kobayashi, and M. Someno, J. Am. Ceram. Soc. 62:443 (1979).

20. E. Iguchi and K. Yajima, J. Phys. Soc. Japan 32:1415 (1972).
21. O. W. Johnson, Phys. Rev. 136:A284 (1964).
22. O. W. Johnson, S. H. Paek, and J. W. Deford, J. Appl. Phys. 46:1026 (1975).
23. K. Hoshino and N. L. Peterson, to be published.
24. J. Yahia, J. Phys. Rev. 130:1711 (1963).
25. V. J. Rudolph, Z. Naturforschg. 14a:727 (1959).
26. R. D. Shannon and C. T. Prewitt, Acta Cryst. B25:925 (1969).

DIFFUSION OF ^{51}Cr TRACER IN Cr_2O_3 AND THE GROWTH OF Cr_2O_3 FILMS

A. Atkinson and R.I. Taylor

Materials Development Division
Building 552
AERE Harwell
Didcot, Oxon. OX11 ORA. U.K.

ABSTRACT

The tracer self diffusion coefficient of Cr in Cr_2O_3 single crystals has been measured at temperatures of 1100 and 1300°C and for oxygen activities ranging from the Cr/Cr_2O_3 equilibrium to air. The dependence of D*(Cr) on $a(O_2)$ is consistent with diffusion by vacancies at high $a(O_2)$ and by interstitials at low $a(O_2)$. The activation energy of D*(Cr) in the vacancy region at constant $a(O_2)$ is 6.0 eV.

All the tracer penetration profiles show a strong contribution from diffusion along dislocations, probably in the form of low angle boundaries. The effective radius of the dislocation is determined to be ~ 0.4 nm and the dependence of the dislocation diffusion coefficient on $a(O_2)$ is similar to that observed for lattice diffusion which suggests that similar defects are involved. The present lattice diffusion coefficients are several orders of magnitude smaller than those of earlier studies which suggests that the earlier measurements were mainly controlled by short-circuit diffusion.

The lattice diffusion coefficients are far too low to account for the rate of oxidation of Cr and it is concluded that grain boundary diffusion must be responsible for Cr_2O_3 film growth.

INTRODUCTION

Cr_2O_3 is an extremely important oxide because it grows as a film on stainless steels and other technological alloys at elevated temperatures and endows the alloy with considerable resistance against rapid oxidation and corrosion. Consequently, the transport properties of Cr_2O_3 and their relationship to Cr_2O_3 film growth have long been a subject of interest in this field.

The early tracer diffusion studies of Cr and O in Cr_2O_3 have been summarised by Kofstad[1]. Cr diffusion in sintered polycrystalline material has been measured by Lindner and Åkerstrøm[2] and Hagel and Seybolt[3] at temperatures between 1000 and 1500°C. Walters and Grace[4] measured the diffusion of Cr in single crystals of Cr_2O_3 as a function of oxygen activity, $a(O_2)$, at 1300°C. Kofstad and Lillerud[5] considered all these diffusion data in relation to the rate of growth of Cr_2O_3 films by thermal oxidation. They concluded that the most likely mechanistic interpretation of the diffusion data is that the majority point defects in Cr_2O_3 are Cr interstitials and that on this basis the diffusion data can satisfactorily account for the observed rate of growth of Cr_2O_3 films in the temperature range 800-1400°C. However, Atkinson et al.[6] have demonstrated that in the oxidation of Ni at temperatures below approximately 1100°C it is diffusion of Ni along oxide grain boundaries which controls film growth. This conclusion should be valid quite generally for slow-growing crystalline oxides and it is surprising that such a protective oxide as Cr_2O_3 appears to be growing by lattice diffusion.

The aim of the work described here was to study Cr diffusion in Cr_2O_3 single crystals at as low a temperature as possible and over a range of oxygen activity in order to explore the point defect structure of Cr_2O_3 ad the relationship between Cr diffusion and the rate of Cr_2O_3 film growth.

EXPERIMENTS

The single crystals of Cr_2O_3 which were used in this study were obtained from Cristal Tec (Grenoble, France) and had been prepared from 'Specpure' grade starting materials. The crystals were of approximate dimension 4 x 4 x 3 mm and the largest surfaces had been cut perpendicular to the 'c' crystallographic axis. (Diffusion measurements were made from this surface i.e. diffusion was parallel to the c-axis.) The crystals were ground and polished on this surface to a mirror finish (1 μm diamond on a fibre lap) and then annealed to stabilise polishing damage for 2 h at 1500°C in an Al_2O_3 furnace tube in flowing CO_2 containing one volume percent CO. This gas mixture was chosen because annealing in air or oxygen caused considerable etching of the

polished surface due to the volatilisation of CrO_3 gas. In all
anneals the crystals were supported in a sintered Cr_2O_3 crucible
with lid (Specpure grade) to minimise contamination and loss of
volatile species.

Each crystal was then pre-annealed under the same conditions
of temperature and oxygen activity as the intended diffusion
anneal. In experiments at 1100°C (except in air) the oxygen
activity was obtained using a metal/metal oxide couple and CO_2 in
a closed silica system with two furnaces[7]. One furnace
maintained the metal/oxide couple at one temperature to establish
the required CO_2/CO ratio, whilst the other maintained the crystal
at 1100°C. The couples which were used for different $a(O_2)$ are
given in Table 1. In the single experiment at 1300°C an Al_2O_3
tube and flowing CO_2/CO mixture was used.

After the pre-anneal the surface of each crystal was
carefully inspected. Often the surfaces were observed to have
been etched and if this occurred the experiment was restarted.
(The reason for the etching was not discovered.) ^{51}Cr tracer was
applied to the diffusion surface by vacuum evaporation of $^{51}CrCl_3$
and the chloride was converted to oxide by heating in air for
30 mins at 500°C. The crystal was then given a diffusion anneal
(Table 1). Initial experiments showed that considerable loss of
tracer could occur at the higher oxygen activities. Therefore
under these conditions two single crystals were used with their
diffusion surfaces in contact and held together using Pt wire.

After the diffusion anneal some material was ground from the
back and sides of the crystal to eliminate errors from transport
by surface or vapour diffusion. The tracer penetration profile
was then determined by sputter-sectioning[7]; the maximum
penetration depth in all experiments being less than 1.4 μm.

RESULTS

Some of the measured penetration profiles are shown in
Figures 1 and 2. The profiles in Figure 1 illustrate the effect
of temperature on diffusion. Those in Figure 2 illustrate the
effect of anneal time and also show how close are the results for
two crystals annealed face to face under the same conditions. All
the profiles have the same general shape in that there is a
prominent 'tail' in which the logarithm of the mean concentration
$\bar{c}$ is proportional to depth, y. (The quantity plotted is $\bar{c}/K$,
where K is the tracer activity per unit area on the initial
surface before diffusion.) Such tails have been observed in
similar experiments on NiO[8] and are characteristic of diffusion
along dislocations. The appropriate method for analysing such
tails depends on whether the dislocations are randomly distributed
or whether they are arranged in arrays forming low angle

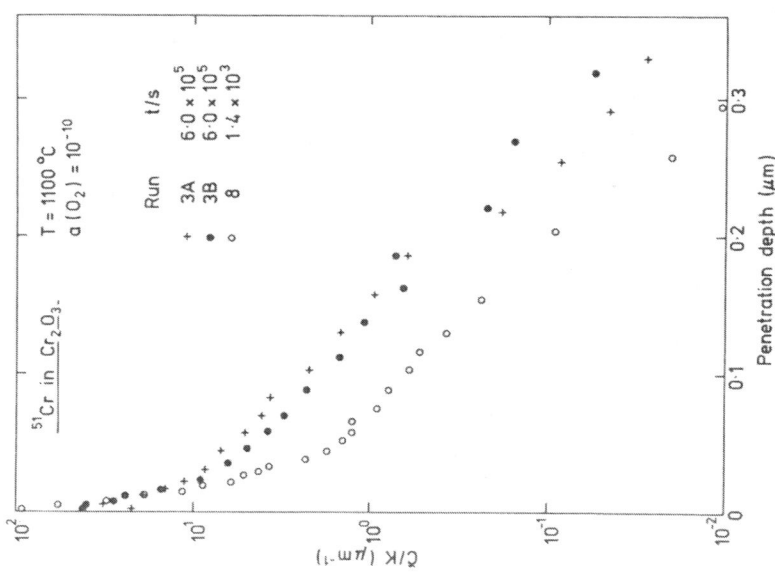

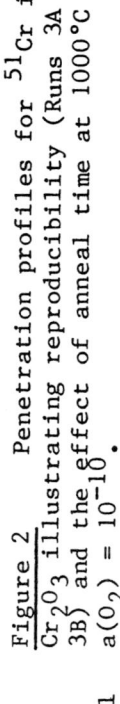

Figure 1 Penetration profiles for ^{51}Cr diffusing in Cr_2O_3 single crystals showing the effect of temperature at approximately constant oxygen activity. The curves have no theoretical significance.

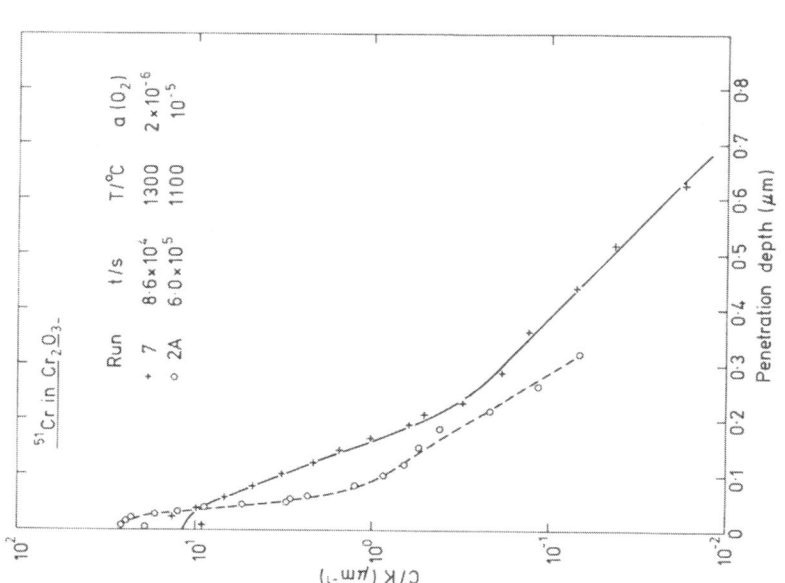

Figure 2 Penetration profiles for ^{51}Cr in Cr_2O_3 illustrating reproducibility (Runs 3A and 3B) and the effect of anneal time at 1000°C and $a(O_2) = 10^{-10}$.

boundaries. In the case of NiO the latter case was shown to be the more appropriate and we will assume that this is also the case in these experiments. Therefore the profiles were analysed as described previously for NiO to give the parameters in Table 1 in which D is the lattice diffusion coefficient, D' is the grain boundary diffusion coefficient for the low angle boundaries and δ is the boundary width. Also shown in Table 1 is the lattice penetration distance $(Dt)^{\frac{1}{2}}$ which is seen to be extremely small in some experiments, particularly runs 3 and 5. The profile for run 8, in which the anneal time was very short, shows that even when there is negligible lattice diffusion ($(Dt)^{\frac{1}{2}}$ is estimated to be only 0.4 nm in this experiment) there is an apparent lattice diffusion profile which is caused by the sputtering process and corresponds to $(Dt)^{\frac{1}{2}} \sim 5$ nm. Therefore the results from runs 3 and 5 are subject to large errors since they are close to the minimum diffusion which can be detected by the experimental methods. Also listed in Table 1 is the parameter β $(=D'\delta/(2D(Dt)^{\frac{1}{2}}))$ which should be greater than 10 for a good determination of $D'\delta$.

The results for diffusion as a function of oxygen activity at 1100°C are shown in Figure 3. For runs in which two crystals were used, two points have been plotted, except for lattice diffusion in run 3. In this particular case the lattice diffusion contribution is so small (Figure 2) that an error bar is shown to indicate the probable uncertainty in D.

DISCUSSION

Lattice Diffusion

The defect structure of Cr_2O_3 is still a subject of controversy and has recently been surveyed by Kröger[9]. He concluded that Cr_2O_3 is an instrinsic semiconductor (equal concentrations of electrons and holes) at temperatures above 1250°C and may be p-type or n-type at lower temperatures depending on doping. Greskovitch[10] has measured the deviation from stoichiometry at 1100°C and for $a(O_2)$ in the range 10^{-4} to 10^{-1}. He concludes that Cr_2O_3 is cation deficient with a deficit that increases as $a(O_2)^{1/8}$. The simplest point defect reaction for deficiency on the cation sublattice is

$$\tfrac{3}{2} O_2 \rightarrow 2 V_{Cr}''' + 3O_O^x + 6 \overset{\bullet}{h}. \qquad (1)$$

This reaction predicts p-type semiconductivity with $3[V_{Cr}'''] = [\overset{\bullet}{h}]$ and a concentration of vacancies proportional to $a(O_2)^{3/16}$. At

Table 1   Details of diffusion runs and results

Run	T (°C)	a(O$_2$)	Method	t (s)	D (cm^2 s^{-1})	D'δ (cm^3 s^{-1})	(Dt)$^{\frac{1}{2}}$ (nm)	β
1 A	1100	0.2	air	6.0 x 10^5	1.2 x 10^{-16}	3.8 x 10^{-20}	86	18
1 B				6.0 x 10^5	1.2 x 10^{-16}	1.9 x 10^{-20}	86	9
2 A	1100	10^{-5}	Cu at 950°C	6.0 x 10^5	4.7 x 10^{-18}	3.8 x 10^{-22}	17	24
2 B					3.7 x 10^{-18}	1.4 x 10^{-22}	15	13
3 A	1100	10^{-10}	Co at 1100°C	6.0 x 10^5	1.2 x 10^{-18}	1.2 x 10^{-22}	8.6	55
3 B						8.2 x 10^{-23}		32
4	1100	10^{-13}	Fe at 970°C	6.0 x 10^5	2.7 x 10^{-18}	9.6 x 10^{-23}	13	14
5	1100	10^{-15}	Graphite at 1100°C	6.0 x 10^5	1.1 x 10^{-18}	4.2 x 10^{-23}	7.9	31
6	1100	5 x 10^{-20}	Cr at 1100°C	5.2 x 10^5	9.6 x 10^{-18}	3.4 x 10^{-22}	22	8
7	1300	2 x 10^{-6}	CO$_2$ + 1% CO	8.6 x 10^4	2.8 x 10^{-16}	2.8 x 10^{-20}	49	10
8	1100	10^{-10}	Co at 1100°C	1.4 x 10^3	D_d 2.6 x 10^{-14}	–	0.4 (estimate)	–

the present time this would appear to be the most likely defect structure at high $a(O_2)$ at 1100°C. The broken line in Figure 3 has the theoretical slope of $^3/16$ for diffusion via vacancies created in this way and the data points are not inconsistent with this mechanism considering the experimental uncertainties. Greskovitch also measured the chemical diffusion coefficient at $a(O_2) = 2 \times 10^{-4}$ to be 4.6×10^{-12} cm^2 s^{-1} and combined this with data on deviation from stoichiometry at the same oxygen activity to predict a tracer diffusion coefficient of 1.6×10^{-17} cm^2 s^{-1}, in good agreement with the present measured value.

Hoshino and Peterson[11] have measured tracer diffusion of Cr in Cr_2O_3 (in crystals from the same source as those used in this work), but at higher temperatures (1490 and 1570°C). Their data are in extremely good agreement with the relationship $D \propto a(O_2)^3/16$. This is somewhat surprising since if Cr_2O_3 is an intrinsic semiconductor at such high temperature we would expect $D \propto a(O_2)^3/4$ from equation (1). The diffusion data therefore seem to be at variance with earlier electrical conductivity measurements at high temperatures (> 1250°C).

Kofstad and Lillerud[5] suggested that point defects on the Cr sublattice are $\overset{.}{Cr}$ interstitials formed by the reaction

$$Cr_2O_3 \rightarrow 2Cr_i^{\cdot\cdot\cdot} + 3/2O_2 + 6\,e\,. \tag{2}$$

This reaction predicts n-type semiconductivity with $3[Cr_i^{\cdot\cdot\cdot}] = [e^-]$ and $D \propto a(O_2)^{-3}/16$. As can be seen in Figure 3 there is some indication of a changeover to interstitial diffusion at low $a(O_2)$ at 1100°C. This is analogous to the established defect structure for cations in Fe_3O_4[12].

The present diffusion data (at $a(O_2) = 10^{-5}$) are plotted in Arrhenius form in Figure 4. (The data point at 1300°C has been adjusted slightly by assuming $D \propto a(O_2)^3/16$. They form a reasonably self-consistent set when combined with the measurements of Hoshino and Peterson and the activation energy (for the combined set) is approximately 5.5 eV. This set of results is between 4 and 6 orders of magnitude lower than any of the other earlier measurements. The reason for this is not clear. Hoshino and Peterson point out that Hagel and Seybolt's penetration profiles are not inconsistent with diffusion along grain boundaries in their polycrystalline samples. Whilst this may account for their relatively fast diffusion coefficients the same explanation cannot be applied to the data of Walters and Grace, who used single crystals.

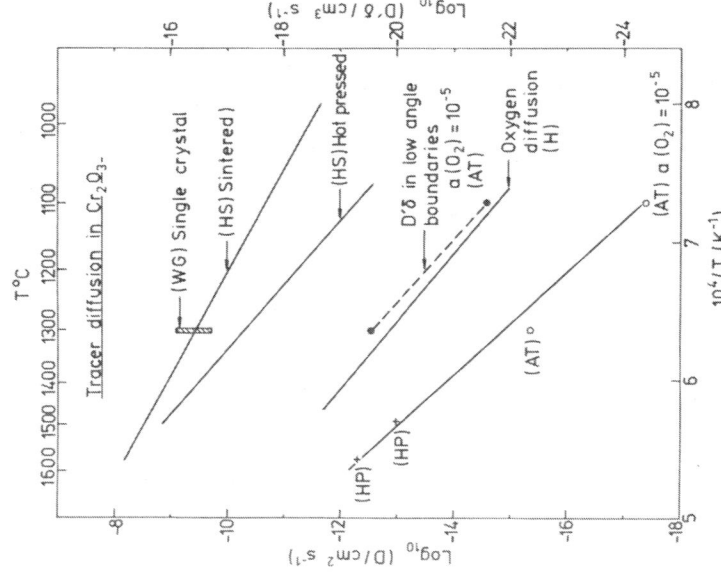

Figure 4   Arrhenius plot showing the relationship between the present measurements at $a(O_2) = 10^{-5}$ (labelled AT) and the data of other workers labelled by authors' initials (e.g. HP ≡ ref. 11).

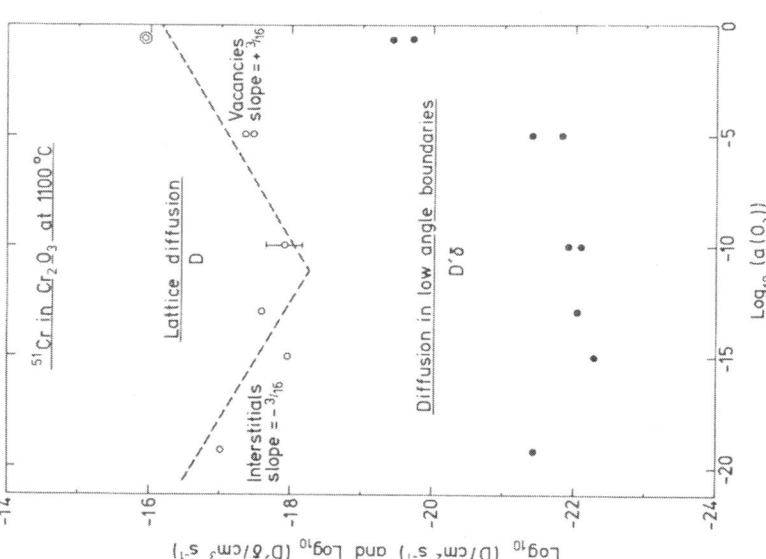

Figure 3   Diffusion of ^{51}Cr in the Cr_2O_3 lattice (D) and along low angle boundaries ($D'\delta$) as a function of oxygen activity at 1100°C.

Diffusion Along Low Angle Boundaries

The parameter $D'\delta$ for diffusion of Cr along low angle boundaries in Cr_2O_3 is plotted as a function of $a(O_2)$ in Figure 3 and shows a dependence which is similar to that found for lattice diffusion. (There is also a tendency for any errors in D to be reflected in $D'\delta$ since D is required in order to evaluate $D'\delta$.) This similarity between D and $D'\delta$ was also observed for Ni diffusion along dislocations in NiO[8] and most probably indicates that the mechanism of dislocation diffusion involves something similar to a lattice point defect i.e. vacancies at high $a(O_2)$ and interstitials at low $a(O_2)$ in Cr_2O_3.

Run 8 was designed to have negligible lattice diffusion so that the dislocation diffusion coefficient would be determined directly, D_d, and a value for the dislocation radius estimated as done previously for NiO[8]. At $a(O_2) = 10^{-10}$ (and T = 1100°C) D_d is measured to be 2.6×10^{-14} cm^2 s^{-1} which is approximately 2×10^4 times faster than lattice diffusion. For a low angle boundary the relationship between boundary diffusion and dislocation diffusion is

$$D'\delta = \pi a^2 D_d/\ell \qquad (3)$$

where ℓ is the dislocation spacing. We have not measured ℓ in the boundaries in these crystals, but if we assume a spacing of 10 nm (as was measured previously in NiO) then a is deduced to be 0.4 nm, which is a reasonable value and is of the same order as that estimated for NiO.

The data for $D'\delta$ at 1100°C and 1300°C are also plotted in Figure 4 and the scale of $D'\delta$ has been chosen so that if δ is assumed to be 1 nm the boundary diffusion coefficient is given directly on the diffusion coefficient scale.

Cr_2O_3 FILM GROWTH

Wagner's[13] theory of oxide film growth by ambipolar diffusion enables the parabolic rate constant for growth ($k_p = X^2/t$, where X is film thickness) to be related to the tracer self-diffusion coefficients in the oxide. When this is applied to Cr_2O_3 (assuming dominant electronic conductivity and neglecting correlation effects) the appropriate expression for k_p is

$$k_p = \int_{a(O_2)'}^{a(O_2)''} (1.5\,D^*(Cr) + D^*(O))\,d\ln a(O_2) \qquad (4)$$

where the limits of integration correspond to the inner and outer
surfaces of the film. According to the data presented in Figure 4
oxygen lattice diffusion[14] is faster than Cr lattice diffusion
in Cr_2O_3. In view of the difference between the present and
earlier data for Cr diffusion, the oxygen diffusion data must be
considered suspect. It has been concluded from many marker
studies that Cr_2O_3 grows by Cr diffusion outwards and the most
recent oxygen tracer studies in growing films[15] confirm that
this is so. We will therefore only consider the contribution from
Cr diffusion. The contribution from Cr lattice diffusion at
1100°C has been estimated by evaluating equation (4) using the
present diffusion data. The result is shown in the Arrhenius plot
of Figure 5 and compared with the range of k_p observed in
oxidation studies, compiled by Lillerud and Kofstad[16]. The
comparison shows that lattice diffusion is too slow by at least
three orders of magnitude to account for the observed rate of film
growth. Also shown in Figure 5 is k_p calculated by Lillerud and
Kofstad from the earlier diffusion measurements. As they point

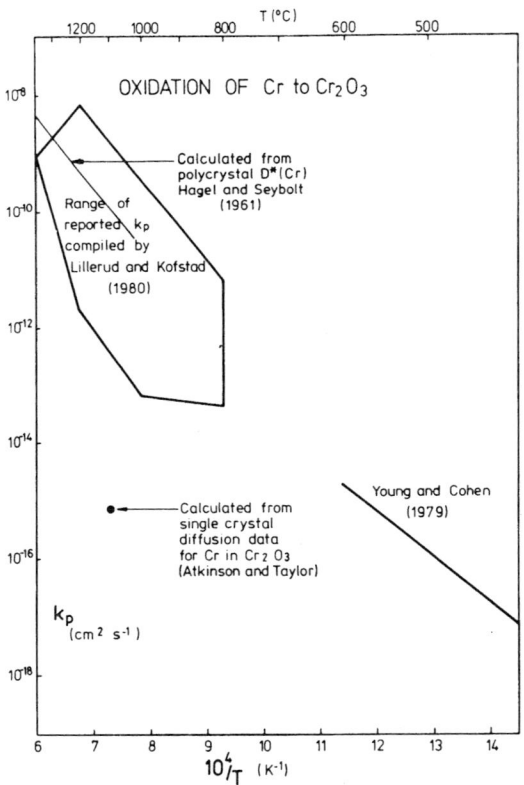

Figure 5   Arrhenius plot of the parabolic rate constant for
Cr_2O_3 growth by the oxidation of Cr. The values of k_p predicted
by the current and previous tracer diffusion data for Cr in Cr_2O_3
are compared with the measured values.

out, the earlier diffusion data can account for the rate of film growth, but in view of the present diffusion data, this agreement appears to be fortuitous. Furthermore, the observed values of kp cannot be accounted for even if diffusion of Cr along dislocations is included. We estimate that to account for the lower bound of measured k_p would require a dislocation density of approximately 3×10^{12} cm^{-2} which is extremely high. Whilst this may be acceptable to account for the lower bound of k_p, it would be unreasonable to account for the higher values on this basis. It must therefore be concluded that the most likely transport process which controls Cr_2O_3 film growth is Cr diffusion along oxide grain boundaries.

ACKNOWLEDGEMENT

The authors are grateful to C. Greskovitch for permission to refer to his results prior to publication.

REFERENCES

1. P. Kofstad, "Nonstoichiometry Diffusion and Electrical Conductivity in Binary Metal Oxides", (Wiley, New York) (1972).
2. R. Lindner and Å. Åkerstrøm, Z. Phys. Chem. N.F., 6:162 (1956).
3. W.C. Hagel and A.U. Seybolt, J. Electrochem. Soc., 108:1146 (1961).
4. L.C. Walters and R.E. Grace, J. Appl. Phys., 8:2331 (1965).
5. P. Kofstad and K.P. Lillerud, J. Electrochem. Soc., 127:2410 (1980).
6. A. Atkinson, R.I. Taylor and A.E. Hughes, Philos. Mag., A45:823 (1982).
7. A. Atkinson, M.L. O'Dwyer and R.I. Taylor, J. Mater. Sci., 18:2371 (1983).
8. A. Atkinson and R.I. Taylor, Philos. Mag., A39:581 (1979).
9. F.A. Kröger in: "High Temperature Corrosion", Ed. R.A. Rapp, (NACE 6, Houston), p. 89 (1983).
10. C. Greskovitch, to be published.
11. K. Hoshino and N.L. Peterson, J. Amer. Ceram. Soc., 66:C-202 (1983).
12. R. Dieckmann and H. Schmalzried, Ber. Bunsenges. Phys. Chem., 81:414 (1977).
13. C. Wagner, Z. Phys. Chem., B21:25 (1933).
14. W.C. Hagel, J. Amer. Ceram. Soc., 48:70 (1965).
15. D.F. Mitchell, R.J. Hussey and M.J. Graham, J. Vac. Sci. Technol. A., 1:1006 (1983).
16. K.P. Lillerud and P. Kofstad, J. Electrochem. Soc., 127:2397 (1980).

DEFECT STRUCTURE AND TRANSPORT PROPERTIES OF THE MANGANESE OXIDES

MANGANOSITE ($Mn_{1-\Delta}O$) AND HAUSMANNITE ($Mn_{3-\delta}O_4$)

Rüdiger Dieckmann and Michael Keller

Institute for Physical Chemistry and Electrochemistry
University of Hannover, Callinstrasse 3-3a, D 3000
Hannover 1, Federal Republic of Germany

ABSTRACT

The effect of temperature and oxygen partial pressure on the nonstoichiometry of two manganese oxides, manganosite ($Mn_{1-\Delta}O$) and hausmannite ($Mn_{3-\delta}O_4$), has been studied by thermogravimetry between 900 and 1400°C. It is concluded that cation vacancies and holes are the dominant defect species in manganosite over at least most of its stability range at high temperatures. For the low and also for the high temperature phase of hausmannite it was found that cation vacancies predominate at high oxygen activity, while manganese interstitials are the prevailing point defects at low oxygen activity. The relationships between the defect structure of manganosite and its transport properties are discussed briefly.

INTRODUCTION

At high temperature and atmospheric pressure, manganosite ($Mn_{1-\Delta}O$), hausmannite ($Mn_{3-\delta}O_4$) and bixbyite ($Mn_2O_{3-\varepsilon}$) are the three thermodynamically stable oxide phases of the manganese-oxygen system (Fig. 1). Manganosite crystallizes in the sodium chloride structure. Hausmannite occurs as a tetragonal low temperature phase ($\alpha-Mn_{3-\delta}O_4$) and as a cubic high-temperature phase ($\beta-Mn_{3-\delta}O_4$) with spinel structure. Bixbyite has a cubic crystal symmetry. At high temperatures, these three oxides are nonstoichiometric compounds. At constant total pressure, the diviations from stoichiometry, Δ, δ and ε, depend on temperature and oxygen partial pressure, and are determined by the concentrations of the different point defects present in the crystals at thermodynamic equilibrium. As long as impurities, dislocations and grain boundaries can be neglected, transport processes in these oxides, such as the electrical conduction and ionic diffusion, are closely related to the concentration and to the mobility of the different equilibrium point defects.

The goal of this paper is to present new experimental results on the nonstoichiometry of manganosite ($Mn_{1-\Delta}O$) and hausmannite ($Mn_{3-\delta}O_4$) and to interpret them in terms of defect structure and transport properties of these oxides.

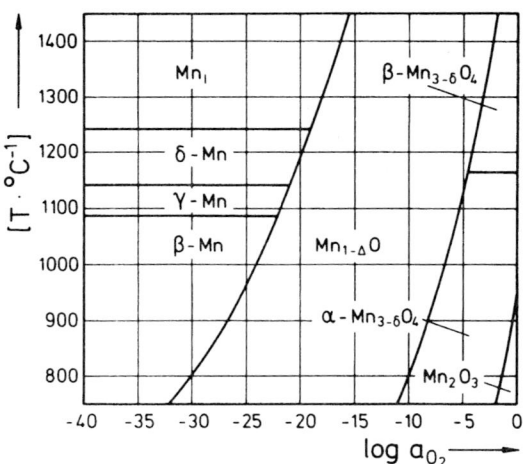

Fig. 1. The manganese-oxygen system between 750 and 1450°C and at one atmosphere[1].

MANGANOSITE ($Mn_{1-\Delta}O$)

Nonstoichiometry and Point Defect Structure

Although the nonstoichiometry of manganosite is already widely documented[2-16], a few problems remain concerning the defect structure of this material and the exact value of Δ. To answer these questions, we have performed new thermogravimetric measurements on manganosite. It is generally agreed that the disorder in manganosite affects essentially only the cation sublattice and that at high oxygen activities, cation vacancies and holes are the dominant defects, corresponding to a cation deficit. However, it is still somewhat unclear, whether or not manganese interstitials become the dominant defect species at low oxygen activities, as suggested in the literature[15,17].

Cation vacancies and holes can be formed by an incorporation of oxygen in manganosite, according to the reaction

$$1/2 \ O_2 + (Mn_{Mn}^{2+}2+)^x \ \rightleftharpoons \ (V_{Mn}2+)'' + 2 \ h^\cdot + MnO. \qquad (1)$$

With increasing defect concentrations, singly charged and neutral vacancies may be formed to reduce the electrostatic interaction between differently charged point defects. The corresponding reactions are:

298

$$(V_{Mn}2+)" + h\cdot \rightleftharpoons (V_{Mn}2+)', \qquad (2)$$

$$(V_{Mn}2+)" + 2 h\cdot \rightleftharpoons (V_{Mn}2+)^x. \qquad (3)$$

The thermal ionic disorder equilibria of Frenkel- and Schottky-type, respectively, are:

$$(Mn_{Mn}^{2+}2+)^x + (V_I)^x \rightleftharpoons (V_{Mn}2+)" + (Mn_I^{2+})^{\cdot\cdot}, \qquad (4)$$

$$(Mn_{Mn}^{2+}2+)^x + (0_2^{2-})^x \rightleftharpoons (V_{Mn}2+)" + (V_02-)^{\cdot\cdot} + MnO. \qquad (5)$$

Furthermore, the thermal electronic disorder equilibrium is formulated:

$$0 \rightleftharpoons h\cdot + e'. \qquad (6)$$

The concepts of point defect thermodynamics can be applied to Eqs. (1) to (6) to formally model the oxygen activity dependence of experimentally obtained Δ-values. In doing so one often assumes an ideal solution of point defects in $Mn_{1-\Delta}O$. However, this assumption, commonly made when the exact activity coefficients are unknown, is not necessarily quite realistic.

Based on theoretical calculations it has also been suggested[18-20] that, for large deviations from stoichiometry, defect clusters consisting of interstitials and cation vacancies should occur. Up to now there is no clear experimental evidence for the existence of such clusters in manganosite and they will therefore not be considered here.

Our very precise thermogravimetric measurements were carried out on dense, large grained, polycrystalline samples, between 900 and $1400^\circ C$ and over a very broad oxygen activity range. CO/CO_2 and N_2/O_2 gas mixtures, flowing between 0.3 and 0.6 atm total pressure, were used to fix the oxygen activity in the system. Under these conditions the resolution and the noise level were approximately one microgram. More details will be reported elsewhere[1]. Our experimental data points for Δ as a function of oxygen activity ($a_{O_2} = P_{O_2}/1$ bar) are reported in Fig. 2.

Our data show that the deviation from stoichiometry Δ increases with increasing oxygen activity, in general agreement with the observations reported in the literature. No inflection point can be seen in our Δ versus oxygen activity curves for the oxygen activity range covered here. Such an inflection point would indicate stoichiometric manganosite, the point from which to lower oxygen activities cation interstitials and/or anion vacancies become the dominant point defects. The oxygen activity dependence of Δ can be formally calculated using Arrhenius-type expressions for the equilibrium constants of reactions (1) to (3) and by neglecting the thermal disorder described by reactions (4) to (6). More details and numbers will be given elsewhere[1]. The fit of these calculated values with our experimental results is excellent, as shown in Fig. 2. In Fig. 3 the corresponding calculated lines are plotted using a logarithmic scale for Δ. All isotherms are bent upwards, indicating, according to our model that the part of nonstoichiometry due to doubly charged vacancies decreases with increasing oxygen activity, while that due to neutral vacancies increases. The fraction of Δ due to singly charged vacancies is always smaller than that due to doubly charged or neutral ones and goes through a maximum at higher temperatures. The defect model involving neutral clusters at high defect concentrations, as proposed by Catlow and Stoneham[20], cannot be fitted to our results. Only if clusters were charged, such a model could be compatible with the observed oxygen activity dependence of Δ.

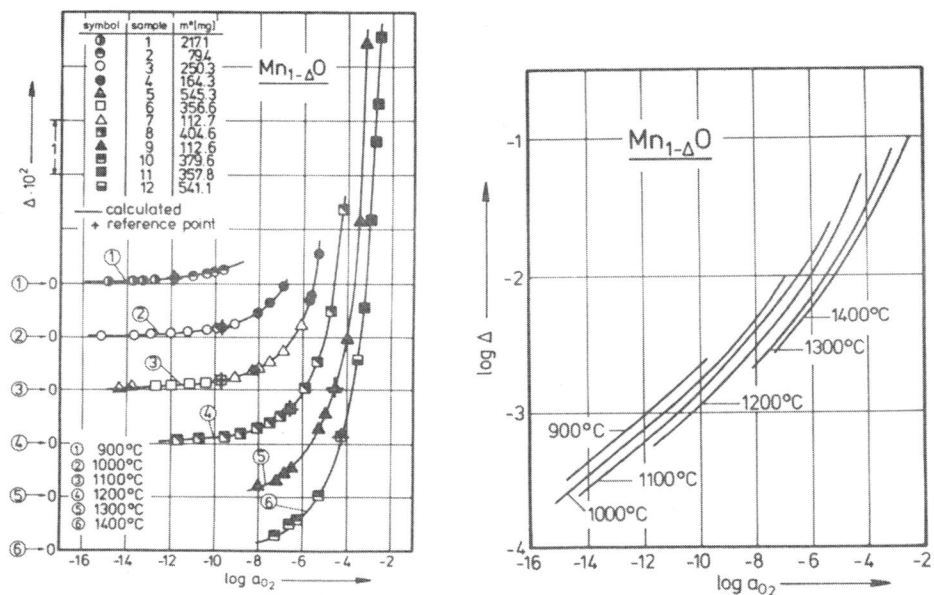

Fig. 2. (left) Thermogravimetrically measured values for Δ in $Mn_{1-\Delta}O$ as a function of oxygen activity at temperatures between 900 and 1400°C. The lines are calculated as described in the text.

Fig. 3. (right) Logarithmic representation of the deviation from stoichiometry Δ in $Mn_{1-\Delta}O$ as a function of the oxygen activity between 900 and 1400°C.

Our data are compared with other experimental data for Δ[3-15] somewhere else[1]. At higher temperatures, our results are in good agreement with the results of Bransky and Tallan[9] who also investigated over large temperature and oxygen activity ranges. At lower temperatures this agreement is not as good, probably due to an insufficient equilibration of Bransky and Tallan's samples.

Using our nonstoichiometry data we have also worked out the stability field of manganosite in the manganese-oxygen phase diagram for temperatures between 850 and 1450°C. The result is shown in Fig. 4. The maximum deviation from stoichiometry, Δ, significantly increases with temperature. Exact values of Δ for manganosite in equilibrium with manganese phases, are still unknown, but any cation excess or cation deficit can only be very small. This statement may be inferred from (1) the probably most reliable electrical conductivity results obtained for $Mn_{1-\Delta}O$ by Eror and Wagner[10], (2) electron mobility data from literature[21-23] and (3) values of hole concentrations derived from our nonstoichiometry data. Previous reports that significant cation excess should occur in manganosite at low oxygen activities are therefore to some extent[15] or totally[24,25] in error.

It has been claimed several times[17,26-28] that carbon from the CO/CO_2 gas mixtures used during the experiments could dissolve in manganosite and significantly influence its defect structure. The straight forward experiment of Grabke and Wolf[29] has recently disproved this claim. Using the radiotracer carbon-14, these authors have found that the bulk solubility of carbon in several oxides is extremely small, namely, less than 10^{-6} weight percent in manganosite, in a 1:1 CO/CO_2 mixture at 1100°C and 0.66 atm total pressure. We are therefore confident that our present results cannot have been influenced by carbon dissolution.

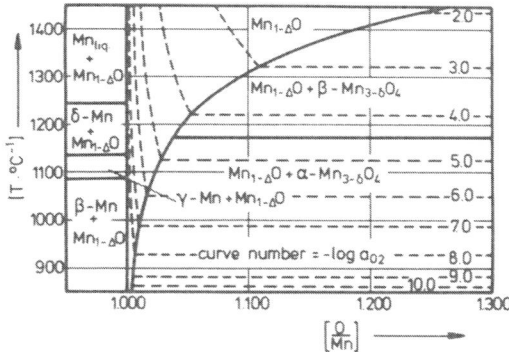

Fig. 4. The manganosite field of the manganese-oxygen phase diagram with oxygen isoactivity lines.

Electrical Conductivity

The oxygen activity dependence of the electrical conductivity of manganosite at high temperatures is of interest for the formulation of the defect structure of this material and has been studied in several instances[4,10,30-36]. It is generally accepted that, at high temperatures, the electrical conduction in manganosite is electronic. According to an estimate made by Eror and Wagner[10], the ionic transference number is less than 10^{-3} at about 1030°C and at $\log a_{O_2} = -11$. The electrical conductivity therefore is due to holes and electrons and is, independently from the accurate conduction mechanisms, given by:

$$\sigma = \sigma_{h\cdot} + \sigma_{e'} = F\,(c_{h\cdot}\,u_{h\cdot} + c_{e'}\,u_{e'})\cdot \qquad (7)$$

$u_{h\cdot}$ and $u_{e'}$ are the electrochemical mobilities of holes and electrons, respectively. Because the hole concentration decreases with decreasing oxygen activity and the electron concentration does the opposite, a minimum should be observed in the electrical conductivity of manganosite as oxygen activity is varied at constant temperature, if $c_{h\cdot}u_{h\cdot}$ becomes equal to $c_{e'}u_{e'}$ within the stability range of $Mn_{1-\Delta}O$. Depending on the ratio $u_{e'}/u_{h\cdot}$, the oxygen activity corresponding to this minimum of electrical conductivity may be different from the oxygen activity for which Δ would become zero. In $Mn_{1-\Delta}O$, the mobility ratio $u_{e'}/u_{h\cdot}$ is much larger than unity, so that any conductivity minimum would occur at an oxygen activity higher than that where Δ would become zero, provided that manganosite would still be stable at this value. In addition to the electronic mobilities, the different equilibrium constants for the formation of point defect in manganosite will fix the position of this minimum, as well as the dependence of the conductivity on oxygen activity close to stoichiometry. In order to discuss different limiting cases for the oxygen activity dependence of σ when the composition is not far from $\Delta=0$, a schematic plot (Fig. 5.) has been derived for the concentration of doubly charged cation vacancies, manganese interstitials, holes and electrons and for the resulting electrical conductivity as a function of oxygen activity. For the sake of simplicity, singly charged and neutral cation vacancies have been neglected as well as oxygen vacancies. From Fig. 5, it follows that the oxygen activity dependence of σ in the vicinity of $\Delta=0$ depends on the values of the equilibrium constants K_4 and K_6 corresponding to the defect formation reactions (4) and (6) relative to K_1 for the component activity dependent disorder reaction (1). When $4\,K_4 = K_6$, and only then, no change in the oxygen activity dependence of σ is expected in the vicinity of $\Delta=0$.

All the systematic investigations[4,10,30-36] on the dependence of the electrical conductivity of manganosite on the oxygen activity at constant temperature agree on the existence of the conductivity minimum described above. From the observed oxygen activity dependences of σ, it is concluded that the electronic conduction of $Mn_{1-\Delta}O$ is due to holes at high oxygen activities and to electrons at low oxygen activities. Absolute values for the electrical conductivity of dense manganosite as a function of oxygen activity are available from two sources [10,35]. In only one of these studies[10] were single crystals used in order to exclude any contribution from grain boundaries to the electrical conduction. Consequently, to compare the different results, we have normalized all the conductivity data to the conductivities at the minimum from the different studies. A compilation of such normalized conductivity data for 1000 and 1100°C is shown in Fig. 6. The position of the conductivity minima with respect to a_{O_2} does not vary much from one study to another. However, the oxygen activity dependences disagree to some extent, expecially at high oxygen activities. This scatter prevents the determination of the true oxygen activity dependence and, as a consequence, these data cannot contribute to the development of the appropriate, quantitative defect model for manganosite. Furthermore, none of the data sets of Fig. 6 shows any evidence for a change in the oxygen activity dependence due to $\Delta = 0$, as discussed above. The conductivity data obtained on single crystals by Eror and Wagner[10] are probably the most reliable, but unfortunately, they do not cover the range of higher oxygen activities. Some of these results are displayed in Fig. 7 together with conductivity data by O'Keeffe and Valigi[35]. The oxygen activity dependences at the highest oxygen activities investigated by Eror and Wagner[10] are in good agreement with those derived for the hole concentration from our fit for the nonstoichiometry data of manganosite. By combining Eror and Wagner's conductivity data with the hole concentrations from our fit, the electrochemical mobility of holes in $Mn_{1-\Delta}O$ can be determined. Furthermore, by using literature data[21-23] for the electron mobility in manganosite and the conductivities at the conductivity minima, the equilibrium constant K_6 for the recombination reaction of holes and electrons can be estimated as a function of temperature. Finally, from the temperature dependence of K_6, the band gap in $Mn_{1-\Delta}O$ can be calculated to be 2.34 eV, in agreement with values reported by Eror and Wagner[10] (2.3 eV) and by O'Keeffe and Valigi[35] (2.2-2.25 eV).

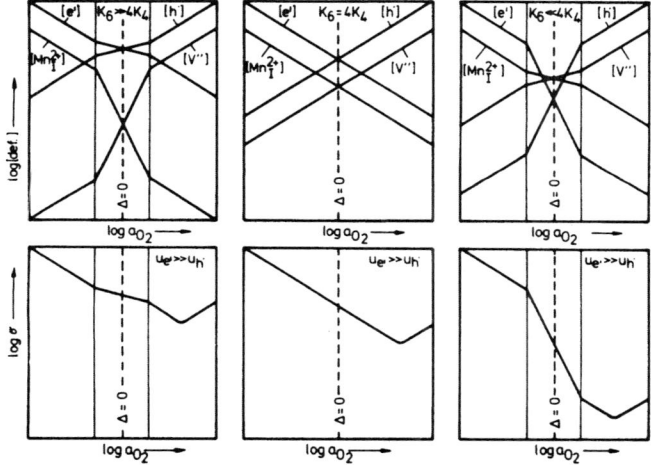

Fig. 5. Schematic representation of the concentrations of point defects and the electrical conductivity as a function of oxygen activity for various limiting cases in the vicinity of the hypothetical stoichiometric point of manganosite.

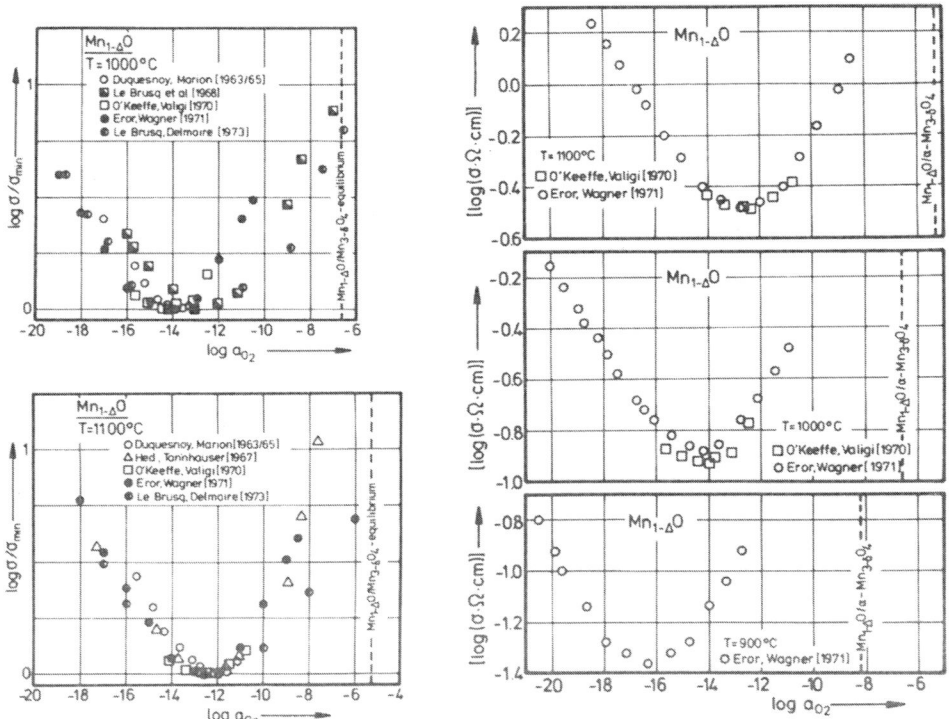

Fig. 6. (left) Literature data for the electrical conductivity of manganosite as a function of oxygen activity at 1000 and 1100°C. The conductivity values are displayed relative to the conductivity minima reported from the different investigations.

Fig. 7. (right) Absolute values from literature for the electrical conductivity of manganosite as a function of oxygen activity for temperatures between 900 and 1100°C.

Manganese Tracer Diffusion

The manganese tracer diffusion coefficient D^*_{Mn} is related to the manganese self-diffusion coefficient D_{Mn} according to

$$D^*_{Mn} = D_{Mn} \cdot f^*, \tag{8}$$

where f^* is the correlation factor. f^* takes into account the fraction of ineffective elementary jumps for the manganese tracer transport in the sense of the mean square displacement. The correlation factor is a rather simple quantity as long as only one type of point defects is involved in the cation diffusion. It is more complicated in the case of tracer diffusion when different defect species must be considered. If cation vacancies and manganese interstitials are involved in the manganese diffusion, the manganese diffusion coefficient is the sum of two partial diffusion coefficients, $D_{Mn(V)}$ and $D_{Mn(I)}$, describing the diffusion of manganese via cation vacancies and via interstitials, respectively:

304

$$D_{Mn} = D_{Mn(V)} + D_{Mn(I)} \qquad (9)$$

Because the dependences of the concentrations of cation vacancies and of manganese interstitials on oxygen activity are opposite, the manganese self-diffusion coefficient is expected to go through a minimum at a certain oxygen activity, if the dominant defect type controlling the cation diffusion changes. If different types of cation vacancies contribute to the partial diffusion coefficient $D_{Mn(V)}$, then $D_{Mn(V)}$ is the sum of the partial diffusion coefficients specific of the different cation vacancies multiplied by the fraction of these vacancies:

$$D_{Mn(V)} = \sum_i \frac{c_V i}{\Sigma c_V i} \cdot D_{Mn(V^i)}. \qquad (10)$$

i represents the vacancy charge.

Only three studies of the dependence of manganese tracer diffusion on oxygen activity have been published[33,37,38]. In addition, Kofstad[17,27] has derived from kinetic measurements somewhat speculative data for manganese self-diffusion in $Mn_{1-\Delta}O$. The manganese tracer diffusion data obtained by Peterson and Chen[38] are probably the most precise and are reported in Fig. 8. No minimum is observed in the D^*_{Mn} versus log a_{O_2} curves in the range of oxygen activities investigated. Such a minimum would indicate a change in the controlling defects for manganese diffusion from cation vacancies to manganese interstitials, as the oxygen activity decreases. In Fig. 8, the oxygen activity dependences of D^*_{Mn} measured by Peterson and Chen[38] are compared with the oxygen activity dependence of our Δ-values. Surprising enough, these two sets of oxygen activity dependences practically coincide, up to very high Δ-values in $Mn_{1-\Delta}O$ (about 0.1). Only at 1000°C and high oxygen activities, this agreement is not as good. Comparing the experimental procedures described by Peterson and Chen[38] with data for the reequilibration kinetics of $Mn_{1-\Delta}O$, discussed in the next section, it may be concluded that the tracer diffusion coefficients obtained at 1000°C and high oxygen activities are somewhat in error due to an incomplete sample equilibration before the diffusion runs. The agreement between the oxygen activity dependences of both D^*_{Mn} and Δ suggests that the partial self-diffusion coefficients $D_{Mn(V^i)}$ must be quite similar and that the correlation factor f* cannot vary significantly with the vacancy concentration in $Mn_{1-\Delta}O$. On the one hand, defect clusters suggested by theoretical calculations[18-20] would affect the mobility and, consequently, the oxygen activity dependence of D^*_{Mn}. No such effect is observed. On the other hand, measurements of the isotope effect in cation tracer diffusion experiments by Peterson and Chen[38] show an influence of Δ on the isotope effect. Based on this result, Peterson and Chen[38] suggest that defect clustering plays some role for cation diffusion in manganosite. These conflicting observations forbid any conclusion about the existence of defect clusters in $Mn_{1-\Delta}O$ at high Δ-values, and about their possible role in diffusion. The agreement between the oxygen activity dependences of

D_{Mn}^* and of Δ suggests further investigations of the temperature dependence of the ratio D_{Mn}^*/Δ. This ratio is the product of a correlation factor by a vacancy diffusion coefficient. An appropriate analysis yields an Arrhenius dependence of D_{Mn}^*/Δ on temperature; the activation energy is 167 kJ/mole.

Recently Kofstad[28] has suggested that the tracer diffusion results obtained from experiments carried out in carbon containing atmospheres at constant oxygen activity, will differ from that derived from carbon-free runs because carbon containing defect clusters are formed. Following this suggestion, Peterson[39] measured manganese tracer diffusion coefficients in either Ar/O_2 or CO/CO_2 atmospheres at log $a_{O_2} = -5$ at 1300 and 1400°C. He found no effect of the gas mixture used on the diffusion coefficient. Furthermore, Peterson's measurements are in good agreement with the earlier diffusion results of Peterson and Chen[38]. Consequently, Kofstad's suggestion is not relevant to the case of manganosite.

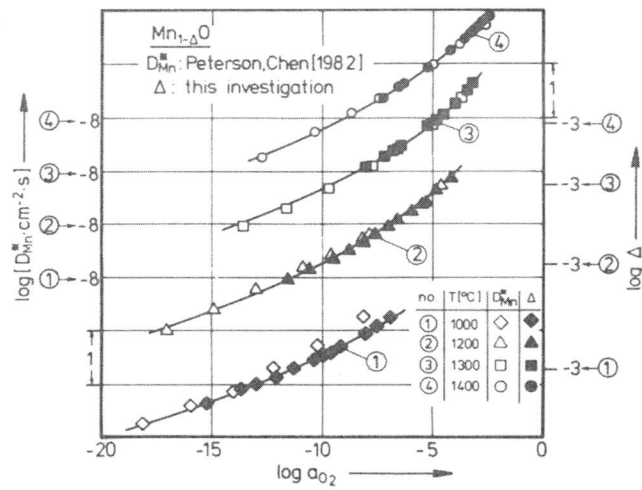

Fig. 8. Comparison of cation tracer diffusion data by Peterson and Chen[38] with our results for Δ in $Mn_{1-\Delta}O$ for temperatures between 1000 and 1400°C.

Reequilibration Kinetics

During our studies on the nonstoichiometry of $Mn_{1-\Delta}O$, we have observed that the kinetics of the sample reequilibration after a sudden jump in oxygen activity between two oxygen activity values depend on the jump direction, when CO/CO_2 mixtures are used. An example of this behavior is shown in Fig. 9a for a dense polycrystalline sample at 1000°C. Such a decrease of the reequilibration rate with increasing oxygen activity has also been observed in other studies[40-45] of the reequilibration of $Mn_{1-\Delta}O$ samples in CO/CO_2 mixtures. The reequilibration kinetics have sometimes[40-44] been interpreted as being due to diffusion processes in the bulk. Accordingly, chemical diffusion coefficients, $\tilde{D}$, strongly decreasing with increasing oxygen activity, have been derived. We do not believe that these $\tilde{D}$-data are representative of actual diffusion processes in $Mn_{1-\Delta}O$. Firstly, the agreement between the oxygen activity dependences of D^*_{Mn} and of Δ as presented above, do not allow such a strong oxygen activity dependence of $\tilde{D}$. Secondly, Bransky and Tallan[44] and Wimmer[45] have observed that a reduction of the total pressure of a CO/CO_2 mixture at constant oxygen activity or the dilution of a CO/CO_2 mixture with Argon, lead to a decrease of the reequilibration rate. Some of Wimmer's[45] results are displayed in Fig. 9b. From this it is clear that in addition to bulk diffusion processes, the rate of oxygen transfer between $Mn_{1-\Delta}O$ and CO/CO_2 gas plays a role in the reequilibration kinetics. Independently from the elementary-rate-determining step of the oxygen exchange reaction, the rate of oxygen transfer between the gas phase and the surface is well described by:

$$\frac{1}{A} \cdot \frac{d\,n_O}{d\,t} = P_{tot} \cdot N_{CO_2} \cdot k(a^*_{O_2}) \cdot [1 - (a^*_{O_2}/a_{O_2}(gas))^{1/2}] \cdot \quad (11)$$

A is the surface area, P_{tot} is the total pressure and N_{CO_2} the mole fraction of CO_2 in the gas. $k(a^*_{O_2})$ is the rate constant of the reaction for oxygen exchange; it is dependent on the oxygen activity $a^*_{O_2}$ established in $Mn_{1-\Delta}O$ near the surface. The kind as the reequilibration kinetics depend on the direction of the oxygen activity jump suggests that $k(a^*_{O_2})$ decreases with increasing oxygen activity $a^*_{O_2}$. Grabke[46] reached this conclusion by performing isotope exchange experiments for the oxygen exchange between $Mn_{1-\Delta}O$ and CO/CO_2 mixture at 900, 950 and 1000°C. Grabke's (still unpublished) results are displayed in Fig. 9c as reaction rate constant $k(a^*_{O_2})$ versus oxygen activity.

HAUSMANNITE ($Mn_{3-\delta}O_4$)

Nonstoichiometry and Point Defect Structure

Literature data[7,16,47-51] on the nonstoichiometry of $\alpha-$ and $\beta-Mn_{3-\delta}O_4$ are rather scarce and are in poor agreement. No reliable systematic data set on the deviation from stoichiometry in the hausmannite phases has yet been published. The recent results of Terayama et al.[16] are erroneous as discussed somewhere else[52]. In order to determine the point defect structure of the two manganosite phases it was, therefore, necessary to investigate systematically the temperature and oxygen activity dependence of δ in $Mn_{3-\delta}O_4$. Thermogravimetry measurements have been made between 1000 and 1130°C for the low temperature phase and between 1200 and 1350°C for the high temperature phase. The detailed results of this study will be reported elsewhere[53], whereas Fig. 10 shows δ-values at 1100°C for $\alpha-Mn_{3-\delta}O_4$ and at 1200°C for $\beta-Mn_{3-\delta}O_4$. The S-shaped curves for δ as a function of oxygen activity are quite similar to those found for magnetite ($Fe_{3-\delta}O_4$)[54]. By analogy with the defect structure found for magnetite, in α-$Mn_{3-\delta}O_4$ and in β-$Mn_{3-\delta}O_4$, cation vacancies are believed to be the dominant defect species at high oxygen activities, while manganese interstitials are thought to prevail at low oxygen activities. Because the temperature dependence of the cation distribution in α- and β-$Mn_{3-\delta}O_4$ is still unknown, the defect equilibria can only be formulated by considering a unique cation sublattice. Then, vacancies form according to the reaction

$$3\ Mn_{Mn}^{2+} + 2/3\ O_2 \rightleftharpoons 2\ Mn_{Mn}^{3+} + V_{Mn} + 1/3\ Mn_3O_4, \qquad (12)$$

and manganese interstitials form in the following way:

$$Mn_{Mn}^{n+} + V_I \rightleftharpoons Mn_I^{n+} + V_{Mn}. \qquad (13)$$

Using point defect thermodynamics, the oxygen activity dependence of the deviation from stoichiometry δ in $Mn_{3-\delta}O_4$ can be derived. According to analogous formulations made for magnetite, the result for δ is:

$$\delta = (K_V/4) \cdot a_{O_2}^{2/3} - 4\ K_I \cdot a_{O_2}^{-2/3}. \qquad (14)$$

K_V and K_I are equilibrium constants for the formation of cation vacancies and of manganese interstitials, respectively. The fit between Eq. (14) and experimentally obtained δ-values is excellent, as shown in Fig. 10. This agreement indicates that the defect model selected for hausmannite is appropriate. The equilibrium constant K_V and K_I obey Arrhenius-type equations, for α- and β-$Mn_{3-\delta}O_4$, respectively. Based on these data it is possible to derive the stability field of hausmannite in the manganese-oxygen phase diagram of Fig. 11. This diagram is the first of this type ever reported for the hausmannite phases. It shows that the nonstoichiometry variation in hausmannite increases with increasing temperature for α- and β-$Mn_{3-\delta}O_4$. However, at the phase transformation this nonstoichiometry variation is smaller in the high temperature phase than in the low temperature phase. In Fig. 11 the line representing the α/β-transformation is only an approximation. Details of this area of the phase diagram will be reported elsewhere[53].

Transport Properties

No data are available for the diffusion of cations and anions in hausmannite,, but some results have been reported[32,55-57] for the electrical conductivity of polycrystalline hausmannite. The most extensive measurements were made by Metselaar et al.[57] who found that the electrical conductivity, σ, is of the order of 100 $\Omega^{-1}cm^{-1}$ at 1340°C and of 2 $\Omega^{-1}cm^{-1}$ at 1010°C. These authors have also reported that, at constant temperature, the electrical conductivity is approximately constant; it only slightly decreases at low oxygen activities. This oxygen activity dependence has not yet been explained. It may be that this behavior does not reflect bulk properties, since the samples were polycrystalline. This possible explanation is supported by the observation by Metselaar et al.[57] that the conductivity values for melt grown samples differ from those obtained for sintered samples. A similar decrease of the electrical conductivity at low oxygen activity has also been observed for polycrystalline magnetite[58], but not for single crystalline samples[59].

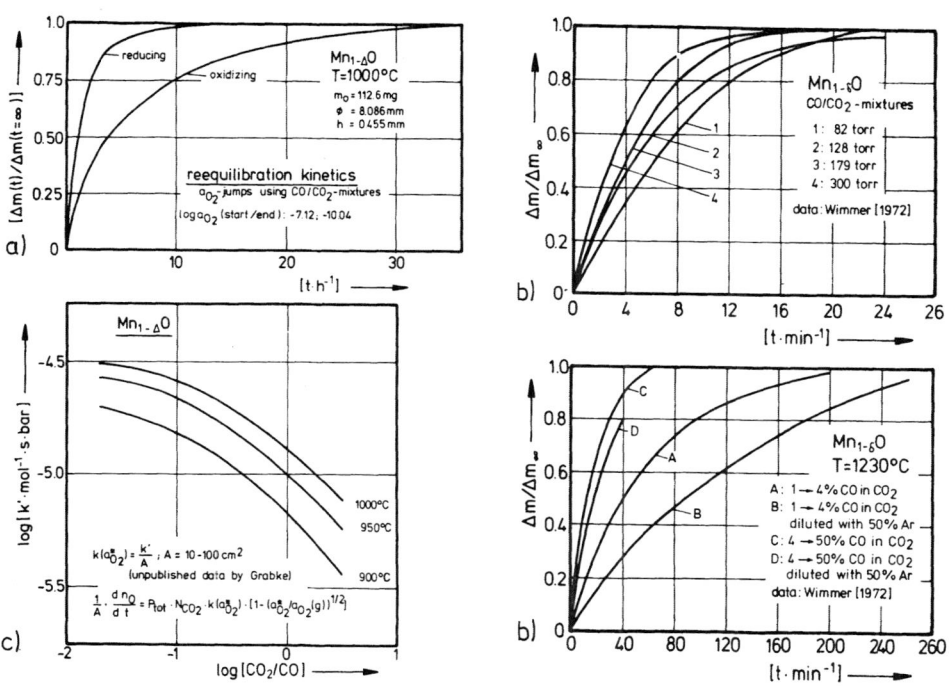

Fig. 9. (a) Reequilibration kinetics of a manganosite sample at 1000°C after a sudden increase or decrease of oxygen activity in CO/CO$_2$ atmospheres.

Fig. 9. (b) Results obtained by Wimmer[45] for the influence of the absolute pressure and the dilution of CO/CO$_2$ mixtures, and of the jump direction on the reequilibration kinetics of manganosite.

Fig. 9 (c) Unpublished results from isotope exchange experiments performed by Grabke[46]. The rate constant of the oxygen exchange between CO/CO$_2$ mixtures and manganosite is reported as a function of the CO$_2$/CO ratio between 900 and 1000°C.

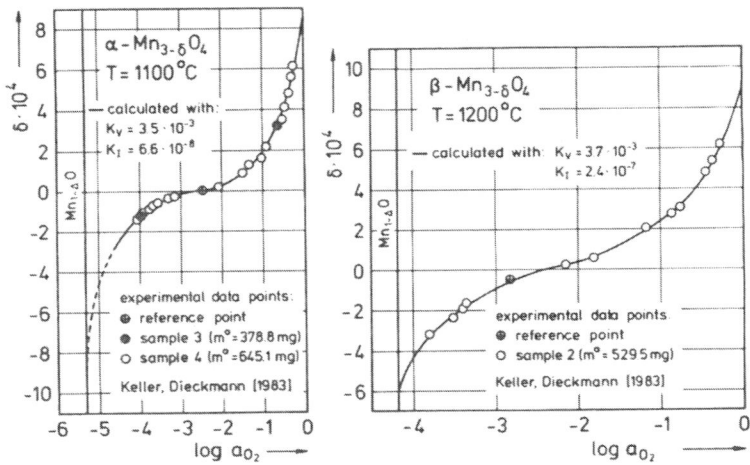

Fig. 10. Thermogravimetrically determined values of the deviation from stoichiometry δ in $\alpha\text{-Mn}_{3-\delta}O_4$ at 1100°C and in $\beta\text{-Mn}_{3-\delta}O_4$ at 1200°C as a function of oxygen activity.

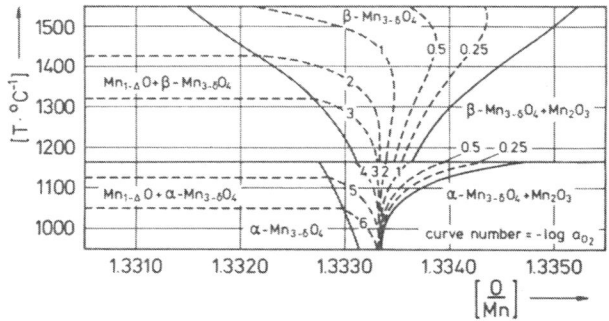

Fig. 11. The hausmannite field in the manganese-oxygen phase diagram with oxygen isoactivity lines.

ACKNOWLEDGEMENTS

We are indebted to the "Deutsche Forschungsgemeinschaft" for financial support and to Prof. Schmalzried for his supporting interest in our work. Thanks are due to Prof. Grabke for allowing us to make use of his unpublished experimental results and to Dr. Peterson for his communication concerning the effect of carbon in manganosite. Thanks are also due to Dr. D. Ricoult for his assistance with the manuscript.

REFERENCES

1. M. Keller and R. Dieckmann, Ber. Bunseges, Phys. Chem. (to be published).
2. M. W. Davies and F. D. Richardson, Trans. Faraday Soc. 55:604 (1959).
3. A. Z. Hed and D. S. Tannhauser, J. Electrochem. Soc. 114:314 (1967).
4. A. Z. Hed and D. S. Tannhauser, J. Chem. Phys. 47:2090 (1967).
5. K. Schwerdtfeger and A. Muan, Trans. AIME 239:1114 (1967).
6. N. G. Schmahl and D. Hennings, Z. Phys. Chem. NF 63:111 (1969).
7. N. G. Schmahl and D. F. K. Hennings, Arch. Eisenhuttenwes. 40:395 (1969).
8. B. E. F. Fender and F. D. Riley, Thermodynamic Properties of $Mn_{1-x}O$, in: "Chemistry of Extended Defects in Non-Metallic Solids", Proc. Advan. Study Inst., Scottsdale, Ariz., 1969 (1970).
9. I. Bransky and N. M. Tallan, J. Electrochem. Soc. 118:788 (1971).
10. N. G. Eror and J. B. Wagner, J. Electrochem. Soc. 118:1665 (1971).
11. F. Schaberg, Dissertation TU Clausthal (1973).
12. G. Tromel, W. Fix, K. Koch and F. Schaberg, Erzmetall 29:234 (1976).
13. C. Picard and P. Gerdanian, J. Solid State Chem. 11:190 (1974).
14. B. Touzelin, These Orsay, Ser. A., n°d' ordre 1302 (1974).
15. J. Couzin and A. Duquesnoy, C. R. Acad. Sc. Paris, Ser. C 281:259 (1975).
16. K. Terayama, M. Ikeda and M. Taniguchi, Trans. Jap. Inst. Met. 24:24 (1983).
17. P. Kofstad, J. Phys. Chem. Solids 44:129 (1983).
18. D. G. Muxworthy, "The Defect Properties of Transition Metal Oxides", UKAEU Dept. No. AERF-TP 665 (1976).
19. C. R. A. Catlow, B. Fender and D. G. Muxworthy, J. Phys. (Paris) Colloq. 7:67 (1977).
20. C. R. A. Catlow and A. M. Stoneham, J. Am. Ceram. Soc. 64:234 (1981).
21. H. J. DeWit and C. Cerevecoeur, Phys. Lett. 25A:393 (1967).
22. M. Gvishi, N. M. Tallan and D. S. Tannhauser, Solid State Commun. 6:135 (1968).
23. T. G. M. Kleinpenning, J. Phys. Chem. Solids 37:925 (1976).
24. C. Carel, On a New T-P-X Diagram of Manganese Monoxide, in: "Preprints 9th Int. Symp. Reactivity of Solids", Cracow, Poland, Sept. 1-6, 1980", 420 (1980) and in: "Reactivity of Solids", Proc. 9th Int. Symp. React. Solids, Cracow, Poland, Sept. 1-6, 1980, eds. K. Dyrek, J. Haber and J. Nowotny, Elsevier Sci. Publ., Amsterdam and New York, 2:596 (1982).
25. C. Carel, C. R. Acad. Sc. Paris, Ser. II. 295:853 (1982).

26. J. M. Pope and G. Simkovich, *J. Electrochem. Soc.* 116:292C (1969).
27. P. Kofstad, *J. Phys. Chem. Solids* 44:879 (1983).
28. P. Kofstad, *Oxid. Met.* 19:129 (1983).
29. H. J. Grabke and I. Wolf, Private Communication (1984).
30. A. Duquesnoy and F. Marion, *C. R. Acad. Sc. Paris, Ser. C,* 256:2862 (1963).
31. A. Duquesnoy, *Rev. Hautes Temp. Refract.* 3:201 (1965).
32. J. - J. Oehlig, H. Le Brusq, A. Duquesnoy and F. Marion, *C. R. Acad. Sc. Paris, Ser. C,* 265:421 (1967).
33. J. - P. Bocquet, M. Kawahara and P. Lacombe, *C. R. Acad. Sc. Paris, Ser. C.* 265:1318 (1967).
34. H. Le Brusq, J. - J. Oehlig and F. Marion, *C. R. Acad. Sc. Paris, Ser. C,* 266:965 (1968).
35. M. O'Keeffe and M. Valigi, *J. Phys. Chem. Solids* 31:947 (1970).
36. H. Le Brusq and J. - P. Delmaire, *Rev. Int. Htes. Temp. Refract.* 10:15 (1973).
37. J. B. Price and J. B. Wagner, Jr., *J. Electrochem. Soc.* 117:242 (1970).
38. N. L. Peterson and W. K. Chen, *J. Phys. Chem. Solids* 43:29 (1982).
39. N. L. Peterson, Private Communication (1983).
40. J. B. Price, Jr. "Chemical and Radiotracer Diffusion in MnO_{1+x}" Ph.D. Thesis, Northwestern University, Evanston, Ill. (1968).
41. P. E. Childs, Diffusion in Chromium-Doped Manganous Oxide, *in:* "Proc. Thomas Graham Memorial Symposium on Diffusion Processes, Glasgow, Scotland, Sept. 22-24, 1969" Gordon and Breach, London 437 (1970).
42. P. E. Childs and J. B. Wagner, Jr., Chemical Diffusion in Wustite and Chromium-Doped Manganous Oxide, *in:* "Heterogeneous Kinet. Elevated Temp.", Proc. Int. Conf. Mat. Mater. Sci., Philadelphia, 1969, eds. G. R. Belton and W. L. Worell, Plenum Press, New York 269 (1970).
43. P. E. Childs, L. W. Laub and J. B. Wagner, Jr., *Proc. Brit. Ceram. Soc.* 19:29 (1971).
44. I. Bransky and N. M. Tallan, A System for the Determination of Oxidation-Reduction Kinetics in Nonstoichiometric Metal Oxides, *in:* "Vacuum Microbalance Techniques", Plenum Press, New York-8:29 (1970).
45. J. M. Wimmer, "Chemical Diffusion in Cobalt (II) Oxide, Manganese (II) Oxide and Iron (II) Oxide", Ph.D. Thesis, Marquette Univ., Milwaukee, Wis. (1972).
46. H. J. Grabke, Private Communication (1983).
47. M. LeBlanc and G. Wehner, *Z. Phys. Chem. A* 168:59 (1934).
48. C. H. Shomate, *J. Am. Chem. Soc.* 65:785 (1943).
49. T. E. Moore, *J. Am. Chem. Soc.* 72:856 (1950).
50. W. C. Hahn, Jr and A. Muan, *Am. J. Sci.* 258:66 (1960).
51. A. Schmier and G. Sterr, *Z. Anorg. Allg. Chem.* 346:181 (1966).

52. M. Keller and R. Kieckmann, *Trans. Jap. Inst. Met.* 24:650 (1983).
53. M. Keller and R. Kieckmann, *Ber. Bunsenges, Phys. Chem.* (to be published).
54. R. Dieckmann, *Ber. Bunsenges. Phys. Chem.* 86:112 (1982).
55. F. C. Romeijn, *Philips Res. Rep.* 8:304 (1953).
56. E. M. Logothetis and K. Park, *Solid State Commun.* 16:909 (1975).
57. R. Meselaar, R. E. J. VanTol and P. Piercy, *J. Solid State Chem.* 38:335 (1981).
58. T. O. Mason and H. K. Bowen, *J. Am. Ceram. Soc.* 64:237 (1981).
59. R. Dieckmann, C. A. Witt and T. O. Mason, *Ber. Bunsenges. Phys. Chem.* 87:495 (1983).

THE THERMODYNAMIC FACTORS FOR THE CATION DIFFUSION IN A P-TYPE

OXIDE AO DOPED WITH A MONOVALENT IMPURITY

Francesco Gesmundo

Istituto di Chimica Fisica Applicata dei Materiali

Lungobisagno Istria, 34 - Genova (Italy)

SUMMARY

The thermodynamic factors for the cation diffusion in a p-type semiconducting oxide of a divalent metal containing a monovalent impurity are evaluated by means of thermodynamic considerations, assuming that the defect structure of the base oxide is known. It is shown that these parameters are not only functions of the oxygen activity and of the dopant concentration, but depend also strongly on the ratio between the gradients of these two variables. Limiting expressions corresponding to the presence of only one gradient are obtained and evaluated approximately for very small or large impurity concentrations. It is found in particular that under constant oxygen activity the two thermodynamic factors are always close to one but change in opposite directions with the impurity concentration. On the contrary, under constant impurity concentration the thermodynamic factors are both very large and decrease as the oxygen activity increases. A general relationship between the two thermodynamic factors is obtained by means of the Gibbs-Duhem equation applied to the relevant ternary system. Finally, the theoretical equations are used to calculate the thermodynamic factors in NiO doped with a monovalent impurity at 1000°C, using a detailed model to represent its defect structure.

INTRODUCTION

The thermodynamic factor is a parameter relating the intrinsic diffusion coefficient of a given species to the corresponding tracer diffusion coefficient in the same system.[1-3] In a binary system the thermodynamic factors of the two diffusing species are also involved in the relationship between the interdiffusion coefficient of the system and the tracer diffusion coefficients of the two species through Darken's equation:[1,2,4] the same type of relationship has also been extended to oxide solid solutions of the (A,B)O type. [5,6,7] The thermodynamic factor for the diffusion of the cations in a pure oxide containing metal vacancies as prevailing defects and in solid solutions between two oxides with the same metal/oxygen ratio has already been considered elsewhere.[8] The present paper is concerned instead with the calculation of the thermodynamic factors for the diffusion of the cations in a p-type oxide doped with a monovalent impurity.

The oxide AO is assumed to be a p-type semiconductor and to contain metal vacancies of different effective charges as prevailing lattice defects, electrically compensated by an equivalent concentration of electron holes. It is further assumed that the oxygen sublattice is practically undefected and that oxygen does not diffuse significantly in presence of oxygen activity gradients applied to the sample. Finally, interactions between defects are disregarded.

THEORY

Activities of the Metal Ions and of the Oxide Components in the Mixed Oxides

The calculation of the thermodynamic factors for the two ions in a mixed oxide requires a knowledge of the activities of the two cations as well as of those of the two oxides in the solid solution. For this, a correct definition of the composition of a doped sample of AO is also needed. The number of oxygen ions per unit volume is denoted by c_O (moles cm^{-3}), while c_A and c_M denote the corresponding concentrations of A and of the impurity M and c_V the overall concentration of the metal vacancies, irrespective of their charges. The mole fractions of the various species in the same system can be defined in two slightly different ways, i.e. as

$$\xi_A = c_A/(c_A+c_M) \; ; \quad \xi_M = c_M/(c_A+c_M) \; ; \quad \xi_V = c_V/(c_A+c_M) \qquad (1)$$

with $\xi_A+\xi_M=1$, or as

$$x_A = c_A/c_O \; ; \quad x_M = c_M/c_O \; ; \quad x_V = c_V/c_O \qquad (2)$$

Since the oxygen sublattice is assumed to be perfect, one has also $c_O=c_A+c_M+c_V$, so that $x_A+x_M+x_V=1$. In view of the previous definitions, the two concentrations for A, M and the vacancies are related by the equations

$$\xi_A = x_A/(1-x_V) \; ; \quad \xi_M = x_M/(1-x_V) \; ; \quad \xi_V = x_V/(1-x_V) \qquad (3)$$

316

Finally, the overall deviation from stoichiometry δ as defined by the formula $A(1-\delta)O$ as commonly used for a pure oxide of the type considered here, when extended to the case of doped oxides, is evidently given by $c_O/(c_A+c_M)=1/(1-\delta)$. Use of the previous definitions shows that ξ_V is equal to δ so that it will be denoted in this way in the following, while ξ_A and ξ_M will be indicated as $1-\xi$ and ξ respectively.

The next step is represented by the derivation of an expression for the activities of A, M, AO and $MO_{0.5}$. For AO, the activities of A, O and AO are related through the condition of equilibrium for the formation of AO according to the reaction

$$A + \tfrac{1}{2} O_2 = AO \tag{a}$$

with an equilibrium constant

$$K_{AO} = a_{AO}/(a_A a_O) \tag{4}$$

In case of ideal behaviour of the oxide components in the solid solution one has[9]

$$a_{AO} = 1 - \xi \tag{5}$$

so that, according to Eq.(4), one has also

$$a_A = (1-\xi)/(K_{AO} a_O) \tag{6}$$

In a similar way, the activity of a monovalent impurity M is related to that of the corresponding oxide $MO_{0.5}$ through the condition of equilibrium for the reaction

$$M + 1/4 O_2 = MO_{0.5} \tag{b}$$

with an equilibrium constant

$$K_{MO_{0.5}} = a_{MO_{0.5}}/(a_M a_O^{1/2}) \tag{7}$$

The activity of $MO_{0.5}$ dissolved in AO will be a function of the concentration of M. This dependence can be obtained by considering the reaction of incorporation of $MO_{0.5}$ into AO, i.e.

$$MO_{0.5} + 1/4 O_2 = M' + O^X + h^{\bullet} \tag{c}$$

where $h^{\bullet}$ is an electron hole , with an equilibrium constant

$$K_s = \xi p/(a_{MO_{0.5}} a_O^{1/2}) \tag{8}$$

where p is the concentration of the electron holes (mole fraction). Elimination of the activity of $MO_{0.5}$ from Eqs.(7) and (8) yields

$$a_M = \xi p/(K(MO_{0.5}) K_s a_0) \qquad (9)$$

while substitution of Eq.(9) into Eq.(7) gives

$$a(MO_{0.5}) = \xi p/(K_s a_0^{1/2}) \qquad (10)$$

The previous expressions show that the activities of M and of $MO_{0.5}$ will depend both on a_0 and on ξ not only directly but also indirectly through p. In fact p is related to a_0 and to the impurity concentration through the condition of electroneutrality which in this case has the form[10]

$$p = \xi + n + \left[V'\right] + 2\left[V''\right] \qquad (11)$$

where n is the concentration of the electrons (mole fraction), usually negligible in these cases: the presence of the impurity concentration in Eq.(10) is due to the fact that the M ions carry a unit negative effective charge in substitutional sites in the AO lattice. The concentrations of the vacancies are in turn functions of a_0 and p through the conditions of equilibrium for their reactions of formation, and introduction of these expressions into Eq.(11) leads finally to a cubic equation in p which can be solved numerically for assigned values of a_0 and ξ to give the concentrations of all the defects.[10] Thus, the activities of M and $MO_{0.5}$ can be calculated for any condition provided that the appropriate equilibrium constants are known.

Thermodynamic factors

As for the case of a pure oxide or of a solid solution of the (A,B)O type,[8] the thermodynamic factor for the diffusion of a cation in a mixed oxide can be defined as the ratio between the intrinsic diffusion coefficient of that species D^1 and the corresponding tracer diffusion coefficient D.

The diffusion flux of A in AO either pure or doped can be expressed in the form

$$J_A = - c_A D_A (\partial \ln a_A / \partial x) \qquad (12)$$

where D_A is the diffusion coefficient of A in the mixed oxide: according to Wagner[11] D_A can be either the self- or the tracer diffusion coefficient of A, since correlation effects have been explicitly disregarded in his treatment. To make the following expression of the thermodynamic factors consistent with their general definition, it will be assumed here that D_A is the tracer diffusion coefficient. According to the definition of the intrinsic diffusion coefficient,[1-3] the same diffusion flux can also be expressed as

$$J_A = - D_A^i (\partial c_A / \partial x) \tag{13}$$

by comparison between the two expressions of J_A one obtains

$$D_A^i = D_A (\partial \ln a_A / \partial x)/(\partial \ln c_A / \partial x) \tag{14}$$

or also

$$\vartheta_A = D_A^i / D_A = d \ln a_A / d \ln c_A \tag{15}$$

In the same way, the flux of M through the doped oxide takes the form[11]

$$J_M = - c_M D_M (\partial \ln a_M / \partial x) \tag{16}$$

where D_M is the tracer diffusion coefficient of M in the oxide, or also

$$J_M = - D_M^i (\partial c_M / \partial x) \tag{17}$$

Again, comparison of the two expressions for J_M yields

$$\vartheta_M = D_M^i / D_M = d \ln a_M / d \ln c_M \tag{18}$$

Use of these definitions, taking into account that all the parameters involved depend on both a_O and ξ , yields for the two thermodynamic factors the following general expressions

$$\vartheta_A = \left[(\partial \ln a_A / \partial \xi)_{a_O} \, d\xi + (\partial \ln a_A / \partial \ln a_O)_\xi \, d \ln a_O \right] / \tag{19}$$
$$/ \left[(\partial \ln c_A / \partial \xi)_{a_O} \, d\xi + (\partial \ln c_A / \partial \ln a_O)_\xi \, d \ln a_O \right]$$

and

$$\vartheta_M = \left[(\partial \ln a_M / \partial \xi)_{a_O} \, d\xi + (\partial \ln a_M / \partial \ln a_O)_\xi \, d \ln a_O \right] / \tag{19'}$$
$$/ \left[(\partial \ln c_M / \partial \xi)_{a_O} \, d\xi + (\partial \ln c_M / \partial \ln a_O)_\xi \, d \ln a_O \right]$$

Simpler limiting expressions for ϑ_A , ϑ_M are obtained when diffusion occurs in presence of gradients of only one variable. In fact, under constant a_O one obtains

$$\vartheta_A (a_O = \text{const.}) = \vartheta_A (I) = (\partial \ln a_A / \partial \xi)_{a_O} / (\partial \ln c_A / \partial \xi)_{a_O} \tag{20}$$

and

$$\vartheta_M (a_O = \text{const.}) = \vartheta_M (I) = (\partial \ln a_M / \partial \xi)_{a_O} / (\partial \ln c_M / \partial \xi)_{a_O} \tag{20'}$$

while under constant ξ one has

$$\vartheta_A(\xi=\text{const.})=\vartheta_A(II)=(\partial\ln a_A/\partial\ln a_0)_\xi/(\partial\ln c_A/\partial\ln a_0)_\xi \qquad (21)$$

and

$$\vartheta_M(\xi=\text{const.})=\vartheta_M(II)=(\partial\ln a_M/\partial\ln a_0)_\xi/(\partial\ln c_M/\partial\ln a_0)_\xi \qquad (21')$$

When instead both gradients are present simultaneously, then division of Eqs.(19) and (19') by d ln a_0 or by dξ shows that both ϑ_A and ϑ_M will depend also on the ratio between the gradients of these two variables.

For brevity sake, only the simpler expressions applying in case of presence of only one gradient will be considered in the following. These can be put into more explicit forms by using the equations given earlier for the activities and concentrations of the two metals in the solid solution. Considering first the case of presence of gradients of impurity concentration only, one has for A

$$(\partial\ln a_A/\partial\xi)_{a_0} = -\,1/(1-\xi)\approx -1 \qquad (22)$$

since $\xi\ll 1$, and

$$(\partial\ln c_A/\partial\xi)_{a_0} = -1/(1-\xi) -\left[1/(1-\delta)\right](\partial\delta/\partial\xi)_{a_0}\approx -(1+d_1) \qquad (23)$$

with $d_1=(\partial\delta/\partial\xi)_{a_0}$, since $\delta\ll 1$, so that

$$\vartheta_A(I) \approx 1/(1+d_1) \qquad (24)$$

In a similar way one finds for M

$$(\partial\ln a_M/\partial\xi)_{a_0} = 1/\xi + (\partial\ln p/\partial\xi)_{a_0} = 1/\xi + d_2 \qquad (25)$$

with $d_2=(\partial\ln p/\partial\xi)_{a_0}$, and

$$(\partial\ln c_M/\partial\xi)_{a_0} = 1/\xi -\left[1/(1-\delta)\right](\partial\delta/\partial\xi)_{a_0}\approx 1/\xi - d_1 \qquad (26)$$

so that

$$\vartheta_M(I) \approx (1+\xi d_2)/(1-\xi d_1) \qquad (27)$$

Consideration of the effect of the addition of a monovalent impurity to an oxide AO of p-type according to the procedure developed elsewhere[10] shows that d_1 changes from about $-1/(z+1)$ at low ξ to very small values at large ξ values, while d_2 changes from $\left[p(z+1)\right]^{-1}$ to about $1/\xi$ in the same range of ξ values. Thus one may predict that $\vartheta_A(I)$ will change from $(z+1)/z$ to about 1 as ξ increases while $\vartheta_M(I)$ will change from nearly one at small ξ values (since $\xi\ll p$ under these conditions) to about 2 at large ξ values. These predictions are in agreement with the results of a direct calculation presented later.

In case of presence of gradients of oxygen activity only, one obtains instead for A

$$(\partial \ln a_A / \partial \ln a_O)_\xi = -1 \tag{28}$$

and

$$(\partial \ln c_A / \partial \ln a_O)_\xi = -\left[1/(1-\delta)\right](\partial \delta / \partial \ln a_O)_\xi \approx - d_3 \tag{29}$$

with $d_3 = (\partial \delta / \partial \ln a_O)_\xi$, so that one obtains

$$\vartheta_A(II) \approx 1/d_3 \tag{30}$$

For the impurity M one gets instead

$$(\partial \ln a_M / \partial \ln a_O)_\xi = -1 + (\partial \ln p / \partial \ln a_O)_\xi = -1 + d_4 \tag{31}$$

with $d_4 = (\partial \ln p / \partial \ln a_O)_\xi$, and

$$(\partial \ln c_M / \partial \ln a_O)_\xi = -\left[1/(1-\delta)\right](\partial \delta / \partial \ln a_O)_\xi \approx - d_3 \tag{32}$$

so that one obtains finally

$$\vartheta_M(II) \approx (1-d_4)/d_3 \tag{33}$$

An approximate evaluation of the two thermodynamic factors can be made also in this case. In fact, when ξ is considerably smaller than the concentration of defects in the pure oxide, assuming that the prevailing lattice defects are metal vacancies with an effective charge z, one has[8,10]

$$\delta \approx \left[V^{z'}\right] \propto a_O^{1/(z+1)} \quad \text{and} \quad p \approx z\left[V^{z'}\right] \propto a_O^{1/(z+1)} \tag{34}$$

so that $d_3 \approx \delta/(z+1)$ and $d_4 \approx 1/(z+1)$. Thus one obtains

$$\vartheta_A(II)(\xi \to 0) \approx (z+1)/\delta \quad \text{and} \quad \vartheta_M(II)(\xi \to 0) \approx z/\delta \tag{35}$$

When instead ξ is large, one obtains approximately[10] $d_3 \approx \delta$ and $d_4 \approx 0$, so that this gives

$$\vartheta_A(II)(\xi \text{ large}) \approx 1/\delta \quad \text{and} \quad \vartheta_M(\xi \text{ large}) \approx 1/\delta \tag{36}$$

In addition, since in these oxides δ increases with a_O under a constant ξ, both thermodynamic factors will decrease when a_O will increase under a constant impurity concentration. On the contrary, the values of the two thermodynamic factors $\vartheta_A(II)$ and $\vartheta_M(II)$ calculated for a constant value of a_O and different values of ξ will increase with ξ since δ is a decreasing function of ξ.[10]

Finally, it seems interesting to derive a general relationship between the thermodynamic factors of the two cation species. This can be obtained by means of the Gibbs-Duhem equation, which, when applied to the ternary A,M,O system, takes the form

$$(1-\xi) \, d \ln a_A + \xi \, d \ln a_M + d \ln a_O - \xi \, d \ln p = 0 \qquad (37)$$

as shown elsewhere.[12] Division of Eq.(37) by $d \ln c_A$, recalling that, according to the definitions of c_A, c_M given previously one has

$$(1-\xi)(1-\delta) \, d \ln c_A + \xi(1-\delta) \, d \ln c_M + d\delta = 0 \qquad (38)$$

and taking into account the general definitions of the two thermodynamic factors given above, neglecting δ,ξ with respect to one since they are both small, yields finally

$$(1-\xi) \, \vartheta_A - \left\{ \xi / \left[(\xi/\vartheta_M) + (d\delta /d \ln a_M) \right] \right\} + d \ln a_O/d \ln c_A -$$

$$- \xi(d \ln p/d \ln c_A) = 0 \qquad (39)$$

It is important to recall that all the derivatives in Eq.(39) will depend on the ratio between the gradients of $\ln a_O$ and of ξ unless one of them is equal to zero. In fact for a mixed oxide such as considered here, the variables p, δ, a_M and c_A will all be functions of both a_O and ξ. Taking this fact into account one can express for example the term $(d\ln p/d \ln c_A)$ in the form

$$d \ln p/d \ln c_A = \left[(\partial \ln p/\partial \xi)_{a_O} \, d\xi + (\partial \ln p/\partial \ln a_O)_\xi \, d \ln a_O \right] /$$

$$/ \left[(\partial \ln c_A/\partial \xi)_{a_O} \, d\xi + (\partial \ln c_A/\partial \ln a_O)_\xi \, d \ln a_O \right] \qquad (40)$$

Division of the numerator and the denominator of the right-hand member of Eq.(40) by $d \ln a_O$ or $d\xi$ shows that the ratio $(d \ln p//d \ln c_A)$ will depend on the ratio between the gradients of $\ln a_O$ and ξ when they differ both from zero. When instead only the gradient of one variable is present, the left-hand member of Eq.(40) becomes equal to the corresponding partial derivative. The same conclusion applies also to the other similar terms in Eq.(39).

NUMERICAL APPLICATIONS

The treatment developed in the previous section is applied here to a specific system, i.e. to NiO doped with a monovalent impurity at 1000°C, for which the relevant data are reasonably well established. The defect structure of NiO is represented by means of the detailed model proposed by Osburn and Vest,[13] which considers the presence of metal vacancies of different electric charges. At 1000°C the equilibrium constants for the formation of the three types of vacancies are

$$K(V^x)=5.20 \times 10^{-6} \; ; \; K(V')=2.92 \times 10^{-3} \; ; \; K(V'')=5.33 \times 10^{-5}$$

while the constant for intrinsic ionization takes the value[13]

$$K_i = 1.12 \times 10^{-16}$$

The effect of the addition of a monovalent impurity on the concentration of the various defects in NiO has already been considered quantitatively elsewhere[10] and is therefore only briefly recalled here. As implied by the reaction of dissolution of $MO_{0.5}$ in AO (reaction (c)), the addition of $MO_{0.5}$ to NiO produces an increase in the concentration of the electron holes under a constant oxygen activity, which in turn produces a decrease in the concentration of the charged vacancies. When the impurity concentration is sufficiently large to dominate the concentration of the electron holes, the approximate condition of neutrality becomes $p \approx \xi$, while one has also $\delta \ll p$.

The thermodynamic factors for Ni and M in NiO doped with a monovalent impurity at 1000°C calculated according to the equations presented above and to the present model for the defect structure of NiO under conditions of constant oxygen activity are shown in Figs. 1 and 2 as functions of the impurity concentration for two different values of a_O. The thermodynamic factor for Ni decreases as ξ increases and tends to one when ξ is sufficiently large. This limiting value corresponding to large impurity concentrations is reached earlier at low a_O values since in these conditions the concentration of the intrinsic defects is lower and the condition of control of the defects concentration by the impurity is easier. On the contrary, the thermodynamic factor for M increases with the impurity concentration and ranges from about 1 at small ξ to about 2 at high ξ values: the change occurs again earlier at low oxygen activities for the same reasons examined above. The dependence of the two thermodynamic factors on the oxygen activity under a constant value of the impurity concentration is shown more clearly in Figs. 3 and 4 and is in agreement with the previous conclusion. In fact, at low oxygen activities the change to the condition of control of the defects concentration by the impurity occurs earlier: therefore under a fixed ξ value the situation is closer to the limiting case corresponding to large impurity concentrations at low than at large oxygen activities. Thus, an increase of a_O under a fixed ξ is qualitatively equivalent to a decrease of ξ under a fixed a_O and is expected to produce an increase of the thermodynamic factor of Ni and a simultaneous decrease of that of M.

The thermodynamic factors of Ni and M under conditions of constant impurity concentration are shown in Figs. 5 and 6 as functions of a_O for $\xi=10^{-6}$ and $\xi=10^{-3}$ respectively. In the former case the

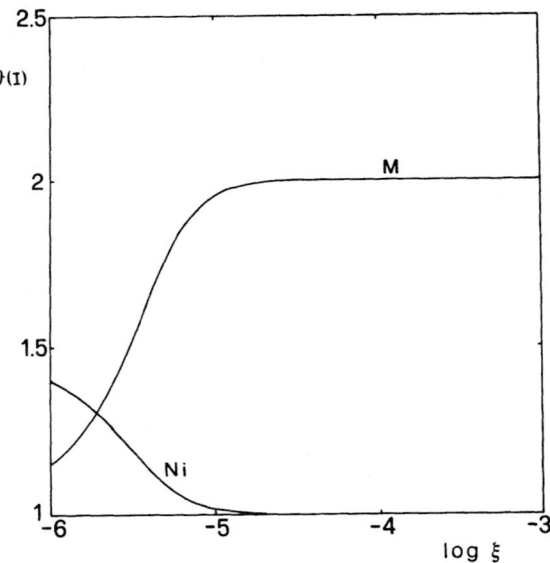

Fig. 1. Thermodynamic factors for Ni and a monovalent impurity in
NiO at 1000°C in presence of gradients of ξ only as functions of ξ at $a_0 = 10^{-5}$.

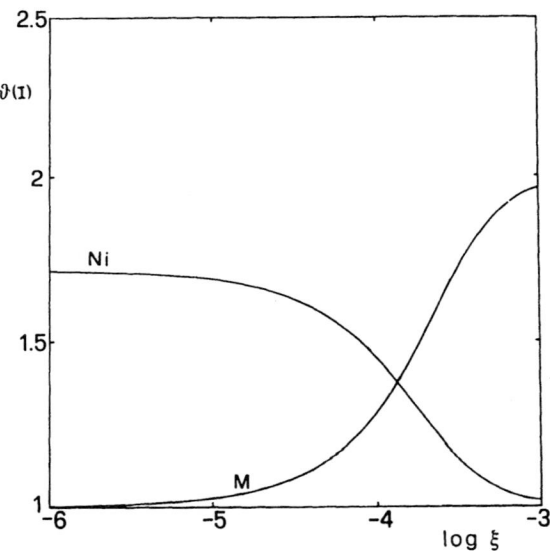

Fig. 2. Thermodynamic factors for Ni and a monovalent impurity in
NiO at 1000°C in presence of gradients of ξ only as functions of ξ at $a_0 = 1$.

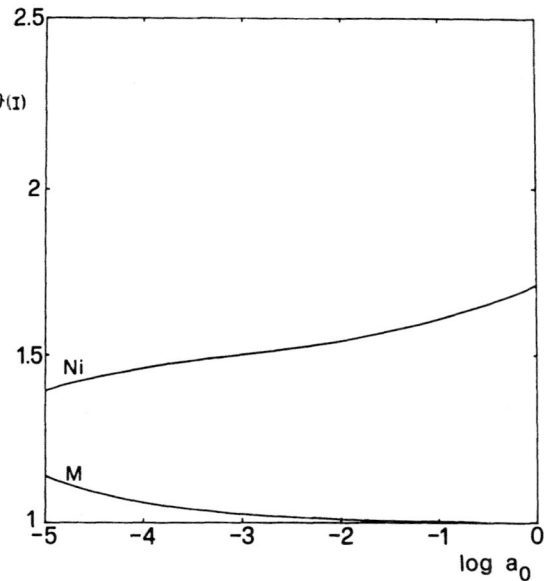

Fig. 3. Thermodynamic factors for Ni and a monovalent impurity in NiO at 1000°C in presence of gradients of ξ only as functions of a_0 at $\xi = 10^{-6}$.

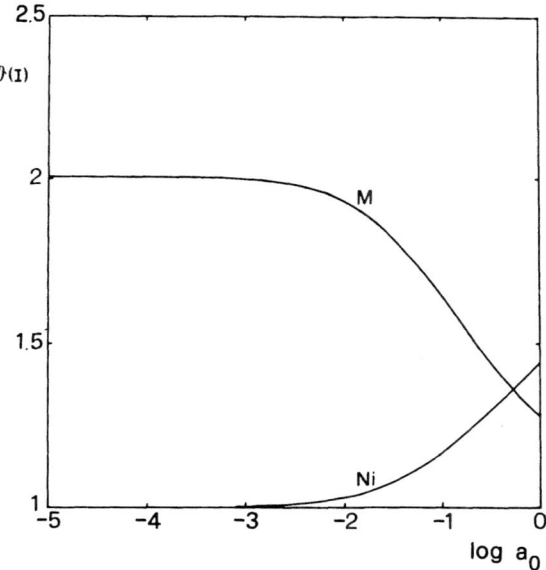

Fig. 4. Thermodynamic factors for Ni and a monovalent impurity in NiO at 1000°C in presence of gradients of ξ only as functions of a_0 at $\xi = 10^{-4}$.

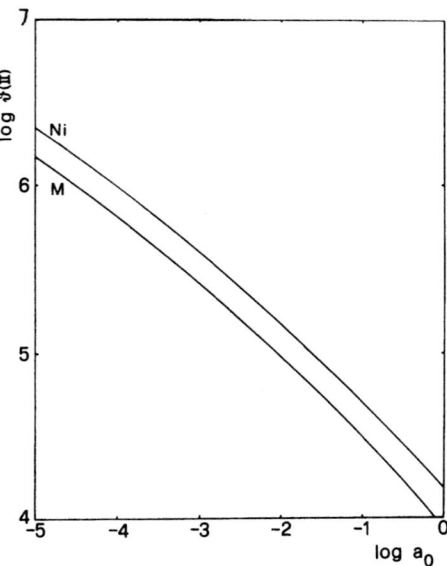

Fig. 5. Thermodynamic factors for Ni and a monovalent impurity in NiO at 1000°C in presence of gradients of a_0 only as functions of a_0 at $\xi = 10^{-6}$.

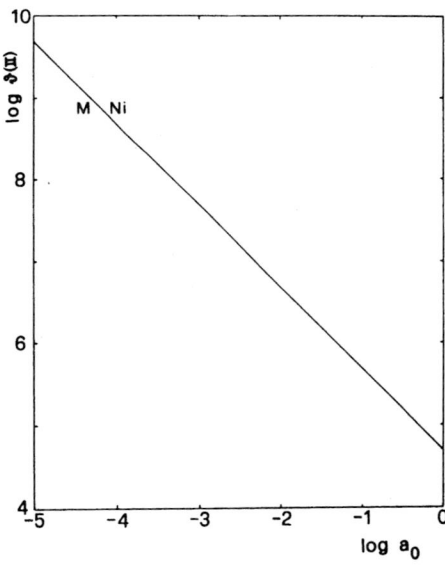

Fig. 6. Thermodynamic factors for Ni and a monovalent impurity in NiO at 1000°C in presence of gradients of a_0 only as functions of a_0 at $\xi = 10^{-3}$.

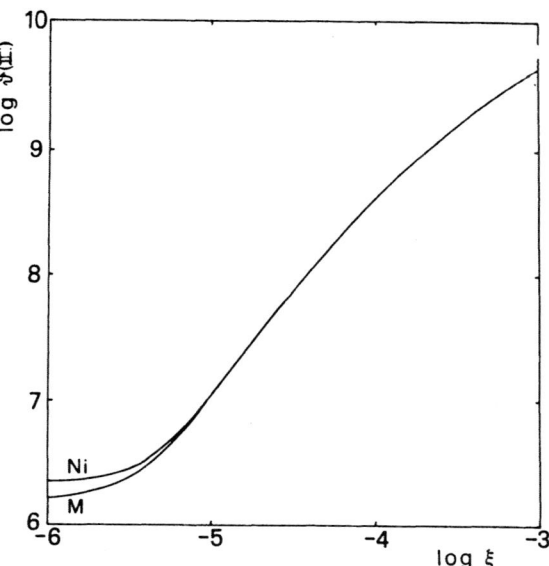

Fig. 7. Thermodynamic factors for Ni and a monovalent impurity in
NiO at 1000°C in presence of gradients of a_0 only as functions of ξ at $a_0=10^{-5}$.

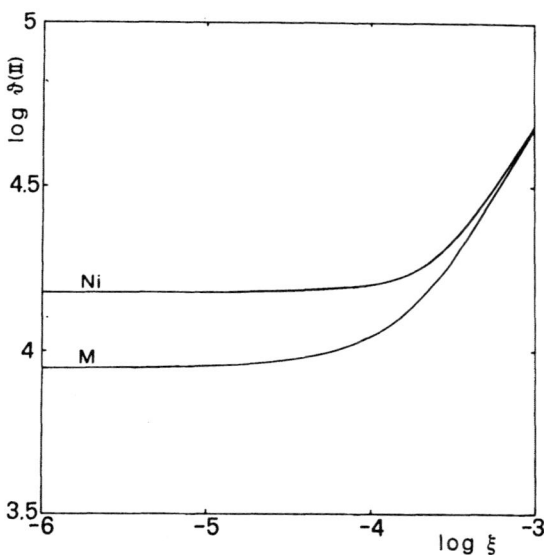

Fig. 8. Thermodynamic factors for Ni and a monovalent impurity in
NiO at 1000°C in presence of gradients of a_0 only as functions of ξ at $a_0=1$.

327

thermodynamic factor for Ni is slightly larger than that of M while in the latter case they are about the same. In addition, both parameters decrease strongly as a_O increases, as expected. Finally, the two thermodynamic factors increase significantly with the impurity concentration. This condition is shown more clearly in Figs. 7 and 8 where the two thermodynamic factors are reported as functions of the impurity concentration under a fixed value of the oxygen activity. This increase is related to the effect that a monovalent impurity has on the concentration of the defects in NiO and is much stronger at low than at high oxygen activities.

CONCLUSIONS

The analysis of the thermodynamic factors for the cation diffusion in a p-type oxide doped with a monovalent impurity developed in the present paper leads to the following conclusions.

1. The thermodynamic factors of the cations in a doped oxide are not only functions of the oxygen activity and of the impurity concentration, but depend as well on the ratio between the gradients of these two variables.

2. The thermodynamic factors of the two ions are connected by a relationship obtained from Gibbs-Duhem equation for the ternary system involved and in general will differ from each other.

3. In case of presence of gradients of impurity concentration under a constant oxygen activity the two thermodynamic factors are both close to one but they change in a different way with the impurity concentration.

4. In case of presence of gradients of oxygen activity under a constant impurity concentration the two thermodynamic factors are again close to each other but they are very large and decrease both as the oxygen activity increases.

REFERENCES

1. P. G. Shewmon, "Diffusion in Solids," McGraw-Hill, New York (1963).
2. J. R. Manning, Theory of Diffusion, in: "Diffusion,", H. I. Aaronson, ed., American Society for Metals, Metals Park, Ohio (1973).
3. A. D. Le Claire, Diffusion, in: "Treatise on Solid State Chemistry," N. B. Hannay, ed., Plenum Press, New York (1976).
4. H. Schmalzried, "Solid State Reactions," Academic Press, New York (1974).

5. G. J. Yurek and H. Schmalzried, Ber. der Bunsenges. Phys. Chem. 78:1379 (1974).
6. G. J. Yurek and H. Schmalzried, Ber. der Bunsenges. Phys. Chem. 79:225 (1975).
7. W. K. Chen and N. L. Peterson, J. Phys. Chem. Solids 34:1093 (1973).
8. F. Gesmundo, J. Phys. Chem. Solids 44:819 (1983).
9. F. A. Kroger, "The Chemistry of Imperfect Crystals," Vol. II, North-Holland, Amsterdam (1973).
10. F. Gesmundo, J. Phys. Chem. Solids, in press.
11. C. Wagner, Corros. Sci. 9:91 (1969).
12. F. Gesmundo, Oxid. Met., submitted for publication.
13. C. M. Osburn and R. W. Vest, J. Phys. Chem. Solids 32:1343 (1971).

RADIATION ENHANCED DIFFUSION IN NUCLEAR CARBIDES

Hj. Matzke

Commission of the European Communities
Joint Research Centre, Karlsruhe Establishment
European Institute for Transuranium Elements
Postfach 2266, D-7500 Karlsruhe
Federal Republic of Germany

ABSTRACT

The radiation enhanced diffusion (coefficient $D*$) of Pu-238 and U-233 in polycrystalline and single crystalline UC and in sintered (U,Pu)C with 3 and 15 % Pu was measured in a nuclear reactor during fission. Conventional diffusion sandwiches with thin tracer layers and a RF furnace for temperature variation (130 to 1400 °C) were used. $D*$ was found to be completely athermal below 1100 °C and to increase linearly with fission rate, $\dot{F}$, according to $D* = A\dot{F}$ with $A = (2.5 \pm 0.5) \times 10^{-30} cm^5$. The mechanism of enhancement is discussed as well as the implications on technologically important processes. The results are related to recent fission damage studies and compared with those obtained on other nuclear fuels, UO_2 and UN.

INTRODUCTION

The monocarbide of uranium, UC, or the mixed carbide (U,Pu)C, are considered as advanced nuclear fuels for liquid metal fast breeder reactors. At present, a heavy water reactor, the WR-1 reactor in Canada, uses UC as fuel, and the Indian fast breeder, FBTR, scheduled to become operative in 1985, uses (U,Pu)C as fuel. The response to radiation of nuclear carbides is therefore of technological and of basic interest. Since nuclear carbides have a much higher thermal conductivity than today's conventional nuclear fuels, UO_2 and $(U,Pu)O_2$, they remain colder during operation and, therefore, radiation enhanced diffusion is anticipated to be effective in a large fraction of the fuel. The thermally activated self-diffusion of both carbon, uranium and plutonium in UC and (U,Pu)C has been measured extensively [1,2]. It is quite well understood and was, in fact, treated in some detail at the 2nd Conf. on Transport in Non-Stoichiometric Compounds in a more general review on carbides[3].

Carbon diffusion in nuclear carbides is much faster than U or Pu diffusion (e.g. $D^C/D^M \sim 10^6$ in UC at 1700 °C). Consequently, the less mobile metal atoms (M) will be the rate-controlling species for other processes of technological interest (densification, sintering, grain growth, creep etc.). Therefore, the effect of radiation on <u>metal</u> atom diffusion was studied. To better understand the mechanism of radiation enhancement, the contributions of displacement spikes and of fast neutron induced atom mixing were estimated, a consideration of the fission fragment impact was made, and the suggested mechanisms are related to the thermal and electrical conductivity of the matrix. The results are compared to those of similar experiments on two other ceramic fuels, UO_2 and UN (or (U,Pu)N), and they are related to results of fission damage studies in UC where fission-induced changes of lattice parameter, volume, lattice strain and electrical resistivity were measured[4].

EXPERIMENTAL

The materials used and the irradiation conditions are summarized in Tables 1 and 2. 3 mm diam. specimens were cut from larger

Table 1. Materials used

Material	Composition	Density (% of theor.)	irradiated at T,°C
UC single crystals	stoichiometric (<30 ppm metal impur.)	100	130, 1120, 1300
UC arc cast	stoichiometric ($\sim$120 ppm metal impur.)	100	400, 800*, 900, 1120, 1300
(U, 3% Pu)C, sinter	$(U,Pu)C_{0.96} - -O_{0.04}N_{0.01}$	95	400, 800*, 1120, 1300, 1400
(U, 15% Pu)C, sinter	$(U,Pu)C_{0.97} - -O_{0.05}N_{0.01}$	92	130, 700+, 900, 1120, 1300

* neutron flux dependence measured at 800 °C
+ time dependence measured at 700 °C

332

samples, one face was polished and coated with a thin tracer layer of U-233 by flash evaporation or with monolayers of Pu-238 with an electron beam evaporator using (U,Pu-238)C as evaporating target. The small dimension of the samples ensured that radial temperature gradients in the carbide samples during irradiation due to fission -induced heat were acceptable. Specially produced W-containers were filled with 4 specimens (2 diffusion couples) each, closed by electron beam welding and inserted into a Nb-container. Figures of the arrangement have been given before[5]. Four thermo-couples were used to measure temperatures. The temperature differences ΔT between central and surface temperature of the small specimen discs varied with Pu-content and neutron flux (hence with fission rate $\dot{F}$).

Table 2. Irradiation Conditions

T	t	flux	rate $\dot{F}$, 10^{13}f cm^{-3}s^{-1}			dose F, 10^{18}f cm^{-3}		
°C	h	ncm^{-2}s^{-1} x 10^{13}	0% Pu	3% Pu	15% Pu	0% Pu	3% Pu	15% Pu
130	120	2.9	0.38	–	7.7	1.64	–	33.3
400	143	1.4	0.18	1.04	–	0.93	5.35	–
700[+]	56	0.81	–	–	2.17	–	–	4.37
700	168	0.75	–	–	2.03	–	–	12.3
700	334	0.75	–	–	2.03	–	–	24.4
800[x]	168	0.70	0.09	0.54	–	0.54	3.27	–
800	189	2.65	0.34	2.00	–	2.31	13.6	–
800	173	9.7	1.26	7.36	–	7.85	45.8	–
880	113	11.7	1.5	–	–	6.10	–	–
900	159	2.2	–	–	5.95	–	–	34.1
1120	123	3.4	0.44	2.56	9.1	1.95	11.3	40.3
1300	115	3.6	0.47	2.72	9.7	1.95	11.3	40.3
1400	155	7.2	0.92	5.44	–	5.13	30.3	–

*Temperature of RF furnace. The real specimen temperature is higher by (most frequently) 5 to 50 °C because of ΔT_{gap} and ΔT_{pellet} (see text).

+T = const. to study possible time dependence.

xT = const. to study dependence on n-flux or on fission rate.

The $\dot{F}$ and T-values are calculated for specimen surfaces, corrected for neutron absorption in furnace, container and capsule. The flux depression in the samples is negligible for UC but important for $(U_{0.85}Pu_{0.15})C$. The average fluxes, in fraction of the surface flux, are 0.994, 0.951 and 0.77 for 0, 3 and 15 % Pu, the corresponding central fluxes are ∿0.98, ∿0.91 and ∿0.59 of the surface flux.

Parallel experiments were performed with as-pressed green pellets of starting densities of 61 % of theoretical, in order to investigate radiation-enhanced densification. For those, the $\dot{F}$ and F-values were lower in the ratio of the densities.

With UC, this ΔT_{pellet} was always <10 °C, with (U, 15% Pu)C it amounted up to 50 °C. An average temperature

$$T_{av} \sim T_{irr} + \Delta T_{gap} + \frac{1}{3} \Delta T_{pellet}$$

was calculated from the furnace temperature T_{irr}, the temperature gradient in the pellet (taken to 1/3 since the tracer layer was smaller than the pellet surface) and the temperature drop ΔT_{gap} within the gap of $\sim$ 100 μm between specimen and W-capsule. These temperatures were used in the Arrhenius diagrams shown below.

The diffusion profiles were measured following irradiation applying high-resolution α-spectroscopy. This method and the algorithms to deduce D-values as low as 10^{-18} cm^{2}s^{-1} were described before[6,7]. The method is non-destructive and yields depth profiles. It uses the energy loss of the α-particles emitted from U- or Pu-atoms diffused into the specimen to determine their diffusion distance. As shown before[5], an effect of β-, γ-irradiation of the highly radioactive specimen in temporarily deteriorating the resolution of the spectrometer can exist, in particular for the highest fission doses used. However, in these cases, a sufficiently long waiting time between irradiation and measurement of the tracer concentration profile guarantees a very low limit of detection in radiation enhanced diffusion coefficients, D*, of $\leq 1 \times 10^{-17}$ cm^{2}s^{-1}.

Diffusion sandwiches identical to those used for irradiation experiments were also annealed in a laboratory furnace at 700, 900, 1100 and 1300 °C. Therefore, the determination of enhancement factors due to irradiation was based on direct comparison and not on literature data alone.

The irradiations were performed in the SILOË reactor at CEA Grenoble. A total of 13 irradiations was performed (see Table 2). The irradiation series was named RADIF, for radiation enhanced diffusion. The total fission doses were chosen such that the total amount of metallic fission products produced during irradiation was less than the level of pre-existing impurities in the (U,Pu)C, except at the highest burn-up achieved (0.26 at.%).

RESULTS

Figs. 1 and 2 show the results for UC and for (U,Pu)C as function of temperature in Arrhenius-diagrams. Full symbols give results under irradiation, open symbols show furnace anneals and the steep lines labelled "thermally activated diffusion" represent literature data[1] for self-diffusion. For UC in Fig. 1, two lines are shown, one for the very pure single crystals that were also used in the present study, and another one for less pure arc-cast UC where impurity effects cause a lower activation enthalpy, ΔH,

below ∿2000 °C. This latter line only is shown for UC in Fig. 2
since Fig. 2 shows results for sintered specimens of (U,Pu)C of
comparatively high impurity contents (∿1300 ppm). The equations
describing thermally activated diffusion are summarized in Table 3.
Generally, the smaller Pu-atoms show a slightly higher D-value than
the larger U-atoms. (D^{Pu}/D^{U} ∿ 3 to 5 in UC). Also, diffusion in
(U,Pu)C with its somewhat bigger lattice spacing is faster than diffusion in UC. Note also, that all (U,Pu)C investigated so-far was

Table 3: Arrhenius equation for metal atom diffusion in nuclear
carbides[1] (ΔH in kcal mol^{-1}, thermally activated diffusion).

UC single crystals	$D^U = 11.7 \exp(-141.9/RT)$ cm^{2}s^{-1}
UC arc cast	$D^U = 6.9 \exp(-141.0/RT)$ {intrinsic} $+ 3.6 \times 10^{-5} \exp(-84.5/RT)$ cm^{2}s^{-1} {extrinsic, impurity controlled}
$(U_{0.85}Pu_{0.15})C$	$D^{Pu} = 0.013 \exp(-96.0/RT)$ cm^{2}s^{-1}

(typical result; no data exist for (U,3%Pu)C)

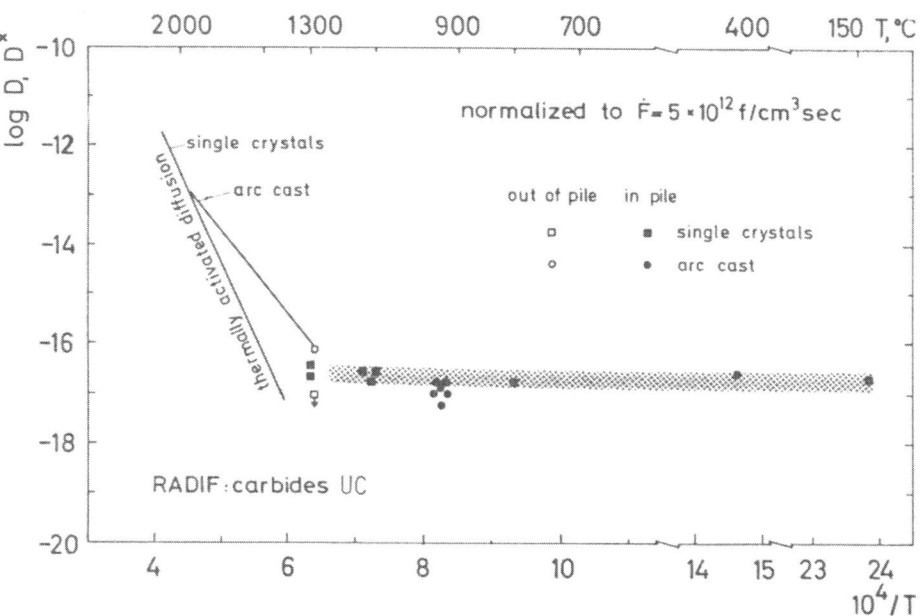

Fig. 1: Radiation enhanced diffusion of U and Pu in UC in an
Arrhenius diagram (full symbols), normalized to a fission
rate of $\dot{F} = 5 \times 10^{12}$ f cm^{-3}s^{-1}
(Note that the temperature scale is broken twice).

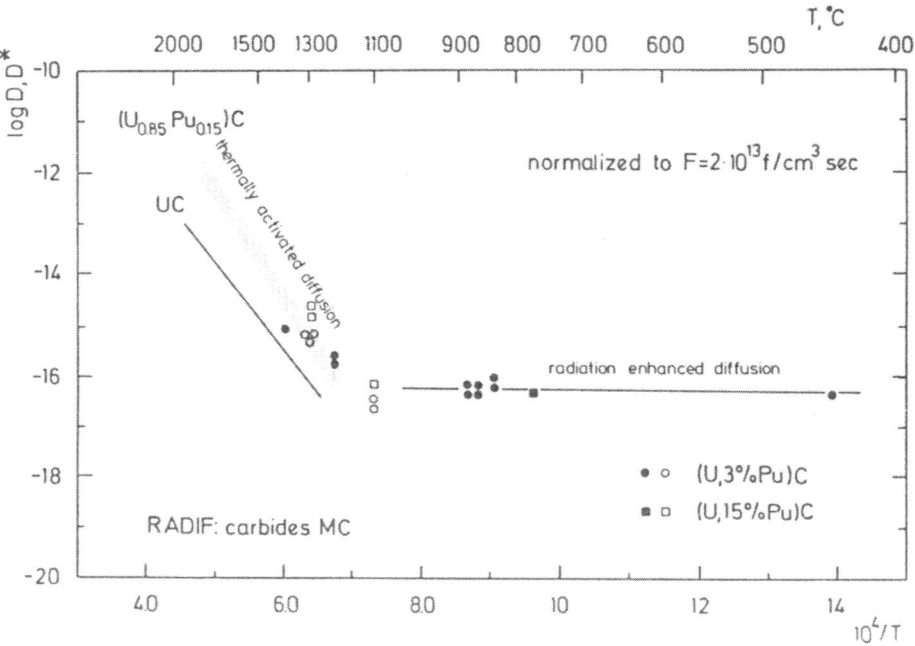

Fig. 2: Radiation enhanced diffusion of Pu in (U,Pu)C with 3 or
15 % Pu (full symbols). As in Fig. 1, open symbols are for
laboratory anneals. The dotted band represents the data on
thermally activated Pu-diffusion in a series of different
batches of (U,Pu)C with 15 or 20 % Pu.

sintered, had lower densities, smaller grain sizes and higher impurity contents than the corresponding UC compounds. All this acce-
lerates thermally activated diffusion.

The laboratory anneals (open symbols in Figs. 1 and 2) showed
absence of any measurable thermally activated diffusion at 700 °C
and 900 °C. At 1100 °C, no measurable diffusion was detected in
UC, whereas D-values of $3 - 7 \times 10^{-17} \mathrm{cm^2 s^{-1}}$ were measured for (U,Pu)C.
At 1300 °C, an upper limit of $10^{-17} \mathrm{cm^2 s^{-1}}$ was obtained for UC single
crystals (see small arrow at data point in Fig. 1), whereas the expected higher values were observed for arc-cast UC and for (U,Pu)C,
in agreement with the literature data. These combined results on
thermally activated diffusion represent thus a reliable basis for
estimating radiation effects.

The radiation enhanced diffusion coefficients D*, in contrast,
were above the detection limit at all temperatures, even at a
temperature as low as 130 °C. They were completely temperature independent, or athermal below ∿ 1100 °C, and even up to 1300 °C for
UC single crystals. Fig. 3 shows that D* is roughly proportional to
fission rate, Ḟ, within the investigated range. The slope is appro-

336

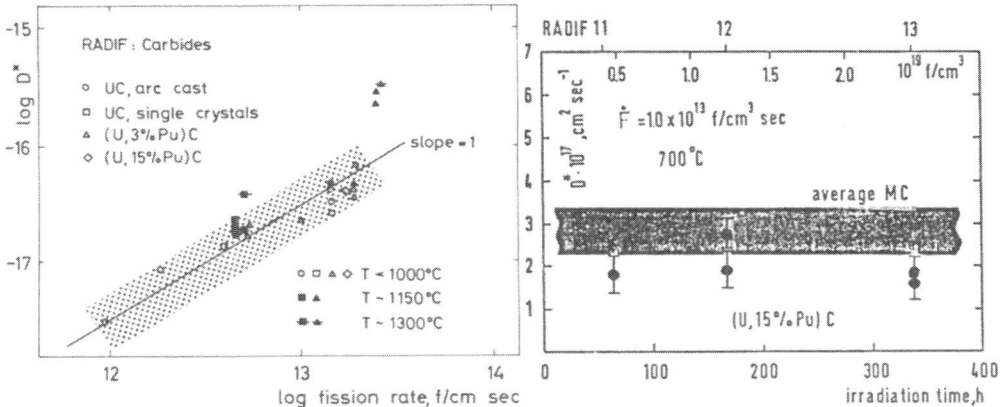

Fig 3: Fission rate dependence of D* in nuclear carbides. Only data for T ≳ 1100 °C do not fall in the dotted area of slope 1.

Fig. 4 : Results for D* at 700 °C for different irradiation times for constant Ḟ. No obvious effect of dose F or time t is seen.

ximately 1. The data points representing temperatures ≳ 1100 °C are expected to be high and, therefore, tend to fall above the band describing the results for T < 1100 °C. The resulting relation is

$$D* = A\dot{F} = (2.5 \pm 0.5) \times 10^{-30} \dot{F} \text{ cm}^5.$$

Fig. 4, finally, shows that the measured D*-values do not depend on total dose, or time, within the range investigated.

Some thermal anneals in a laboratory furnace were performed on RADIF samples irradiated at 130 or 400 °C. Anneal temperatures were 700 and 900 °C, i.e. temperatures where metal point defect become or are mobile. On average, the tracer atoms migrated at the most very small distances of < 50 Å during the recovery of the excess defects that survived irradiation.

DISCUSSION AND CONCLUSIONS

The reported experiments show that metal atom diffusion in nuclear carbides is enhanced during reactor irradiation by very high factors indeed (e.g. at 1000 °C: D*/D ∿ 10^3 and this ratio increases drastically with decreasing temperature). The resulting D*-values are completely athermal below 1100 °C and proportional to fission rate Ḟ. The exact temperature where D ∿ D* depends on Ḟ and on C/M-ratio since thermally activated metal diffusion in nuclear carbides depends strongly on C/M-ratio[1]. There are different possible means of enhancing D during fission. Most of these can be discarded. The following cannot be the main reasons for enhancement:
- neutrons: atom mixing by neutrons is expected to give a D* ∿ 10^{-20} cm^{2}s^{-1}. Also, D* is proportional to Ḟ and not to n-flux
- direct collisions with fission fragments. This would yield, for

the $\dot{F}$ range used, D*-values of the order of 5 x 10^{-20}cm^{2}s^{-1}
- point defect mobility or recovery of surviving excess point defects following irradiation . No significant effect was seen in subsequent laboratory anneals.
- the lattice expansion that occurs in nuclear fuels during irradiation [4],[9-14] (with maxima of 0.1 to 0.2 % at about 10^{18}f cm^{-3}s^{-1} - see Fig. 5 - and subsequent decrease due to point defect cluster formation). No effect of total dose (or irradiation time) was seen.

To obtain a better understanding of the mechanism operative for D*, it is illustrative to compare the carbides with other ceramic fuels, i.e. oxides and nitrides. Fig. 5 shows fission damage ingrowth in these 3 materials. Maxima in lattice expansion are obtained at about the same total dose, but the size of the effect differs. Above a dose of 10^{18}f cm^{-3}, cluster formation causes the lattice to recover. The curves of Fig. 5 can be taken as an indication that less damage recovery occurs in materials with higher thermal (or electrical) resistivity. The thermal recovery of the surviving defects, e.g.[4],[12-14],[16],[17] occurs between about 200 - 800 °C, and therefore well below the 1100 °C until which D* shows no T-dependence. The reason for the observed radiation enhancement must therefore mainly be sought in the fusion event itself. Figs. 6 and 7 show that most of the fission energy is lost in electronic stopping, and only a small fraction of about 3-6 % is transformed into nuclear stopping, i.e. in atomic collisions and colli-

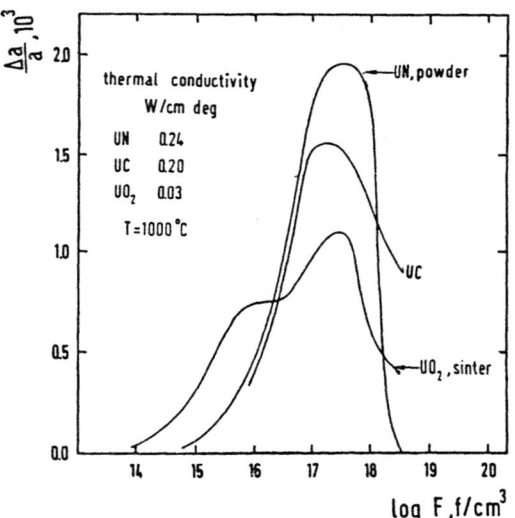

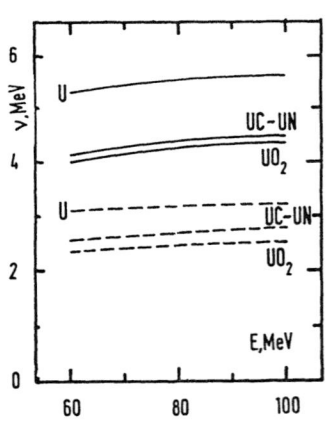

Fig. 5: Lattice parameter changes in UC, UO$_2$ and UN as function of fission dose.[4],[9-14]

Fig. 6: Fraction ν of energy lost by nuclear stopping of a fission fragment (heavy fragments = full line; light fragments = dashed line) for U, UC or UN, and UO$_2$, as function of fission energy.[15]

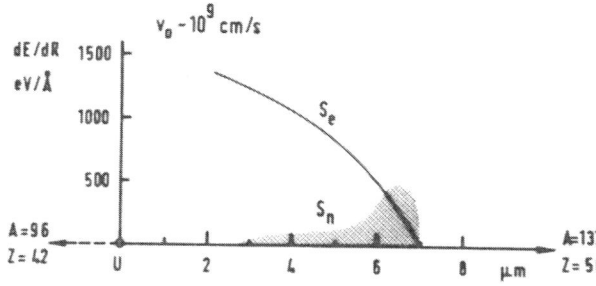

Fig.7: Energy loss of a fission fragment along its track, split up in electronic stopping power, S_e, and nuclear stopping power, S_n, for a median heavy fission fragment in UO_2

sion cascades. These nuclear collisions occur predominantly towards the end of the fission tracks (s. Fig. 7). Calculations on the total number of displaced atoms per fission exist, predominantly, however, for UO_2, where Soullard[18] determined $\sim$ 27000 displaced uranium and $\sim$ 73000 displaced oxygen atoms (total 100000) to be formed per fission. 20 % of the U-defects were determined to survive the fission event. First calculations for UC exist also[19]. Any difference between UC, UN and UO_2 in terms of number of displaced U-atoms can be assumed to be much less than the observed differences in D*. These were (normalized to 1 for UC)

$$D*(UO_2) : D*(UC) : D*(UN) = 5 : 1 : 0.5$$

These differences can thus not be explained in terms of nuclear collisions. As shown in Fig. 5, the thermal conductivities of these three materials show the inverse relation. If UC is again normalized to 1, then the relation (at 1000 °C) is UO_2 : UC : UN = 0.15 : 1 : 1.2 . Blank and Matzke[20] have argued that fission spikes in UO_2 involve significant thermoelastic pressure fields. Volumes of $\sim 1.5 \times 10^{-16} cm^3$ per fission event were predicted to be heated above the melting point, causing even bigger volumes ($\sim 2.3 \times 10^{-16} cm^3$) to be under high pressures (> 5×10^3 atm). The atomic mixing in these volumes is very effective. Brucklacher and Dienst[22,23], to explain their creep results (Figure 8) discuss a "thermal rod", i.e. a molten zone along the fission track existing for $\sim 2 \times 10^{-11}$ sec with a D as in the melt. With this assumption, results of creep under irradiation could be well described. The predicted D*-values were still smaller than the measures ones[21]. In contrast, direct atomic collisions, the effect of fast neutrons and conventional displacement and thermal spikes were calculated to yield D* values which are lower by a factor >100 than the experimental results. Therefore, an even more effective mixing than in the thermal rod picture must take place in UO_2; one possibility for this more effective mixing might be a larger than expected separation of Frenkel defects (interstitials and vacancies) due to the large pressure gradients along the fission track. Such processes are expected to be important in UO_2 because of its low thermal and electrical condictivity. They would be predicted to be largely absent in the more metallic substances UC and UN. In these materials, the distribution of energy originating from the electronic stopping over larger volumes is much easier than in UO_2, since their electrical and thermal conductivities are much higher. Whereas thus the volume V_d affected by displacements should be similar in the three materials, the temperature volume V_T and the pressure volume V_p are

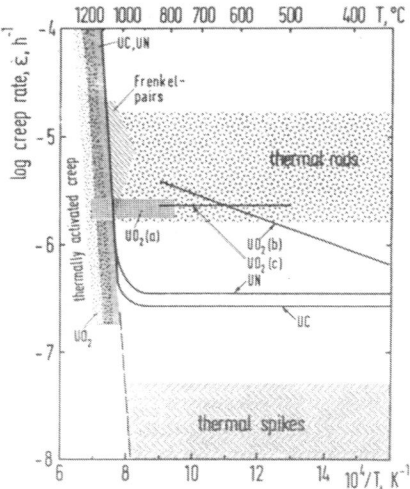

Fig. 8: Temperature dependence of thermally activated and radiation enhanced creep[22-25] (for 3 at.% burn-up, a stress of 2 kp mm^{-2} and a fission rate of~3 x 10^{13}f cm^{-3}s^{-1}) for UO_2, UC and UN, together with predictions of Brucklacher and Dienst[22,23] for contributions of thermal spikes , thermal rods and Frenkel pair defect annealing (see text). Radiation enhances creep drastically below ~1000 °C, and radiation enhanced creep in UC is completely athermal, just as radiation enhanced diffusion.

much larger in UC than in UO_2, with simultaneously much smaller center values of T and p. This explains the smaller effect of radiation on diffusion and creep in nuclear carbides than in nuclear oxides. However, since the observed values are larger than expected from displacement phenomena alone, some additional though small effect of the temperature and/or pressure connected with fission spikes probably exists in the carbides as well. Also, a thermal migration of point defects in the time between the spikes (typically some 30 min) might exist causing $D_v \cdot c_v = \text{const}$ (c = concentration, v = vacancy) thus cancelling the opposite T-dependences of c and D and yielding the observed athermal D*-values.

The pronounced enhancement of metal atom diffusion by radiation has consequences on processes of technological importance. The most obvious one is densification[26]. The irradiation of green pellets of UC (and also UO_2 and UN, as well as of the corresponding Pu-containing compounds) confirmed this fact. Even at 130 °C, densification was measurable, and, as expected, it was much more pronounced in the oxides than in the carbides. Detailed results on radiation enhanced densification will be reported elsewhere.

The fact, that D*(carbide) << D*(oxide), has also technologically important consequences. Since the same relation holds for creep, accommodation of fission-gas swelling by creep into pre-existing porosity will exist in carbides, but to a smaller extent than in oxides. Also because of the smaller creep rates under irradiation, the mechanical interaction between the (relatively cold) outer rim of the fuel and surrounding metal tube (the clad) will be more severe with carbides than with oxides. Finally, an important question in operating fuel elements is the formation of gas-filled bubbles by the precipitation of rare gases, Kr and Xe. These occur in fission with a high yield, have a very small solubility, and tend therefore to precipitate. The pronounced temperature and pressure effects connected with fission spikes in oxides cause a radiation-induced "re-solution" of precipitated fission gas atoms back into the lattice, which is larger than that expected by simple collisional knock-on processes [20,27,28]. In fact, a single fission event can destroy a small fission gas bubble in UO_2 completely. In contrast, the fact that these phenomena are much smaller in nuclear carbides implies that the re-solution probability will also be much smaller approaching that due to atomic collisions alone. This contributes to the known fact that carbides swell more under irradiation than oxides.

In summary, measurements of radiation enhanced diffusion of U and Pu in nuclear carbides show that an important effect with interesting implications on problems of their technological application exists. The enhancement is smaller than in oxides. It is connected with fission spikes and directly proportional to fission rate, and it is completely athermal, temperature-independent below ~ 1100 °C.

References

1. Hj. Matzke and M.H. Bradbury, Euratom Report EUR 5906 EN (1978)
2. H. Matsui and Hj. Matzke, J. Nucl. Mater. 88:41 (1980).
3. Hj. Matzke, Solid State Ionics 12:25 (1984).
4. Hj. Matzke, Radiation Effects 64:3 (1982).
5. A. Höh and Hj. Matzke, J. Nucl. Mater. 48:157 (1973).
6. F. Schmitz and R. Lindner, J. Nucl. Mater. 17:259 (1965).
7. A. Höh and Hj. Matzke, Nucl. Instrum. Methods 114:459 (1973).
8. Hj. Matzke, J.L. Routbort and H.A. Tasman, J. Appl. Phys.
9. N. Nakae, A. Harada, T. Kirihara and S. Nasu, J. Nucl. Mater. 71:314 (1978).
10. T. Kirihara, N. Nakae, H. Matsui and M. Tamaki, in "Plutonium (1975) and other Actinides", North Holland Publ.Co., Amsterdam p. 903 (1976).
11. M. Tamaki, A. Ohnuki, H. Matsui, G. Matsumoto and T. Kirihara,
12. M. Tamaki, S. Matsumoto, K. Ishimaru, G. Matsumoto and T. Kirihara, J. Nucl. Mater. 108 & 109:671 (1982).
13. M. Tamaki, PhD Thesis, University of Nagoya (1983).
14. H. Matsui, PhD Thesis, University of Nagoya (1975).
15. J. Lindhard and P.V. Thomsen, in "Radiation Damage in Solids", IAEA, Vienna 1:65 (1962) and P.V. Thomsen, private communication (1972).
16. W. Dienst, German Report KfK-1215 (1970).
17. H. Matsui and Hj. Matzke, J. Nucl. Mater. 89:41 (1980).
18. J. Soullard, French Report CEA-R-4882 (1977).
19. D. Leseur, Phil. Mag. A44:905 (1981).
20. H. Blank and Hj. Matzke, Radiation Effects 17:57 (1978).
21. Hj. Matzke, Radiation Effects 75:317 (1983).
22. D. Brucklacher and W. Dienst, J. Nucl. Mater. 42:285 (1972).
23. D. Brucklacher, PhD-Thesis, University of Karlsruhe, FRG (1981); and in "Physical Metallurgy of Reactor Fuel Elements", Eds. J.E. Harris and E.C. Sykes, The Metals Society, London, p.118 (1975).
24. D.J. Clough, J. Nucl. Mater. 65:24 (1977).
25. P. Zeisser, G. Marianiello and C. Merlini, J. Nucl. Mater. 65:48 (1977).
26. Hj. Matzke, in "Fission Gas Behavior in Nuclear Fuels", Eds. C. Ronchi, Hj. Matzke, J.v.d. Laar and H. Blank, Report EUR-6600 EN and Europ. Appl. Research Reports 1:289 (1979).
27. R.S. Nelson, J. Nucl. Mater. 31:153 (1969).
28. H. Blank, in "Fission Gas Behavior in Nuclear Fuels", Eds. C. Ronchi, Hj. Matzke, J.v.d. Laar and H. Blank, Report EUR-6600 EN and Europ. Appl. Research Reports 1:307 (1979).

DIFFUSION PROPERTIES - PRIMARILY INTERFACIAL

VOLUME, GRAIN BOUNDARY AND SURFACE DIFFUSION OF AN

ISOTOPE IN SOME REFRACTORY OXIDE SYSTEMS: A COMPARISON

V. S. Stubican

Department of Materials Science and Engineering
The Pennsylvania State University
University Park, PA   16802

INTRODUCTION

It is the purpose of this paper to compare some recent results obtained on volume, grain boundary diffusion and surface diffusion of ^{51}Cr in refractory oxides such as MgO, Al_2O_3 and $MgAl_2O_4$. In these materials the concentration of intrinsic point defects is very small and diffusion transport phenomena are very sensitive to the concentration of aliovalent impurities. However, from the technological point of view the knowledge of transport phenomena in these materials is important to explain sintering, grain growth, mechanical, thermal and electrical properties. To avoid uncontrolled influence of impurities carefully doped materials can be used. The doping with aliovalent ions is particularly interesting because the presence of such ions determines the concentration of point defects and allows experiments to be performed with well defined materials.

DIFFUSION EXPERIMENTS

All diffusion experiments were performed with single crystals. Single crystals of MgO and Cr_2O_3 doped MgO crystals were grown by the arc melting technique, while Al_2O_3 single crystals were grown by using the Czochralski method and $MgAl_2O_4$ crystals by using fusion cast technique. All specimens were analyzed for impurity concentration by semiquantitative emission spectroscopy and the following results were obtained: MgO contained Fe 10-30 ppm, Al 3-25 ppm, Ca 450-500 ppm, Si 200 ppm; Al_2O_3 contained Ca 150-250 ppm, Si 400-500 ppm, Fe < 5 ppm, Ti < 5 ppm, $MgAl_2O_4$ contained Ca 150-250 ppm, Si 150-200 ppm Fe < 10 ppm, Cu < 5 ppm. All impurity concentrations are given in parts per million by weight. Wet chemical analysis was used to determine the MgO:Al_2O_3 ratio of the $MgAl_2O_4$ single crystals. The analysis gave a molar ratio of 1:1. The preparation and the application of the isotope ^{51}Cr for volume, grain boundary and surface diffusion as well as sectioning technique and counting technique were described in previous papers[1,2,3].

VOLUME DIFFUSION

A comparison of the lattice diffusion coefficients of ^{51}Cr for the temperature 1300-1600°C is shown in Fig. 1. The temperature dependence of the lattice diffusion coefficient of ^{51}Cr is given by $D = 1.02 \times 10^{-3}$ exp-(293 ± 40 KJ/mole/RT), $D = 2.58 \times 10^{-3}$ exp -(306 ± 59 KJ/mole/RT) and 2.56×10^{-2} exp -(337 ± 40 KJ/mole/RT) for MgO, Al_2O_3 and $MgAl_2O_4$ respectively[4]. Diffusion of ^{51}Cr in these oxides is by the vacancy mechanism and the activation energy is the energy of motion for an associated impurity. It seems very probable that in all of these materials diffusion is dominated by silicon impurity and it is extrinsic to the melting point[1,5]. Similar conclusion was reached by Lloyd and Bower[6], who studied iron tracer diffusion in alumina.

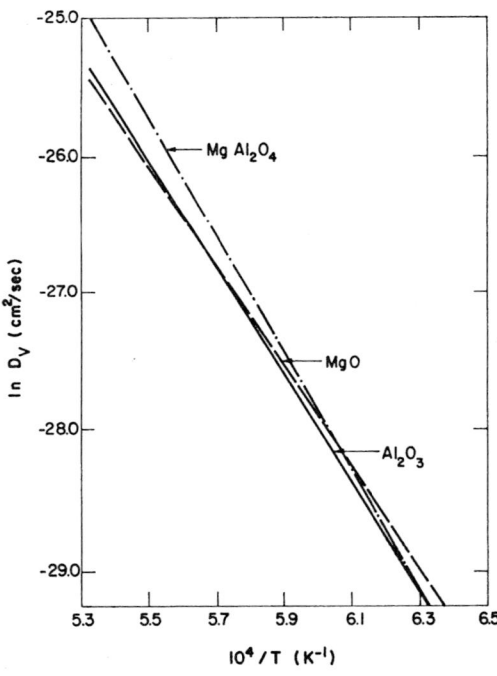

Figure 1. Arrhenius plots for ^{51}Cr lattice diffusion in MgO, Al_2O_3 and $MgAl_2O_4$.

To avoid uncontrolled influence of impurities on the diffusion in highly refractory oxides it is useful to fix the concentration of cation vacancies by doping material with the known concentration of an aliovalent ion. Weber et al.[1] have studied diffusion of ^{51}Cr in Cr-doped MgO single crystals at three temperatures 1383, 1444 and 1495°C. An approximately linear relationship was found between $D(Cr*)$ and the concentration of dopant (Cr^{3+})[1].

According to Lidiard's[7] model in which the complexes are regarded as diffusing species, the intrinsic diffusion coefficient, $D(Cr)$, of the impurity is related to the diffusion coefficient of the complex as follows:

$$D(Cr) = D_o \left[\frac{d(pc)}{dc}\right] \tag{1}$$

where D_o is the diffusion coefficient of the complex, p is degree of association, c is the total impurity (dopant) concentration $[Cr_{tot}]$ expressed as atomic fraction. For diffusion of Cr* tracer in uniformly doped crystals p is constant along the diffusion profile so that:

$$D(Cr*) = D_o \cdot p \tag{2}$$

Perkins and Rapp[8] following Lidiard's theory developed an expression for p:

$$p = 3/4 + 1/cA - (1/16 + 3/4\ cA + 1/4\ c^2A^2)^{\frac{1}{2}} \tag{3}$$

This yields for $D(Cr*)$ in the limit of low concentration:

$$D(Cr*) = D_o \cdot Ac/2 \tag{4}$$

where D_o = const. exp $(-E_m/kT)$; E_m is the activation energy for motion of the Cr-vacancy complex, $A = 12$ exp (E_A/kT), E_A is the association energy of the Cr-vacancy complex and c is total Cr-ion concentration expressed as atomic fraction. For the above conditions the slope of lnD* vs 1/T would yield the difference in the energy of association and energy of motion:

$$E = E_m - /E_A/ \tag{5}$$

In our observations, the diffusion data were well represented as being proportional to c but did not extrapolate to zero as predicted by the model[7]. The extrapolated values for $D(Cr*)$ to zero concentration of dopant (Cr^{3+}) were consistent in order of magnitude with the values obtained in measurements on similar but undoped MgO crystals. It is postulated that the diffusivity observed in the undoped MgO and the non-zero extrapolated values of $D(Cr*)$ for the doped MgO crystals result from the contribution of the background impurities to the diffusivity. Assuming for example that the significant impurity is Si^{4+}, that at the temperature of interest Si^{4+} is sufficiently associated such that it does not affect the Cr ion-vacancy association equilibrium, the contribution of the Si^{4+} to the observed $D(Cr*)$ would be then independent and additive to the Lidiard expression. The data for the undoped MgO crystals (Fig. 1) were therefore used to determine a $D(Cr*)$ Si, which is the tracer diffusion coefficient of Cr^{3+} in MgO controlled by the Si^{4+} impurity concentration, which was subtracted from the measured data for the doped crystal. The results obtained are shown in Fig. 2. The Arrhenius presentation of the data, Fig. 3, yielded an average energy of 82.0 ± 7.1 KJ/mole. The activation energy determined from the doped crystal corrected for the background impurity contribution is then the difference between the

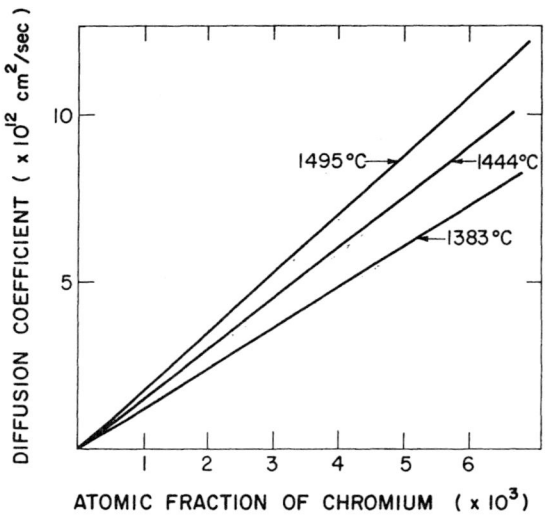

Figure 2. Corrected values for the lattice diffusion coefficient of ^{51}Cr in Cr-doped MgO as the function of the concentration of dopant and temperature.

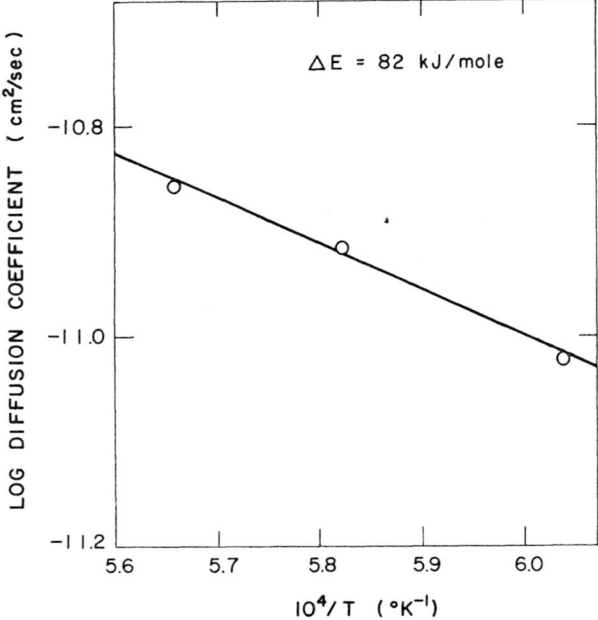

Figure 3. Log D vs 1/T for the diffusion of ^{51}Cr in Cr-doped MgO obtained from Fig. 2.

energy of motion E_m and the energy of association E_A of a Cr-vacancy complex:

$$E = E_m - /E_A/ = 82.0 \pm 7.1 \text{ KJ/mole}$$

According to Glass and Searly[9] the energy for motion of the Cr-vacancy complex in MgO is 166.94 KJ/mole. If we take this value to be correct then according to our results the energy for association of Cr-vacancy pairs is 84.9 ± 12.6 KJ/mole (0.88 ± 0.13 eV).

The anisotropy of the volume diffusivity of ^{51}Cr in Al_2O_3 was also investigated (Fig. 4). Within the experimental error, anisotropy was not observed, which may be explained by the close packed structure of oxygens. This is in agreement with the results obtained by Lloyd and Bowen[6] who studied volume diffusion of iron in Al_2O_3.

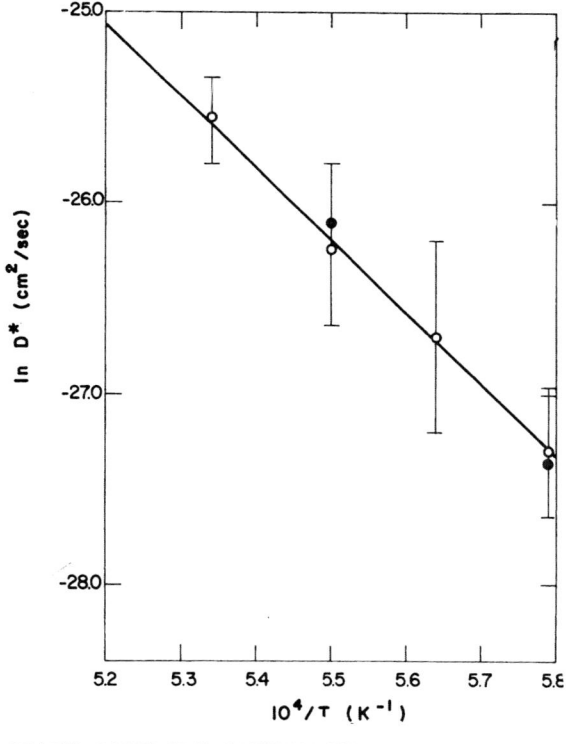

Figure 4. Arrhenius plot for ^{51}Cr lattice diffusion in Al_2O_3. The open circles are the data measured normal to a 6° $(01\bar{1}2)$ plane, the closed circles are the data measured parallel to this plane.

GRAIN BOUNDARY DIFFUSION

In view of the poor understanding of grain boundary diffusion in ionic systems, the study was initiated to determine the effect of anisotropy and doping with aliovalent ions on the grain boundary diffusion of ^{51}Cr in MgO, Al_2O_3 and $MgAl_2O_4$. It was hoped that investigation of anisotropy would contribute to the better understanding of the mechanism of the grain boundary diffusion in ionic crystals and that the doping with aliovalent ions will show if the change in the concentration of point defects has an influence on the grain boundary diffusion. For the experiments bicrystals were cut out of the group of crystals, which were grown as described previously. The chemical analysis of crystals used for this study was the same as those used for volume diffusion studies. The values for the grain boundary diffusion parameter, $\alpha D'\delta$, where α is the isotope segregation factor, D' is the grain boundary diffusion coefficient and δ is the effective grain boundary thickness, were determined by using Whipple's[10] solution. The grain boundary diffusion parameter was determined as a function of orientation both parallel ($\alpha D'\delta_{11}$) and perpendicular to the growth direction ($\alpha D'\delta_{\perp}$). The results obtained could be combined in one plot, Fig. 5, which is a plot of $D'_{11}/D'_{\perp}$ versus tilt angle for three investigated oxides. It is evident that the ratio $D'_{11}/D'_{\perp}$ which reflects the anisotropy of grain boundary diffusion is constant for tilt angles above ~10°, and increase sharply if tilt angles become smaller than ~10°. It was found that large decrease in the $D'_{11}/D'_{\perp}$ ratio is due to a large increase in $D'_{\perp}$ with the increase in tilt angle[2].

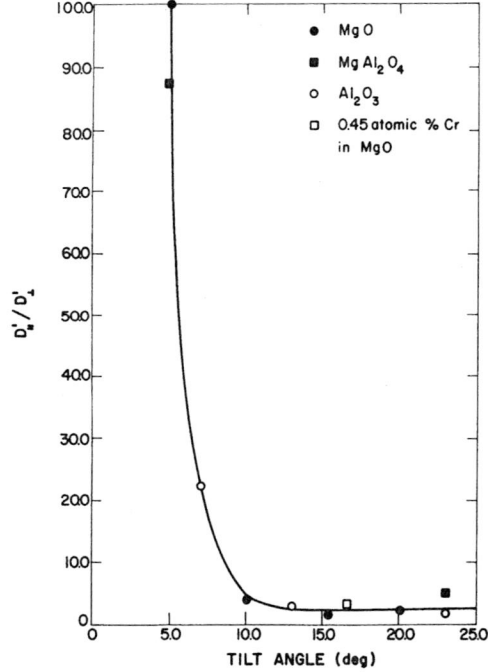

Figure 5. Dependence on the anisotropy of the grain boundary diffusion on tilt angle Θ for ^{51}Cr diffusion in refractory oxides.

In their work on grain boundary diffusion Turnbull and Hoffmann[11] proposed a pipe diffusion mechanism. Their model is based on the dislocation model of a small angle grain boundary. Our results indicate that this model can be applied to ionic crystals and that the mechanism of the grain boundary diffusion

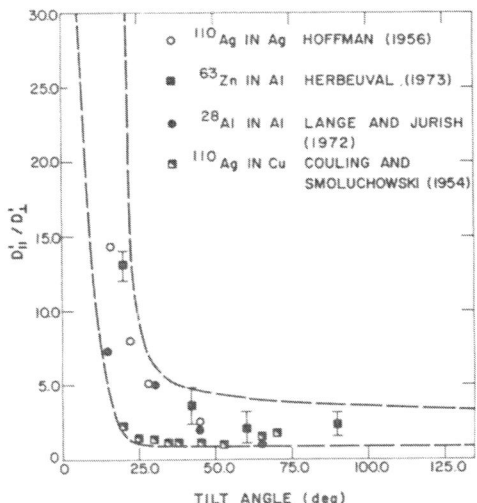

Figure 6. Dependence of
the anisotropy of the grain
boundary diffusion on tilt
angle Θ for metal systems.
Refs. 12, 13, 14, 15.

in ionic crystals is a dislocation pipe mechanism. Our results
also indicate grain boundaries with the tilt angles less than ~10°
can be considered low angle grain boundaries with oxides. It is
interesting to compare anisotropy of grain boundary diffusion in
oxide systems with the anisotropy of grain boundary diffusion in
metallic systems (Fig. 6). In metallic systems the anisotropy of
diffusion becomes large at tilt angles below ~20° [12,13,14,15] and
in oxides at tilt angles below ~10°. Possible reasons for this
difference may be due to the charge effect in grain boundaries of
oxides and/or to the fact that edge dislocations in oxides extend
over two planes of atoms, compared to one in metals. In metals
as well as in oxides a small anisotropy of grain boundary diffusion
can be observed with large angle grain boundaries indicating that
large angle boundaries must have some structure to produce
anisotropy.

The activation energies of diffusion of ^{51}Cr in tilt
boundaries of MgO, Al_2O_3 and $MgAl_2O_4$ are independent of tilt angle
and are 180 ± 20, 170 ± 15 and 185 ± 20 KJ/mole respectively.

The influence of the concentration of the metal vacancies
(V_{Mg}) on the ^{51}Cr grain boundary diffusion was studied by using
Cr-doped MgO bicrystals. Fig. 7 shows that the grain boundary
diffusion parameter, $\alpha D'\delta_{11}$, increases almost linearly with the
increase of the concentration of Cr-ions in the bulk of the bi-
crystal. This result indicates that magnesium vacancies play an
important role in diffusion along dislocations in grain boundaries.

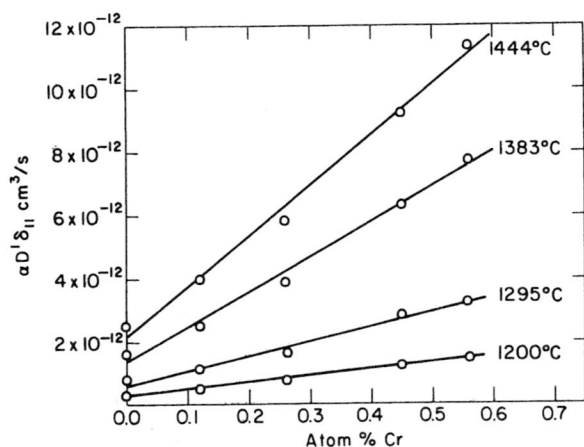

Figure 7. Dependence of the grain boundary diffusion parameter, $\alpha D' \delta_{11}$, on the concentration of Cr^{3+} ions in the bulk of MgO bicrystals. Diffusion along $\Theta = 15°(100)$ tilt boundary.

SURFACE DIFFUSION

Diffusion of ^{51}Cr isotope on (100) MgO, (0001) Al_2O_3, and (111) $MgAl_2O_4$ surfaces was investigated by using the edge source method[16]. The surface diffusion parameter, $\alpha D_s \delta$, where α is the segregation factor, D_s the surface diffusion coefficient, δ the thickness of the high diffusivity layer, was determined, for the temperature region 650-1250°C. For calculation of experimental results Whipple's solution, which contains factor 1/2, was used.

Figs. 8, 9 and 10 show the Arrhenius plots for the surface diffusion of ^{51}Cr on the (100) plane of MgO, the (111) plane of $MgAl_2O_4$ and (0001) plane of Al_2O_3 respectively. It is evident that there is a strong change of slope in Arrhenius plots at ~1050°C for MgO and ~900°C for $MgAl_2O_4$. This change of slope does not occur, however, for Al_2O_3 at least not in the temperature region 650-1250°C. It was noticed that all diffusion profiles for MgO above 1050°C showed substantial tails and the residual activity increased with the temperature. The same was true for $MgAl_2O_4$ above ~900°C. However, for Al_2O_3 residual activity after approx. 30-40 μm were removed was negligible in the entire temperature region ~650-1250°C. This indicated that the vapor transport of ^{51}Cr become an important mechanism of diffusion at high temperatures for MgO and $MgAl_2O_4$ surfaces. To verify this assumption, one surface of a MgO crystal was coated with ^{51}Cr isotope. This crystal was suspended above Al_2O_3 and MgO crystals in a Pt-crucible, which was then heated for 9 hours at 750°C and for 2 hours at 110°C. After the heat

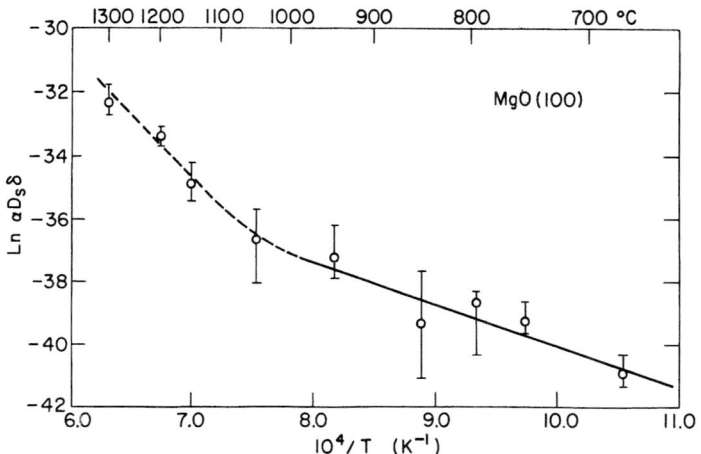

Figure 8.   Arrhenius plot for the surface diffusion of ^{51}Cr on
a (100) MgO plane.

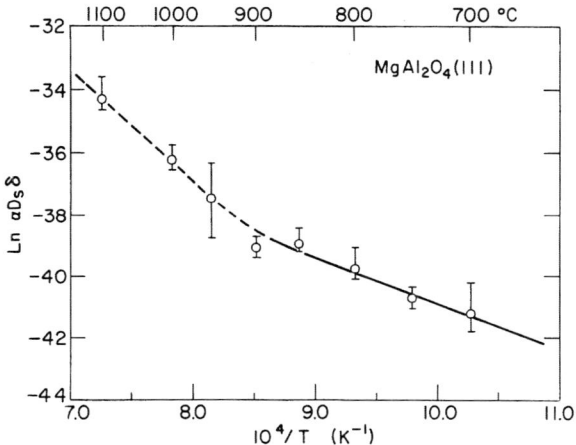

Figure 9.   Arrhenius plot for the surface diffusion of ^{51}Cr on a
(111) MgAl$_2$O$_4$ plane.

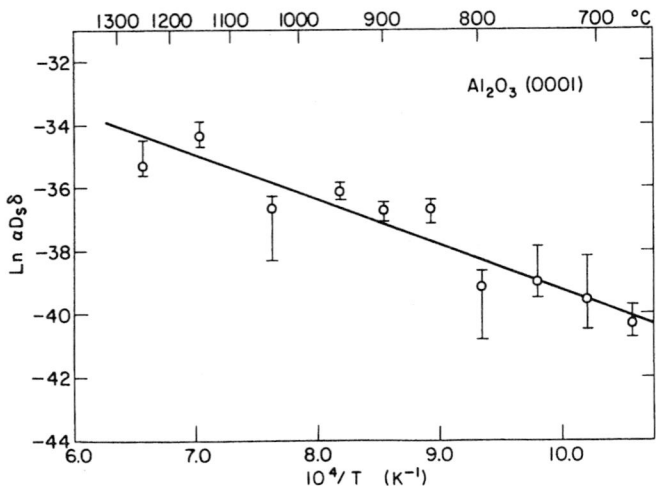

Figure 10. Arrhenius plot for the surface diffusion of ^{51}Cr on a (0001) Al$_2$O$_3$.

treatment at 750°C the radioactivity of both MgO and Al$_2$O$_3$ uncoated crystals was negligible. After heat treatment at 1100°C MgO crystals showed very strong activity and Al$_2$O$_3$ displayed negligible activity.

In the case of Al$_2$O$_3$ the sticking coefficient for Cr-vapor species must be low so that in the entire temperature region 650-1250°C the surface diffusion seems to occur predominantly by the localized motion in the high diffusivity layer.

The apparent activation energies for surface diffusion of ^{51}Cr on (100) MgO, (111) MgAl$_2$O$_4$ and (0001) Al$_2$O$_3$ surfaces are 110 ± 12, 121 ± 12 and 119 ± 12 KJ/mole respectively. These activation energies are for the temperature region where the localized motion in the high diffusivity layer is the predominant mechanism of the transport. Since the values for the activation energies are close and do not depend strongly on the material investigated it is quite possible that these values are extrinsic.

LEED, Auger and SIMS investigation indicate that this may be the case. Only MgO crystals were investigated by LEED. No LEED pattern was found for any of the single crystals which were studied. These single crystals were annealed at 1200°C for 4 hours and quenched. If no LEED pattern is found, then the surface is nonperiodic. There are two possibilities for this "amorphous" nature of the surface, (i) it may be due to mechanical damage which occurs during crystal preparation, or (ii) it may be due to impurity segregation. To determine which of these two possibilities is more probable, single crystals of MgO were annealed at 1470°C for seven days and quenched. This anneal should be sufficient to recrystallize the surfaces. A single crystal of MgO was also chemically polished with hot phosphoric acid at 150°C for 30 min. which removed ~75 μm of material. The samples were then annealed at 1450°C for seven days and quenched. All of the single crystals were then analyzed with LEED and subsequently used for diffusion anneals. No LEED patterns were obtained from these crystals, indicating that the nonperiodic nature of the surfaces is probably due to the segregation of impurities. Also, within experimental error, no difference in the diffusion behavior was found with the crystals prepared in several different ways. Auger spectra taken from the (100) MgO surface of the specimen heated at 1200°C for 2 hrs showed small Ca and Si signals. An Auger spectrum taken after 1.0 nm of MgO was removed showed no Si and Ca peaks indicating that Ca^{2+} and Si^{2+} segregated to depths no greater than 1.0 nm. Single crystals of Al_2O_3 were heated at 1250°C for 2 hr and at 750°C for 8.5 hr and the surfaces were analyzed by SIMS. The segregation of Si^{4+} and Ca^{2+} occurred at both temperatures. Si^{4+} segregated to the distance of ~1-1.5 nm and Ca^{2+} segregated to the distance of 5 nm. Some of the single crystals of Al_2O_3 which contained higher concentration of Mg^{2+} (300 ppm) and Fe^{2+} (40 ppm) showed also segregation of these impurities to the surface.

COMPARISON

Considering the volume diffusion of ^{51}Cr isotope in the investigated high refractory oxides it is evident that mechanism of diffusion is a vacancy mechanism and the obtained results show relatively small difference in the values for preexponential factor, D_0, and for the activation energy ΔQ for oxides with considerable different structure like MgO and Al_2O_3. It seems that in all investigated materials volume diffusion is extrinsic and is dominated by the aliovalent impurities. The measured activation energy is then the energy for motion of an associated impurity probably Si-vacancy complex.

This conclusion is further supported by the diffusion results obtained with the Cr-doped MgO crystals. The activation energy for the ^{51}Cr diffusion in Cr-doped MgO corrected for the background impurity contribution was found to be only 82 ± 7.1 KJ/mole. It is very probable then this value is the difference between the energy of motion and the energy of association of a Cr-vacancy complex. No anisotropy for the volume diffusion of ^{51}Cr was found with Al_2O_3 crystals which can be explained by the close packing of oxygens in Al_2O_3 structure.

A strong anisotropy in the grain boundary diffusion of the investigated oxides was found, which indicates the mechanism of diffusion is a dislocation pipe mechanism. The observed anisotropy is consistent with Turnbull and Hoffmann's model[11] for diffusion in low angle grain boundaries. In tilt boundaries of investigated oxides with the tilt angle below ~10° the anisotropy is high, however, some anisotropy of diffusion is still present in high angle grain boundaries. From the Table I it can be seen the apparent activation energy of the grain boundary diffusion is 0.56-0.61 times the activation energy for volume diffusion. Furthermore, it was found the activation energy for grain boundary diffusion is independent of the tilt angle and direction of diffusion[2].

Table I. Measured activation energies (ΔQ) for the volume, grain boundary and surface diffusion of ^{51}Cr in MgO, Al_2O_3 and $MgAl_2O_4$.

^{51}Cr DIFFUSION

MATERIAL	VOLUME	GRAIN BOUNDARY	SURFACE
	ΔQ	ΔQ	ΔQ
MgO	293 ± 42	180 ± 20	110 ± 12
Al_2O_3	306 ± 59	170 ± 15	121 ± 12
$MgAl_2O_4$	337 ± 40	180 ± 20	119 ± 12

ΔQ in KJ/mole

Diffusion of ^{51}Cr in both grain boundaries and lattice depends on the concentration of cation vacancies in a qualitatively similar way (Figs. 2 and 7). Enhancement of the grain boundary diffusion with an increase in the concentration of point defects indicates that vacancies play an important role in grain boundary diffusion. Atkinson and Taylor[17] who studied the diffusion of ^{63}Ni along the grain boundaries in NiO as a function of oxygen partial pressure observed an enhancement of diffusion with an increase in the cation vacancy concentration. Roubort and Matzke[18] have also observed the influence of stoichiometry on the grain boundary diffusion of ^{233}U in UC.

Surface diffusion data obtained in this study are some of the first data obtained on the surface diffusion of isotopes on ionic surfaces. The only other data on surface diffusion of an isotope on ionic surfaces were reported by Marlowe and Kaznoff[19] and Zhou and Orlander[20]. They studied surface diffusion of U-isotope on UO_2. The former authors did not account for simultaneous volume and surface diffusion during tracer spreading and obtained very high activation energy of 450-540 KJ/mole. The same authors claimed that the results obtained were in good agreement with the results of mass transport experiments. Zhou and Orlander, however, accounted for volume diffusion and obtained an activation energy of 301 ± 63 KJ/mole and a preexponential factor 5×10^6 cm^2/β for the temperature region of 1760 - 2110°C. These authors tried to explain very high value for the preexponential factor by using Bonzel's theory[16] of "nonlocalized" surface migration of UO_2 molecules.

It seems from our results that at low temperatures we have "localized" surface migration with the relatively low apparent activation energy of ~110 - 120 KJ/mole (Table I). These activation energies are 1.5-1.6 times smaller than for the grain boundary diffusion of ~2.5-2.8 times smaller than for the volume diffusion. However, due to the strong segregation of impurities to the surface and structural changes of the surface caused by impurities, it is quite possible that the activation energies which are very close for all three investigated materials are extrinsic.

ACKNOWLEDGMENTS

Supported by US Department of Energy Contract No. DE-AC02-78ER04998.

REFERENCES

1. G. W. Weber, W. R. Bitler and V. S. Stubican, J. Phys. Chem. Solids 41:1355 (1980).
2. V. S. Stubican and J. W. Osenbach, Solid State Ionics 12:375 (1984).
3. V. S. Stubican, G. Huzinec and D. Damjanovic, J. Am. Ceram. Soc., in press.
4. J. W. Osenbach, Ph.D. Thesis, The Pennsylvania State University, University Park, PA (1982).
5. B. J. Wuensch, W. C. Steel and T. Vasilos, J. Chem. Phys. 58:5258 (1973).
6. I. K. Lloyd and H. K. Bowen, J. Am. Ceram. Soc. 64:744 (1981).
7. A. B. Lidiard, "Handbuch der Physik," S. Fluegge, Ed., Springer Verlag, Berlin (1957).
8. R. A. Perkins, and R. A. Rapp, Met. Trans. 4:193 (1973).
9. A. M. Glass and T. M. Searly, J. Chem. Phys. 48:1420 (1968).
10. R. T. Whipple, Phil. Mag. A45:1220 (1954).

11. D. Turbull and R. Hoffmann, Acta Met. 2:419 (1954).
12. R. E. Hoffman, Acta Met. 4:97 (1956).
13. I. Herbeuval, M. Biscondi and C. Goux, Mem. Sci. Rev. Met. 70:39 (1973).
14. W. Lange and M. Jurisch, quoted in H. Gleiter and B. Chalmers, Prog. Mater. Sci. 16 (1972).
15. S. R. L. Couling and S. Smoluchowski, J. Appl. Phys. 25:1538 (1954).
16. See H. P. Bonzel, 1983, in "Surface Mobilities on Solid Materials", Vu Tieu Binh, Ed., Plenum Press, New York, N.Y.
17. A. Atkinson and R. J. Taylor, Phil. Mag. 43:979 (1981).
18. J. L. Routbort and H. J. Matzke, J. Am. Ceram. Soc. 58:81 (1975).
19. M. O. Marlowe and A. I. Kaznoff, J. Nucl. Mater. 25:328 (1968).
20. S. Y. Zhou and D. R. Olander, Surface Sci. 136:82 (1984).

DIFFUSION OF ^{18}O TRACER IN NiO GRAIN BOUNDARIES

A. Atkinson and F.C.W. Pummery

Materials Development Division
Building 552
AERE Harwell
Didcot, Oxon. OX11 ORA. U.K.

C. Monty

Laboratoire de Physique des Matériaux
CNRS
1, Place Aristide Briand
92190 Meudon, France

ABSTRACT

The diffusion of ^{18}O in polycrystalline NiO has been studied at temperatures in the range 1100-1600°C and an oxygen activity of 0.2. The penetration profiles of tracer were determined using the ^{18}O $(p,\alpha)^{15}$N nuclear reaction and the lateral distribution of tracer was observed by SIMS imaging. The penetration profiles were analysed by computer fitting to the appropriate theoretical solution to determine the apparent lattice diffusion coefficient and the product $D'\delta$ for grain boundary diffusion. The grain boundary diffusivity of oxygen in NiO is approximately equal to the lattice diffusivity of Ni in NiO under the conditions of these experiments.

INTRODUCTION

The study of oxygen self-diffusion in oxides has always lagged behind that of the metal ion because of the lack of a suitable radioactive tracer isotope. The non-radioactive isotopes ^{17}O and ^{18}O are available, but until recently they could only be analysed conveniently by mass spectrometry of the vapour phase. Consequently, the early studies of oxygen diffusion relied on following exchange kinetics (between ^{18}O in the gas and ^{16}O in the oxide) by mass-spectrometric analysis of the gas (e.g. Oishi and Kingery[1]).

The most serious disadvantage of this approach is that only the total exchange is monitored, but this may be controlled by more than one diffusion process (lattice, grain boundary, dislocation) or even perturbed by non-diffusive processes such as evaporation and surface exchange rate. Consequently, the method tends to be unreliable and alternative methods in which the depth distribution of exchanged isotope is determined are to be preferred. This has become possible during the last five years or so as the techniques of secondary ion mass spectrometry (SIMS) and nuclear depth profiling have become more refined.

Oxygen self-diffusion may control, or contribute to, many important technological properties of oxides which depend on mass transport. The work described in this paper relates to NiO which is typical of oxides in which oxygen is less mobile in the lattice than is the metal ion. In such oxides it is the slowest moving species (i.e. usually oxygen) which controls sintering, creep and grain growth. Furthermore, in many mechanistic interpretations of these processes it is diffusion in grain boundaries which is the controlling factor and not lattice diffusion. The oxidation of metals is another area for which knowledge of oxygen diffusion is required. Whilst it is known that outward diffusion of Ni is the dominant process controlling the rate of growth of NiO films, the inward diffusion of oxygen may contribute to the growth of two-layered (duplex) films and the generation of growth stresses by countercurrent diffusion and reaction (lateral growth).

In the case of NiO, data now exist for diffusion of Ni in the lattice[2,3] and along dislocations[3] and grain boundaries[4], and for diffusion of O in the lattice[5]. NiO is therefore the oxide for which knowledge of different diffusion pathways is currently the most comprehensive. The aim of the experiments reported here was to extend this knowledge even further by measuring oxygen grain boundary diffusion using a depth profiling technique. (The only other oxide in which oxygen grain boundary diffusion has been measured directly by a profiling technique is Al_2O_3[6].)

EXPERIMENTAL METHODS

An essential prerequisite for grain boundary diffusion studies is a supply of suitable polycrystalline specimens. A

360

method of fabricating polycrystalline NiO specimens suitable for Ni grain boundary diffusion studies had already been developed based on the complete oxidation of Ni foils supported on sintered NiO substrates[4]. The same method was used to prepare specimens of approximate dimensions 1 cm x 1 cm x 2 mm for the oxygen diffusion measurements. The specimens had a columnar structure parallel to the diffusion direction with a grain size of approximately 10 µm. The characterisation of such specimens has been reported previously[4].

The diffusion anneals were carried out in a vertical Al_2O_3 tube furnace which was connected to a vacuum system and reservoir of $^{18}O_2$-enriched gas. The specimen was contained within a sintered NiO crucible with lid to minimise contamination from the furnace tube. The apparatus was the same as had been used in the earlier study of ^{18}O diffusion in NiO single crystals[5]. The specimen was first pre-annealed in oxygen of natural isotopic abundance (0.204% $^{18}O_2$) at a pressure of 0.2 atm for a time at least as long as the intended diffusion anneal. The furnace was then quickly evacuated and refilled to the same pressure with oxygen gas enriched in ^{18}O (approximately 50% $^{18}O_2$), which was allowed to exchange with the ^{16}O in the specimen. After the diffusion anneal the labelled gas was recovered into the reservoir (sorption pump).

Since the diffusion profiles which are reported here were obtained exclusively using the nuclear reaction technique, this will be described in some detail. The method is based upon the ^{18}O (p,α)^{15}N reaction which has good specificity for ^{18}O. The experiments were carried out using the 'IBIS' accelerator and nuclear microprobe at Harwell. The specimen was bombarded with a beam of protons of incident energy 845 keV and the beam was scanned electrostatically over an area of approximately 0.6 mm x 0.6 mm on the specimen surface. The energy spectrum of α particles emerging at an angle of 135° was measured and compared with the spectrum obtained from a standard specimen of NiO containing a known uniform distribution of ^{18}O. From the comparison, and knowing the rate of energy loss with depth of the α particles in NiO, the depth distribution of ^{18}O in the specimen was obtained[7]. Depth distributions were always measured at two different places on the specimen in order to check uniformity. In no case did we observe any large differences between the two profiles and therefore the two profiles were combined and averaged for subsequent analysis.

The nuclear reaction technique suffers from two main disadvantages when used in this way. The first is that the maximum depth which can be analysed is limited to approximately 2 to 3 µm, chiefly because of interference from backscattered protons. To circumvent this problem, profiles to depths greater

361

than 2 μm were analysed in two stages. After analysing the first 2 μm we removed approximately 2 μm from part of the surface by radiofrequency sputtering. The depth removed was determined from the loss in weight during sputtering. Analysis of the sputtered surface then enabled the next 2 μm of the depth profile to be determined. Therefore, profiles which extend to depths greater than 2 μm are seen to be composed of two parts (e.g. Figures 1 and 2).

The second problem concerns the depth resolution of the technique. The resolution is limited by intrinsic characteristics of the method i.e. the straggling of incident protons and emerging α-particles; the latter being the most important. In addition, there are other experimental sources of degraded resolution such as the energy spread of incident protons, the resolution of the detector system and variable lengths of emergent paths as a result of surface roughness. In an earlier study Atkinson et al.[7] attempted to both calculate and measure the eventual practical resolution of the technique. They concluded that the measured resolution was approximately eight times worse than expected and that this was due to non-uniform layers of ^{18}O on the test specimens, which had been produced by thermal oxidation. We have made a second assessment of the resolution of the technique in this study using sputtered $Ni^{18}O$ films on various substrates. Two film thicknesses (nominal thickness 0.2 μm and 0.5 μm) were combined with three substrates; polished Ni, polished NiO single crystal and the polycrystalline NiO as used in the present diffusion experiments. The resolution was estimated from the depth profiles as one half the difference in depth between concentrations of 0.12 and 0.88 of the surface concentration. This is equivalent to the half-width at half maximum of a delta function which has been degraded to a gaussian distribution and is approximately equal to (1.17 times) the standard deviation. The results are shown in Figure 3 and confirm the earlier conclusion that the experimental resolution is always worse than estimated theoretically. Since the resolution is significantly worse on the unpolished surfaces of the type used in the present diffuson experiments we conclude that surface roughness is the principal cause of the poorer-than-optimum resolution. However, it should be noted that even for rough surfaces the resolution should approach the optimum ($\pm$ 0.015 μm) as the penetration depth tends to zero (provided that the surface contours do not have angles steeper than 45° such that the detector would be obscured from the 'valleys' in the surface). This therefore represents an absolute limit for detectable diffusion (($Dt)^{\frac{1}{2}} \sim$ 15 nm) irrespective of the surface roughness. However, the profile at greater depth will be more degraded on a rough surface than on a smooth one and the effect will be greatest when the depth is of the same order as the surface roughness.

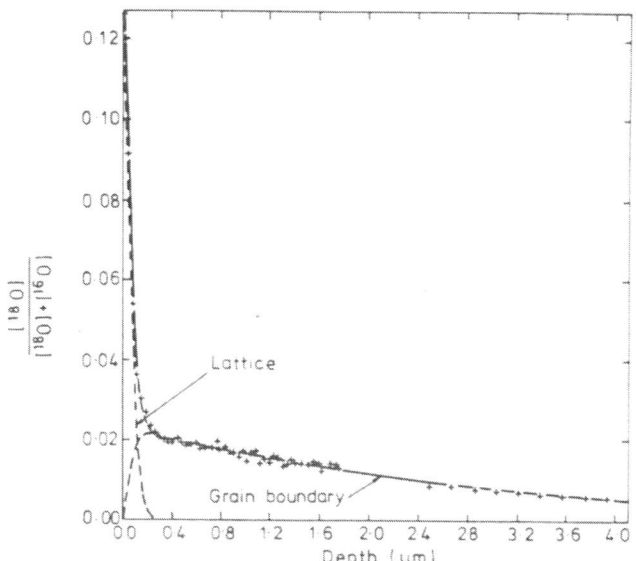

<u>Figure 1</u> ^{18}O penetration profile measured using the nuclear reaction technique after diffusion into polycrystalline NiO for 8 h at 1300°C (Run 5). The solid curve is the best fit to the theoretical solution of the diffusion equations and the broken lines are the individual contributions from lattice and grain boundary diffusion.

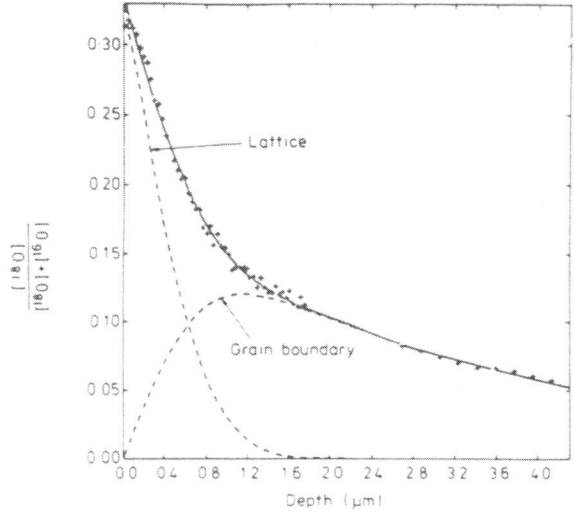

<u>Figure 2</u>   As for Figure 1 but after a longer anneal time of 145 h at 1300°C (Run 7).

RESULTS

The depth profiles have been interpreted as being controlled by only diffusion processes with no contribution from the rate of exchange at the specimen surface. The solution of the diffusion equations for combined lattice and grain boundary diffusion appropriate to the experimental boundary conditions (constant surface concentration) has been given by Whipple[8] and is used here in the form suitable for a polycrystalline material[9]

$$\bar{C}\ (y,t)\ =\ C(0)\ \{ \text{erfc}\ (\eta/2)\ +\ \bar{C}_I/n\}. \tag{1}$$

In this equation $\bar{C}$ is the average concentration of tracer (^{18}O) at depth y after time t and $C(0)$ is the concentration at the surface which is maintained constant by equilibrium with the gas. The dimensionless quantity $\eta = y/(Dt)^{\frac{1}{2}}$, and therefore the first term in brackets in equation (1) is the normal depth profile for lattice diffusion. The second term is the contribution from grain boundary diffusion with $n = g/2(Dt)^{\frac{1}{2}}$ where g is the grain size. The quantity $\bar{C}_I$ cannot be expressed algebraically, but must be evaluated from the integral[9]

$$\bar{C}_I\ =\ \frac{4z}{\pi^{\frac{1}{2}}}\ \int_0^{\beta^{\frac{1}{2}}}\ \exp\ (\frac{-z^2\tau^2}{4})\ [\frac{1}{\pi^{\frac{1}{2}}}\ \exp\ (-V^2)-V\text{erfc}(V)]d\tau \tag{2}$$

with $V = (\frac{1}{\tau^2} - \frac{1}{\beta})/2$, $\quad \beta = \frac{D'\delta}{2\ D(Dt)^{\frac{1}{2}}}$ and $z = \eta/\beta^{\frac{1}{2}}$.

The parameter characterising grain boundary diffusion is the product of the boundary diffusion coefficient, D', and the effective boundary width, δ. $\bar{C}_I$ has been evaluated and tabulated by Suzuoka[9] for certain values of β and z.

There is no general way of extracting the required diffusion parameters (D and $D'\delta$) from a simple analysis of the tracer penetration profile. If conditions are such that the lattice and grain boundary contributions are well separated then the lattice diffusion coefficient can be extracted in the normal way and $D'\delta/D^{\frac{1}{2}}$ obtained from the slope of a plot of log $\bar{C}$ against $y^{6/5}$[10]. However, this was found to be inappropriate for many of the profiles in these experiments and was only used to obtain initial estimates. These estimates were then refined by least squares computer fitting of the profiles (after subtracting the natural background concentration) to the full solution. The computer programme minimised the sum of squared residuals function with respect to the variables $C(0)$, g, D and $D'\delta$. Since, from equation (1), $\bar{C}$ is linearly dependent on both $C(0)$ and g, the sum of squared residuals can be readily minimised with respect to

Table 1. Summary of diffusion anneals and fitted parameters

Run No.	T (°C)	t (h)	D (cm² s⁻¹)	$D'\delta$ (cm³ s⁻¹)	g (μm)	C(0)	β	$D'\delta$ (using single crystal D) (cm³ s⁻¹)
1	1100	24	3.9×10^{-16}	1.1×10^{-17}	3.7	0.21	2400	2.2×10^{-19}
17	1100	100	1.0×10^{-15}	2.9×10^{-17}	3.0	0.31	740	3.5×10^{-19}
15	1200	24	1.4×10^{-15}	1.3×10^{-16}	2.9	0.26	4000	6.1×10^{-18}
3	1300	2.25	1.3×10^{-15}	1.4×10^{-17}	0.2	0.06	1700	3.1×10^{-18}
5	1300	8	1.1×10^{-15}	3.0×10^{-17}	1.6	0.15	2300	6.9×10^{-18}
6	1300	71	2.4×10^{-15}	2.6×10^{-17}	3.0	0.29	210	4.0×10^{-18}
4	1300	100	2.3×10^{-15}	2.9×10^{-17}	3.3	0.19	210	4.6×10^{-18}
7	1300	145	3.4×10^{-15}	2.1×10^{-17}	3.6	0.33	74	2.8×10^{-18}
18	1400	12	1.7×10^{-14}	1.0×10^{-16}	3.3	0.25	120	2.2×10^{-17}
19	1500	1	4.0×10^{-14}	5.1×10^{-16}	0.8	0.27	520	2.0×10^{-16}
20	1600	1	1.8×10^{-13}	2.9×10^{-16}	1.4	0.19	31	1.3×10^{-16}

these two variables, but a function minimisation subroutine (steepest descent) was used to fit with respect to D and D δ using equations (1) and (2). Since the data were obtained by particle counting procedures, the individual residual from a given point on the penetration profile was given a weight proportional to the square root of number of counts which contributed to that point.

Two examples of experimental profiles are shown in Figures 1 and 2 and have been chosen to illustrate the effect of anneal time at constant temperature. The solid curves are the theoretical profiles (equation (1)) which give the best fit to the data points and the two dashed curves in each profile are the individual contributions from lattice and grain boundary diffusion (the first and second terms in equation (1)). The details of all the diffusion anneals and the fitted parameters are given in Table 1 and the values for D and D δ are plotted in Arrhenius form in Figure 4. The scales for D and D δ in Figure 4 have been offset such that if we assume δ = 1 nm, as has been shown to be the case for Ni diffusion in NiO grain boundaries[4], then D$'$ can be read directly on the same scale as D.

The nuclear reaction technique which was used for depth profiling in these experiments did not prove to be suitable for studying the lateral distribution of tracer. However, this information was obtained using SIMS (Cameca IMS 300). Figure 5 shows a series of images of the same area formed with the mass 18 signal at the indicated depths below the original surface. The bright regions are therefore regions rich in ^{18}O. The micrographs clearly demonstrate a uniform lateral distribution at the surface and preferential penetration along grain boundaries becoming more prominent at greater depths.

DISCUSSION

Lattice Diffusion

The lattice diffusion coefficients for O measured in this work are shown in Figure 4 compared with the data of Dubois et al.[5], which were measured in single crystals. The present measurements are consistently higher by approximately a factor of 4 at 1600°C to perhaps as much as 10^3 at 1100°C. The reason for this difference is not clear, but there are several possible sources of error which may have contributed. One is the limited resolution of the profiling technique discussed previously. For t = 100 h and a flat surface the limiting value of D which is measurable is ~ 10^{-17} cm^2 s^{-1} and for the rough surface of these specimens is, at worst, ~ 10^{-16} cm^2 s^{-1}. The high values cannot therefore be entirely attributed to the analysis techniques.

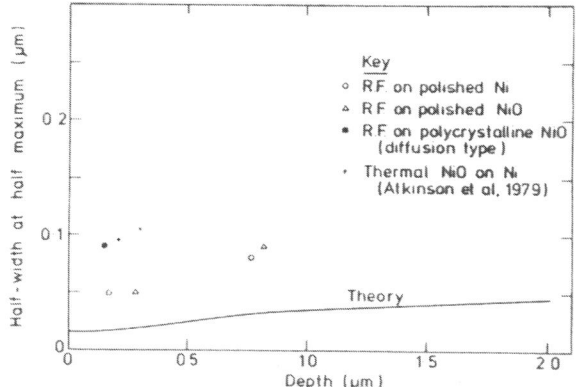

<u>Figure 3</u>   The resolution of the nuclear profiling technique (using the ^{18}O $(p,\alpha)^{15}N$ reaction) measured as a function of depth using sputtered films on various substrates and estimated theoretically. The resolution is approximately equal to one standard deviation (see text).

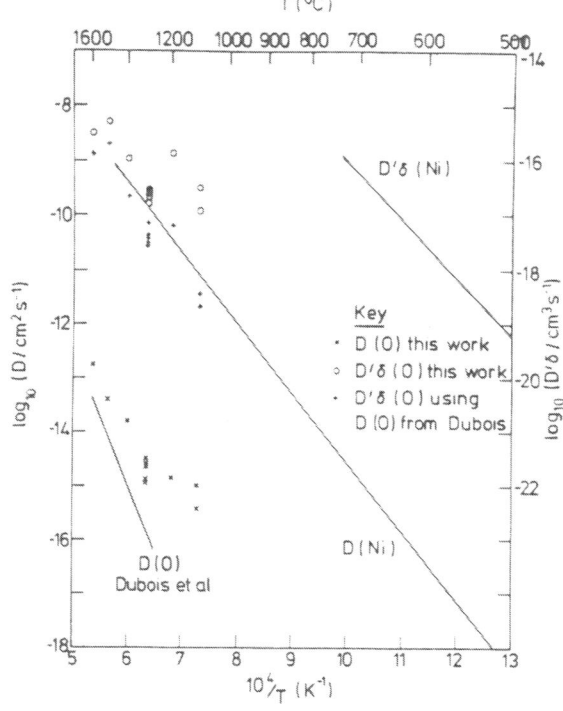

<u>Figure 4</u>   Arrhenius plot of the oxygen diffusion parameters measured in this work compared with previously published data for lattice diffusion of oxygen and nickel and grain boundary diffusion of nickel.

Another possibility is a contribution from dislocation diffusion. Such a contribution was observed in Ni tracer diffusion experiments in specimens fabricated in the same way as the ones used in this study. However, the effect was small and it would appear to be insufficient to explain the current observations.

Grain Boundary Diffusion

The ambiguity in D leads to a related ambiguity in $D'\delta$ since D is required in order to extract $D'\delta$ from the penetration profiles. We do not know a priori whether the appropriate lattice diffusion coefficients are those measured in the polycrystalline specimens or those measured in single crystals since the higher (polycrystalline) values may be the result of some experimental artefact which is not yet evident. We have therefore listed in Table 1, and plotted in Figure 4, a second set of values for $D'\delta$ which have been deduced from the grain boundary portion of the profiles using D from the single crystals. This alternative interpretation results in less scatter on the Arrhenius plot and is probably the correct one.

The apparent grain size which is deduced from fitting the penetration profiles to equation (1) is also given in Table 1. This effective grain size is approximately one third of the microscopic grain size and has a definite tendency to increase with increasing anneal time (e.g. in the experiments at 1300°C). These effects may have been caused by some grain growth during both the pre-anneal and diffusion anneals (n.b. a smaller grain size results in a larger contribution from grain boundary diffusion). The specimens were examined by optical microscopy at the end of the experiments. No significant grain growth was detected in specimens annealed at up to 1400°C, but after annealing at 1500°C (Run 19) the grain size had increased to approximately 20 μm and at 1600°C (Run 20) to approximately 30 μm.

Also plotted in Figure 4 are data for lattice and grain boundary diffusion of Ni in NiO. Whichever set of data is used for lattice diffusion of oxygen, the grain boundary diffusion coefficient for oxygen is approximately equal to that for lattice diffusion of Ni. Thus the more mobile ion in the lattice (i.e. Ni) is also the more mobile in grain boundaries. Furthermore, the activation energy of $D'\delta$ for oxygen (using the preferred data set) is approximately equal to that for lattice diffusion of Ni (~ 2.5 eV). This similarity illustrates the danger of trying to identify a rate-controlling species merely from the activation energy of some transport controlled process. For example, Notis et al.[11] concluded that sintering of NiO was controlled by Ni lattice diffusion because the activation energy for densification

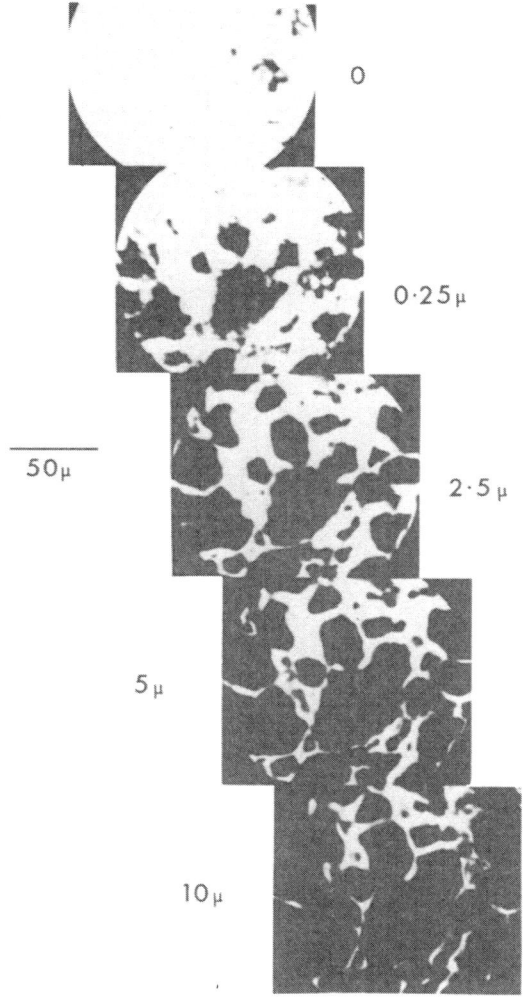

50μ

0

0·25μ

2·5μ

5μ

10μ

Figure 5   Mass 18 SIMS images as a function of depth after diffusion into a polycrystalline specimen for 71 h at 1300°C (Run 6). The bright regions are rich in ^{18}O and the depth below the original surface is indicated beside each micrograph.

was ~ 2eV. It is evident from the present data that the process controlling densification could equally well be grain boundary diffusion of oxygen (indeed, this is much more plausible).

The present data also indicate that oxygen diffusion, even along grain boundaries, can make no significant contribution to the rate of growth of NiO by the oxidation of Ni. Thus the countercurrent diffusion model of NiO growth suggested by Rhines and Wolf[12] is untenable and inward diffusion of oxygen along oxide grain boundaries cannot account for the penetration of NiO films by ^{18}O tracer which is observed under conditions where two-layered films are formed.

ACKNOWLEDGEMENTS

The authors wish to thank J.W. McMillan for advice concerning nuclear analysis, C. Dubois for SIMS analysis, R. Atkinson for computer programming and R.I. Taylor for fabricating the specimens. The authors are also grateful to NATO for the award of a travel grant which has made this collaborative research possible.

REFERENCES

1. Y. Oishi and W.D. Kingery, J. Chem. Phys., 33:480 (1960).
2. M.L. Volpe and J. Reddy, J. Chem. Phys., 53:1117 (1970).
3. A. Atkinson and R.I. Taylor, Philos. Mag., A39:581 (1979).
4. A. Atkinson and R.I. Taylor, Philos. Mag., A43:979 (1981).
5. C. Dubois, C. Monty and J. Philibert, Philos Mag., A46:419 (1982).
6. K.P.R. Reddy, PhD. Thesis, Case Western Reserve University, Cleveland, USA (1979).
7. A. Atkinson, R.I. Taylor and P.D. Goode, Oxidation of Metals, 13:519 (1979).
8. R.T.P. Whipple, Philos. Mag., 45:1225 (1954).
9. T. Suzuoka, J. Phys. Soc., Japan, 19:839 (1964).
10. A.D. LeClaire, Brit. J. Appl. Phys., 14:351 (1963).
11. M.R. Notis, P.A. Urick and R.M. Spriggs in "Sintering and Related Phenomena", Ed. G.C. Kuczynski (Mat. Sci. Res. Vol. 6) Plenum, New York, p. 409 (1973).
12. F.N. Rhines and J.S. Wolf, Metall. Trans., 1:1701 (1970).

NEAR – SURFACE MASS TRANSPORT IN NICKEL OXIDE

Guy Dhalenne and Alexandre Revcolevschi

Laboratoire de Chimie Appliquée
Université Paris-Sud – Bâtiment 414
91405 Orsay Cédex, France

Claude Monty

Laboratoire de Physique des Matériaux, C.N.R.S.
1, place Aristide Briand
92195 Meudon Principal Cédex, France

INTRODUCTION

It is well known that the evolution of the topography of a
surface at high temperature may result from a near-surface mass
transport phenomenon involving simultaneous processes of surface
diffusion, volume diffusion and gas phase diffusion. It has been
shown that the study of the kinetics of thermal grooving of grain
boundaries makes possible the determination of volume and surface
diffusion coefficients[1].

The knowledge of these two coefficients is of great importance
for the understanding of phenomena such as sintering, Herring-Nabarro
creep, precipitate coalescence, pore annealing, and bubble migration,
in which mass transport takes place.

A very large amount of information concerning diffusion processes in metals has been amassed so far and indicates in particular some scatter in the surface diffusion data[2-6]. In the case of
oxides, most of the results concern only volume diffusion[7,8] with
the exceptions of UO_2, MgO and Al_2O_3[9] for which surface diffusion
data have been determined. Besides it can be noted that in the case
of compounds, several species can migrate and very often data relative to only one of them can be determined.

371

The present contribution is an extension of earlier work[10] concerning the study of the kinetics of thermal grooving of grain boundaries in bicrystalline nickel oxide, concentrating more on the surface mass transport phenomena taking place at high temperature in this oxide.

EXPERIMENTAL

Theory of thermal grooving of grain boundaries

The kinetics of the thermal grooving of grain-boundaries has been analyzed by Mullins[1],[11]. In the case of a boundary normal to the surface, high-temperature annealing leads to the formation of a groove which results from near-surface mass transport occurring by surface, volume or gas phase diffusion processes and which corresponds to an equilibrium between intergranular and surface tensions.

Mullins has shown that the width of the groove can be expressed as a function of the time t of annealing by a relation of the form :

$$w = a(Kt)^{1/n} \tag{1}$$

where a is a constant and n and K are coefficients characteristic of the mechanisms to which the mass transport can be attributed.

These coefficients take the following values :

- in the case of surface diffusion :

$$a = 4.6 \quad n = 4 \quad \text{and} \quad K = \frac{D_s \gamma_s \eta \Omega^2}{kT}$$

- for volume diffusion :

$$a = 5.0 \quad n = 3 \quad \text{and} \quad K = \frac{D_v \gamma_s \Omega}{kT}$$

- for gas-phase diffusion :

$$a = 5.0 \quad n = 3 \quad \text{and} \quad K = \frac{\rho_o \gamma_s D_g \Omega^2}{(kT)^2}$$

where γ_s is the surface free energy, D_g, D_v and D_s the diffusion coefficients in the gas phase, the volume and on the surface, respectively ; Ω is the molar volume of the solid, k the Boltzmann constant, η the surface concentration of the diffusing species, ρ_o the equilibrium vapour pressure of the solid and T the absolute temperature.

According to this the graphical representation of log w vs
log t will be a straight line, the slope of which will be 1/4 in the
case of a surface diffusion mechanism and 1/3 in the case of a
volume or gas phase diffusion mechanism.

In the case of metals, Mullins and Shewmon have shown that, by
measuring n, it was possible to separate the contributions of volume
and surface diffusion[12]. When both mechanisms contribute to the
groove formation, the slope of the line has a value between 1/3 and
1/4.

The Mullins theory has been developed for monatomic solids. In
the case of compounds, the diffusion coefficients which have to be
taken into account are effective diffusion coefficients, which are
functions of the diffusion coefficients of the different chemical
species. The three diffusion mechanisms being independent, the
fastest will determine the kinetics, but for a given mechanism the
necessity of keeping the composition constant makes the slowest
species control the kinetics. Dhalenne et al.[10] have shown that in
the case of compounds it is possible to separate the relative contributions of each mechanism and to determine surface and volume
diffusion coefficients. It was shown in particular in the case of
NiO that the study of thermal grooving could lead to the volume
and surface diffusion coefficients of nickel, the oxygen transport
occurring through the gas phase.

Experimental study of the grooving of NiO bicrystals

The experimental details of the techniques used in this study
have already been described in detail elsewhere[10,13]. We shall
mention here only some essential points.

Nickel oxide bicrystals having symmetrical <011> tilt boundaries
were grown by the flame fusion method using 99.999 Johnson Matthey
NiO. The grown boules which have a diameter of about 10 mm and which
are 50 mm long were annealed at 1600°C for 24 hours in order to
relieve stresses and reduce the brittleness during subsequent ope-
rations.

The bicrystals were then cut into slices 2 mm thick with a wire
saw, perpendicular to the grain boundary. After annealing at the
temperature at which the grooving experiments had to be carried out,
one face of each slice was carefully polished.

The slices were then annealed in air in a lanthanum chromite
resistor furnace in which temperatures up to 1800°C could be reached.
The samples were placed inside closed sintered nickel oxide cruci-
bles, thus avoiding possible contamination. Temperature was controlled and regulated with a (Pt-6 Rh/Pt-30 Rh) thermocouple.

After annealing, the groove widths were measured by optical interferometry.

RESULTS

In order to relate the results of the present study, carried out on <011> tilt axis boundaries, to previous data obtained on <001> tilt samples[10], we analyzed the influence of crystallographic characteristics (bicrystalline misorientation angle and tilt axis) on the grooving process.

Grooving kinetics experiments carried out on <011> samples at 1520°C, i.e., a temperature at which previous experiments had been performed, indicated that the width w of the groove is independent of the misorientation angle θ^{14}, except for small misorientations (0-10° and 170-180°) and for angular values corresponding to lattice coincidences, e.g., $\theta \sim 129°$. In these two particular cases, the widths which were measured were slightly smaller, by about 20 %, than the values relative to general orientations. Except for these particular cases, the value of w was very nearly equal to that measured in the case of <001> tilt axis samples, as can be seen in fig. 1 in which identical slopes are observed for the two series of samples. At higher temperatures a similar phenomenon was observed, with an increase of the groove width difference for the two sets of samples. Data relative to samples corresponding to small misorientation angles were discarded in this study.

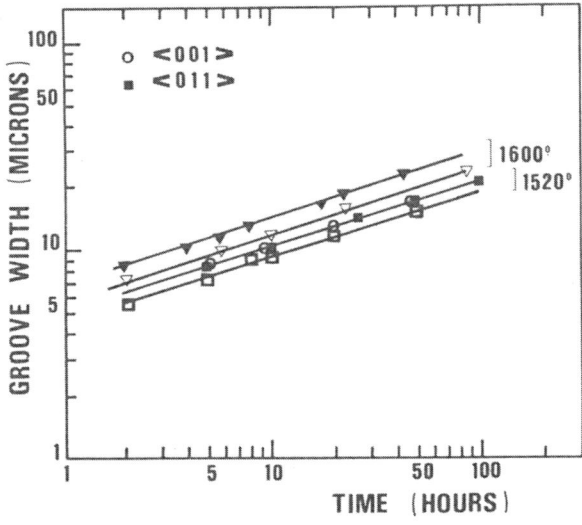

Fig. 1. Grooving kinetics at 1520 and 1600°C. (▼, ■ and ▽, □ correspond to large and small misorientations, respectively, in <011> tilt axis samples ; open circles are relative to <001> tilt axis samples).

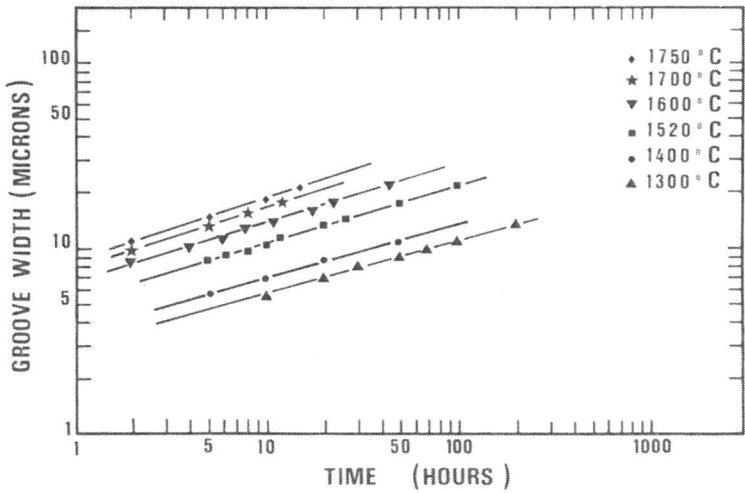

Fig. 2. Variation of the grain boundary groove width as a function
of annealing time in air at various temperatures. Slope
values are given in table 1.

After establishing this relationship between <011> and <001>
samples we carried out thermal grooving experiments in the temperature range 1300-1750°C on the samples with a <011> tilt axis, and
determined the slopes of the corresponding log w vs log t lines by
linear regression of the experimental values (fig. 2).

This graphical representation leads for each temperature to a
value of the coefficient n. This value is in fact a mean value $\bar{n}$
(table 1), as evidenced by a small, but systematic variation of n
as a function of annealing time (fig. 3). This observation can be
explained by the fact that the Mullins theory which we have used in
our study can be applied only to the case of small surface perturbations, whereas in our experiments for long annealing periods the
grooving is quite large.

Furthemore, it can be seen that the variation of n with annealing time becomes more pronounced with temperature. This observation
may be compared with recent results[15,16] of periodic scratch-smoo-
thing experiments carried out at high temperature on alumina, which
have shown an acceleration of the grooving kinetics associated with
a facetting of the sample surface. This phenomenon has apparently
not been reported previously.

Table 1. Experimental values of the coefficient n
of the Mullins relation (1) (see text).

Temperature (°C)	$\bar{n}$	n_o
1300 (0.70 Tf)	3.88 ± 0.08	3.91
1400 (0.75 Tf)	3.46 ± 0.1	3.53
1520 (0.80 Tf)	3.26 ± 0.05	3.35
1600 (0.84 Tf)	3.18 ± 0.07	3.30
1700 (0.89 Tf)	2.90 ± 0.1	3.06
1750 (0.91 Tf)	2.80 ± 0.1	3.06

Taking these observations into account, the curves representing
for different temperatures the variation of n with annealing time t
have been extrapolated to t = 0 ; the corresponding n_o values of n
(table 1) have been considered for the interpretation of our results.

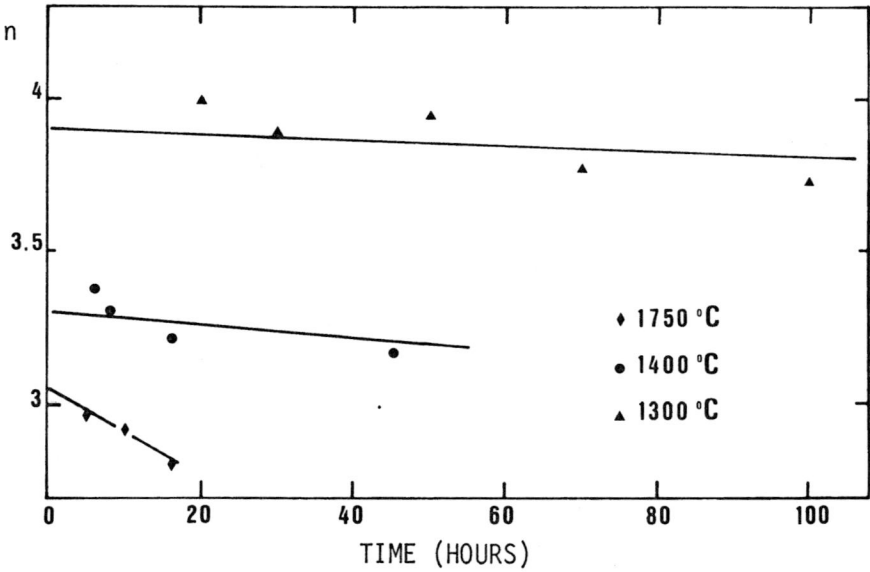

Fig. 3. Variation of the coefficient n of the Mullins relation, as
a function of time, at different temperatures.

We shall recall here that, as set out in an earlier paper[10], our treatment is based on an analysis which postulates a coupling of distinct grooving rates pertaining to each species, which maintains the composition of the solid constant : the contribution of each diffusion mechanism can be separated and volume and surface diffusion coefficients can be determined.

Our results can be summarized as follows :

1. At 1300°C, the log w bs log t slope is close to 1/4 (n = 3.91). This suggests that at this temperature surface diffusion is the predominant mechanism of mass transport in the thermal grooving of NiO. The application of our analysis separates the contribution of volume and surface diffusion.

2. Between 1400°C and 1600°C, the slopes have values between 1/3 and 1/4. This implies that mass transport involves a simultaneous contribution from both surface and volume diffusion.

3. At temperatures of 1700°C and 1750°C, the slope is close to 1/3 which indicates that volume diffusion is the predominant grooving mechanism. Nevertheless we could determine both the D_s and D_v coefficients. It has to be noted that at these two temperatures a tendency to surface facetting appears for annealing times greater than 15 hours. This facetting leads to parallel striations on some areas of the surface of the sample. These facetted areas were ignored for the groove width measurements.

The values of the diffusion coefficients obtained for the different temperatures are presented in table 2.

In the various calculations, the molar volume Ω was taken as equal to $1.81.10^{-23}$ cm^3 per mole of NiO; and for the surface energy γs, we have ssumed a value of 1555 mJ.m^{-2} at 0 K[14,17]. This value is different from that proposed by Stewart and Mackrodt[18] and Duffy and Tasker[19]. For the high temperatures of our study, an estimated entropy contribution of 0.2 mJ.m^{-2}.K^{-1} has been taken into account[20].

The variation of the surface diffusion coefficient of nickel as a function of temperature is represented on an Arrhenius type plot (fig. 4).

In fig. 5, we have plotted the variation with temperature of the volume diffusion coefficients for nickel which we have determined ; each of these points, as well as those in fig. 4, represent average values corresponding to about fifty measurements. We have reported on the same figure the self-diffusion volume coefficient of nickel (D_{Ni}^{*}) determined by radiotracer methods[21,22]. For this comparison our experimental values have been corrected by

Table 2. Volume and surface diffusion coefficients of nickel in NiO in air.

Temperature (°C)	D_v^{Ni} (cm^2.s^{-1})	D_s^{Ni} (cm^2.s^{-1})
1300	1.58 10^{-10}	1.8 10^{-6}
1400	3.46 10^{-10}	4.41 10^{-6}
1520	1.78 10^{-9}	1.35 10^{-5}
1600	4.04 10^{-9}	2.89 10^{-5}
1700	1.13 10^{-8}	5.1 10^{-5}
1750	1.64 10^{-8}	7.2 10^{-5}

a correlation factor f defined by $D_v^{*}/f = D_v$ (f = 0.7815 for a nickel vacancy diffusion mechanism) (table 2). The diffusion coefficients may be described by $D = D_0 \exp(-Q/kT)$ in cm^2.s^{-1}, Q being the activation energy and k the Boltzmann constant.

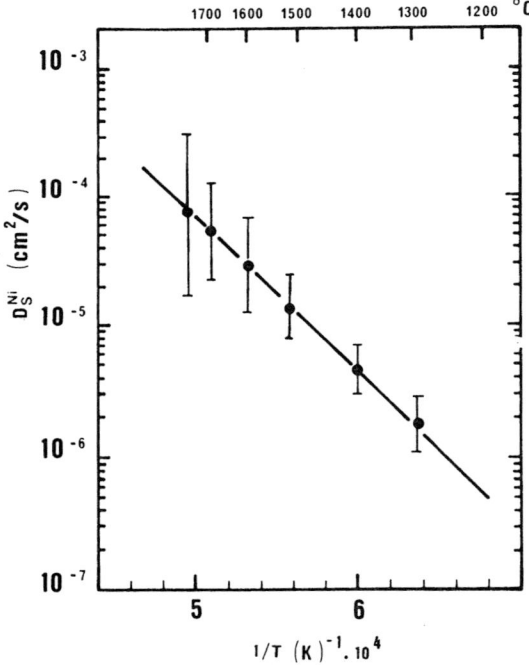

Fig. 4. Arrhenius plot of nickel surface diffusion coefficient in NiO.

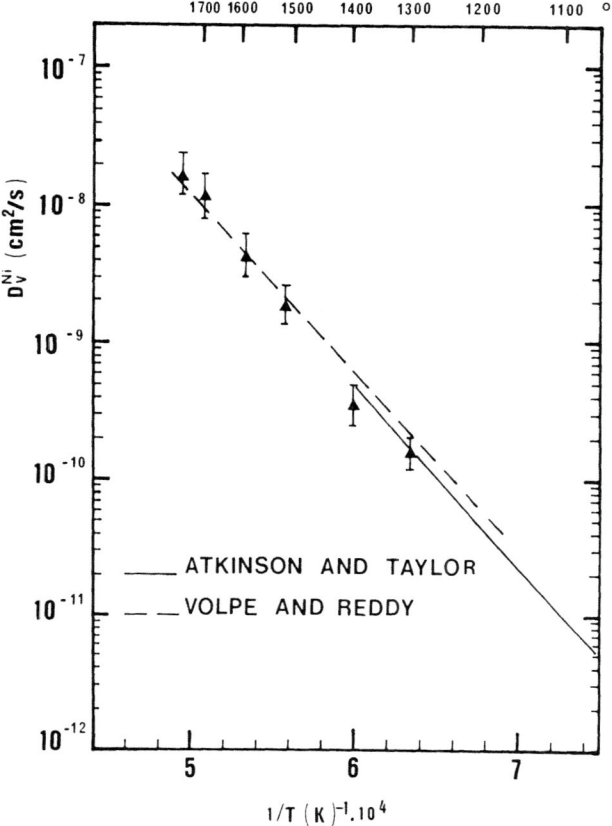

Fig. 5. Comparison of volume self-diffusion coefficients of nickel in NiO determined by the tracer method with those calculated in the present work from groove data[1].

The volume diffusion coefficient of nickel, calculated by linear regression of the experimental points is given by

$$D_v^{Ni} \ (cm^2.s^{-1}) = 3.50 \ 10^{-1} \ exp \left[- \frac{2.94 \pm 0.20 \ (eV)}{kT} \right]$$

The surface diffusion coefficient of nickel may be expressed by :

$$D_s^{Ni} \ (cm^2.s^{-1}) = 3.55 \ 10^{1} \ exp \left[- \frac{2.29 \pm 0.10 \ (eV)}{kT} \right]$$

INTERPRETATION AND DISCUSSION

Our results may be compared with those determined on other ceramic materials by similar techniques. In all cases[9,23-26] the use of the Mullins theory for the analysis of the kinetics of the thermal grooving, has shown only one simple mechanism (either surface or volume diffusion) to be responsible for the mass transport. One can note however a few cases for which both mechanisms have been observed, but each in a distinct temperature range. Thus, in the particular case of CoO, Maiya et al. have shown that at temperatures lower than 1084°C, surface diffusion is the controlling mass transport mechanism, whereas above 1140°C volume diffusion prevails[28]. A similar change in the mass transport mechanism has been observed for MgO at a temperature of 1500°C[28].

We shall also note that our results differ from those obtained on NiO by Readey and Jech[29] which suggested, in particular at 1535°C a single volume diffusion mass transport mechanism. However a large scatter of the experimental data of these authors could be observed ; if these results are analyzed taking into account an additional surface diffusion contribution, one finds for D_s a value in good agreement with our own determination.

A very good fit between our nickel volume diffusion coefficients and those determined by both Volpe and Reddy[21] and Atkinson and Taylor[22], can also be noted. This fit which is obtained considering that the above mentioned γ_s values are nearly identical for (001) and (011) planes is self-consistent with this approximation. The activation energy we measured for the volume diffusion (2.9 eV) is higher than that determined by these authors (2.56 eV and 2.64 eV respectively). This difference is about equal to the uncertainty attached to the groove width measurements. This uncertainty can be estimated from the linear regression of the diffusion coefficient values, at about 0.2 eV.

We shall note that the values obtained for the surface and volume diffusion activation energies, Q_s and Q_v respectively, are quite near to one another. This situation can be compared to that found for other oxides, or metals.

The comparison with other oxides is difficult, since the activation energies for D_v are known with quite a good accuracy, whereas those for D_s are rather scattered : 4.05 to 5.29 eV for Al_2O_3, 2.48 to 3.90 eV for MgO and 3 to 5 eV for UO_2[9,26,30,31].

Concerning now metals and more particularly the surface diffusion coefficients D_s, we note that Gjostein[32] has collected many data for f.c.c. elements and has shown that a correlation could be established between this coefficient and the reduced temperature Tf/T, where T and Tf are the temperature of the experiment and the melting temperature respectively.

Distinct relations have been proposed for the following two temperature ranges :

for $T/Tf > 0.75$:

$$D_s \ (cm^2.s^{-1}) = 7.4 \ 10^2 \ \exp \left[- \frac{30 \ Tf \ (cal.mole^{-1})}{RT} \right]$$

for $T/Tf < 0.75$:

$$D_s \ (cm^2.s^{-1}) = 1.4 \ 10^{-2} \ \exp \left[- \frac{13 \ Tf \ (cal.mole^{-1})}{RT} \right]$$

As for the volume self-diffusion coefficient of metals it has been shown[33] that in most cases the semi-empirical relation $Q_v = 34 \ Tf$ was obeyed.

It is then interesting to note that the ratio of the activation energies relative to volume and surface diffusion mechanisms in NiO is nearly the same as that of metals at high temperature.

This leads to the possibility of considering that at high temperatures the mechanisms accounting for surface diffusion are similar for both oxides and metals.

Several interpretations have been given in order to explain in the case of metals the rather high values of the surface diffusion activation energy at high temperatures. One of them considers a contribution of an evaporation-condensation process above 0.75 Tf[34-36].

Another one suggested by Bonzel et al. takes into account the contribution of polydefects (divacancies or biadatoms)[37], i.e. the occurrence of a new elementary surface diffusion mechanism. This assumption is supported by computer simulations using molecular dynamics which have shown that surface migration by jumping processes extending over a few atomic distances could contribute notably to high temperature surface diffusion[38-39].

CONCLUSION

This work confirms the possibility of determining volume and surface diffusion coefficients from high temperature mass-transport experiments, by separating the contributions of volume and surface processes. It seems that at high temperature surface diffusion takes place through a similar mechanism for both oxides and metals. We shall also note that within the accuracy available in the type of experiments used in this study we did not observe any influence of the orientation of the reference planes (001) and (011), or of the crystallographic directions, on the D_s values.

REFERENCES

1. W. W. Mullins, J. Appl. Phys. 28:333 (1957).
2. Y. Adda, J. Philibert, "La diffusion dans les solides", P.U.F., (1966).
3. G. Neumann and G. M. Neumann, Surface self-diffusion of metals, in: "Diffusion Monograph Series", F. H. Wohibier, ed., Bay Village (1972).
4. G. E. Rhead, Surface Sci. 47:207 (1975).
5. G. Martin, B. Perraillon, "La diffusion dans les milieux limites", 19e Colloque de Metallurgie, I.N.S.T.N. Saclay (1976).
6. G. L. Kellog, T. T. Tsong, and P. Cowan, Surface Sci. 70:485 (1978).
7. P. J. Harrop, Surface Sci., 3:206 (1968).
8. R. Freer, J. Mat. Sc. 15:803 (1980).
9. W. M. Robertson, J. Nucl. Mater. 30:36 (1969).
10. G. Dhalenne, A. Revcolevschi, and C. Monty, Phys. Stat. Sol. (a) 56:623 (1979).
11. W. W. Mullins, Trans. AIME 218:354 (1960).
12. W. W. Mullins and P. G. Shewmon, Acta Met. 7:163 (1959).
13. G. Dhalenne, A. Revcolevschi, and A. Gervais, J. Cryst. Growth 44:297 (1978).
14. G. Dhalenne, M. Dechamps, and A. Revcolevschi, "Energy of tilt boundaries and mass transport mechanisms in nickel oxide", in: "Character of grain boundaries", M. F. Yan and A. H. Heuer, eds., The Am. Ceram. Soc. , Inc. (1983).
15. C. Monty, J. Le Duigou, High Temp.-High Pres., 14:709 (1982).
16. P. Braudeau, These 3e Cycle, Parix VI (1983).
17. G. Chalenne, M. Dechamps, and A. Revcolevschi, Experimental study of the relative energy of symmetrical <110> tilt boundaries in semiconductors", H. J. Leamy, G. E. Pike and E. H. Seager, eds., North Holland (1982).
18. R. F. Stewart and W. C. Mackrodt, J. Phys. (Paris) Colloq., C7:248 (1976).
19. D. M. Duffy and P. W. Tasker, Phil. Mag. 48:155 (1983).
20. W. D. Kingery, J. Am. Ceram. Soc. 37:42 (1959).
21. M. L. Volpe and J. Reedy, J. Chem. Phys. 53:1117 (1970).
22. A. Atkinson and R. I. Taylor, Phil. Mag. 39:581 (1979).
23. K. Kitazawa, T. Kuriyama, K. Fueki, and T. Mukaibo, J. Am. Ceram. Soc. 60:363 (1977).
24. G. L. Reynolds, J. Nucl. Mater. 24:69 (1967).
25. K. Kitazawa, I. Komaki, K. Matsukawa, and K. Fueki, Surface mass transport study on alumina and magnesium aluminium spinel by multiple scratch smoothing method, Yogyo-Kyokai-shi, 88:42 (1980)
26. S. A. Lytle and V. S. Stubican, J. Am. Ceram. Soc. 65:210 (1982).
27. P. S. Maiya, W. K. Chen, and N. L. Peterson, Metallurg. Trans. 1:801 (1970).
28. W. M. Robertson, "Sintering and related phenomena", G. C. Kuczynski, N. A. Hooton and C. F. Gibbon, eds., N.Y. (1967).

29. D. W. Readey and R. E. Jech, J. Am. Ceram. Soc. 51:201 (1968).
30. A. B. Lidiard, J. Nucl. Mater. 19:106 (1966).
31. Hj. Matzke, Chap. IV in: "Nonstoichiometric oxides", O. T. Sorensen, ed., Academic Press (1981).
32. N. A. Gjostein, "Surfaces and interfaces, I", J. J. Burke, N. L. Reed and V. Weiss, eds., Syracuse Univ. Press (1967).
33. P. Guiraldenq, J. Phys. (Paris) Colloq. C4:201 (1975).
34. B. Perraillon, R. Peix, and V. Levy, Script. Met. 5:1001 (1971).
35. V. T. Binh, A. Piquet, H. Roux and R. Uzan, Surface Sci. 44:598 (1974).
36. J. Friedel, Ann. Phys. 1:257 (1976).
37. H. P. Bonzel and N. A. Gjostein, "Molecular Processes in Solid Surfaces", McGraw Hill series in Materials Science and Engineering (1969).
38. G. DeLorenzi, G. Jacucci, and V. Pontikis, Surface Sci. 116:391 (1982).
39. D. Ghaleb, These d'Etat, Orsay (1984).

DIFFUSION IN Cr_2O_3 VIA INITIAL SINTERING EXPERIMENTS

Ming-Yih Su, Huan-Yeong Chang and George Simkovich

The Metallurgy Program
Department of Materials Science & Engineering
The Pennsylvania State University
University Park, PA   16802

INTRODUCTION

Chromium oxide, Cr_2O_3, has been successfully sintered to >98% of theoretical density by Ownby and Jungquist.[1] It was shown in their study that the sintering kinetics of Cr_2O_3 were strongly dependent upon the oxygen partial pressure. Hagel et al.[2] studied the initial sintering of Cr_2O_3 in an Ar atmosphere from 1050° to 1300° and concluded that volume diffusion of oxygen ions controlled the sintering mechanism. However later studies by Halloran and Anderson[3] and Neve and Coble[4] indicated that the oxygen pressure utilized in the study by Hagel et al.[2] was not properly controlled due to the presence of a graphite sample holder employed in their experiments. More recently, Atkinson and Taylor[5] in a tracer diffusion study of Cr in Cr_2O_3 pointed out that grain-boundary diffusion may be important in Cr_2O_3.

In view of the apparent dependence of sintering rates upon oxygen partial pressure and the fact that chromium diffusion occurs readily via grain boundaries it was decided to conduct carefully controlled sintering experiments on high purity Cr_2O_3. Additionally, in view of the possibility of various sintering mechanisms the results obtained in this study were analyzed in terms of the multi-path sintering model briefly described in the following section.

INITIAL SINTERING MODELS

The mass transportant mechanisms occurring during the initial stage of sintering have been considered by many during the past

30-40 years. Based on the neck growth theory various mechanisms may in principle take place: viscous flow, evaporation-condensation, surface diffusion, lattice diffusion, etc.[6-11]. However, it is generally agreed that diffusion is the most important transport mechanism which can explain shrinkage. Although several diffusion paths may be used for mass transport, grain-boundary diffusion and volume diffusion from the boundary to the neck are the major mechanisms causing shrinkage of the compact whereas surface diffusion does not contribute to the shrinkage. From sintering shrinkage studies values of volume and grain-boundary diffusion coefficient may be estimated.

Assuming only one mechanism controls the transport leading to shrinkage, the classical sintering models[6-11] result in equations of the form:

$$Y = \frac{\Delta L}{L_o} = [\frac{K\gamma\Omega D}{kT\ a^P}]^m t^m \qquad (1)$$

where Y = fractional shrinkage; K = geometrical factor; γ = surface free energy, Ω = vacancy volume of controlling species; D = self-diffusion coefficient of diffusion species, bD_B, for grain boundary control, and D_V, for volume control; a = particle (or sphere) radius; m = constant, 0.31 - 0.33 for grain boundary control and 0.46 - 0.50 for volume control; P = constant, 4 for grain boundary control and 3 for volume control; and, t = time. When log Y is plotted against log t, the controlling mechanism may be determined by comparing the slope with the predicted values, and the diffusion coefficient of the controllng species may be estimated from the intercept of such a plot.

If more than one mechanism contribute to the sintering process or if a change in the controlling mechanism occurs, interpretation of the kinetics by a single mechanism alone is no longer valid, and the diffusivity calculated from eqn. (1) will not represent the correct value of the self-diffusion coefficient of the rate controlling species. To overcome this difficulty, Berrin and Johnson[12] and Johnson[13] have developed models for the initial sintering stage in which simultaneous diffusion contributions may occur.

Assuming that all the material in the neck volume results from transport from the grain-boundary region the shrinkage as a function of time is expressed as[12,13]

$$Y^{2.06}\dot{Y} = [\frac{2.63\gamma\Omega D_v}{kT\ a^3}] \ Y^{1.03} + [\frac{0.7\gamma\Omega bD_B}{kT\ a^4}] \qquad (2)$$

386

where Y = fractional shrinkage ; $\dot{Y}$ = rate of fractional shrinkage, dY/dt; b = width of enhanced diffusion near the boundary; D_V = volume self diffusion coefficient; D_B = grain boundary self diffusion coefficient; and the other parameters have been defined following eqn. (1).

The volume diffusion coefficient D_V and the grain boundary diffusion coefficient bD_B may be obtained simultaneously, if both paths are operative, from the slope and intercept respectively, of a $Y^{2.06} \dot{Y}$ vs $Y^{1.03}$ plot. If only volume diffusion from the grain boundary to the neck is operative, eqn. (2) becomes[8]

$$Y \cong [\frac{5.34\gamma\Omega D_V}{kT \ a^3}]^{0.49} \ t^{0.49} \tag{3}$$

Similarly, for only grain-boundary diffusion[8] the following relation holds:

$$Y \cong [\frac{2.14\gamma\Omega bD_B}{kT \ a^4}]^{0.33} \ t^{0.33} \tag{4}$$

EXPERIMENTAL PROCEDURE

High purity (99.999%) Cr_2O_3 powder used in this study was purchased from Johnson Matthey Chemicals Limited with listed impurity contents of 1 ppm Ag and less than 1 ppm of Al, Ca, Cu, Fe, Mg and Si. The raw powder was ground lightly to pulverize loosely aggregated lumps and uniaxially pressed at 3.45×10^8 N/m^2 (50,000 psi) into pellets of 7.5 mm in diameter and 7.5 mm in height without using any binder. The green density of the compacts was 50% of theoretical density. Average particle radius of the ground powder, 0.75 μm, was calculated from the data measured with a Sedigraph 5000D Particle Size Analyzer from Micromeritics Instrument Co., Norcross, Georgia, USA.

Sintering of the compacts was carried out over the temperature range of $1250°$ to $1600°C$ in a horizontal Al_2O_3 tube furnace which was molybdenum-wire wound and hydrogen protected. A quartz cover window on one end of the reaction tube was sufficiently transparent to permit photographing the sample during shrinkage. The P_{O_2} was controlled by using CO/CO_2 gas mixtures with flow rates that gave a velocity of gas of 1 cm/sec calculated on room temperature volumes.

During the experiment, the Cr_2O_3 sample was placed on a sintered Cr_2O_3 disc which, in turn, was contained in an alumina boat. This was done to minimize possible contamination from the boat. The alumina boat was carefully inserted into the hot zone with the furnace at a pre-set temperature and allowed to sit in the air for two minutes to attain thermal equilibrium. No shrinkage was observed during this two-minute period. The CO/CO_2 gas mixture was then passed through the reaction tube. Photographs of the sample were taken intermittently throughout the experiment. A typical sequence of photographs of a sample at 1250°C and P_{O_2} of 1×10^{-16} atm is shown in Figure 1. Zero time was always referred to the moment when the flowing CO/CO_2 gas mixture just reached the sample position, that is, 30 seconds after starting the gas flow. The relative diameter change, $\Delta L/L_o$, of the sample was measured on the projection of these photographic films with a magnification of about 10.

RESULTS AND DISCUSSION

The shrinkage of the Cr_2O_3 compacts sintered in various P_{O_2}'s at 1250°, 1400° and 1500°C is plotted as log Y versus log t in Figure 2. Two dashed lines are drawn to indicate the extremes of the classical sintering model; the upper line has a slope of 0.5 as predicted for a volume diffusion model while the lower line has a slope of 0.33 according to a grain boundary control mechanism. As may be seen in Figure 2, when the P_{O_2} is decreased at a given temperature, the measured shrinkage increases for a constant sintering time. This phenomenon is consistent with Ownby and Jungquist's results that high temperatures and low P_{O_2}'s ($>Cr/Cr_2O_3$ equilibrium P_{O_2}) increase the sintering rate and the final density.[3]

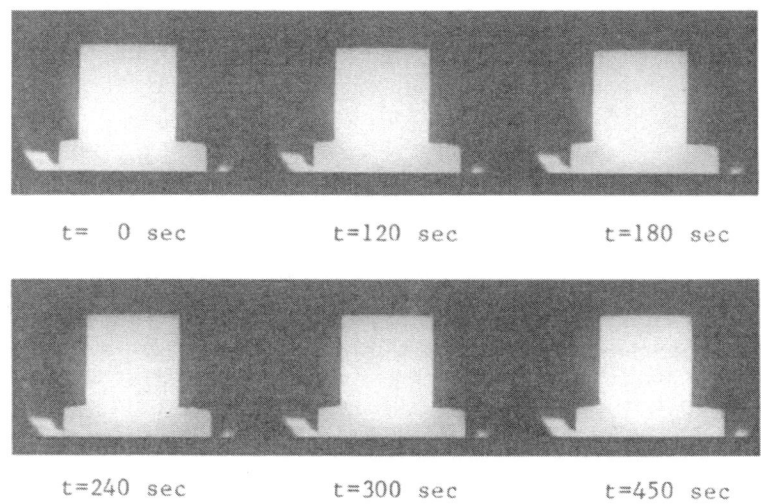

| t= 0 sec | t=120 sec | t=180 sec |

| t=240 sec | t=300 sec | t=450 sec |

Figure 1. Time sequence of Cr_2O_3 pellets at t=1250°C and P_{O_2} = 10^{-16} atm.

388

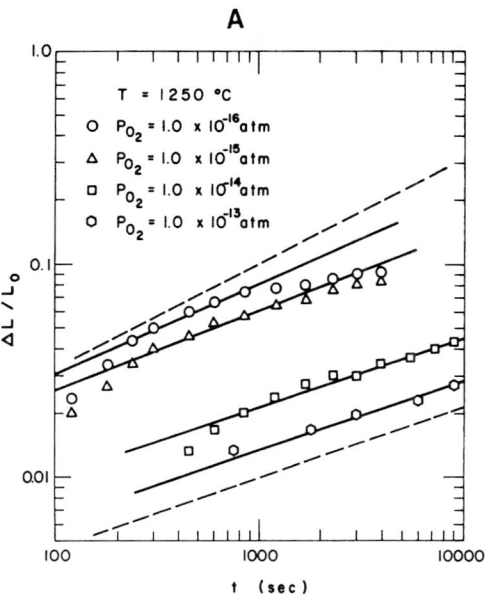

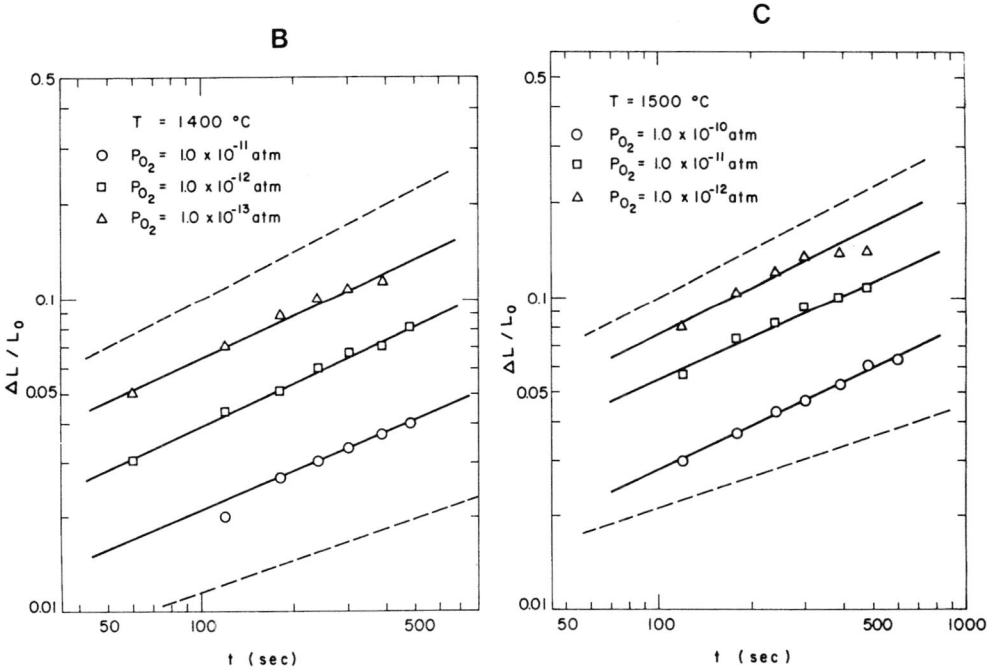

Figure 2. Initial shrinkage of pure Cr_2O_3 at various P_{O_2}'s and temperatures: A = 1250°C; B ≅ 1400°C; and, C ≅ 1500°C.

However, the fact that the slopes of the log Y - log t plots change as a function of P_{O_2} indicates that a single sintering mechanism does not adequately describe the sintering of Cr_2O_3 over the temperatures and P_{O_2} region studied.

Due to the apparent change in the controlling mechanism the data in Figure 2 were plotted and analyzed according to Johnson's multiple sintering mechanism, as previously discussed. Thus, the relation as described in eqn. (2)

$$Y^{2.06} \dot{Y} = [\frac{2.63\gamma\Omega D_V}{kT\ a^3}] Y^{1.03} + [\frac{0.7\gamma\Omega b\ D_B}{kT\ a^4}] \qquad (2)$$

was utilized. Shrinkage rates, Y, were calculated by differentiating the equation of best fit of the form $y = a\ x^b$ to the data. When $Y^{2.06}\dot{Y}$ is plotted against $Y^{1.03}$, the slope will be proportional to D_V and the intercept to bD_B if eqn. (2) is valid. Figure depicts the $Y^{2.06}\dot{Y}$ vs $\dot{Y}^{1.03}$ plots.

In Figure 3, at $1250^{\circ}C$ the two nearly horizontal lines at the P_{O_2}'s of 1×10^{-13} and 1×10^{-14} atm indicate that $D_V \tilde{\ } 0$ or that $D_B \gg D_V$, that is, diffusion in the grain boundaries controls the sintering rate. For such cases the simplified equation, eqn. (4), was used to calculate $b\ D_B$. At $1500^{\circ}C$, the fact that intercepts of all three lines are nearly zero implies volme diffusion control, that is $D_V \gg D_B$ or $D_B \tilde{\ } 0$. Eqn. (3) was applied to calculate D_V although similar values of D_V may still be obtained from the slope of a $Y^{2.06}\dot{Y}$ us $Y^{1.03}$ plot. At $1400^{\circ}C$ for all P_{O_2}'s studied and at $1250^{\circ}C$ for the two lower P_{O_2}'s studied, 1×10^{-15} and 1×10^{-16} atm, both the intercepts and the slopes have finite values which indicate contributions to the sintering process from both grain-boundary diffusion and volume diffusion. Eqn. (2) was then utilized to calculate both D_V and bD_B.

It is apparent from Figure 3 that at a constant temperature both the slope and the intercept increase as P_{O_2} is decreased. This trend indicates that the diffusion processes in both the bulk lattice and the grain boundries are more rapid when the oxygen partial pressure is lowered. By substituting the appropriate numbers (including $\gamma = 1000$ erg/cm, $\Omega = 1.60\times10^{-23}$ cm^3 and a = 0.75 μm) into one of the eqns. (2), (3), or (4), the apparent diffusion coefficients (D_V and bD_B) were calculated and are plotted as a function of P_{O_2} in Figure 4. A comparison line with the slope of -0.5 based on the following discussion is also shown.

In order to aid in the interpretation of the effect of P_{O_2} upon the sintering rates of Cr_2O_3 electrical conductivities of dense Cr_2O_3

390

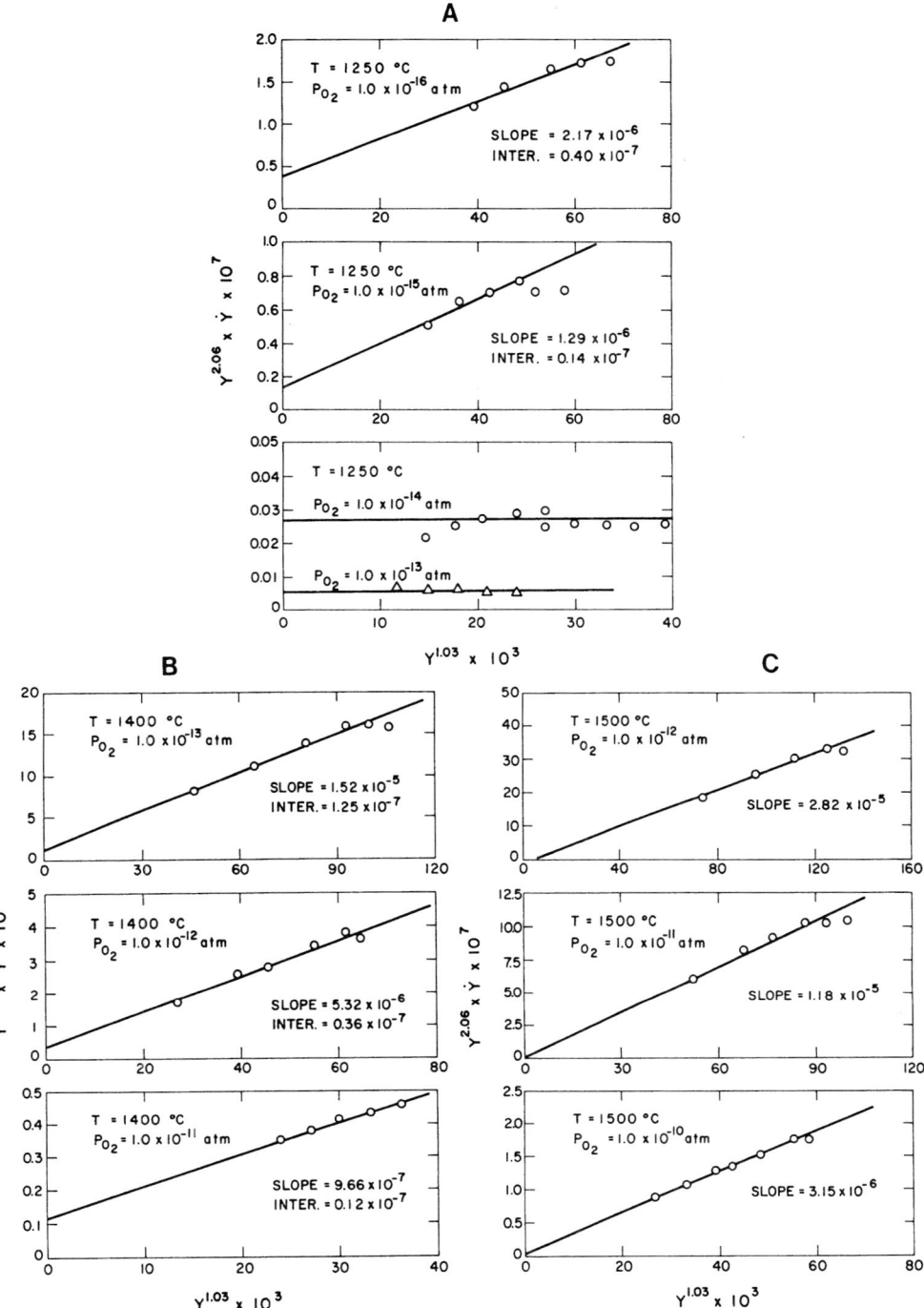

Figure 3. Shrinkage of Cr_2O_3 at various P_{O_2}'s and $1400°C$ and $1500°C$ temperatures plotted according to multiple shrinkage model (eqn. 2): $A = 1250°C$; $B = 1400°C$; and $C = 1500°C$.

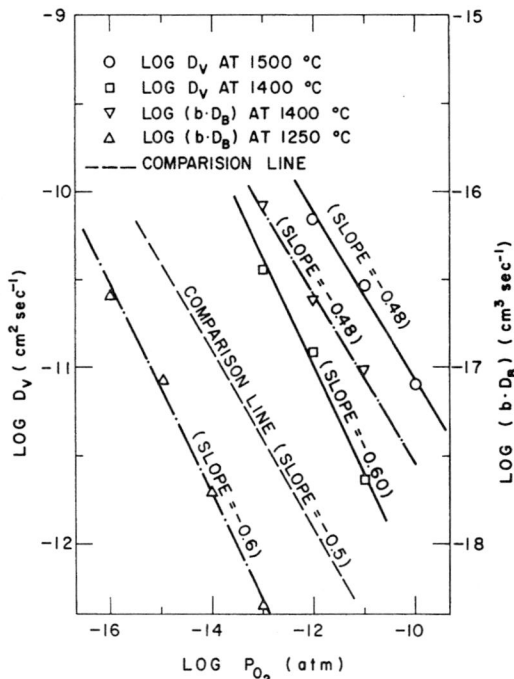

Figure 4. Calculated apparent diffusion coefficient as a function
of oxygen potential at T = 1250°C, 1400°C, and 1500°C.

pellets were made utilizing the 4-point probe D.C. method. These
results will be presented in more detail in a future publication,
however, in respect to the temperatures and oxygen pressures utilized
in this sintering study it was found that the electrical conductivities
of the dense Cr_2O_3 samples remained essentially constant as P_{O_2} was
varied at a particular temperature. Similar results were found by
Crawford and Vest[14] on single crystals of Cr_2O_3 at temperatures
comparable to this study but at higher P_{O_2}'s.

From these results and the knowledge that Cr_2O_3 is a semiconductor one must conclude that over the range of temperatures and
oxygen potentials utilized in these sintering studies that the
Cr_2O_3 is behaving as an intrinsic semiconductor or that the impurity
levels are so high that variations in oxygen potentials do not affect
the electronic defect concentrations. Since the Cr_2O_3 utilized in
this study is of relatively high purity we assume that the Cr_2O_3
employed is an intrinsic semiconductor.

Next one must consider the motion of the chromium and oxygen
ions in relation to the sintering behavior as a function of P_{O_2}
and the intrinsic semiconductor behavior of the Cr_2O_3. If one

assumes that the motion of a chromium ion is rate controlling in the sintering process then one may obtain that the diffusivity of chromium interstitials is given as

$$D_{Cr_i^{n\bullet}} \quad \alpha \quad P_{O_2}^{-3/4} \tag{5}$$

where $Cr_i^{n\bullet}$ indicates a chromium interstitial ion carrying an excess charge of n value, however, the slopes of the diffusivities in Figure 4 are much closer to $-1/2$ rather than $-3/4$. Thus, one may postulate that oxygen ion motion is rate controlling in the sintering process. The formation of oxygen ion vacancies, $V_O^{\bullet\bullet}$, is given as

$$O_O^x = V_O^{\bullet\bullet} + 2e + 1/2 \ O_2 \tag{6}$$

with the equilibrium condition, at a constant electron concentration (intrinsic semiconductor behavior), of

$$K_6 \cdot P_{O_2}^{-1/2} = [V_O^{\bullet\bullet}] \tag{7}$$

Since the diffusivity of oxygen ions may be anticipated to be proportional to the concentration of oxygen ion vacancies one may conclude that the "apparent" diffusion coefficients depicted in Figure 4 relate primarily to oxygen motion. It is perhaps worthy of mention at this point that a relation similar to eqn. (7) is obtained for oxygen vacancies of single charge and and also of neutral charge since the electronic defects remain essentially constant as a function of P_{O_2} and therefore do not alter the exponent of the P_{O_2} term.

Although one should not attach too great a significance to the actual numerical value of the deduced diffusion coefficients obtained from sintering experiments one may, nevertheless, consider that the variation of the coefficients with P_{O_2} is reasonable because most of the parameters entering the calculation will remain constant from one experiment to another with the possible exception of the surface energy. Thus the results depicted in Figure 4 deserve further comment in view of the fact that both grain-boundary and volume diffusion vary in the same manner with oxygen pressure variations.

These results indicate that the vacancies in the grain boundaries are directly proportional to the vacancies in the bulk crystal, that is,

$$V_O^{\cdot\cdot}(\text{Volume}) = V_O^{\cdot\cdot}(\text{Grain Boundary}) \tag{8}$$

with a "distribution" ratio

$$K_8 = \frac{[V_O^{\cdot\cdot}(\text{GB})]}{[V_O^{\cdot\cdot}(\text{Vol})]} \tag{9}$$

existing between these vacancy concentrations. In fact, one may also postulate that a similar distribution exists between surface (the gas-solid interface) and bulk vacancies, that is, that the relation

$$K = \frac{[V_O^{\cdot\cdot}(\text{surface})]}{[V_O^{\cdot\cdot}(\text{Vol})]} \tag{10}$$

also holds. Eqns. (9) and (10) are certainly tentative postulates which require additional experimental results prior to their general acceptance. Nevertheless, for such cases one would obtain essentially similar behavior to the variation of a parameter such as P_{O_2} for both grain boundary and bulk transport properties and such is the case for Cr_2O_3 under the sintering conditions of this study.

CONCLUSIONS

1. The initial sintering rate of Cr_2O_3 is a function of T and P_{O_2}. High temperatures and low P_{O_2}'s enhance the initial sintering rate.
2. Grain-boundary diffusion is important at low temperatures($\simeq 1250^\circ$C) and high oxygen potentials ($P_{O_2} > 10^{-14}$ atm at 1250°C).
3. Volume diffusion becomes rate controlling at higher temperatures (T = 1500°C) and low oxygen potentials, ($P_{O_2} \leq 10^{-10}$ atm at 1500°C).
4. Electrical conductivities at all temperatures studied were constant as a $f(P_{O_2})$ over the low pressure P_{O_2}'s studied.
5. Calculated apparent diffusion coefficients indicate that both bulk and grain-boundary diffusion are via oxygen vacancies.
6. Both grain-boundary and bulk diffusion coefficients vary as a function of P_{O_2} in essentially an identical manner.

ACKNOWLEDGEMENTS

The authors gratefully acknowledge the support of the Applied
Research Laboratory in the Pennsylvania State University. One of
the authors (H.Y.C.) wishes to thank China Steel Corporation in
Taiwan, R. O. C., who supported his two-year graduate study at The
Pennsylvania State University.

References

1. P. D. Ownby and G. E. Jungquist, J. Am. Ceram. Soc. 55:433
    (1972).
2. W. C. Hagel, P. J. Jorgensen, and D. S. Tomalin, J. Am. Ceram.
    Soc. 49:23 (1966).
3. J. W. Halloran and H. W. Anderson, J. Am. Ceram. Soc. 57:150
    (1974).
4. J. M. Neve and R. L. Coble, J. Am. Ceram. Soc. 57:274 (1974).
5. A. Atkinson and R. I. Taylor, in: this conference.
6. J. Frenkel, J. Phys. USSR 9:385 (1945).
7. G. C. Kuczynski, Trans. AIME 185:169 (1949).
8. G. C. Kuczynski, J. Appl. Phys. 21:632 (1950).
9. W. D. Kingery and M. Berg, J. Appl. Phys. 26: 1205 (1955).
10. R. L. Coble, J. Am. Ceram. Soc. 41:55 (1958).
11. D. L. Johnson and I. B. Cutler, J. Am. Ceram. Soc. 46:541
    (1963).
12. L. Berrin and D. L. Johnson, in: "Sintering and Related
    Phenomena," G. C. Kuczynski, N. A. Hooton and C. F. Gibbon,
    ed., Gordon and Breach, N.Y. (1967).
13. D. L. Johnson, J. Appl. Phys. 40:192 (1969).
14. J. A. Crawford and R. W. Vest, J. Appl. Phys. 35:2413 (1964).

THE ATOMISTIC SIMULATION OF IMPURITY SEGREGATION AT THE SURFACES

OF MgO AND CaO

E.A. Colbourn and   P.W. Tasker
W.C. Mackrodt

New Science Group   Theoretical Physics Division
I.C.I. plc, Runcorn   AERE Harwell
Cheshire, England   Oxon, England

INTRODUCTION

For the most part the atomistic simulation of ceramics has
been concerned with bulk material and, in particular, with
structural features and defect properties that are well-
established, at least in principle, if not always known in
quantitative detail [1]. The simulation of ceramic surfaces on
the other hand, in most cases breaks new ground, for here
knowledge and understanding at an atomic level are still both
sparse and rudimentary [2]. For example, it is only very recently
that information on the crystal structure and relative stability
of the lowest index surfaces of α-alumina has become available,
solely from simulation studies [3], although the (bulk) properties
of alumina have been investigated extensively for many years. The
same is true for impurities. They are known to exert an
important, often controlling, influence on the properties of
ceramics, yet phenomena such as surface segregation remain largely
unexplored. The present chapter, then, gives a brief review of
some recent atomistic simulations of impurity segregation at the
lowest index surfaces of MgO and CaO. The aim here is to
illustrate the variety of behaviour, the level of detail and the
unexpected nature of the results that such studies can sometimes
reveal. In so doing it might also prompt further experimental
investigation into this important aspect of ceramic materials.

THEORETICAL MODEL

The atomistic simulations reviewed here are based on a simple
theoretical approach which has been described in full elsewhere
[4,5]. It is assumed, with some justification [6,7], that MgO and

CaO are essentially ionic materials, for which interatomic potentials, including those for impurities, are derived from density-functional methods [8]. Electronic polarisation of the lattice is accounted for by means of the shell-model introduced some time ago by Dick and Overhauser [9] and used subsequently in the majority of recent studies [1]. Simulations of the type reported here are carried out within the static lattice approximation; furthermore, all surfaces are assumed to be planar and infinite in two dimensions. The essence of the method is to calculate the <u>potential</u> energy difference at 0^oK and constant (bulk) lattice parameter between a non-defective and defective lattice, <u>both at zero strain.</u> Kinetic energy differences are neglected. The term defective lattice here refers to bulk and surface lattices containing a single impurity ion (infinite dilution limit) or infinite planes with fractional occupancy by impurity. The zero strain configuration is obtained by relaxing the lattice containing the defect/impurity so that the net force on each ion is zero [4]. Thus in the case of impurity segregation the thermodynamic quantity derived is the difference in the internal energy per impurity ion, Δh, between the bulk and surface. To this can be added the configurational entropy contribution, Δs^c, to give an approximate value for the change in the total free energy of segregation per ion, Δg. At present, the vibrational contribution, Δs^v cannot be estimated with any certainty. Atomistic simulations also give surface (and bulk) crystal structures directly, an example of which is given later in this paper.

COVERAGE-DEPENDENT HEATS OF SEGREGATION

The heat of segregation, Δh, is the quantity which is most directly obtained from simulation studies. It is necessarily coverage-dependent, though prior to recent theoretical studies [10-12] the form of this dependence seems to have been unknown. Figure 1 illustrates the variety of coverage-dependence that such studies have revealed for even the simplest of systems. Coverages here are up to a complete monolayer of the (001) surface. Starting with BeO in MgO, which is probably the least reliable example due to uncertainties in the beryllium-oxygen bonding, we find Δh varying from -0.2 ev at zero coverage to +0.4 ev at full (monolayer) coverage, from which we deduce a limit to the surface concentration of approximately a third of a monolayer. For BaO in MgO, on the other hand, Δh is calculated to be negative for all coverages, as shown in Figure 1(b): it is also highly non-linear with a maximum at about 50% occupancy. The values of Δh, which range from -3.9 ev to -1.4 ev, suggest a strong tendency to segregate. Similar results are found for CaO and SrO in MgO, but with reduced heats of segregation. In the case of CaO the calculated range is -1.1 ev at zero coverage to -0.5 ev for a monolayer, which compares well with the measured value of -0.52 ev

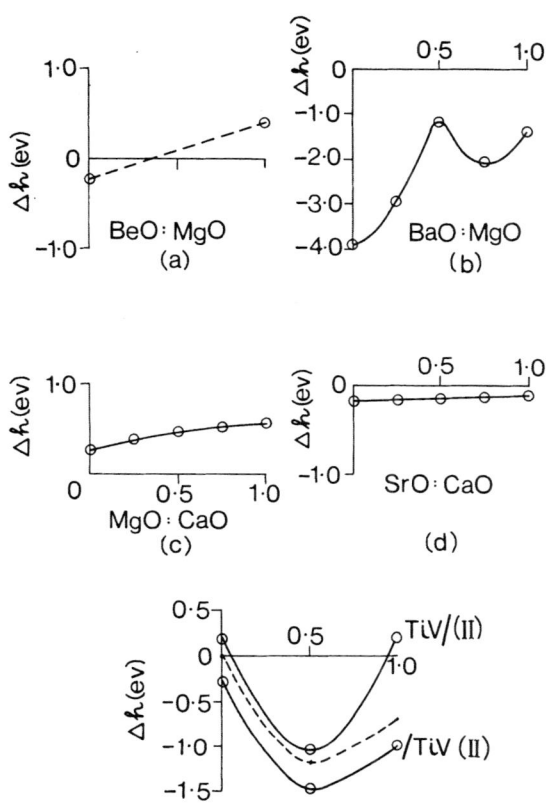

Figure 1 Calculated monolayer coverage-dependent heats of segregation at the (001) surfaces of MgO and CaO

reported recently by McCune and Wynblatt [13]. Turning now to isovalent impurities in CaO, Figure 1 (c) shows that the calculated heat of segregation of MgO in CaO is <u>positive</u> from zero to full coverage, which suggests that Mg^{2+} ions <u>will deplete</u> the (001) surface, while that for SrO (in CaO) is close to zero throughout the coverage range, as shown in Figure 1(d). Thus for the very simplest isovalent impurities in essentially ionic f.c.c. oxides we find examples of each type of possible behaviour, namely, impurity concentration at the surface (BaO in MgO), depletion of impurity at the surface (MgO in CaO), uniform distribution (SrO in CaO) and partial surface coverage (BeO in MgO). Furthermore for the single case where there is experimental data, theory and observation seem to be in good accord.

Aliovalent impurities, in general, lead to charged defects in ionic materials with the possibility of complicated space-charge effects at surfaces. However, for TiO_2 in both MgO and CaO, calculations indicate that the titanium will exist almost exclusively as 4+ ions bound to cation vacancies giving <u>neutral</u> defect pairs, which, in some respects are comparable isovalent impurities of the type referred to previously. Figure 1(e) shows a plot of Δh against surface coverage for TiO_2 in MgO based on this defect model. The abbreviations TiV/(‖) and /TiV(‖) refer to parallel-aligned defect pairs in the surface and one plane below respectively. The dotted plot corresponds to an infinite plane of the corresponding defects in the bulk. For isolated pairs (zero coverage/infinite dilution) the heat of segregation is found to be approximately −0.4 ev. As the concentration/coverage of titanium is increased calculations indicate a strong tendency to form an ordered (two-dimensional) second-phase both in the bulk and at the surface, the composition of which is approximately $TiMg_2O_4$. Furthermore this tendency to aggregate ($\Delta H = -1.5ev$ at 50% coverage) is much greater than that to segregate. What the simulations have shown is that above a certain concentration of TiO_2 in MgO, ilmenite, $TiMgO_3$, will be formed as a second-phase and that this second-phase formation will occur both in the bulk and at the surface. Since the calculations have been restricted to 2-dimensional planes of impurity associates, the predicted second-phase does not correspond to ilmenite exactly; however, the important features of the transformation seem to have been established.

SURFACE STRUCTURE

Earlier in this chapter we have alluded to the fact that atomistic simulations can provide structural information about surfaces (and the bulk) directly. We now consider two aspects of this related to impurity segregation, with particular reference to calcium, strontium and barium ions at the (001) surface of MgO. Figure 1(b) shows a typical plot of Δh against surface occupancy

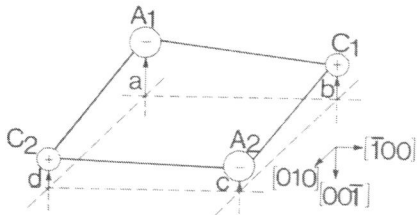

Figure 2a   Heat of segregation per ion, Δh, for multi-layer
segregation of CaO at the (001) surface of MgO

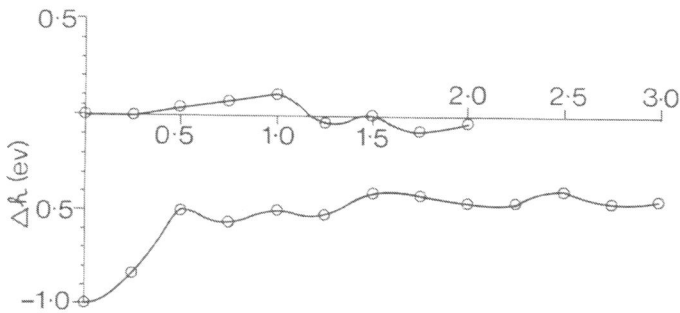

Figure 2b   Reconstructed unit cell of the heavily segregated
(001) surface of MgO. Values of a.....d are given in Table 1.

for <u>monolayer</u> coverage, from which we deduce a strong tendency for barium (calcium and strontium) to segregate at the (001) surface. Two questions that might be asked with regard to this are, first, is the <u>surface</u> monolayer the lowest in energy, and second, is <u>segregation</u> restricted to monolayer coverage, or can multiple-layer segregation take place? Figure 2(a) shows a plot of Δh against coverage for sub-surface and multiple-layer segregation of calcium ions at the (001) surface of MgO. The lower curve corresponds to multiple-layer segregation starting with the surface layer (0-1.0), followed by the plane below the surface (1.0-2.0) and so on; the upper curve corresponds to the same process, but starting one plane below the surface. Clearly surface coverage is all important, for without it the tendency to aggregate below the surface is negligible, as shown by the upper curve. There is a gradual increase in Δh as multiple layers are formed with a very approximate extrapolation to about 8-10 layers of impurity at which point Δh is zero. Thus our calculated heats of segregation suggest multiple-layer segregation, at least in the case of calcium in MgO. The segregation of calcium, strontium and barium ions at the surface of MgO is driven essentially by size effects. The impurity ions are bulkier than the host cation and the resulting elastic strain is minimised at the surface. What effect, then, does segregation have on the crystal structure of the (001) surface? The non-defective surface contains two ions per (surface) unit cell and is very close to a straightforward termination of the bulk structure [5]. At full monolayer coverage, however, we find a 2 x 2 reconstruction of the surface shown in Figure 2(b). Cations are indicated by C, anions by A and the displacements a, b, c, d for the three impurities are given in Table 1. Reconstructions of this sort should be detectable by surface-sensitive techniques such as LEED, the appropriate patterns of which will be quite different from that of the unconstructed non-defective surface.

TEMPERATURE EFFECTS

In the discussion so far there has been no mention of temperature; yet the high-temperature properties of ceramics are among those

Table 1. Perpendicular displacements of impurity ions in the reconstructed surface of MgO

Displacement*	Ca^{2+}	Sr^{2+}	Ba^{2+}
a	0.39	0.75	0.79
b	0.21	0.68	1.21
c	0.05	0.14	0.79
d	0.21	0.24	0.38

---

* in units of a_o, the host cation-anion lattice spacing

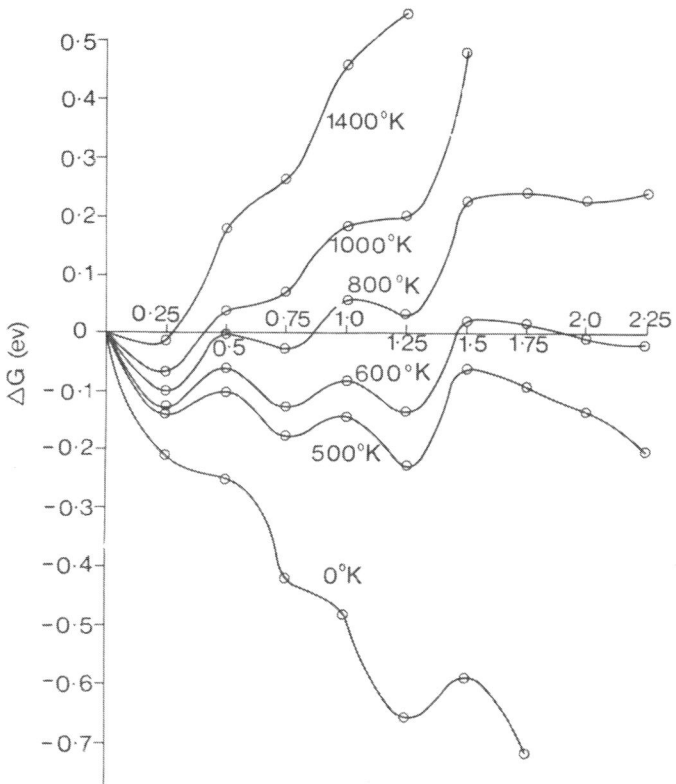

Figure 3 Total free energy of segregation, ΔG, as a function of surface coverage, y, for a bulk concentration of 220 ppm of calcium in MgO

which are most valuable from a technological point of view. From classical thermodynamics we have

$$G = H - TS$$

so that temperature effects are closely associated with entropy changes. In solids there are essentially two contributions to the classical entropy, a configurational entropy, S^c, and a vibrational contribution, S^v. At present the calculation of S^v for surfaces is uncertain; furthermore, there is some evidence from bulk studies which suggests that vibrational entropy differences, such as those involved in the relative stability of alternative defect configurations, or in defect migration, are small [14]. To a first approximation then we neglect the vibrational entropy contribution to surface segregation. It can be shown that the configurational contribution per impurity ion, Δs^c, is given by:

$$\Delta s^c = -k \left\{ \ln(x/y) + (1-y)/y.\ln[(1-x)/(1-y)] \right\}$$

in which x and y are the bulk and surface impurity concentrations respectively. Provided y is not too small, this reduces to

$$\Delta s^c \approx -k\ln(x/y)$$

so that the total free energy of segregation, ΔG, corresponding to a surface coverage, y, is given by,

$$\Delta G(x,y,T) \approx y[h(y)-kT\ln(x/y)]$$

It is the effective driving force for segregation at finite temperatures and is a function of both the bulk and surface concentrations of the impurity. The calculation of ΔG enables the equilibrium surface concentrations ($\partial \Delta G/\partial y=0$) to be determined at a given temperature. We note that this cannot be obtained from the usual Langmuir isotherm due to the complex coverage dependence of Δh. Figure 3 shows ΔG as a function of surface coverage, y, for 220 ppm of calcium in MgO, the concentration used by McCune and Wynblatt in their recent study [13]. At 0°K, ΔG decreases to a minimum value ($\partial \Delta G/\partial y=0$) at approximately 5 layers of impurity, so that in principle this would be the equilibrium coverage of the (001) surface. As the temperature is increased, the coverage (number of layers) at which ΔG is a minimum decreases. Thus at roughly 600°K the maximum equilibrium coverage is one and a quarter (surface) monolayers with further (metastable) equilibria at one quarter and three quarters surface coverage. At approximately 800°K the most stable equilibrium is now at one quarter coverage and at this temperature kT is sufficiently high to de-populate the metastable equilibria at three-quarters and one and a quarter monolayer coverage.

At still higher temperatures the equilbrium coverage falls below one quarter of a monolayer. In the range 1200-1600°K, for example, calculations predict a reduction in coverage from 20% to 10%. McCune and Wynblatt have reported coverages of about 20% of a monolayer of calcium at the (001) surface of MgO at around 1200°K [13] which seems to be in good agreement with the calculated value. In the case of strontium and barium segregation we obtain plots which are broadly similar to that shown in Figure 3; however, the temperatures at which particular surface coverages are stable are much higher. Thus for bulk concentrations of 100 ppm, we find equilibrium coverages of approximately a quarter of a monolayer of strontium up to ~2000°K and a similar coverage of barium stable to beyond the melting point of MgO (~3080°K).

SURFACE ENERGIES

The surface energy of a material is normally thought of as an inherently positive quantity: it represents the energy required to create surface from bulk material. Impurity segregation lowers the surface free energy. For barium and calcium in MgO the calculated heat, $\Delta h(y)$, and total free energy, $\Delta G(x,y,T)$, respectively have been referred to in previous sections and estimates given of their magnitudes. This raises an interesting question: could the lowering in free energy associated with impurity segregation be sufficiently large to reduce the surface energy to a negative value? For if it could reduce the surface energy to a negative value or even reduce it substantially, this would have important implications for the (high temperature) properties of a material. Figure 4 shows the calculated reduction in the surface energy of the (001) surface of MgO, $\delta\gamma_{001}$, as a function of monolayer coverage of calcium, strontium and barium at low temperatures. Also shown as a dotted line is the surface energy of the pure surface, γ_{001} (= 1.06Jm^{-2}): thus for values of $-\delta\gamma_{001}$ greater than γ_{001}, the energy of the segregated surface, $\gamma_{001} + \partial\gamma_{001}$, is negative. In the case of calcium this is not quite achieved, although the energy for full monolayer coverage is only 0.2Jm^{-2}. For strontium and barium on the other hand, calculations predict negative surface energies for monolayer coverages greater than ~70% and ~10% respectively.

Up to what temperatures then might we expect negative surface energies to occur? Calculations suggest that barium coverage of a quarter of a monolayer will lead to negative (001) surface energies up to ~1000°K. At 2000°K it increases to 0.31 Jm^{-2} and at the melting point of MgO (3080°K) to 0.46 Jm^{2}, which is less than half the value for the pure surface. For strontium coverages of a full monolayer negative surface energies are predicted to occur up to ~700°K. At 1000°K it increases to 0.41 Jm^{-2} and at 1500°K to 1.04 Jm^{-2} which is very close to the pure surface energy

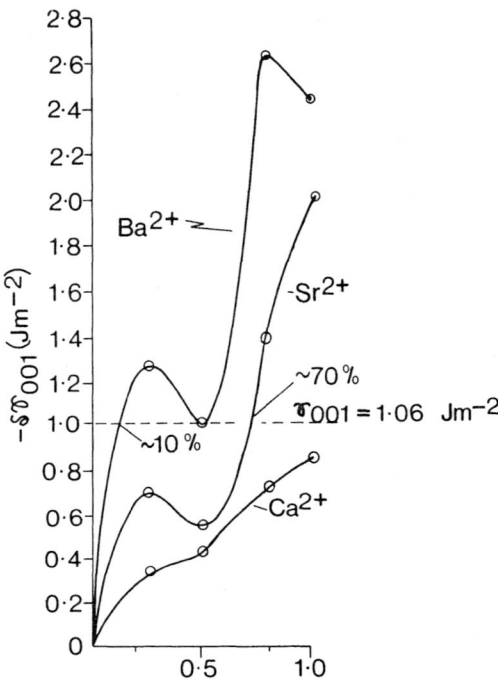

Figure 4 The reduction in surface energy, $\delta\gamma_{001}$ as a function of surface coverage due to the heat of segregation of calcium, strontium and barium at the (001) surface of MgO

(1.06 Jm^{-2}). However, it is important to note that we have not included the vibrational entropy contribution in our calculations. It has been suggested [15] that the vibrational entropy will increase on segregation, in which case the temperatures up to which negative surfaces are predicted to occur will also increase.

Now negative surface energies imply a driving force for the spontaneous creation of surface; but this seems unlikely on kinetic grounds. However, an even more important implication is that negative surface energies represent a (thermodynamic) barrier to sintering, which is the destruction of surface. Although the limiting temperatures for negative surface energies given above are well below the temperature at which sintering might normally be expected to occur (roughly two thirds the melting point) the large reduction in surface energy we predict in the case of barium at 2000 $^{\circ}$K should have a substantial effect on the sintering of MgO at this temperature.

REFERENCES

1. W. C. Mackrodt, Defect Calculations for Ionic Materials, in "Computer Simulation of Solids", edited by C. R. A. Catlow and W. C. Mackrodt, Springer-Verlag, Berlin-Heidelberg-New York (1982).
2. A. M. Stoneham, J. Am. Ceram. Soc. 64:54 (1981).
3. P. W. Tasker, The Surfaces of Magnesia and Alumina, Ceramic Advances, Vol. 12, Edited by W. D. Kingery, The American Ceramic Society, Inc., Columbus, Ohio (1984).
4. C. R. A. Catlow and W. C. Mackrodt, Theory of Simulation Methods for Lattice and Defect Energy Calculations in Crystals, in "Computer Simulation of Solids", edited by C. R. A. Catlow and W. C. Mackrodt, Springer-Verlag, Berlin-Heidelberg-New York (1982).
5. P. W. Tasker, Solid St. Ionics, 8:233 (1983).
6. E. A. Colbourn and W. C. Mackrodt, Surf. Sci., in press.
7. C. Pisani and R. Dovesi - Private Communication
8. W. C. Mackrodt and R. F. Stewart, J. Phys. C12:431 (1979).
9. B. G. Dick and A. W. Overhauser, Phys. Rev. 112:90 (1958).
10. E. A. Colbourn, W. C. Mackrodt and P. W. Tasker, J. Mater. Sci. 18:1917 (1983).
11. P. W. Tasker, E. A. Colbourn and W. C. Mackrodt, J. Amer. Ceram. Soc., submitted for publication.
12. W. C. Mackrodt, P. W. Tasker and E. A. Colbourn, Surf. Sci., in press.

13. R. C. McCune and P. Wynblatt, <u>J. Amer. Ceram. Soc</u>. 66:111 (1983).

14. M. J. L. Sangster, D. K. Rowell and A. M. Stoneham, Calculation of Entropies and of Absolute Diffusion Rates in Oxides, <u>Ceramic Advances</u>, Vol. 12, Edited by W. D. Kingery, The American Ceramic Society Inc., Columbus, Ohio (1984).

15. J. M. Blakeley and H. V. Thapliyal, Structure and Phase Transitions of Segregated Surface Layers, <u>Interfacial Segregation</u>, Edited by W. C. Johnson and J. M. Blakeley, American Society for Metals, Ohio (1979).

HIGH TEMPERATURE CORROSION

SIMULTANEOUS EROSION AND OXIDATION OF NICKEL AT HIGH TEMPERATURES

C. T. Kang, F. S. Pettit and N. Birks

Metallurgical and Materials Engineering Department
University of Pittsburgh
Pittsburgh, PA   15261

ABSTRACT

Structural alloys and coating alloys have been developed that successfully resist high temperature oxidation and, to a lesser extent, hot corrosion ⁻ depending on conditions. Such protection depends upon the maintenance of a surface layer of oxide that effectively separates the metal from the reactive environment.

The oxide protective coatings may however be damaged by erosion by solid particles in the gas flow, the effect being worse when the velocity of the gas flow is high. Unfortunately the detailed mechanisms by which the processes of erosion and oxidation interact at high temperatures are not understood. Much of the work carried out so far has involved studies of the erosion of metals and oxidized metals at room temperature under controlled conditions, or has involved tests at high temperature under conditions designed to simulate those in gas turbines or coal combustors. The data obtained from such tests are well suited to producing reliable comparisons of the behavior of different materials but do not readily lead to analysis of the mechanisms by which erosion and corrosion interact at high temperature.

This paper is concerned with the erosion-oxidation of nickel. An apparatus is described in which specimens can be exposed to well defined and controlled conditions. Results are then presented on the erosion-oxidation of nickel and models are developed to describe the observed interaction between these two processes.

411

INTRODUCTION

The high temperature corrosion, or oxidation, of metals and alloys is quite well understood, at least to the extent that interactions between multicomponent alloys and complex gas systems, with and without salt deposition, can be explained in terms of quite detailed mechanisms in a qualitative manner. A general feature of this type of reaction is that the reaction product forms on the surface of the metal and provides some degree of protection from further attack. Thus, where the reaction rate is controlled by the diffusion of species through the scale, the reaction rate is expected to fall as the scale thickness increases, in accordance with the well known parabolic rate law.

The erosion of materials has also been studied, mainly at room temperature and usually in the absence of surface layers - renewable or otherwise. From the work already carried out on erosion, several mechanisms by which material is removed from the surface have been proposed. These have been reviewed by Adler[1]. The mechanisms so far proposed include micromachining[2], plastic deformation[3], ploughing[4], fatigue by repeated impact [5][6], subcutaneous cracking[7] and delamination [8][9]. The predominant mechanism changes according to the experimental parameters such as angle of impingement, degree of ductility shown by the material, the relative hardness of the eroding particle, etc.

When a metal is exposed to a rapidly flowing, particle laden, oxidizing atmosphere, there is a tendency to form a protective oxide layer and, simultaneously, for the incident erosive particles to remove the oxide layer and, perhaps, begin to remove the metal also. The effects of erosion on high temperature oxidation of alloys are not well understood or even characterized. Investigations that have been carried out have largely simulated conditions in gas turbines[10] or coal combustors[11]. Such studies allow the responses of different materials to be compared, but the lack of control and definition of the appropriate experimental variables involved does not allow the observed interactions to be understood in terms of fundamental mechanisms. Laboratory work into the simultaneous erosion and oxidation of metals at high temperature under well defined conditions has been held back by the associated experimental problems. This has restricted the data obtained to a limited range of variables or to that pertaining to the room temperature erosion of specimens oxidized at high temperature.

Such work has, however, indicated that significant interaction occurs between erosion and oxidation occuring simultaneously at high temperature. The present work was undertaken to produce an apparatus to study this interaction under well defined and controlled conditions in order to improve our understanding of the mechanisms by which the interactions occur.

EXPERIMENTAL

The design aims of the apparatus were to supply a stream of
gas at speeds up to 300 ms^{-1} and temperatures up to 900°C. The
gas stream was also to contain a controlled loading of alumina
particles moving at the same speed and at the same temperature as
the gas. A nozzle diameter of about 10mm was chosen in order to
allow reasonably large specimens, 10mm X 15mm X 2mm, to be used in
experiments. This size is well suited to subsequent handling and
examination.

A diagram of the apparatus is shown in Figure 1. The gas used,
compressed air, is cleaned and dried to remove traces of oil, water
and dust. The air then flows through a flowmeter and fluid heater
which preheats it to about 700°C. At this point the alumina
particles are introduced to the gas stream. The particles used are
20µm alumina particles which were chosen since calculations showed
that they would not be deflected by the gas streamlines around the
specimen, also alumina is relatively inert and its properties are
well characterized.

A second flow of compressed air is taken to a Sylco Mark IX
dispenser to produce a particle laden air stream which is passed
into a simple gas flow divider, one tenth being introduced to the
heated air flow to form the erosive stream while the remaining nine
tenths are collected in a filter bag.

The erosive gas stream, already at 700°C, is passed down a
vertical, heated, Inconel tube, 1.5 meters long, to be accelerated
and heated further. Gas and particles emerge at the 9.3mm diameter
nozzle at speeds up to 300ms^{-1} and at temperatures up to 900°C.
The Inconel tube is aluminized to avoid spallation of oxide from
the inner tube surface into the erosive gas stream.

The specimens are held in a holder under spring compression
to allow for differential changes in length due to thermal expansion on heating. The specimen may be held at any angle to the
erosive flow and is heated by contact with this gas stream. In
order to eliminate temperature gradients, the specimen is also
heated by a focussed radiant heater from the back side. This is
also shown schematically in Figure 1 together with the position of
a laser doppler velocimeter used to measure the speed of the
erosive particles.

The laser velocimeter is mounted on a table whose position may
be adjusted accurately in both horizontal directions. This allows
the velocity of erosive particles to be measured at positions over
the entire cross section of the gas stream. An individual analysis
measures the speeds of particles passing through a volume of 0.1mm^{3}
where the twin laser beams cross. The output of data from the

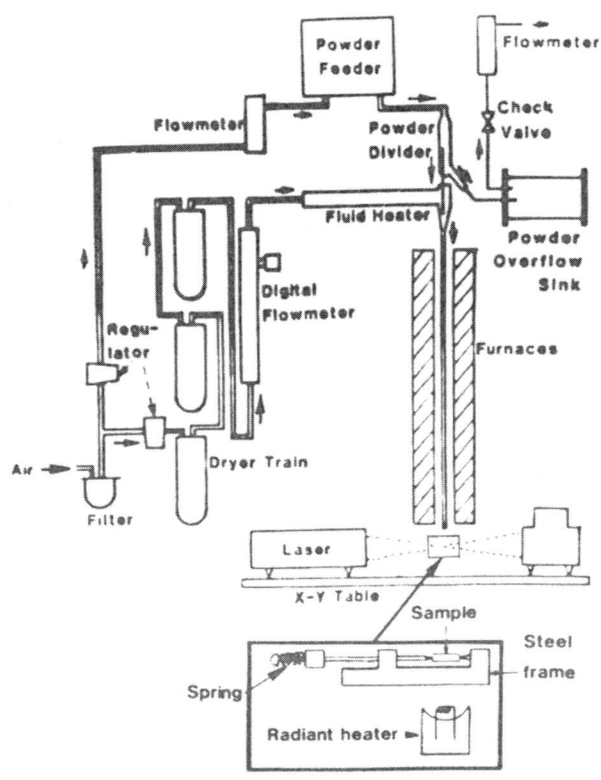

Figure 1. Schematic diagram of apparatus for the simultaneous erosion and oxidation of metals.

velocimeter is fed into an Apple II computer and, in Figure 2, the particle velocities are shown in the form of a frequency distribution which is not symmetrical about the mean velocity. For this gas velocity the peak velocity obtained from the distribution was 85m/s. Excellent agreement has been obtained between measured particle velocities and calculated gas stream velocities, which confirms both the accuracy of the velocimeter and also that most particles reach gas velocity before leaving the Inconel acceleration tube. The velocimeter integrates well with the apparatus and may be used in situ without disturbing the rest of the apparatus to measure velocities of both hot or cold gas streams; so far velocities up to 240 ms^{-1} have been used regularly.

Using this apparatus, specimens can be exposed to conjoint erosion-oxidation over a wide range of well controlled, well defined, particle velocities, temperatures and angles of incidence to the specimen. The gas composition can also be varied and deposits to simulate hot corrosion could be made on the specimen surface.

The specimens are examined after exposure using optical

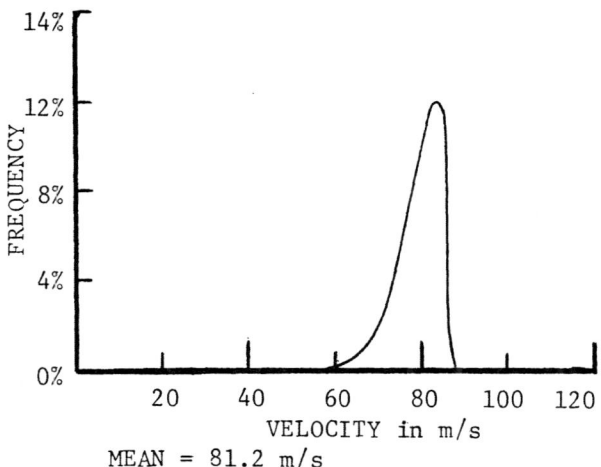

MEAN = 81.2 m/s

Figure 2.   Frequency distribution of particle velocities in the
center of the erosive gas stream. The calculated gas
stream flow rate of 70 ms^{-1} is an average across the
nozzle aperture in the center of the gas stream, the
velocity is expected to be slightly higher in the center
than at the sides, thus the peak velocity here is
about 85 ms^{-1} which is taken to be good agreement between
measurement and calculation.

metallography and scanning electron microscopy with EDX analysis.
Prior to exposure all specimens were polished with 1μm Al$_2$O$_3$
and cleaned by rinsing in acetone.

RESULTS AND DISCUSSION

Specimens of nickel, Ni-270, were exposed in air to different
conditions involving a variety of particle speeds, temperatures
and angles of particle impingement. After exposure, the specimens
were examined using the SEM and an electrolytic scale stripping
procedure to allow the metal and scale surfaces at the metal-scale
interface to be examined. In this paper results will be presented
for temperatures of 800°C, 650°C and 25°C and an impact angle of
90°.

Surface Morphological Features of Nickel

In the case of polished specimens, a thin (1μm) scale of NiO
was observed to form upon the specimens' surfaces under erosion-
oxidation conditions with 90° incidence of the erosion stream. In
addition, this oxide covered surface developed a surface topography
composed of hills and valleys, or moguls, Figure 3a. The moguls
became aligned into ripples at acute angles of incidence such as
45°, Figure 4a and 4b. The oxide scale was composed of nickel
oxide with embedded alumina particles, Figures 3b and 3c. There

was clear evidence of cutting and plastic deformation of the oxide, Figure 4c, and in many cases it appeared that plastic deformation of the metal had also taken place, Figure 5.

Plastic deformation of the metal below the scale is believed to be responsible for mogul formation since this was not evident on specimens with thick scales of NiO formed during preoxidation treatments. The particle impacts are also thought to cause extrusion of both metal and oxide when the oxide scale is thin, thus exposing a greater surface area of both to the gas atmosphere and consequently increasing the oxidation rate.

To attempt to account for mogul formation, via plastic deformation of the substrate, a model, using computer simulation, was constructed based on the formation of a well, or depression, having a raised rim with no material removal at each impact. The computer was used to sum the effects of many such impacts occurring at random over a surface. The resulting surface profile, shown in Figure 6, is very similar to the mogul pattern observed on nickel specimens having thin NiO scales, Figure 3a. Such results confirm the role of plastic deformation in mogul formation by causing

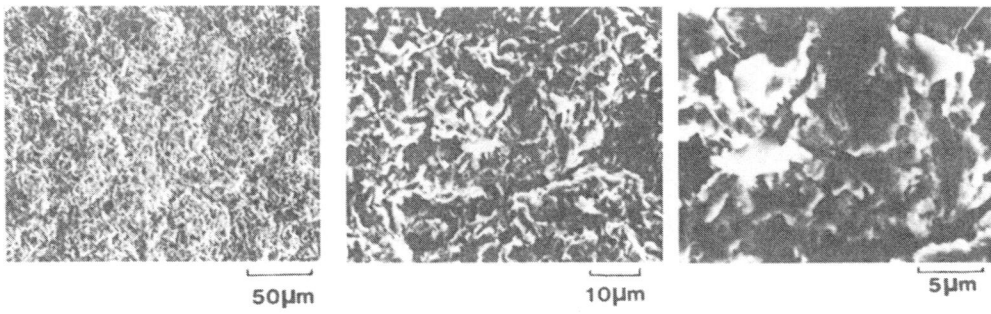

| 50μm | 10μm | 5μm |

(a) Oxide Surface   (b) Oxide Surface   (c)   as (b)
                         with Embedded Particles

Figure 3. Nickel exposed for 8 min. at 810°C at 90° impact angle to 190 ms^{-1} airflow containing 1500 ppm Al_2O_3.

surface rearrangement and also indicate that moguls may form in the absence of material removal. Nevertheless, removal of metal and oxide do occur in the erosion-oxidation process as will be shown subsequently.

The removal of material from specimens during erosion-oxidation is believed to occur by several different mechanisms. Material removal by adhesion to alumina erosion particles has been confirmed by examination of such particles after impact when NiO was seen adhering to the surfaces, Figure 7. Calculations show that about 10% of the material removal might occur by this mechanism. Another perhaps more important, process involves removal of material by plastic deformation of the metal and the thin oxide scale. Typical features are shown in Figures 3c, 4c, and 5b. Repeated deformation of the composite surface (i.e. thin oxide on metal) results in the development of platelets and extrusions of oxide covered metal. Even when a given particle impact fractures the oxide, the exposed

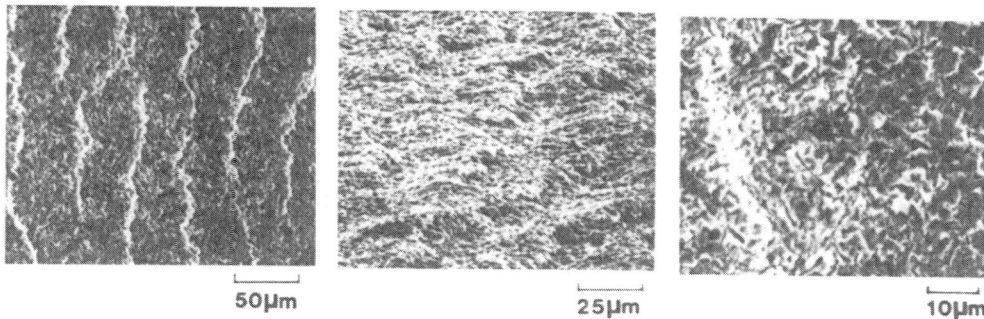

(a) Oxide surface showing ripples

(b) Oblique view of surface in (a)

(c) Surface showing ripple, erosive cuts and embedded Al_2O_3.

Figure 4. Nickel exposed for 30 min. at $730^{\circ}C$ at 45° impact angle to 120 ms^{-1} airflow containing 2400 ppm Al_2O_3.

surface soon becomes oxide covered. Eventually small particles of metal, covered with thin oxide, become detached from the metal surface and excessive deformation results in fracture. It is important to emphasize the composite nature of the surface and that the response of the surface is dependent upon the thickness of the oxide scale present upon the metal.

A typical layer of oxide that formed upon a polished nickel specimen during erosion-oxidation is shown in Figure 8a. The scale was composed of nickel oxide, some extruded nickel, and some embedded alumina particles, Figure 8b. It is worth noting that in all of the experiments performed on the erosion-oxidation of nickel, the erosion rate appeared to be such that the scale approached some constant thickness. For example, when specimens were preoxidized and tested in erosion-oxidation, the scale thickness gradually decreased and approached values close to those observed on polished specimens subjected to no preoxidation treatment.

When experiments were performed at room temperatures or at elevated temperatures in nitrogen, no oxide scale was present except for a small amount of oxidation that occurred as the hot specimen was removed from the nitrogen gas stream. Typical surface morphological features are shown in Figure 9 and 10. Under both conditions substantial amounts of plastic deformation are evident. Most of these morphological features are similar to

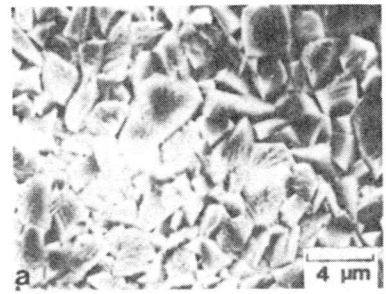 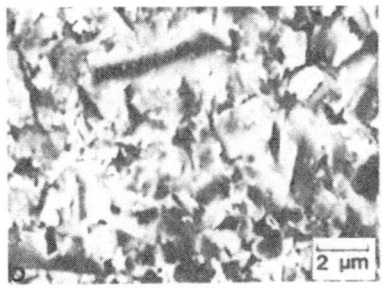

Figure 5. Appearance of scale before corrosion, (a), and (b) after to 74 ms^{-1} for 1 min. showing features consistent with plastic deformation of surface (at temperature of 20°C).

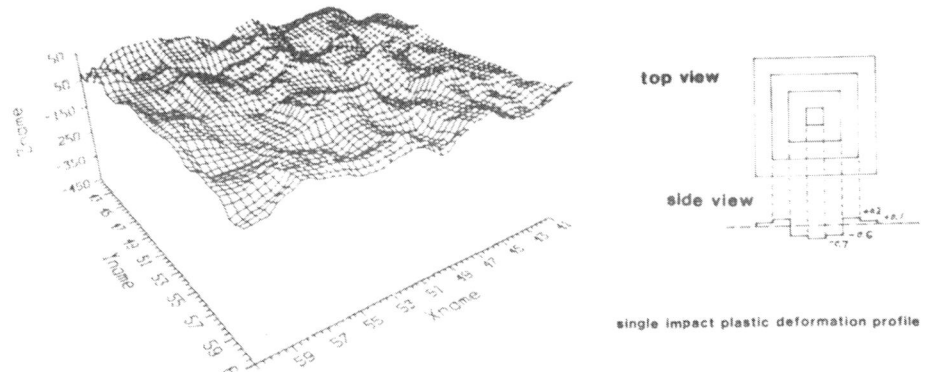

single impact plastic deformation profile

Figure 6. Computer simulation of surface morphology showing mogul formation after 10^{6} random impacts each causing a plastic deformation profile as shown, assuming no material removal occurs.

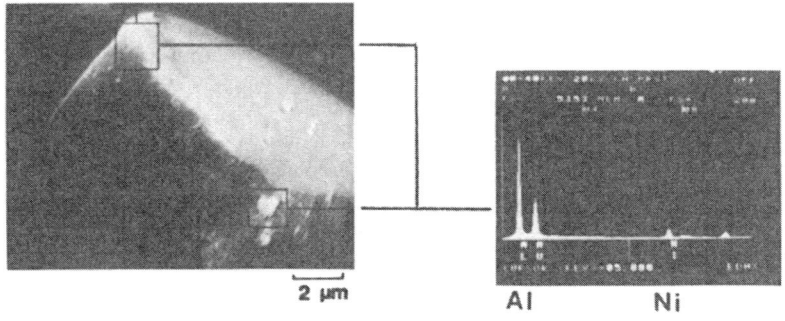

Figure 7. Scanning electron micrograph showing attachment of nickel or nickel oxide to a bombarding Al$_2$O$_3$ particle.

418

those of the oxide covered surfaces except that more deformation
appears to have occurred on the specimens having neglible amounts
of oxidation. It also was observed that more Al_2O_3 particles were
embedded in the sample eroded at 800°C in nitrogen, Figure 9.

Surface Morphological Features of Nickel Oxide

The erosion of nickel oxide was studied by using specimens of
nickel that were covered with thick (i.e. 120µm) layers of NiO
formed by oxidation at 1100°C in air for 72 hrs. Photographs showing the surfaces of typical specimens eroded at 800°C and at room
temperature are presented in Figures 11 and 12 respectively. At
temperatures of about 800°C, features indicative of a ductile response are evident Figure 11b. Nevertheless, some cracking was detected, Figure 11c. At room temperature some features indicating
ductile behavior are evident, Figures 12a and 12b, but cracking
and chipping have also occurred, Figure 12c, suggesting more brittle
behavior.

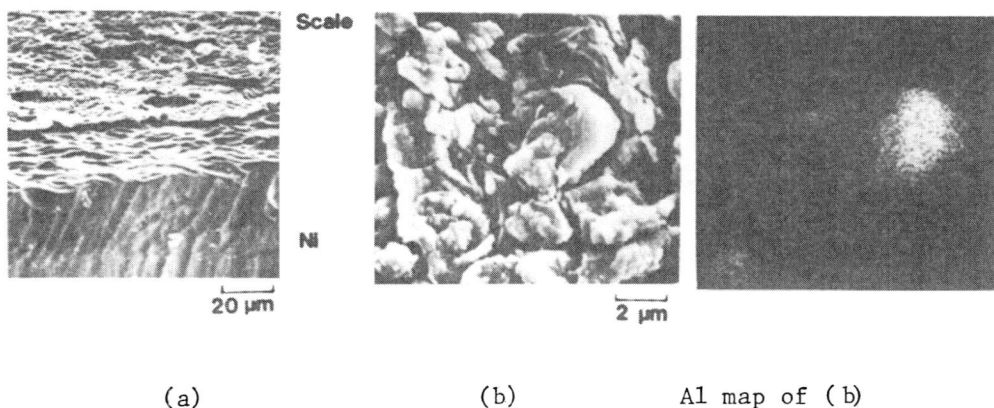

(a)   (b)   Al map of (b)

Figure 8. Nickel exposed for 30 min. at 710°C at 18° impact angle
to 190 ms^{-1} airflow. (a) Removal of scale from metal
facilitated by electropolishing in H_2SO_4-H_2O bath at
1.5 volts for 20 minutes. (b) View of underside of
stripped scale showing embedded Al_2O_3 particles.

Weight Change Measurements of Specimens During Erosion-Oxidation

Weight-change data as a function of time for polished and for
preoxidized samples are presented in Figures 13 and 14, respectively.
These data conform to a linear rate law which is usually preceded
by a nonlinear period. These data are summarized in Table I in
terms of surface recession rates. The data presented in Table I
permits the following statements to be made about the erosion-
oxidation of nickel.

.At 800°C the surface recession rate decreases with particle velocity.

.At a constant particle velocity the surface recession rate decreases as temperature is decreased.*

.At 800°C the surface recession rate decreases dramatically when oxygen is not present.

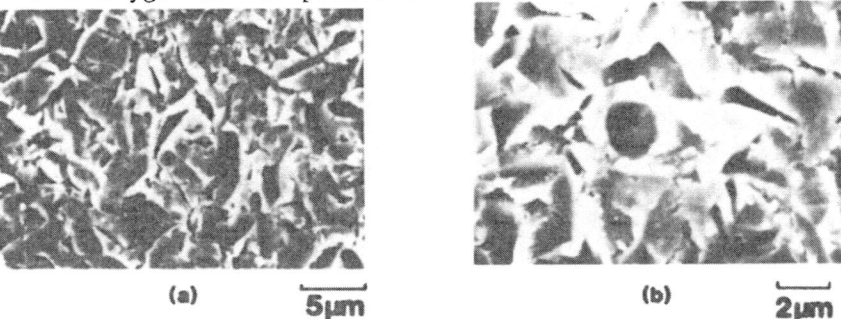

(a) 5μm   (b) 2μm

Figure 9.   Nickel metal eroded at room temperature using 20μm alumina at 73ms $^{-1}$ at 90° for 1 minute.

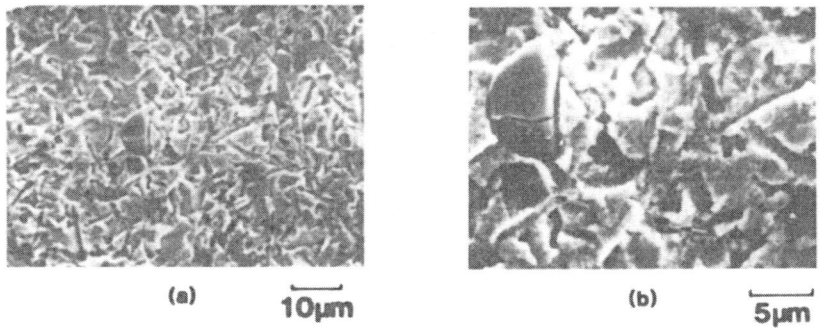

(a) 10μm   (b) 5μm

Figure 10.   Nickel eroded in pure nitrogen flowing at 140 ms $^{-1}$ at 90° at 800° for 1 minute.

The linear material removal rates observed for the specimens undergoing combined oxidation and erosion along with the significant dependence upon oxygen indicate that these two processes are affecting each other. The generation of a thin scale on the nickel specimens subjected to erosion-oxidation conditions is regarded as reflecting the relative rates at which erosion of oxide by the alumina particles and formation of the oxide can occur. It also reflects the kinetic laws followed by the two processes. For instance, if both oxidation and erosion occurred at constant rates,

---

*At a temperature of 850°C and a particle velocity of 125 ms $^{-1}$ the surface recession rate was less than that for 800°C. It is believed at temperatures above 800°C more of the erosive particles are retained in the metal-oxide zone which inhibits the material removal process.

the rate of scale formation could be written, in terms of scale thickness X,

$$\frac{dX}{dt} = k_o - k_e \qquad (1)$$

where k_o and k_e are the rates at which oxidation of nickel and erosion of nickel oxide occur, respectively. Equation (1) would not result in an oxide of constant thickness but rather a scale of either increasing or decreasing thickness.

If the oxide scale is assumed to be protective, the scale thickness will change according to,

$$\frac{dX}{dt} = \frac{k}{X} - k_e \qquad (2)$$

where k is the parabolic rate constant for the oxidation of nickel defined by $X^2 = 2kt$

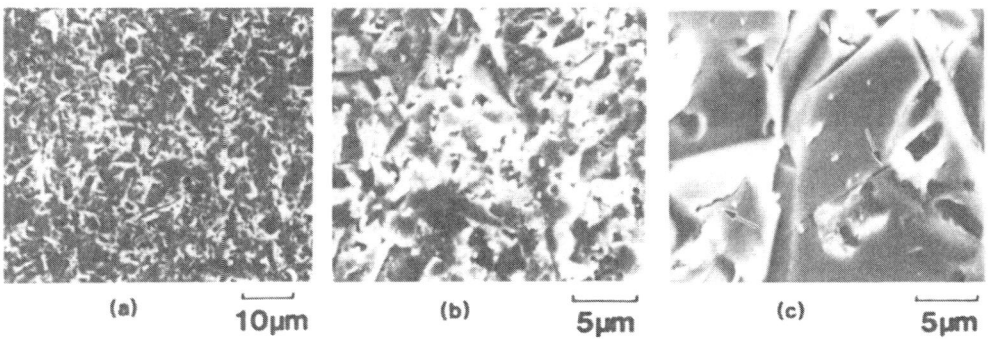

(a) 10μm   (b) 5μm   (c) 5μm

Figure 11. NiO exposed at 800° to 140 ms $^{-1}$ in air at 90° showing plastic deformation in (a) and (b) and cracking or tearing in (c).

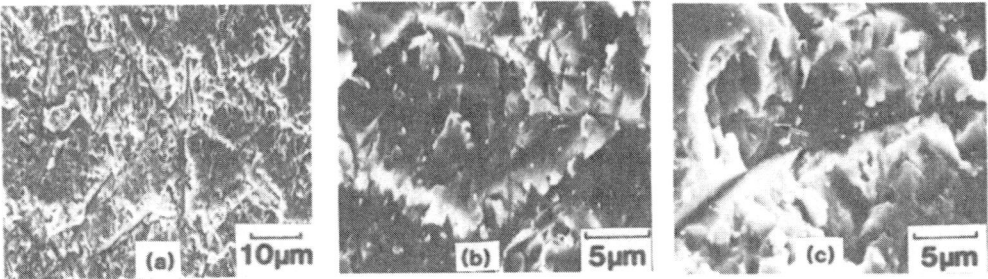

(a) 10μm   (b) 5μm   (c) 5μm

Figure 12. Appearance of NiO after erosion at room temperature in air to 73 ms $^{-1}$ for 30 seconds at 90° showing evidence of plastic deformation in (a) and (b) and cracking or tearing (c).

The interplay between scale growth and erosion processes in this case would lead to the establishment of a steady state where the scale thickness approaches a constant value X* when dX/dt = 0 and hence,

$$X^* = k/k_e \qquad (3)$$

Although the scale thickness is constant, metal removal has not ceased, on the contrary it is being removed at a constant rate given by,

$$\frac{dx}{dt} = \frac{V_M}{V_{MO}} \frac{k}{X^*} = \frac{V_M}{V_{MO}} k_e \qquad (4)$$

where x is the metal surface recession and V_M and V_{MO} are molar volumes of metal and oxide, respectively. While in the case of nickel a constant oxide thickness appears to be formed, the assumption that the scale is continuous and protective does not seem to be justified since it is apparent that the scale is continually being penetrated and made nonprotective, at least in localized areas by the impacting alumina particles. It therefore seems reasonable to suppose that the nickel surface being eroded is a composite consisting of nickel oxide, nickel and embedded alumina particles. The rate of material removal can then be viewed as following the expression,

$$\frac{dx}{dt} = k_e' \qquad (5)$$

where the magnitude of k_e' is determined by the properties of the composite layer as influenced by the particle velocity, impact angle, particle loading, particle size and temperature.

The data presented in Figure 13 and Table I are consistent with such a proposal since,

$$k_e' = \frac{dx}{dt} = \frac{1}{\rho_{Ni}} \frac{d(\Delta M/A)Ni}{dt} \qquad (6)$$

where ρ_{Ni} is the density of nickel and $(\Delta M/A)Ni$ is the weight change per unit area of the nickel specimens. The observed decrease of k_e' with particle velocity at $800^\circ C$ is expected since the higher velocity should produce more damage to the oxide layer and also produce more deformation of the composite layer as observed with metals[1] at room temperature. In air at a fixed particle velocity k_e' decreases as temperature is decreased since the amount of oxidation decreases and, as will be discussed subsequently, oxidation is believed to affect the amount of deformation that the composite layer can withstand prior to fracture. The large reduction in erosion rate on going from air to nitrogen at $800^\circ C$ is also

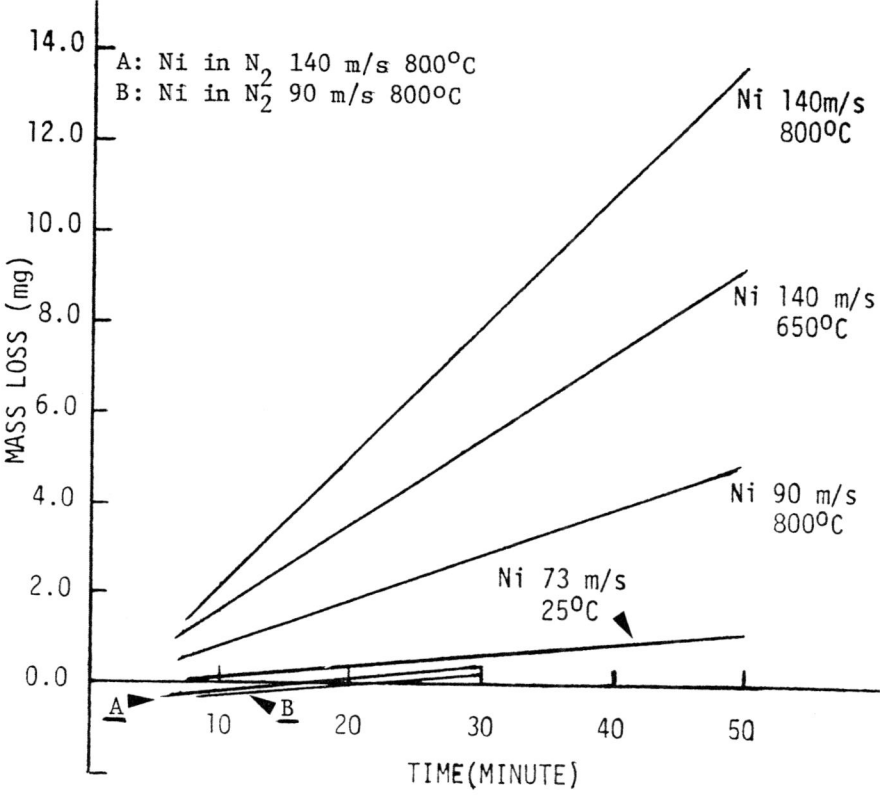

Figure 13.   Mass loss versus time for polished nickel specimens
exposed at 90° impact angle.

reasonable due to the absence of oxidation in nitrogen. Finally,
the observed lower erosion rate at 800°C in nitrogen compared to
25°C in air may be due to work hardening becoming important at 25°C.

TABLE I

SURFACE RECESSION RATES FOR NICKEL AND NiO (cm/s)

EXPERIMENTAL CONDITIONS	NICKEL cm/s X 10[7]	NICKEL OXIDE cm/s X 10[7]
800°C, 140 m/s, air	6.91	8.4
800°C, 90 m/s, air	2.30	3.5
800°C, 140 m/s, N_2	0.71	
800°C, 90 m/s, N_2	0.35	
650°C, 140 m/s, air	4.50	6.2
25°C, 73 m/s, air	0.71	8.4

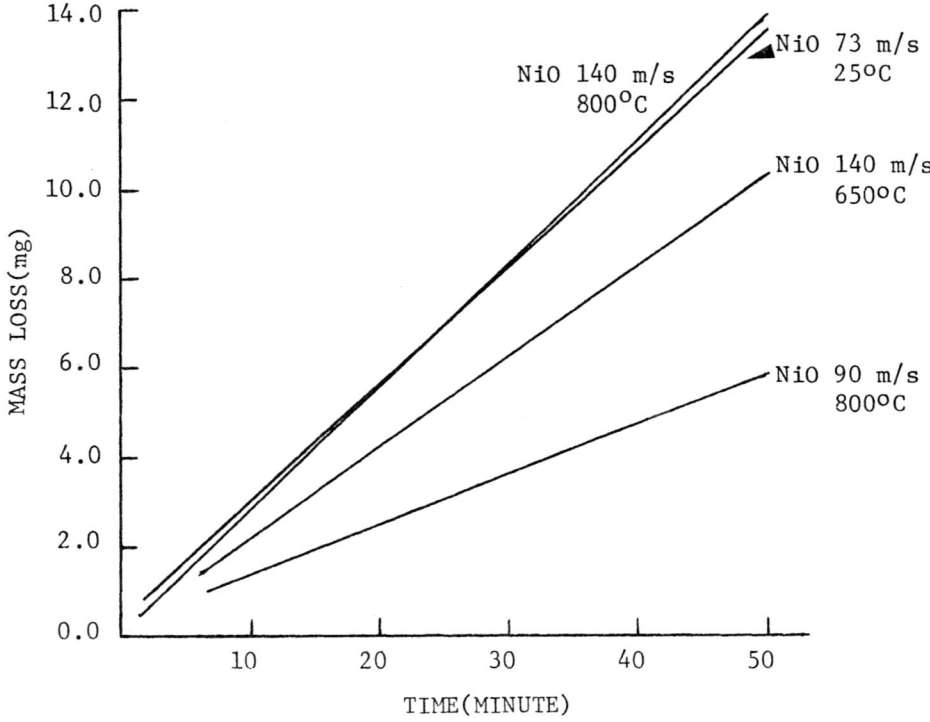

Figure 14. Mass loss versus time for preoxidized nickel specimens
exposed at 90° impact angle.

The data presented in Figure 14 and Table I show that the
erosion rate of nickel oxide decreases with particle velocity at
800°C. This is reasonable since the amount of deformation and
cracking can be expected to decrease with velocity. The dependence
of the erosion of nickel oxide on temperature is not at present understood. The surface recession rates at 800°C and 25°C are about
the same, however, the lower velocity used in the experiment at
25°C suggests a higher erosion rate at 25°C for the same velocity.
This difference could be accounted for by proposing that the erosion
of nickel oxide occurs by two modes, namely, by cracking (brittle
mode) and through repetitive deformation (ductile mode) with more of
the former at room temperature. More recent results obtained at
650°C, however, give an erosion rate that is lower than at 800° and
25°C. This puzzling result is currently being checked. If sub-
stantiated, it may be evidence for a change in fracture mechanism
of the oxide at a temperature below 650°C with the ductile mode
decreasing as temperature is decreased. Finally, it is worth noting
that the erosion of nickel oxide, in terms of surface recession, is
always greater than that for nickel for similar test conditions.
Such results are plausible, since regardless of the mode of erosion

that predominates in nickel oxide, the properties of nickel oxide compared to nickel can be such that higher recession rates occur. For example, erosion of nickel oxide by a ductile mode may lead to the formation of thin sections of displaced oxide that are readily cut and removed by subsequent impacts. In the case of nickel, the metal displaced by plastic deformation may not be removed so readily but rather welded back to the surface by subsequent impacts. This distinction is a direct result of the higher ductility of nickel and is supported by the increased removal rate of nickel at room temperature, compared to 800°C in nitrogen, where work hardening can occur and welding is less likely.

Model for Combined Erosion-Oxidation of Nickel

A schematic model for the combined erosion-oxidation of nickel is presented in Figure 15. When polished nickel is subjected to erosion-oxidation a thin oxide layer is formed, Figure 15a. As erosion occurs, the particles deform the surface and penetrate the oxide scale. For reasons discussed previously, the deformation caused moguls to be developed which results in the particle impact angle being altered. Material begins to be removed by adhering to the ejected alumina particles, Figure 15a. As erosion continues, the exposed metal is oxidized and platelets of metal and oxide begin to extrude out over the surface. Small portions of these platelets eventually become detached from the surface due to subsequent particle impacts cutting them off or by producing deformation to the extent that fracture occurs, Figure 15a. Continuation of this process thus results in the formation of a composite layer consisting of nickel oxide, nickel and parts of embedded particles.

Moreover, this composite layer is maintained at a relatively constant thickness independent of exposure time due to the simultaneous removal of material by excessive deformation and cutting in combination with oxidation of freshly exposed metal surfaces and embedding of additional alumina particles.

In the case of erosion of nickel oxide, and thick layers of nickel oxide on nickel, Figure 15b, the erosion process may involve material removal by both ductile and brittle modes with the ductile mode being more important at 800°C, and perhaps 650°C. In the case of the ductile mode, the mechanisms is similar to that described for nickel with deformation of nickel oxide resulting in very small platelets of oxide being removed. At lower temperatures the brittle mode may become more important with extensive cracking giving rise to the formation of chips of oxide.

For the conditions examined thus far, the erosion rate has always been sufficient to reduce the thicknesses of preformed layers of oxide on nickel to values similar to those developed on polished nickel samples. Hence, for sufficiently long times, preoxidized nickel develops a composite layer virtually identical to that formed on nickel specimens with no preformed nickel oxide.

425

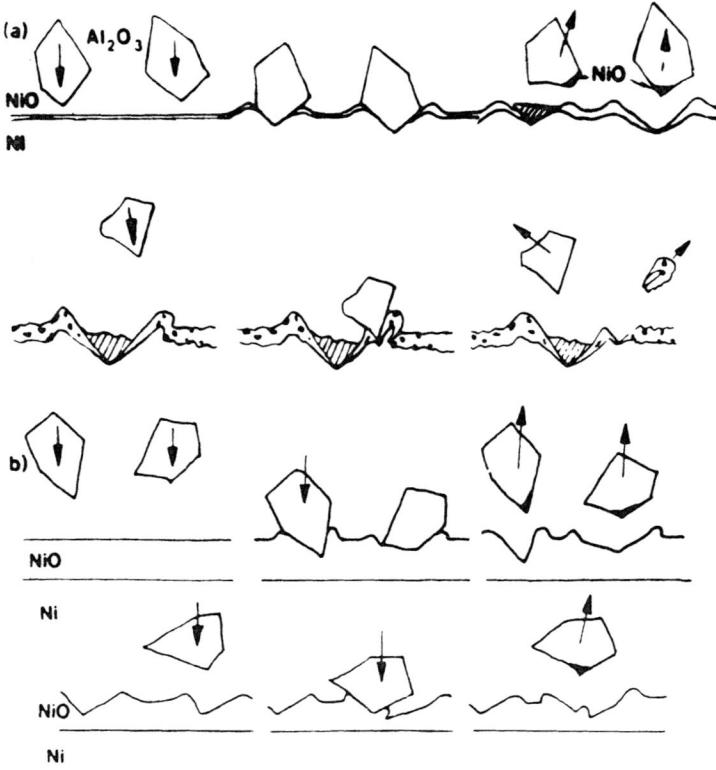

Figure 15. Steps in the erosion-oxidation mechanism of nickel.
(a) clean nickel surface, Al_2O_3 particles embed and
both metal and oxide deform plastically on impact
until fracture. (b) Nickel with thick preoxidized
scale, plastic deformation of oxide and cracking
reduce thickness until conditions of (a) develop.

CONCLUDING REMARKS

The combined erosion-oxidation of nickel has been investigated
at temperatures of 25°, 650° and $800^\circ C$ using alumina particles.
The oxidation process has been found to significantly increase the
erosion process due to the formation of a composite layer consisting of nickel oxide, deformed nickel metal, and embedded
alumina particles. The observed erosion rates have been accounted
for by considering the characteristics of the composite layers
that are formed.

Work is now in progress to investigate the erosion-oxidation
of cobalt which has an oxidation rate substantially greater than
that of nickel.

ACKNOWLEDGEMENTS

This work was supported by the Army Research Office (contract
No. DAAG 29-81-K-0027) under the direction of Dr. R. R. Reeber.
Donation of nickel Ni270 from Williams Co., Pittsburgh, Pa., and
helpful discussions with S. L. Chang are gratefully acknowledged.

REFERENCES

1. W. F. Adler, Assessment of the State of Knowledge Pertaining
to Solid Particle Erosion. Tech. Report, Effects Technology,
Inc. 1979, NTIS, Report No. ETI-CR79-680.

2. I. Finnie, Wear, 3:87 (1960).

3. J. G. A. Bitter, Wear, 6:5 (1963), also 6:169 (1963).

4. R. E. Winter and I. M. Hutchings, Wear, 29:181 (1974).

5. M. M. Mamoun, Mat. Sci. Div. Coal Tech. Third Quarterly Report
No. ANL-75-XX3. Argonne National Lab (1975). Appendix:
Analytical Models for the Erosive-Corrosive Wear Process.

6. P. S. Follansbee, Mechanisms of Erosive Wear of Ductile Metals
Due to Low Velocity, Normal Incidence, Impact of Spherical
Particles. Ph.D. Thesis (1981), Carnegie Mellon University,
Pittsburgh, PA.

7. R. Bellman Jr. and A. Levy, Erosion Mechanism in Ductile Metals.
Tech. Report. Lawrence Berkeley Laboratory, LBL-10289 (1980).

8. N. P. Suhr, Wear, 44:1 (1977).

9. S. Jahanmir, Wear, 309, (1980).

10. Effect of Erosion on Oxidation and Hot Corrosion of Coated and
Uncoated Superalloys. Performed by Pratt and Whitney Aircraft,
East Hartford, CT. for EPRI, Palo Alto, California (R.P. 979-4,
1975).

11. Effect of Alloy Variables on Resistance of a Material to Attack
at High Temperatures where both Erosion by Particulates and
Corrosion can occur. Performed by Battelle - Columbus,
Columbus, OH, for EPRI, Palo Alto, Ca. (RP 589, 1975).

ON THE PHASE TRANSFORMATION DURING REDUCTION OF

MAGNETITE AT HIGH TEMPERATURES

Jan Janowski, Stanislawa Jasieńska, Andrzej Sadowski
and Andrzej Simon

Institute of Metallurgy
University of Mining and Metallurgy
Al. Mickiewicza 30, 30-059 Kraków, Poland

INTRODUCTION

Previous studies on the reduction of iron oxides at temperatures above 1000 K suggest that the process may be particularly dependent on the phase transformations[1,2]. Results reported in papers [3,4] would indicate that the character of those transformations connected with the magnetite-wüstite stage may be determined by the deviation from stoichiometry of the developing wüstite phase.

Despite numerous papers on the subject comparison of results is hardly possible as kinetic investigations were carried out using material with different chemical compositions and physical properties. Furthermore, different methods of modelling and interpretation of results were adopted. Products of reaction usually were not examined structurally.

EXPERIMENTAL

To limit the interpretation problems mentioned above the reduction of pure monocrystalline magnetite (M) has been investigated. Investigations of sintered samples of pure (P-1) and doped (M-1) polycrystalline magnetite were then performed in analogous conditions and referred to the reduction of a monocrystalline sample. Doped magnetite (M-1) was obtained by melting a magnetite ore with total amount of impurities 1.69 pct. In monocrystalline (M) and sintered (P-1) samples only trace concentrations of impurities were found. Contents of impurities in particular materials are listed in Table 1.

Table 1. Contents of impurities in investigated samples

Sample	Concentration of impurities, pct.							
	Ni	Cu	Sr	Cr	Mo	Mn	Sn	Al
$(M)^x$	0.05	0.06	0.01	0.01	0.01	0.01	0.001	0.001
$(P-1)^x$	0.06	0.06	0.01	0.01	0.01	0.02	0.002	0.001
	Al_2O_3	K_2O	Na_2O	MgO	CaO	V_2O_5	MnO	TiO_2	SiO_2
$(M-1)$	0.26	0.01	0.01	0.30	0.03	0.30	0.05	0.30	0.43

x Mg, Ca, Sb, Pb, Ga, W, V, Na, K, Li, B, Ge and Ti were not found

Before each experiment samples were homogenized in the proper gas atmosphere to eliminate FeO and Fe_2O_3. Identification of physico-chemical and structural properties both of reactants and reaction products was made by microscopy, X-ray investigations and measurements of the Mössbauer effect as well as by chemical analysis and porosimetry. Characteristics of the samples investigated are listed in Table 2.

Investigations of the reduction of magnetite to wüstite with different deviations from stoichiometry were carried out isothermically every other 50 K in the temperature range 1123-1373 K using Mettler's thermobalance with a flow system. To ensure the assumed wüstite composition a proper reducing mixture $CO + CO_2$ was chosen according to the phase diagram $Fe - O_2$. Reduction runs were carried out in two series: (i) reduction of magnetite to wüstite with composition $Fe_{0.94}O$ (close to the wüstite-iron interface and (ii) reduction of magnetite to wüstite with composition from $Fe_{0.88}O$ to $Fe_{0.91}O$ (close to wüstite-magnetite interface).

RESULTS AND DISCUSSION

Kinetic data

The results obained were interpreted according to the dominant opinion that high temperature reduction of iron oxides proceeds under mixed control[5-7]. The model adopted was based on one of known equations taking into account chemical reaction at magnetite-

Table 2. Physico-chemical properties of magnetite used in present work

Property	Material		
	(M)	(P-1)	(M-1)
Obtained by method	Verneuille	sintering	melting
Contents of dopes, pct.	0.1	0.1	1.68
Shape of sample	mesh fraction	spherical	mesh fraction
Magnitude of grain, mm	0.70-0.75	0.12-0.15	0.75-0.80
Porosity	0.004		0.004
Specific surface area, m^2/g	0.005	0.02	0.006

wüstite interface and gaseous diffusion across the product layer[5].

Arranging equation (26) from paper[5] and neglecting the gas film diffusion term ($\alpha = \infty$) the following equation was obtained:

$$A\left[3 - 3(1-R)^{2/3} - 2R\right] + C\left[1 - (1-R)^{1/3}\right] = t \quad (1)$$

Parameters A and C were defined as follows:

$$A = \frac{r_o^2 \, R_g \, T \, m_o}{6\left(p_{CO} - p_{CO_2} K_e^{-1}\right) \bar{D}_{eff}} \quad (2)$$

$$C = \frac{r_o \, R_g \, T \, m_o}{k\left(p_{CO} - p_{CO_2} K_e^{-1}\right)} \quad (3)$$

where

$\bar{D}_{eff}$ — mean effective gaseous diffusion coefficient, cm^2/s, defined as:

$$\bar{D}_{eff}^{-1} = \frac{1}{D_{eff}^{CO}} + \frac{1}{D_{eff}^{CO_2} K_e} \quad (4)$$

k — rate constant of chemical reaction at the magnetite-wüstite interface, cm/s

R - reduction degree

r_o - initial grain radius of magnetite particle, cm

m_o - weight of oxygen removed from volume unit, mol/cm^3

T - reduction temperature, K

R_g - gas constant, cm^3 atm/mol K

P_{CO}, P_{CO_2} - partial pressure of CO and CO_2 in gas atmosphere, respectively, atm

K_e - equilibrium constant of reaction

$Fe_3O_4 + CO \quad 3FeO + CO_2$

In the papers mentioned above[5-7] it was assumed that either equation (1) was applicable in the whole range of reduction or the initial stage of the process was controlled solely by chemical reaction, while results obtained in present work showed that in the temperature range investigated reduction was so fast that the process ran initially in conditions of unstable concentration of reducing mixture and/or unstable temperature. Furthermore, processes connected with nucleation of the wüstite phase might occur in this region. From this point of view the essential problem arises how to interpret kinetic data when initial measurement points are neglected. The question may be approached in two ways: (i) one should choose such a point (R_i, t) on the kinetic curve where for values $t_i > t$ experimental disturbances are no longer significant. Several values of R_i can be tested, of course, integrating the kinetic equation in limits from $t = t_i = 0$ to t and from R_i to R. One should however remember that a layer of wüstite was already formed before the reduction degree reached value R_i. Taking into account the influence of this layer on further reduction progress necessitates modification of the kinetic equation. This problem will be a subject of a separate paper. (ii) One should assume that the equation describes the reduction run at a certain further stage of the process. In this case correction of time t_o should be substracted from experimental values of time. It is an additional parameter of the equation (1) chosen to describe in the best way the experimental data according to the least square method. Such a correction compensates to some extent errors in the initial stage of the experiment.

The second approach was applied to interpret results of reduction in the present work. The results were fitted to the equation:

$$A\left[3 - 3(1-R)^{2/3} - 2R\right] + C\left[1 - (1-R)^{1/3}\right] - t_o = t \quad (5)$$

Parameters A, C and T_o were determined by means of the least square method. An agreement between the experimental data and that calculated on the basis of the parameters described above is shown in both the natural coordinates system R vs t (Fig. 1) and one that is linearized (Fig. 2). It may be seen from both plots that

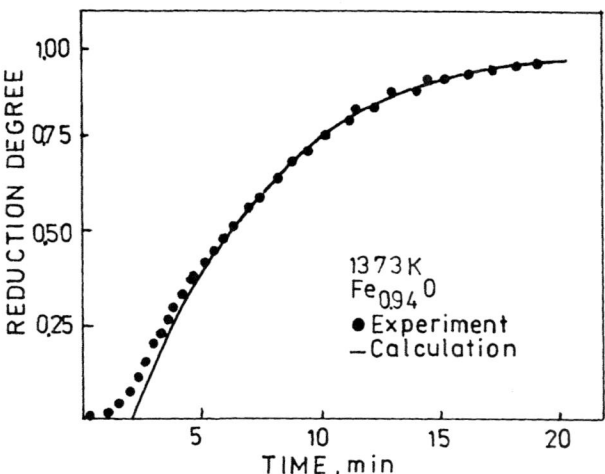

Fig. 1. Kinetic curve for reduction of monocrystalline Fe_3O_4 to $Fe_{0.94}O$ at temperature 1373 K

there is a reasonable agreement except for the beginning and end of the process. For monocrystalline samples the range of the degree of reduction where the equation (5) is valid varies with temperature from 0.20 to 0.80 at 1123 K and from 0.40 to 0.95 at 1373 K.

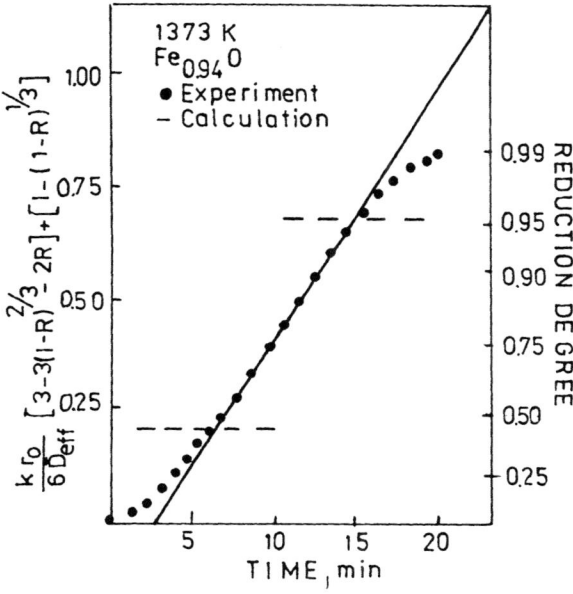

Fig. 2. Reduction progress of monocrystalline Fe_3O_4 to $Fe_{0.94}O$ at temperature 1373 K

Taking advantage of equations (2) and (3) the rate constant of chemical reaction k at the magnetite-wüstite interface was calculated as well as the effective coefficient of gaseous diffusion across product layer $\bar{D}_{eff}$. Calculated values of the rate constant k as a function of temperature are shown in Fig. 3. It was stated that the rate constant k is independent of the stoichiometric composition of wüstite formed in course of the reduction process. The determined value of the activation energy, also independent of the composition of wüstite, amounts to about 100 kJ/mol and is in good agreement with the value reported by Bessieres et al.[7] which is shown in Figure 3 for comparison. Calculated values of the effective diffusion coefficient $\bar{D}_{eff}$ seem to decrease with an increase of deviation from the stoichiometry of wüstite. It may be caused by a difference in texture or a different distribution of pores and consequently different values of the labyrinth coefficient changing with non-stoichiometry of the wüstite obtained.

The first term of the left-handed side of the equation (1) in some way represents the contribution of the gaseous diffusion in the overall reduction process the second one of the interfacial

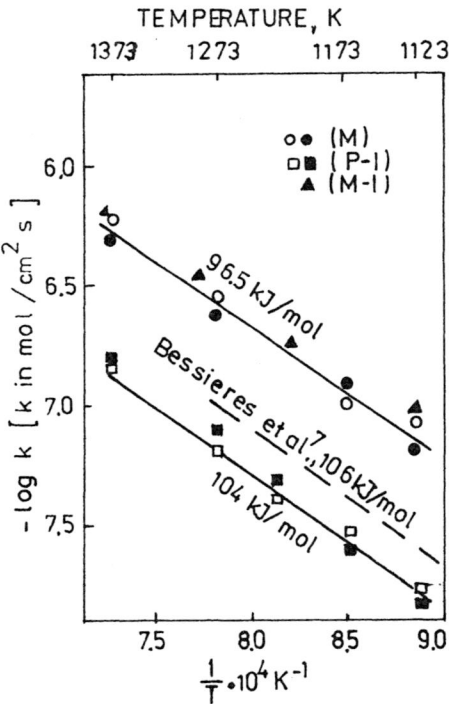

Fig. 3. Arrhenius plot of rate constant of the interfacial reaction vs. temperature. Empty points refer to $Fe_{0.88}O$ to $Fe_{0.90}O$ and dark ones to $Fe_{0.94}O$.

reaction. It was stated that participation of the diffusion (in reduction runs to the same wüstite composition) decreased with an increase of temperature and increased with an increase of deviation from stoichiometry (Fig. 4). Observed relations suggest differences in texture of wüstite samples with different stoichiometry obtained at different temperatures.

In the case of cohesive polycrystalline magnetite microscopic and porosimetric examination showed that samples (M-1) displayed cohesion comparable with that of monocrystalline magnetite. Results of reduction of magnetite (M-1) to wüstite with composition $Fe_{0.94}O$ showed that also in this case the model described above was suitable for description of the process except in the range of the reduction degree R > 0.85. Determined values of the rate constant of chemical reaction and the activation energy did not differ from those for the reduction of monocrystalline magnetite (see Fig. 3).

Analysis of the reduction of sintered polycrystalline Fe_3O_4 (P-1) samples showed that the adopted model did not produce satisfactory results. Calculated kinetic curves in some range of the reduction degree could be interpreted according to the shell-core model taking into account only the chemical reaction. However, in this way it was only possible to estimate values of the rate constant of chemical reaction and the activation energy (see Fig. 3).

Microscopic observations

Microscopic examination of the texture of samples and the morphology of wüstite grains, formed in course of high temperature

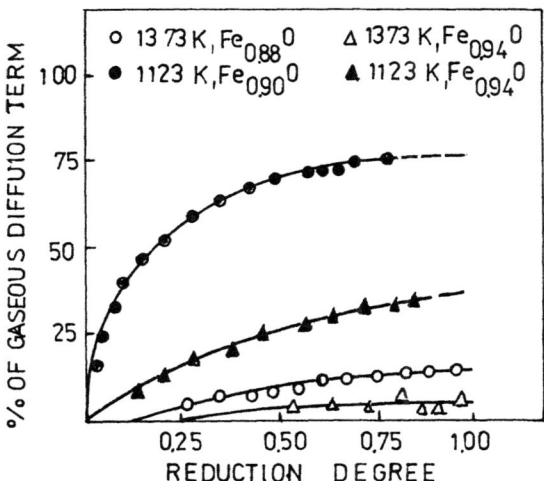

Fig. 4. Contribution of gaseous diffusion term in overall reduction process of monocrystalline Fe_3O_4.

reduction, helped to interpret correctly the above mentioned model of the reduction of magnetite to wüstite.

According to previous investigations[8] the phase transformation of magnetite to wüstite may be divided into the following steps:

(a) Formation of pores
(b) Disappearance of magnetite grains
(c) Growth of anhedral grains of wüstite
(d) Growth of prismatic crystals of wüstite.

Wüstite with composition close to the wüstite-iron interface crystallized in the shape of prisms or plates forming a cohesive texture. In the case of wüstite with a composition close to the magnetite-wüstite boundary the formation of irregularly outlined grains was observed. In contradistinction to the former case the texture was loose.

It was stated on the basis of scanning and optical microscopic investigations the the phase composition and morphology of wüstite grains formed during the reduction of magnetite monocrystalline samples were highly dependent upon both reduction temperature and reducing gas composition and consequently upon deviation of this wüstite from stoichiometry.

Differences of texture and morphology of wüstite with composition $Fe_{0.94}O$ obtained at temperatures 1373 K and 1123 K are visible in scanning micrographs (Figs. 5a and 5b, respectively). Wüstite obtained at 1373 K crystallized in grains shaped like plates and forming a cohesive texture (Fig. 5a). Relicts of magnetite phase may be seen among plates of wüstite. Such a microscopic image testifies that the cohesive texture at this temperature, composed of closely packed plates of wüstite, makes impossible immediate access of the reducing agent.

The texture of wüstite sample with the same composition obtained at 1123 K is distinctly looser (Fg. 5b) and grains with prismatic outlines are less well-shaped and of smaller size.

As the stoichiometric composition of wüstite approaches the magnetite – wüstite interface distinct changes of texture and morphology are observed compared with the ones discussed above. The temperature dependence mentioned is still observed. Wüstite $Fe_{0.88}O$ forms thin leaves with irregular shapes (Fig. 6a) at temperature 1373 K while only original formation of wüstite in form of irregular prisms on surface of a sample may be seen at temperature 1123 K (Fig. 6b). Texture of wüstite with this composition is mixed – more cohesive at surface and looser inside sample.

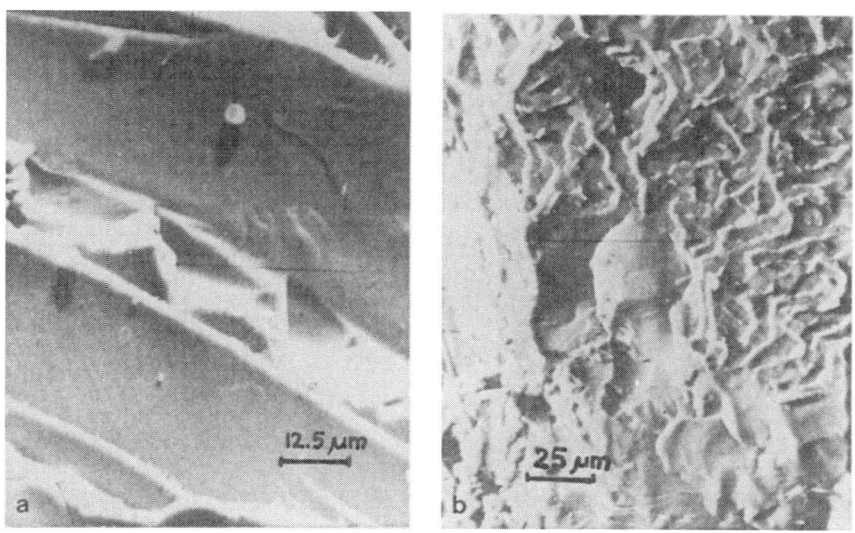

Fig. 5. Scanning micrographs of $Fe_{0.94}O$ formed during reduction of monocrystalline magnetite. Fig. 5a – at 1373 K and Fig. 5b – at 1123 K.

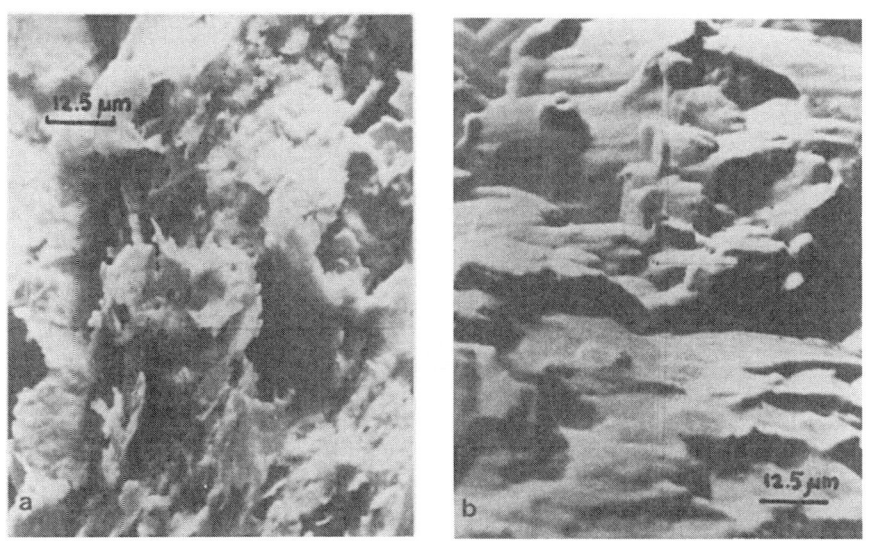

Fig. 6. Scanning micrographs of $Fe_{0.88}O$ formed during reduction of monocrystalline magnetite. Fig. 6a – at 1373 K and Fig. 6b – at 1123 K.

These observations seem to confirm the dependence discussed above (see topic Kinetic data) of the contribution of the gaseous diffusion term from equation (1) to the overall process upon both temperature and stoichiometry of wüstite.

Scanning examination of (M-1) magnetite samples after reduction confirmed the influence mentioned above of reduction temperature on the texture and morphology of grains. At temperature 1123 K wüstite is formed inside magnetite grains as aggregates of small particles (Fig. 7a). In this case the reduction begins with the formation of deep pores in magnetite crystals. The growth of wüstite phase takes place around particular pores. At temperature 1373 K crystallization proceeds in different way. Wüstite crystallizes in the form of plates situated perpendicularly to the outer walls of magnetite crystals. Initially crystals of magnetite are divided into plates. The wüstite phase then arises among them (Fig. 7b). This wüstite phase forms a cohesive layer which contains the rest of the magnetite, one which may be seen in optical micrographs (Figs. 8a and 8b).

Microscopic examination of primary samples of polycrystalline magnetite (P-1) showed the structure to be less cohesive than in the case of monocrystalline magnetite (M-1). Microscopic examination

Fig. 7. Scanning micrographs of $Fe_{0.94}O$ formed during reduction of melted magnetite (M-1). Fig. 7a - at 1123 K and Fig. 7b - at 1373 K.

of partially reduced (P-1) samples showed that in the initial stage
of the reduction a cohesive layer of wüstite was formed on surface
of magnetite grains. As the degree of transformation increased a
grain fell into several pieces which accelerated the process. This
effect was observed in the whole investigated range of temperature.
It was also found that the effect increased with an increase of
temperature. It would suggest a change in the mechanism and rate
determining steps of the process.

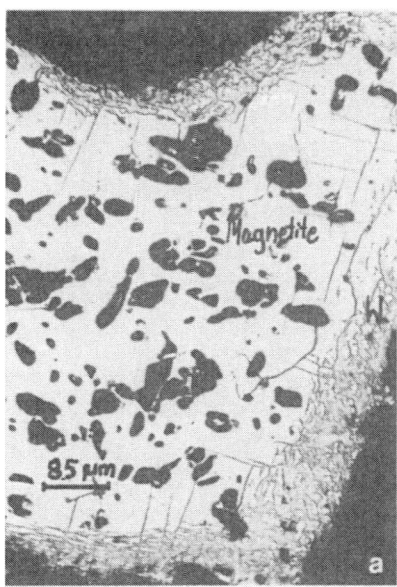 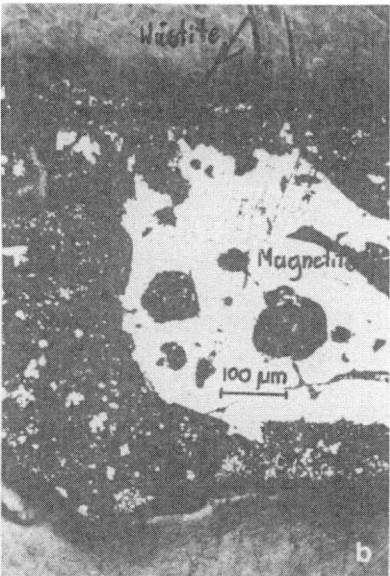

Fig. 8. Optical micrographs of magnetite (M-1) partially reduced
         to wüstite $Fe_{0.94}O$. Fig. 8a - at 1123 K and Fig. 8b -
         at 1373 K.

CONCLUSIONS

     Results of the reduction of magnetite to wüstite were inter-
preted on the basis of the shell-core model taking into account
mixed control.

     It was proved that the model produced staisfactory results to
some extent but several facts suggested that its applicability was
limited:

     (a) Inconsistency of the kinetic equation with experimental
          data in the whole range of the reduction
     (b) The presence of residual magnetite within a layer of
          wüstite formed from this phase.

(c) Cracking of the magnetite core in early stage of the reduction.

(d) Decreasing participation of the gaseous diffusion term from the equation (1) in the overall process with an increase of temperature.

It suggests that applicability of the cracking core model described in papers[9,10] to the process, which is the subject of the present work, should be considered.

REFERENCES

1. J. O. Edström and G. Bitsianes, J. Metals 7 (1955); Trans. AIME 203:760.

2. J. O. Edström, Iron and Steel 26(14):612 (1953).

3. S. Jasieńska, J. Janowski, Mo Ghodsi and G. Naessens, Rev. ATB Met. 23:1 (1983).

4. P. Vallet and C. Carel, Mat. Res. Bull. 14:1181 (1979).

5. R. H. Spitzer, F. S. Manning and W. O. Philbrook. Trans. AIME 236:726 (1966).

6. B. B. L. Seth and H. U. Ross, Can. Met. Quart. 5(4):315 (1966).

7. A. Bessieres, J. J. Heizmann, J. Bessieres and R. Baro, Mem. Sci. Rev. Met. 73:79 (1977).

8. S. Jasieńska, J. Janowski, M. Ghodsi and G. Naessens, Solid State Ionics 12:51 (1984).

9. J. Y. Park and O. Levenspiel, Chem. Eng. Sci. 30:1207 (1975).

10. A. Reizer and A. Barański, Appl. Catal. 9:343 (1984).

THE PREPARATION OF NONSTOICHIOMETRIC

SULPHIDES BY SOLID-GAS REACTIONS

M. Kizilyalli* and H. M. Kizilyalli**

Middle East Technical University

Ankara, Turkey

INTRODUCTION

Copper-iron sulphides were obtained[1] either by heating a quartz ampoule which contains copper, iron and sulphur at high temperatures or by passing hydrogen sulphide H_2S, in a closed system over metallic copper and iron sulphide (pyrite) FeS_2, at temperatures between 1100-1200°C.

On the other hand, if any metal oxide MO, is heated in the flow of either hydrogen sulphide, H_2S, or carbon sulphide, CS_2, the following reactions will take place:

$$MO + H_2S \rightarrow MS + H_2O$$

$$2MO + CS_2 \rightarrow 2MS + CO_2$$

The same kind of reactions may be achieved by either carbonyl sulfide, COS or by a mixture of carbon monoxide, CO, and sulfur

$$MO + COS \rightarrow MS + CO_2$$

$$2MO + S_2 + 2CO \rightarrow 2MS + 2CO_2$$

Welch proposed that a reducing and sulphidizing gas mixture which contains carbon sulphide, carbon monoxide, carbonyl sulphide and a small amount of sulphur and carbon dioxide can be used for the same purpose[2]. This gas mixutre was obtained by passing technical grade SO_2 through activated charcoal heated between

---

* Department of Chemistry
** Department of Physics

700-800°C in the vertical tubular furnace shown in Fig. 1. The change of partial pressures with respect to temperature[3] is shown in Fig. 2.

Fig. 1.
The sulphidizing and reduction chambers.

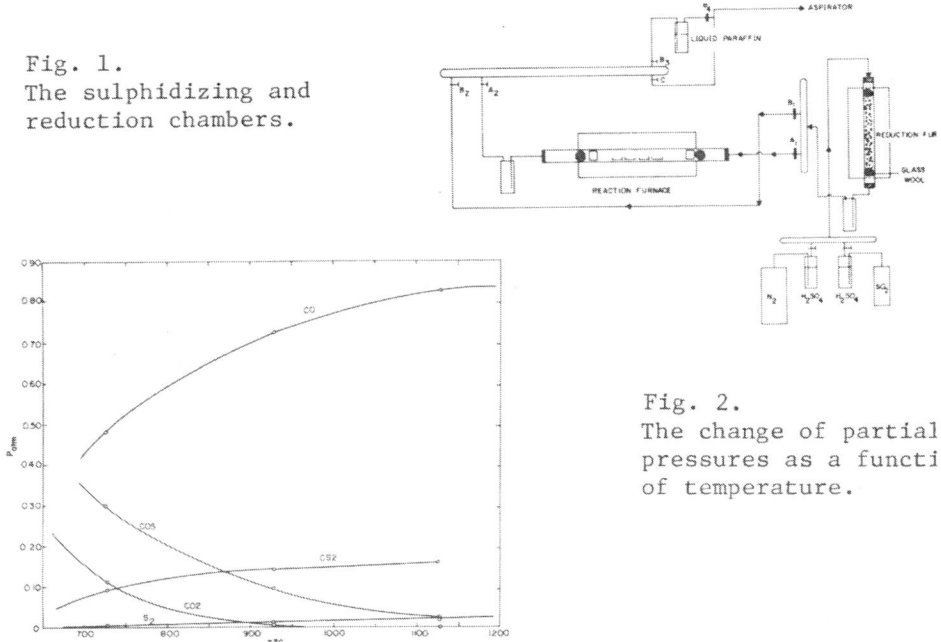

Fig. 2.
The change of partial pressures as a functi of temperature.

The purpose of this work is to prepare stoichiometric and some non-stoichiometric compounds of sulphur, like bornite, cubanite and chalcopyrite by Welch's technique and to examine their crystal structures and changes in the structure due to Cu/Fe mole fraction, by the x-ray powder diffraction method. Roseboom and Kullerud[4] investigated stoichiometric compounds prepared in evacuated closed silica tubes heated between 400-800°C. Kullerud and Yoder[5] performed the same experiments in gold tubes under high pressure.

In this work some stoichiometric and non-stoichiometric compounds of sulphur were prepared by the new technique in the bornite, chalcopyrite and cubanite region and crystal structures of tetragonal bornite and β- chalcopyrite as well as the structural transformations among other structures were investigated. The temperature of reaction, gas flow rate, reaction time and the type of cooling were determined by various preliminary experiments. Afterwards, the mixtures consisting of various ratios of Cu/Fe were prepared and sulphidized.

METHOD

1. The rig and technique used in the synthesis of the copper-iron sulphides:

The vertical tubular furnace is the reduction chamber where SO_2 gas is reduced. The sulphidizing reaction is carried out in the

442

horizontal tubular furnace. The fine powders of Fe_2O_3 + CuO mixtures, in known ratios , have been placed in silica boats and heated in the temperature range of 700-800°C in the reaction chamber. The system has been flushed with nitrogen to be able to remove the oxygen at the beginning. Sulfidizing gas mixture is bubbled through a bubbler, which contains liquid paraffin, with a flow rate of two bubbles per second. The solid-gas reactions in the reaction chamber took 4 to 5 hours.

CuO and Fe_2O_3 were chemically pure and preheated to 800°C then cooled and crushed into fine powder afterwards kept in a desiccator. At the end of the reaction, heating was terminated and the flow of SO_2 gas stopped. The reaction products either cooled slowly in the flow of nitrogen or were cooled rapidly by blowing cold air on the tube which was taken out of the furnace. Four boats which contained four mixtures of different Cu/Fe ratios were sulphidized during the same run. The crystal structures, of about 40 products obtained by about 10 runs, were determined by x-ray powder diffraction technique. FeKα rays (λ = 1.93728 Å) and Guinier camera were used for the experiments.

2. Chemical Analysis:

The sulphide content of compounds was determined[6] by oxidizing it to SO_4^{-2} by bromine in carbon tetrachloride solution and then precipitating as $BaSO_4$ with $BaCl_2$. The formulas calculated through chemical analysis and by theoretical calculations through sulphur gain agreed very well.

RESULTS AND DISCUSSION

In these experiments, as the Cu/Fe ratio, temperature and the degree of reduction were varied, different Cu-Fe-S compounds were obtained which had colors ranging from purple to green. The Cu/Fe ratio versus observed and calculated masses of products (for 1 gr oxide mixtures) are shown in Fig. 3. The comparison of the theoretical and observed masses of the products and the formulas of the products, the investigation of the crystal structures and the unit cell parameters by the x-ray powder diffraction technique have clarified the structure of stoichiometric and non-stoichiometric products in the bornite, cubanite and chalcopyrite regions.

1. Bornite region:

The reactions performed at 800°C/750°C (reducing and reaction furnace temperatures, respectively), 800°C/850°C and 700°C/750°C with a composition corresponding to a ratio of Cu/Fe = 5/1. The formulas of the products were calculated as $CuFeS_{3.84}$, $CuFeS_{3.98}$ and $CuFeS_{3.94}$. The structures agreed with Morimoto and Kullerud's

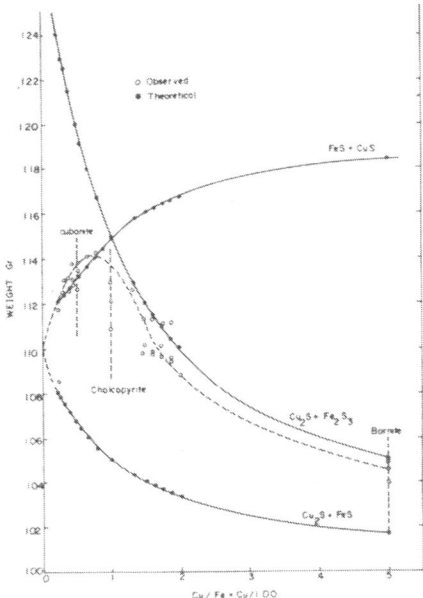

Fig. 3. For 1 gr Fe_2O_3 CuO mixtures, amount of theoretical and observed products versus the Cu/Fe ratio.

low temperature tetragonal bornite form[7] and the unit cell parameters were calculated as "a" ranging between 10.939 and 10.946 Å and c = 21.879 Å. The weights of the three bornite structures were found to be near the Cu_2S + Fe_2S_3 curve in Fig. 3. As the sulphur gain decreases the points move toward the Cu_2S + FeS curve. The presence of a trace amount of FeS in the x-ray diffraction pattern of this product suggested a two phase system of FeS and $Cu_5^{+1}Fe^{+3}S_4$.

If the product were cooled suddenly then a cubic structure was obtained with a = 21.879 Å. Cubic bornite structure was suggested by Morimoto and Kullerud and was explained as the twinning of tetragonal bornite[7]. The high temperature cubic bornite with a = 5.47 Å, found by the same authors, was not observed in this work.

2. <u>Bornite-chalcopyrite transition region</u>:

The product obtained from the mixture of Cu/Fe ratio of 2/1 at 800/750°C by slow cooling in N_2 atmosphere gave an x-ray diffraction pattern of tetragonal α, cubic β-chalcopyrite, and cubic metastable bornite (a = 10.939 Å). The products obtained from the mixtures corresponding to Cu/Fe ratios of 1.875/1, 1.75/1, 1.625/1 and 1.50/1 gave x-ray patterns of tetragonal bornite and β-chalcopyrite. However as the Cu/Fe ratio decreases, the α-chalcopyrite phase decreases and β-chalcopyrite phase increases. Since the system was cooled slowly, the presence of the high temperature β-form below the β to α transition temperature of 550°C, proved the idea of the increasing stability of the β-phase in the iron rich region[10]. On

444

the other hand, when the same products were sulfidized again and cooled slowly, the new products obtained were found to be rich in sulphur. X-ray diffraction analysis of these products showed the disappearance of β-structure and presence of α-structure and bornite structure. Hence by preparing copper-iron sulfides through solid-gas reactions it was proved that α-phase is stable in a sulfur-rich region. This is in agreement with Sukune and Yoshinari's[10] work which was prepared by means of a different reaction.

The bornite structure obtained together with α-phase has a pseudocubic structure with a = 21.879 Å. The presence of forbidden numbers for the cubic system and the presence of doublets suggest a monoclinic bornite[11] which was reported previously, but the x-ray pattern and unit cell parameters were not investigated[12].

3. <u>Chalcopyrite-cubanite region</u>:

In the region where the Cu/Fe ratio changed between 1.33/1 and 0.50/1, α- and β-chalcopyrite structures were obtained upon slow cooling and β and very weak α-chalcopyrite structures were observed upon fast cooling. The formulas of the products obtained, unit cell dimensions and the calculated number of moles of Fe^{+2} and Fe^{+3} are shown in Table I.

Table I

No.	Cu/Fe (mole)	Formulas	Fe^{+2} (mole)	Fe^{+3} (mole)	For β a (Å)	For α* c (Å)
1	1.33/1.00	$Cu_{1.244}Fe_{0.933}S_{2.00}$	0.04	0.89	10.576	10.409
2	1.00/1.00	$Cu_{1.058}Fe_{1.058}S_{2.00}$	0.23	0.83	10.594	--
3	0.80/1.00	$Cu_{0.894}Fe_{1.118}S_{2.00}$	0.25	0.87	10.590	10.423
4	0.66/1.00	$Cu_{0.801}Fe_{1.201}S_{2.00}$	0.40	0.80	10.594	10.439
5	0.57/1.00	$Cu_{0.724}Fe_{1.266}S_{2.00}$	0.52	0.74	10.603	10.457
6	0.50/1.00	$Cu_{0.667}Fe_{1.333}S_{2.00}$	0.67	0.67	10.608	--

*Since some of the lines of α and β structures coincided in some films, only the values that can be calculated are given.

The unit cell dimensions increase as the Cu/Fe ratio decreases which is observed in Table I. Figure 4 shows the changes of some d-spacings as a function of Cu/Fe ratios. The increase in the unit cell dimensions may be explained by the decrease in Fe^{+3}/Fe^{+2} ratio, since the ionic radius of Fe^{+2} is larger than that of Fe^{+3}. On the

445

other hand, one may argue, that the decrease in the ratio of Cu/Fe may reduce unit cell dimensions since the ionic radius of Cu^{+1} is larger than that of Fe^{+2} and Fe^{+3}. However, since the ratio of Cu/Fe decreased by a factor of 2.66 and the ratio of Fe^{+3}/Fe^{+2} decreased by a factor of 20, Fe^{+3}/Fe^{+2} ratio determines the change in the unit cell parameters. Finally, it was observed that the "a" value for the product #2 is larger than expected. This proves that the chalcopyrite is in between $Cu^{+1}Fe^{+3}S_2$ and $Cu^{+2}Fe^{+2}S_2$ configurations as stated by Hall and Steward on single crystal measurements[13]. These different oxidation states explain deviations from each of the calculated three curves corresponding to the Cu/Fe ratio of 1/1 as can be seen in Fig. 3.

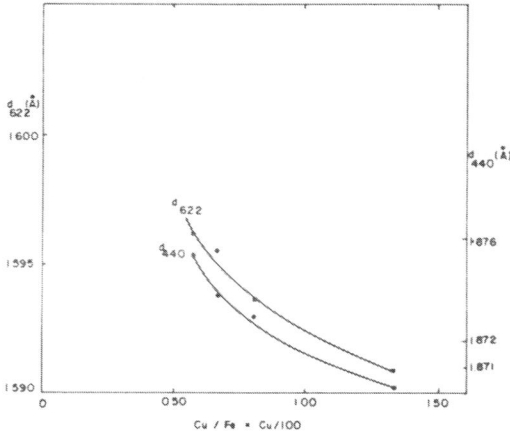

Fig. 4. The spacings of d_{622} and d_{440} lines versus Cu/Fe ratio in non-stoichiometric region.

4. Cubanite Structure:

Another important result obtained in these series of experiments is the equality of the number of moles of Fe^{+2} and Fe^{+3} for the product #6, Cu/Fe ratio of 0.5/1, which corresponds to stoichiometric $CuFe_2S_3$. Following this, the cubanite configuration may be shown as $Cu^{+1}Fe^{+2}Fe^{+3}S_2$, or $Cu_2S.2FeS.Fe_2S_3 = 2CuFe_2S_3$.

Greenwood and Whitfield had earlier suggested that the odd electron per formula unit was being exchanged rapidly between the iron atoms, on the basis of their Mössbauer spectra[14]. On the other hand, Mössbauer spectra obtained at 4°K by Imbert and Wittenberger have not shown any further splitting of lines compared with the room spectra[15]. This excludes the idea of rapid electron exchange and favors the idea of identical oxidation states. In the refinement of the structure of cubanite, Szymanski[16] suggested centrosymmetric space group Pcmn. This implies that the two Fe atoms across the center of symmetry are in identical oxidation states, at least on a time-average basis. Hulliger[17] suggested

446

that the odd electron occupies an incompletely-filled d-band which accounts for the electrical conductivity of cubanite. Townsend et al. reported that cubanite is an n-type semiconductor[18]. Fleet suggested a bonding model for cubanite where both Cu and Fe atoms are in tetrahedral coordination with S, the Fe coordination tetra-hedra are arranged to share a common edge[19]. According to molecular orbital theory, if two valence electrons are allocated to each metal-sσ bond and Cu is assumed to have the $3d^{10}$ configuration, there remain eleven valence electrons per formula unit to be distributed between the d-levels of the two Fe atoms. For the equivalent ionic model, then, one Fe atom per formula unit would be trivalent and one would be divalent. The Mössbauer Spectra of cubanite at 77°K and 295°K are consistent with this model. The chemical isomer shift is intermediate between that expected for Fe^{+2} and Fe^{+3} ions in tetrahedral coordination. However, only a single six-line hyperfine spectrum is observed although two superimposed spectra are to be expected for distinct Fe^{+2} and Fe^{+3} states. It seems that the problem of the location of this odd electron is still unresolved.

In this work, the preparation of colored stoichiometric and non-stoichiometric copper iron sulfides through gas-solid reactions suggested that they could be mixed valence compounds.

5. The definition of non-stoichiometric region:

Figure 5 shows change of cubic β-structure unit cell para-meters with respect to Cu/Fe ratio. These belong to the products prepared at the same temperature and cooled slowly. In the region when the ratio of Cu/Fe changes from 0.25 to 0.50 and between 1.33 and 2.00 the unit cell parameters remain practically constant, however, in the region from 0.50 to 1.33 they gradually decrease. This third region is defined as the non-stoichiometric region for chalcopyrite, although it was rather difficult to describe the boundaries of this region. Better results would have been obtained if the gas-solid reactions could had been repeated under the same conditions, which were rather difficult to obtain[20].

Electrical conductivity is perhaps the most sensitive of all physical parameters; therefore electrical conductivity studies can reveal very small deviations from stoichiometry which are quite undetectable by chemical analysis or by cell parameter measurements through x-ray diffraction, and can also determine in which direction the deviation is occurring.

Since there is a dynamic equilibrium which relates the number of unbalanced defects with the ambient pressure, the magnitude of the electrical conductivity at a given temperature is a sensitive indicator of the extent of the deviation from stoichiometry. This technique is an elegant way of deciding whether a given compound is metal excess or metal deficit by following the changes of

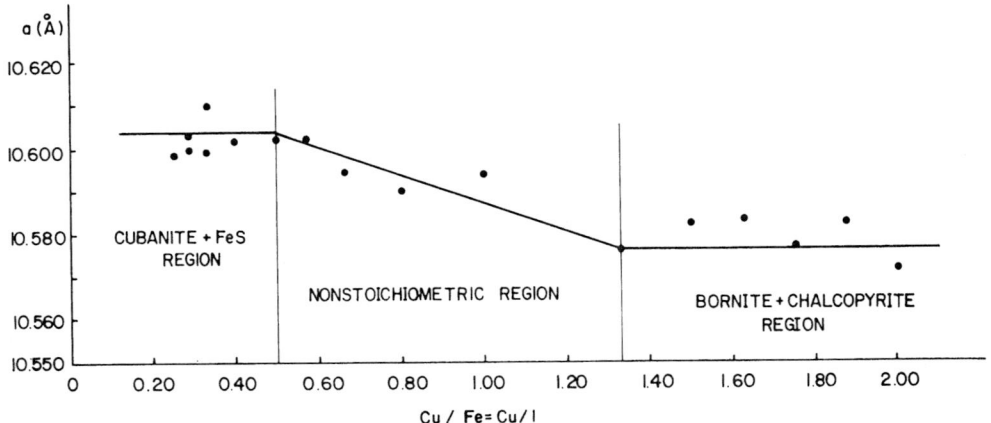

Fig. 5. The change of cell parameter for cubic structure (β)
versus Cu/Fe ratio.

electrical conductivity with pressure at constant temperature. Many
methods are available to measure the conductivity of a material each
having its own merits. The two-terminal method is the simplest,
although surface leakage is one of the drawbacks of this method.
Ohmic contacts are required.

In case it becomes inconvenient or impracticable to follow
the change of the electrical conductivity with ambient pressure of
one of the components, two further electrical effects can be used
to characterize the type of semiconductor and the direction of
stoichiometric imbalance. These are the Seebeck or thermoelectric
effect and the Hall effect. We intend to use these techniques in
the study of the transport properties of non-stoichiometric sulphides.

ACKNOWLEDGMENT

H. M. Kizilyalli would like to thank NATO for financial
assistance in attending the Advanced Research Workshop.
M. Kizilyalli would like to thank Dr. A. J. E. Welch for helpful
discussions and the facilities of the Chemistry Department of
Imperial College, London, and Professor Joshua Bear for
critical reading of the manuscript.

REFERENCES

1. H. E. Mervin and R. H. Lombard, Econ. Geol. 32:203 (1937).
2. A. J. E. Welch, Interim Report, Imperial College, London
   (1970).
3. A. J. Owen, K. W. Sykes and D. J. D. Thomas, Trans. Faraday
   Soc. 47:419 (1951).

4. E. H. Roseboom and G. Kullerud, <u>Carnegie Inst. of Wash. Year Book</u> 57:222 (1957).
5. G. Kullerud and G. Yoder, <u>Econ. Geol.</u> 54:533 (1959).
6. H. I. Vogel, "Textbook of Quantitative Inorganic Analysis", Longmans, London, 467 (1961).
7. N. Morimoto and G. Kullerud, <u>Carnegie Inst. of Washington Year Book</u> 50:118 (1959-60).
8. L. Pauling and L. O. Brockway, <u>Z. Krist.</u> 82:188 (1932).
9. J. E. Hiller and K. Probstain, <u>ibid.</u> 108:108 (1956).
10. T. Sukune and F. Yosinari, <u>J. Fac. Eng. Univ. Tokyo Ser</u> A6:50 (1968).
11. M. Kizilyalli, Report Middle East Technical University, Ankara p. 48 (1977).
12. R. A. Yund and G. Kullerud, <u>J. Petrol.</u> 7:454 (1966).
13. S. R. Hall and J. M. Steward, <u>Acta Cryst. B.</u> 29:579 (1973).
14. N. N. Greenwood and H. J. Whitfield, <u>J. Chem. Soc.</u> A:1697 (1968).
15. P. Imbert and W. Winterberger, <u>Bull. Soc. Franc. Mineral. Cristallogr.</u> 90:299 (1967).
16. J. T. Szymankski, <u>Z. Krist.</u> 140:218 (1974).
17. F. Hulliger, <u>Structure and Bonding</u> 4:83 (1968).
18. M. G. Townsend, J. L. Horwood, and J. R. Gosselin, <u>Canad. J. Physics</u> 51:2162 (1973).
19. M. E. Fleet, <u>Can. Min.</u> 11:901 (1972).
20. M. Kizilyalli, Proc. Science Congress VII. of T.B.T.A.K., Izmir, Turkey, p. 505 (1980).

THE REACTIONS OF COBALT, IRON AND NICKEL IN SO_2

ATMOSPHERES: SIMILARITIES AND DIFFERENCES

Nathan S. Jacobson
NASA-Lewis Research Center
Cleveland, Ohio 44135

Wayne L. Worrell
Department of Materials Science
University of Pennsylvania
Philadelphia, PA 19104

ABSTRACT

The reactions of cobalt, iron and nickel in SO_2 atmospheres are reviewed and compared. A mixed oxide-sulfide product layer is observed in all cases. Cobalt and nickel exhibit similar behavior. The observed rates are near the sulfidation rates, and the reaction rate is strongly influenced by the outward diffusion of metal through an interconnected sulfide network. A continuous interconnected sulfide is not observed in the oxide-sulfide scales formed on iron, and the reaction rates are more difficult to summarize. The differences and similarities among the three metals are explained in terms of the absence of scale-gas equilibrium and the ratio of the metal diffusivity in the corresponding oxide and sulfide.

INTRODUCTION

In recent years, the reactions of cobalt, iron and nickel in SO_2 atmospheres have been studied extensively, and the understanding of the reaction mechanisms has significantly increased. In this paper, the scale growth mechanisms of these reactions are summarized, emphasizing the major similarities and differences. Our summary will consider only the SO_2 or SO_2-Ar atmospheres and studies below the lowest eutectic temperature in each system: 872°C for Co-Co_4S_3[1], 925°C for Fe-FeS-FeO[2], and 637°C for Ni_3S_2[1].

GENERAL THERMODYNAMIC AND KINETIC ASPECTS

A brief summary of the thermodynamic and kinetic aspects of these reactions is helpful before analyzing the mechanisms of scale growth. A thermodynamic stability diagram for each metal-sulfur-oxygen system is used to indicate which phases are stable under the specified conditions of temperature, SO_2 and oxygen pressure[3-5]. For example, the stability diagram for the Ni-O-S system is shown in Fig. 1[3]. The reaction conditions considered in this analysis are temperatures below the eutectic temperature, SO_2 pressures between 0.1 and 1.0 atm, and an estimated oxygen impurity level of 1 PPM. Under these conditions, Fig. 1 indicates that the outer product scale would be nickel sulfate if scale-gas equilibrium were obtained. However it should be emphasized that when nickel is first exposed to SO_2, the nickel activity is very high, and a mixed sulfide-oxide product scale can form according to reaction (1).

$$7/2 \text{ Ni} + SO_2(g) = 2\text{NiO} + 1/2 \text{ Ni}_3S_2 \qquad (1)$$

Thermodynamic calculations[3] indicate that the nickel activity must be greater than 0.01 when the SO_2 pressure is between 0.25 and 1 atm. at 600°C for reaction (1) to proceed. Diagrams similar to Fig. 1 are also available for the Co-S-O[4] and Fe-S-O[5] systems. Under the conditions considered in this analysis, the outer product scale on cobalt should be either cobalt oxide and/or cobalt sulfate, while the equilibrium outer scale for iron is iron oxide. In all three cases, no metal sulfide should be observed at the scale-gas interface, because the sulfur activity in the SO_2 atmosphere is below that necessary for sulfide formation.

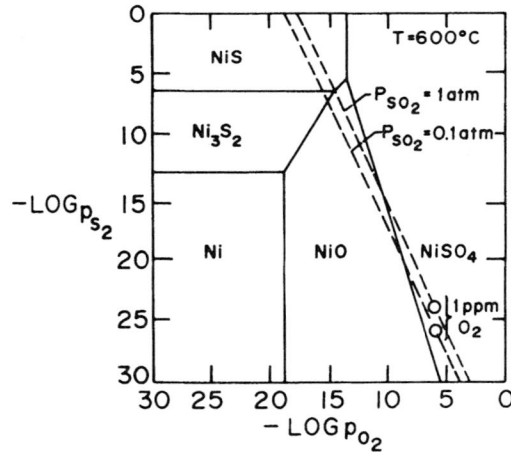

Fig. 1 Stability diagram for the Ni-S-O system at 600°C. The dotted lines are constant SO_2 pressure lines. The o points show the stable solid phase in the indicated SO_2 atmosphere, with an estimated O_2 impurity level of 1 PPM.

452

The observed presence of the metastable sulfide phase[4,6] at the
scale gas interface after extended reaction times is due to the
combination of three factors. The first is that formation of sulfide-
oxide product scales on metal surfaces with high metal activity is
thermodynamically feasible, as discussed for reaction (1). The second
factor is the large difficulty in the formation of sulfate in SO_2
atmospheres. For example, nickel sulfate is observed in several
recent studies[7,8] to form only by reaction of NiO with SO_3, and not
by reactions (2) and/or (3).

$$Ni + SO_2(g) + O_2(g) = NiSO_4 \qquad\qquad (2)$$

$$NiO + SO_2(g) + 1/2 \ O_2(g) = NiSO_4 \qquad\qquad (3)$$

Thus reaction (1) appears to be kinetically favored over reactions
(2) and (3), because of a high activation-energy barrier in the direct
formation of sulfate from solely SO_2 atmosphere. However, experimental
studies using some of the newer surface-characterization techniques,
e.g. in situ Raman spectroscopy, are necessary to elucidate the
kinetic barriers to sulfate formation. Once a sulfide-oxide product
scale is formed, the third factor, metal diffusion through the scale
becomes important. The high metal diffusivities in the sulfides shown
in Table 1 could ensure that an abnormally high metal activity is
maintained at the scale-gas interface and that reaction (1) continues
to be favored over sulfate formation.

TABLE 1.   Self-Diffusion Coefficients $[cm^2 - sec^{-1}]$

Cobalt, 750°C		Iron, 800°C		Nickel, 600°C	
CoS	$2.7 \times 10^{-8}(4,9)^a$	$Fe_{0.9}S^b$	$1.3 \times 10^{-7}(10)$	Ni_3S_2	$1.5 \times 10^{-7}(11)$
CoO^b	$9.2 \times 10^{-11}(12)$	$Fe_{0.9}O$	$1.0 \times 10^{-7}(13)$	NiO	$3.0 \times 10^{-16}(14)$
		Fe_3O_4	$3.3 \times 10^{-11}(13)$		

[a] Reference from which value is obtained.

[b] Single crystal samples, all others polycrystalline

The self-diffusion coefficients for cobalt, iron and nickel in
the pertinent sulfides and oxides are compared in Table 1. For
cobalt and iron, both single-crystal and polycrystalline data are
listed, since grain boundary diffusion in their oxides or sulfides
does not appear to be significant. Since only a polycrystalline
value is available for Ni_3S_2, it is compared with a polycrystalline
value for NiO. Table 1 clearly shows that metal diffusion through
the sulfide is 10^3 and 10^9 times greater than through the oxide
for cobalt and nickel, respectively. However, the iron diffusivity

is almost identical in FeS and FeO at 800°C. Because the product scales from the metal–SO$_2$ reactions contain both oxide and sulfide, the metal diffusivities in these phases must be considered in any kinetic analysis of the SO$_2$ reactions.

The parabolic-rate constants (kp) for oxidation, sulfidation and SO$_2$ reaction for each of the three metals are summarized in Table 2. The values are presented in the units m moles gas^2–cm^{-4}–sec^{-1} to enable a common basis of comparison between the three types of reaction. All the oxidation and sulfidation data are for oxygen and sulfur pressures of 1 atm, respectively, with the exception for nickel where the k_p is for oxidation at 0.1 atm. The k_p (SO$_2$) for nickel is for 0.25 atm, while the values for cobalt and iron are for 0.1 atm SO$_2$ pressure. The temperatures cited in Table 2 are those below the lowest eutectic temperature and having the largest amount of SO$_2$ reaction-rate data for each metal. The ratio of k_p (SO$_2$) to k_p (O$_2$) ranges from 10^3 to 10^7 for cobalt and nickel, respectively. However, the k_p (SO$_2$) for iron is close to the oxidation value.

TABLE 2. Parabolic-Rate Constants, [m moles^2 – cm^4 – sec^{-1}]

Metal	Oxidation	Sulfidation	SO$_2$ Reaction
Cobalt, 750°C	2.2x10^{-7}(15)[a]	3.9x10^{-4}(16)	2.9x10^{-4}(4)
Iron, 800°C	5.2x10^{-5}(13)	8.2x10^{-4}(9)	1.3x10^{-5}(5)
Nickel, 600°C	3.0x10^{-11}(17)	2.1x10^{-4}(18)	1.4x10^{-4}(6)

[a] Reference from which value is obtained.

SCALE GROWTH MECHANISMS IN SO$_2$ ATMOSPHERES
Cobalt

The reaction of cobalt in SO$_2$ atmosphere has been studied by numerous investigators[4,19-22]. Product scales consist of a narrow band of cobalt sulfide adjacent to the metal and an outer two-phase oxide-sulfide layer (Fig. 2). Mechanisms of scale growth in 10% SO$_2$–Ar mixture at temperatures between 650 and 800°C have recently been reported[4]. An inner sulfide layer with an outer porous oxide layer form during the initial stages of reaction. Molecular SO$_2$ can penetrate the porous layer and react to form more oxide and sulfide.

When the oxide pores are filled with the sulfide-oxide reaction product, cobalt can diffuse rapidly through the sulfide phase to the scale-gas interface. At this point the outer two-phase sulfide-oxide scale (Fig. 2) begins to grow, and a parabolic rate low is observed at 700 and 750°C[4]. Cross-sectional micrographs

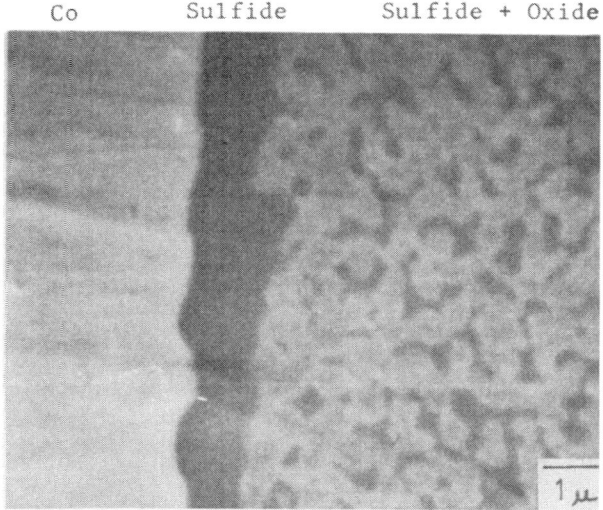

Co   Sulfide   Sulfide + Oxide

1μ

Fig. 2 Cross-sectional view of the product scale formed on cobalt after 2 hr at 750°C under 0.1 atm SO_2[4]. The sulfide is the dark phase. (Electron micrograph).

and electrical resistivity measurements of the product scale indicate that the sulfide phase is interconnected[4]. As shown in Table 1, the cobalt diffusivity in CoS is $\sim 10^3$ times greater than in CoO. This suggests that the interconnected sulfide phase acts as a rapid transport path for cobalt in the oxide matrix. The comparison of the cobalt diffusivities in CoS calculated from straight sulfidation and the SO_2 reactions shown in Table 3 confirms that cobalt diffusion through the sulfide phase is the primary mechanism establishing the scale-growth rate in SO_2 atmospheres. With increasing time the parabolic rates decrease, due to a change in the sulfide distribution in the outer sulfide-oxide scale[4]. The outer scale contains sulfide regions connected by narrow constricted sulfide channels in the oxide matrix, which results in restricted transport paths for cobalt.

IRON

The reaction of iron has been investigated in various SO_2 atmospheres including solely SO_2[5,23,24], $N_2-O_2-SO_2$[25], $CO-CO_2-COS$[25] and 10% SO_2-CO_2[26] gas mixtures. At low SO_2 pressures, the reaction has a linear rate with a clear flow-rate dependence, indicating that diffusion through a gaseous boundary layer at the scale-gas interface is rate limiting[5]. A typical cross-sectional view of the scale observed at 800°C in low pressure SO_2 environments is shown

TABLE 3. Calculated Cobalt and Nickel Diffusivities [cm^2-sec^{-1}] in CoS and Ni_3S_2, respectively from Parabolic Ratio Constants for Sulfidation and SO_2 Reactions.

Sulfide, Temp.	D_M, Sulfidation[a]	D_M, SO_2 Reaction[b]
CoS, 700°C	9.8×10^{-9}	2.2×10^{-8} (0.1 atm SO_2)
CoS, 750°C	2.7×10^{-8}	4.2×10^{-8} (0.1 atm SO_2)
Ni_3S_2, 600°C	1.5×10^{-7}	0.7×10^{-7} (0.25 atm SO_2)
Ni_3S_2, 600°C	1.5×10^{-7}	2.7×10^{-7} (1 atm SO_2)

[a] Sulfidation data from references 4 and 9 for CoS and reference 11 for Ni_3S_2.

[b] SO_2-reaction data from reference 4 for CoS and reference 6 for Ni_3S_2.

in Fig. 3. Although the scale consists of sulfide and oxide, its morphology is very different from that observed with cobalt and nickel. In Fig. 3, a finely dispersed FeO-FeS duplex layer is observed adjacent to the iron instead of the narrow, continuous sulfide layer observed with cobalt and nickel. A lamellar FeO-FeS layer is also observed on top of the inner duplex layer. The observed morphology has been postulated to result from a SO_2 reaction where the formation of FeO increases the sulfur activity to form FeS which in turn increases the oxygen activity to form FeO[5].

At higher SO_2 pressures and extended times, parabolic reaction kinetics have been reported[5,23,24]. The scale morphologies are very complicated, and quantitative interpretation of the observed parabolic rates is not possible. The absence of an interconnected sulfide network in the product scale as observed with cobalt and nickel is presumably related to the similar values for the iron diffusivity in FeS and FeO shown in Table 1. In the oxide-sulfide product scale on iron, an interconnected FeS network in FeO would not enhance the transport of iron to the scale-gas interface. Thus the observed parabolic-rate constants for iron-SO_2 reactions are similar to those observed in the oxidation of iron[5,23]. For example, the value shown in Table 2 for the iron-SO_2 reaction is slightly less than that shown for iron oxidation, while the values for the SO_2 reactions with cobalt and nickel are 10^3 to 10^7 faster than their respective oxidation rates.

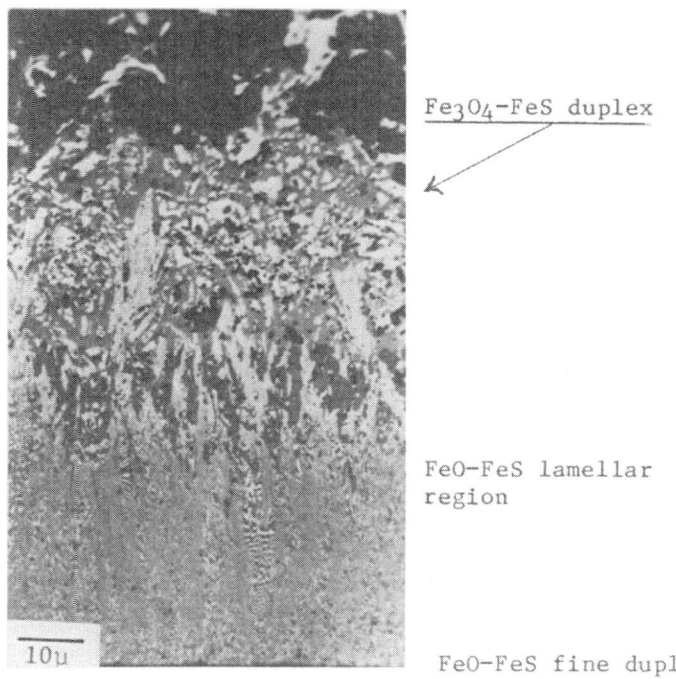

Fe₃O₄-FeS duplex

FeO-FeS lamellar region

FeO-FeS fine duplex

<u>Fig. 3</u> Cross-sectional view of the product scale formed an iron at 800°C under low SO_2 pressures (~0.1 atm)[5]. The sulfide phase is the light region (optical micrograph).

NICKEL

The reaction of nickel in SO_2 atmosphere has been extensively investigated[6,27-32]. The reaction mechanism[6,27,28,31] in solely SO_2 atmospheres and at temperatures below 637°C (the Ni-Ni₃S₂ eutectic temperature) is summarized here. During the early stages of reaction, a porous NiO layer forms on the nickel while the sulfur from the dissociation of SO_2 diffuses through the nickel grain boundaries to form an inner layer of Ni_3S_2[6]. Eventually the pores are filled with the sulfate-oxide product of reaction (1), and nickel can diffuse though the sulfide to the scale-gas interface. After this initial stage, the outer two-phase (NiO-Ni₃S₂) layer begins to grow and parabolic, diffusion controlled kinetics are observed. Inert markers are located between the inner Ni₃S₂ layer and the outer two-phase layer[27-29], confirming that the inner layer grows by inward sulfur diffusion and the outer two-phase layer grows by outward nickel diffusion. The thickness of the inner

Ni Ni$_3$S$_2$ NiO+Ni$_3$S$_2$
↓   ↓

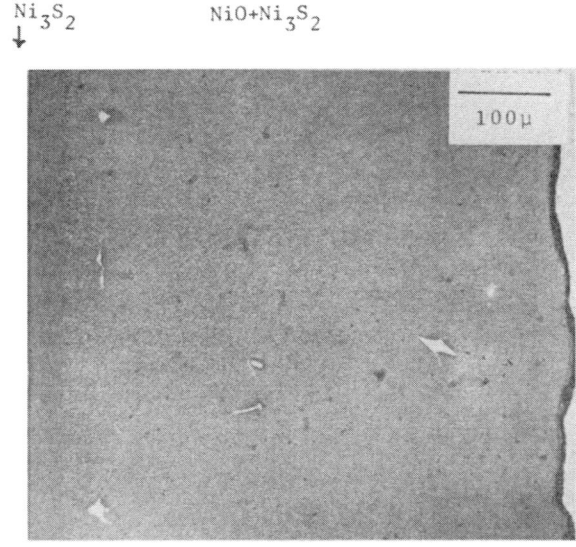

Fig. 4 Cross-sectional view of the scale formed on nickel after 44 min at 603°C under 1 atm SO$_2$[6]. The sulfide is the light phase (optical micrograph).

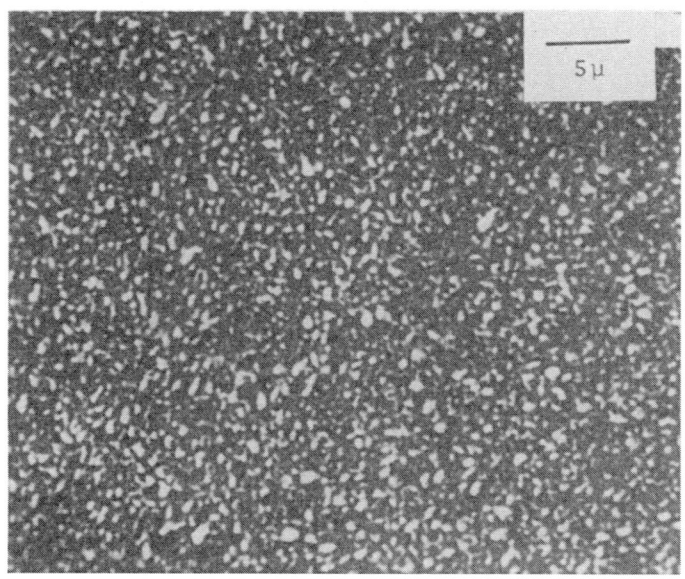

Fig. 5 Higher magnification of the outer two-phase scale shown in Fig. 4[6]. The Ni$_3$S$_2$ (light regions) is fairly uniformly distributed in NiO.

458

Ni_3S_2 layer at 600°C is ~30μ (Fig. 4), while the inner sulfide layer formed on cobalt at 750°C (Fig. 2) is only ~1μ thick. This comparison indicates that the grain-boundary diffusivity of sulfur is significantly higher in nickel than in cobalt.

Figure 5 is a highly magnified picture of the outer two-phase region of the product scale shown in Fig. 4. The lighter areas are the Ni_3S_2 phase, which is rather evenly distributed in the darker oxide matrix. Resistivity measurements indicate that the Ni_3S_2 is interconnected throughout the outer two-phase layer[6]. The nickel diffusivity in Ni_3S_2 is about 10^9 times greater than that in NiO at 600°C (see Table 1), and the Ni_3S_2 phase provides a rapid transport path for nickel diffusion through the outer two-phase scale. Calculated values of the nickel diffusivity from the parabolic rate constants of nickel-SO_2 reactions[6] are shown in Table 3. The agreement between these calculated values and that obtained from sulfidation experiments confirms that nickel diffusion through interconnected Ni_3S_2 is the primary mechanism controlling the growth rate of the outer, two-phase scale. Thus the parabolic growth rates observed for the nickel-SO_2 and the cobalt-SO_2 reactions are controlled by similar mechanisms.

SUMMARY AND CONCLUSIONS

The reactions of cobalt, iron and nickel in SO_2 atmospheres have been compared. In all three metal-SO_2 reactions, a metastable sulfide phase is observed at the scale-gas interface. The presence of the sulfide phase is due to three factors: formation of sulfide from SO_2 is thermodynamically feasible at high metal activities, the large difficulty in the formation of sulfate from SO_2, and rapid metal diffusion through the two-phase product scales maintains a high metal activity at the scale-gas interface.

Cobalt and nickel exhibit similar behavior in that both form an inner thin sulfide layer adjacent to the metal and an outer two-phase layer consisting of an interconnected sulfide phase in an oxide matrix. The metal diffusivity is much greater in the sulfide than in the oxide, and the interconnected sulfide provides a rapid transport path for metal diffusion through the outer two-phase scale. The observed parabolic-rate constants for the SO_2 reactions are 10^3 to 10^7 times greater than those observed for oxidation.

The reaction of iron in SO_2 atmosphere is more complex, because of the influence of gas transport and the similar values of the iron diffusivity in FeS and FeO. The morphology of the scales formed on iron is very complicated, and quantitative interpretation of the parabolic kinetics is not possible. However, the observed parabolic rates are similar to those observed in the oxidation of iron.

ACKNOWLEDGEMENTS

The support of the NSF Materials Research Program under Grant No. DMR-7923647 at the University of Pennsylvania for our investigations of the nickel-SO_2 and cobalt-SO_2 reactions is greatfully acknowledged. The authors also thank Dr. Ajay Misra of NASA Lewis Research Center for helpful discussions.

REFERENCES

1. T. Rosenqvist, J.I.S.I., 176:37 (1954).
2. D. C. Hilty and W. Crofts, Trans. AIME, 194:1307 (1952).
3. N. S. Jacobson and W. L. Worrell in Proceedings of the Symposium on High Temperature Materials Chemistry II, Electrochemical Society, Inc., Pennington, pp. 217-223 (1983).
4. N. S. Jacobson and W. L. Worrell, J. Electrochem. Soc. 131:1182 (1984).
5. T. Flatley and N. Birks, J.I.S.I., 209:523 (1971).
6. K. L. Luthra and W. L. Worrell, Met. Trans. 9A:1055 (1978).
7. W. L. Worrell and B. Ghosal in Proceedings JMIS-3, High Temperature Corrosion, Trans. JIM, Suppl., pp. 419-426 (1983).
8. V. Guerra-Brady and W. L. Worrell, to be published.
9. S. Mrowec and T. Werber, Chemia Analityczna 7:605 (1962).
10. R. H. Condit, R. R. Hobbins, andd C. E. Birchenall, Oxide Met. 8:409 (1974).
11. B. D. Bastow and G. C. Wood, Oxide Met. 9:473 (1975).
12. R. E. Carter and F. D. Richardson, Trans. AIME 200:1244 (1954).
13. L. Himmel, R. F. Mehl, and C. E. Birchenall, Trans. AIME, 197:827 (1953).
14. K. Fueki and J. B. Wagner, J. Electrochem. Soc. 112:284 (1965).
15. H. S. Hsu and G. J. Yurek, Oxid. Met. 17:55 (1982).
16. S. Mrowec and T. Werber, Phys. Met. Metallogr. (Engl. Transl.)8, No. 3, 452 (1959).
17. E. A. Gulbransen and K. F. Andrew, J. Electrochem. Soc. 101:128 (1954).
18. L. Czerski, S. Mrowec, and T. Werber, J. Electrochem. Soc. 109:273 (1962).
19. P. Singh and N. Birks, Oxid. Met. 12:23 (1978).
20. F. Gesmundo and C. deAsmundis in "Behavior of High Temperature Alloys in Aggressive Environments; Proceedings of the International Conference, Petten, The Netherlands, 15-18 October, 1979" pp 435-447, Metals Society, London (1980).
21. K. Holthe and P. Kofstad, Corros. Sci. 20:919 (1980)
22. B. Gillot and D. Garnier, Ann. Chim. Fr. 5:483 (1980).
23. B. Chatterjee and A. D. Dowell, Corros. Sci. 15:637 (1975).
24. F. Gesmundo, C. deAsmundis, S. Merk, and C. Bottino, Werkstoffe and Korrosion, 30:179 (1979).
25. A. Rahmel, Corros. Sci., 13:125 (1973).

26. F. C. Yang and D. P. Whittle in Proceedings of the Symposium on Corrosion in Fossil Fuel Systems, The Electrochemical Society, Pennington, 1983, pp 111-129.
27. C. B. Alcock, M. G. Hocking, and S. Zador, Corros. Sci. 9:111 (1969).
28. M. R. Wootton and N. Birks, Corros. Sci. 12:829 (1972).
29. P. Kofstad and G. F. Akesson, Oxid. Met. 12:503 (1978).
30. K. L. Luthra and W. L. Worrell, Met. Trans. 10A:621 (1979).
31. M. Seierstein and P. Kofstad, Corros. Sci. 12:487 (1982).
32. B. Haflan and P. Kofstad, Corros. Sci. 23:1333 (1983).

THE INFLUENCE OF ION IMPLANTATION OF REACTIVE ELEMENTS ON THE

GROWTH OF Cr_2O_3 SCALES ON Ni-20% Cr AT HIGH TEMPERATURE

F.H.Stott, J.S.Punni, G.C.Wood and G.Dearnaley*

Corrosion and Protection Centre, University of Manchester
Institute of Science and Technology
Manchester M60 1QD UK
* A.E.R.E. Harwell, Oxfordshire, UK

ABSTRACT

    The additions of reactive elements, such as yttrium or cerium,
to nickel-chromium alloys have considerable beneficial effects on
the development, growth and mechanical stability of Cr_2O_3 scales
in oxygen at high temperature. Several mechanisms have been
proposed to account for these effects, involving processes within
the alloy, at the alloy/scale interface or within the bulk scale,
but none is universally accepted. In order to investigate the
phenomenon further, a study of the influence of ion-implanted
reactive elements on the subsequent oxidation behaviour of Ni-20%Cr
has been carried out. Implanted cerium or yttrium are particularly
effective in increasing the rate of establishment of the healing
protective Cr_2O_3 scale and in decreasing its rate of growth. The
latter effect can be associated with a change in diffusion processes in the oxide. The scale on the unimplanted surfaces grows
outwards following diffusion of Cr^{3+} ions from the alloy/scale to
the scale/gas interface. Here, the contribution of oxygen diffusion
is relatively small. However, the scale on the implanted surfaces
grows inwards following diffusion of oxygen from the scale/gas to
the scale/alloy interface. On exposure of the implanted surfaces
to the environment, the implanted reactive element species are
oxidized to form small (5 to 50 nm) particles at or near the metal
surface. As the Cr_2O_3 scale develops, these are incorporated into
it as discrete particles, with only a very small solubility in the
oxide. However, there is a general suppression of outward
diffusion of Cr^{3+} ions through the scale, enabling inward transport
of oxygen to predominate. Possible reasons for these effects are
discussed and related to the distribution of the reactive elements
in the scale.

INTRODUCTION

Many high-temperature alloys have been designed to facilitate development of external Cr_2O_3-rich scales. These scales are reasonably protective, being relatively stoichiometric, pore-free and defect-free. However, their effectiveness can be impaired by a tendency to crack and spall during thermal cycling. This can be a major problem since, in practice, the thermal cycling performance of scales on alloy components is very important and can determine their useful lives.

Empirically, it has been found that additions of reactive elements, such as cerium, hafnium or yttrium, or their stable oxides, to the alloy can influence markedly the development, growth and mechanical stability of Cr_2O_3-rich scales in oxygen at high temperature[1]. Selective oxidation of chromium is promoted[2], enabling the Cr_2O_3-rich scale to develop as a complete layer more rapidly than in the absence of the additions, but, once established, the growth rate of the scales is reduced[3-6]. Also, the adhesion/cohesion of the scale during thermal cycling is much improved[7-11].

Numerous investigations have been undertaken to ascertain the cause of these effects and several, not necessarily mutually-exclusive, mechanisms have been proposed, although none is universally accepted. These have been summarized in various review papers[12,13]. Briefly, the main effects of the reactive element or oxide additions are reported to be as follows.

i) The additions enhance scale plasticity[11,14].

ii) A layer of oxide is formed between the scale and alloy, forming a graded seal[15].

iii) The presence of the additions reduces the development of voids at the scale/alloy interface, such voids forming following condensation of vacancies at the interface. The internal oxide particles of the reactive elements, the reactive element atoms themselves, or the stable oxide dispersions are reported to provide alternative sites for vacancy condensation (vacancy sinks), thus reducing inter-facial porosity and improving scale adhesion[16,17].

iv) The scale is "keyed" or "pegged" to the alloy by the development of oxide stringers of elongated oxide intrusions. Such intrusions form by internal oxidation of the reactive elements and the inward development of the scale to incorporate the internal oxides[8,10,17,18,19]. In addition, establishment of an irregular oxide/alloy interface, with considerable interlocking of alloy and scale, could be important[6,20].

v) The additions affect the growth mechanism of the scale. In particular, in the absence of any additions, the Cr_2O_3-rich scale grows largely by outward diffusion of Cr^{3+} ions and reaction with oxygen at or near the scale/gas interface. Lateral growth of the scale causes large areas to grow out of contact with the alloy substrate. Conversely, in the presence of the additions, the outward transport of the cations is reduced considerably, enabling inward transport of oxygen to make a significant or predominant contribution to scale growth[2,21]. This also leads to total contact between the scale and the alloy[22].

There is evidence to support some of these hypotheses, but the relative importance of any particular theory has not yet been established conclusively. Ion implantation gives a method of examining this process in more detail. Since the technique enables the reactive element to be implanted to a known, shallow depth into the alloy (Ni-20%Cr in this research), the subsequent oxidation behaviour of the implanted specimen should provide further insight into the influence of the additions and should certainly enable the location of the mechanism (i.e. in the alloy, in the scale or at the scale/alloy interface) to be ascertained. In the research reported in this paper, an investigation of the influence of several implanted reactive elements on the oxidation of Ni-20%Cr has been carried out. Particular emphasis has been placed on the effect of the implanted species on diffusion processes in the scale and on its growth mechanism. Although the research has investigated several phenomena, such as surface pretreatment prior to implantation and the influence of a post-implantation internal oxidation treatment, this paper does not consider these in any detail, but is concerned with the general effects of implantation on oxidation at 800° to 1000°C. More specific features will be reported in future publications.

EXPERIMENTAL

The alloy was prepared by vacuum melting nickel (99.97%) and chromium (99.95%) and hot and cold rolling the ingot to 0.75 mm thickness. Specimens (15 x 5 mm) were abraded to 600 grit SiC paper, annealed in an evacuated, sealed capsule containing Cr/Cr_2O_3 powder for 4h at 1050°C to give a relatively large grain size, and electropolished in a perchloric acid (40 ml)/glacial acetic acid (450 ml)/water (15 ml) solution at 40V to remove about 20 μm of metal from each surface. The specimens were then polished on 6 μm diamond paste, degreased, dried and given an appropriate implantation treatment in a Cockcroft-Walton accelerator. An area approximately 10 x 5 mm of each side of the specimen was implanted. The implanted species and conditions are

Table 1

Conditions of Implantation

Implant species	Energy, KeV	Dose, Ion/cm^2	Mean penetration depth, nm
Cr	170	2.10^{16}	100
Al	100	2.10^{16}	100
Si	120	2.10^{16}	100
Sc	156	2.10^{16}	100
Ca	140	2.10^{16}	100
Ce	250	2.10^{16}	67
Y	280	2.10^{16}	100

given in Table 1. Most of these species were selected because they are more reactive with oxygen than chromium. Following implantation, some specimens were given a pre-internal oxidation treatment, to precipitate the reactive element implanted species as its internal oxide prior to the subsequent oxidation procedure. Other specimens were oxidized directly without this pre-internal oxidation treatment.

For the pre-internal oxidation treatment, the specimens were sealed into one arm of an evacuated dumb-bell shaped silica tube, the other arm containing a mixture of chromium and Cr_2O_3 powders. The tube was heated at 1000°C for 5 h. The powder mixture produces an oxygen activity in the tube which is below the level required to form a stable layer of Cr_2O_3 or NiO on the alloy but is sufficient to oxidize internally the more reactive implanted elements.

The specimens were subsequently oxidized in a microbalance without any further surface treatment. The oxidation tests were carried out at temperatures from 800° to 1000°C in pure, dry flowing oxygen. The specimen was suspended from one arm of the microbalance and located in an alumina reaction tube. Following flushing with oxygen, the preheated furnace was raised around the tube so the specimen was located in the hot zone, controlled to ± 2°C. Weight changes during oxidation and subsequent cooling to room temperature were monitored continuously. Following oxidation, the specimens were examined by various electronoptical techniques.

RESULTS

Kinetics of Oxidation

The kinetics of oxidation of the implanted surfaces were
determined from a knowledge of the area of implanted surface (S_1)
and the area of unimplanted surface (S_2) and comparison of the
weight changes of the implanted specimen with those of an unimplanted specimen under otherwise identical conditions, Then,

$$\Delta W_1 = \frac{\Delta W - \Delta W_2 S_2}{S_1}$$

where ΔW_1 is the weight gain per unit area of the implanted region,
ΔW is the total weight gain of the implanted specimen,
ΔW_2 is the weight gain per unit area of the unimplanted
specimen after the same exposure period.

Figure 1 shows typical oxidation kinetics for implanted and unimplanted surfaces at 1000°C (no pre-internal oxidation treatment).
The unimplanted alloys show an initial relatively rapid oxidation
rate due to the formation of transient nickel-rich oxides which
thicken much more quickly than Cr_2O_3. However, after a few hours
exposure, the oxidation rate decreases considerably as a healing
Cr_2O_3 layer develops at the base of the nickel-rich oxide. This
layer is much more protective than the nickel-rich oxide and
thickens at a slower rate.   The reproducibility of the kinetics
for the unimplanted surfaces was quite good, as indicated by the
scatter bands for five runs carried out under identical conditions
(Fig. 1).

Most of the implanted species had an effect on these oxidation
kinetics, although this was most dramatic for yttrium and cerium.
In general, the initial oxidation rate was independent of the
nature of the implanted species and was the same for unimplanted
and implanted surfaces. However, the time to the transition from
the initial rapid oxidation rate to the slower rate associated
with the development of the Cr_2O_3 healing layer was influenced
markedly by the nature of the implanted species. Implantation with
chromium resulted in the changeover occurring in a shorter period
than for unimplanted surfaces, probably due to the higher surface
concentration of chromium in the former case, although the
physical damage associated with the implantation process itself
may have influenced the kinetics to a small degree. Implantation
with cerium, yttrium, calcium, silicon and, to a lesser extent,
scandium and aluminium, resulted in a relatively rapid transition
to the slower oxidation rate.   Cerium and yttrium were
particularly effective in this respect and thus most emphasis has
been placed on studying the influence of these two implanted
species in this research.

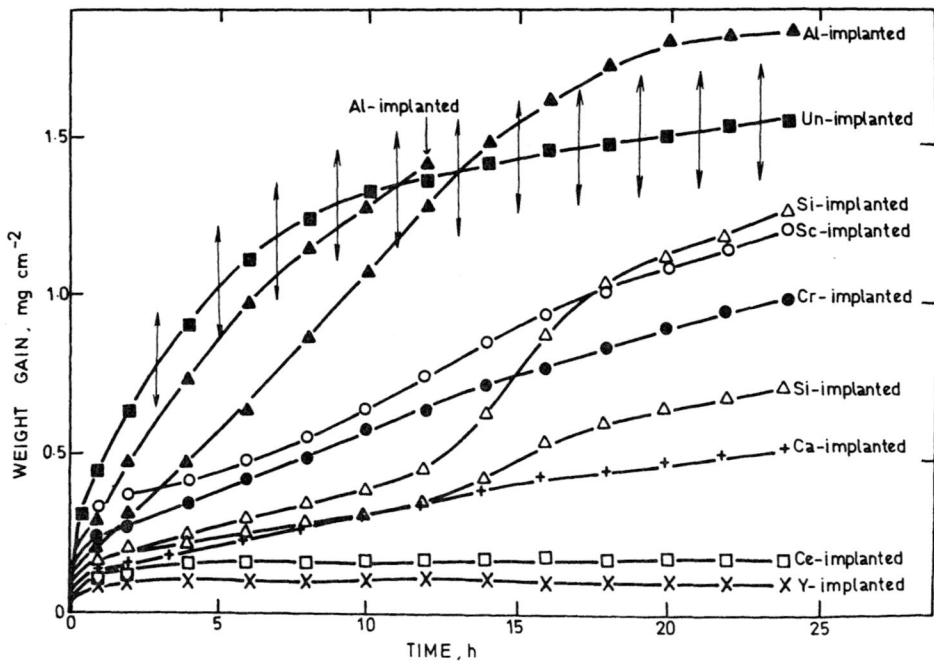

Figure 1. Typical oxidation kinetics for unimplanted and implanted Ni-20%Cr during exposure at 1000°C in 1 atm oxygen (no pre-internal oxidation treatment).

Following the transition, the rates of weight gain for the cerium- and yttrium-implanted surfaces were very low, being significantly less than for the unimplanted surfaces or for the chromium-implanted surfaces. These effects were very reproducible. However, the subsequent kinetics for the silicon-, scandium-, and, particularly, aluminium-implanted surfaces were less reproducible, with slow rates of weight gain after the transition being followed by breakaway-type behaviour. Such irregular kinetic effects were associated with the mechanical integrity of the scales and are not considered further in this paper. Longer-term exposures of the cerium- and yttrium-implanted surfaces (up to 400h) indicated that the very low rates of weight increase were maintained throughout this period.

Tests were also carried out at lower temperatures. However, it was found that the effects of yttrium- and cerium-implantation on the oxidation kinetics became less apparent with decreasing temperature (Fig. 2). At 800° and 900°C, implantation of cerium

468

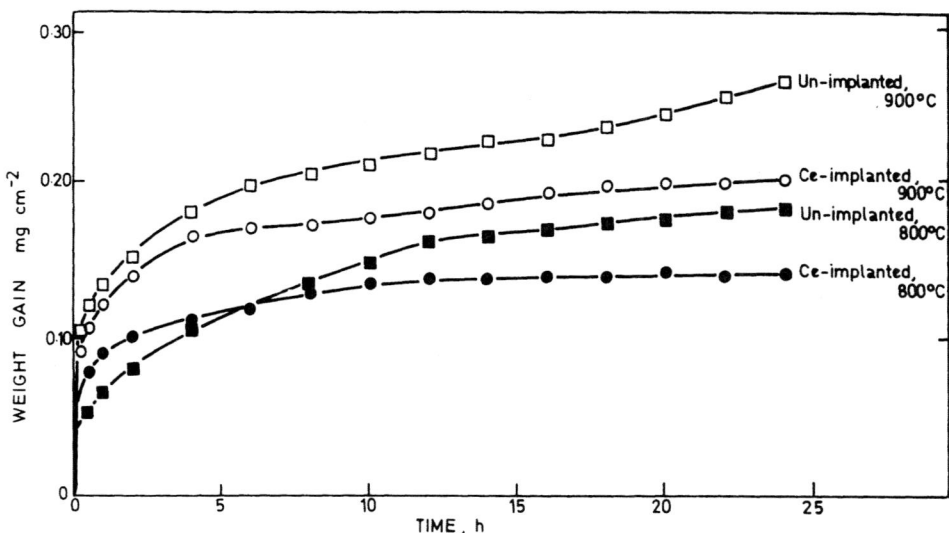

Figure 2. Typical oxidation kinetics for unimplanted and cerium-
implanted Ni-20%Cr during exposure at 800° and 900°C in
1 atm oxygen (no pre-internal oxidation treatment).

or yttrium had relatively little influence on the time to the
transition from the initial rapid oxidation rate to the slower
rate, although the rate of weight gain in the later stages was
lower for the implanted surfaces.

Morphologies and Compositions of Internal Oxides

Following the internal oxidation pretreatment, yttrium- and
cerium-implanted specimens were thinned from the reverse side by
ion beam etching and examined in the analytical transmission
electron microscope. This showed that the reactive elements had
precipitated as internal oxide particles in the alloy, typically
5 to 50 nm in diameter, separated by 7.5 to 100 nm (Fig. 3).
Analysis and electron diffraction confirmed that the precipitates
were Y_2O_3 and CeO_2 respectively. It is expected that similar
internal oxidation would occur for implanted specimens exposed
directly to the oxidation environment without an internal
oxidation pretreatment.

Features of the Oxide Scales

Emphasis has been placed on the scales developed on yttrium-
and cerium-implanted surfaces at 1000°C where the effect of

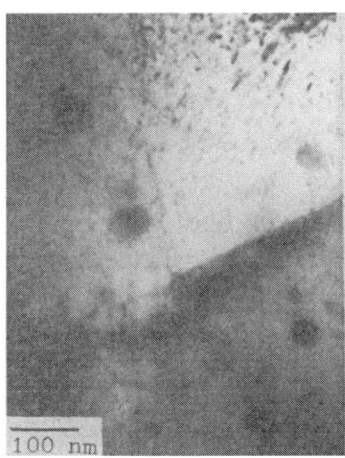

a) Yttrium-implanted   b) Cerium-implanted

Figure 3. Transmission electron micrographs of internal oxide particles in implanted Ni-20%Cr after internal oxidation treatment.

implantation is most dramatic. Examination of unimplanted surfaces after oxidation indicated that significant amounts of scale spallation had occurred on cooling, particularly from over the intersections of the alloy grain boundaries with the surface (Fig. 4(a)). In areas where the scales had not spalled, they had developed a convoluted configuration over the grain boundaries. (Fig. 4(b)). The alloy surfaces exposed by scale spallation were smooth on a microscale but contained undulations and relatively deep voids, while the grain boundaries were deeply grooved, indicating that the scale had not been in contact with the alloy at temperature (Fig. 4(a)), as expected from previous work[23,24]. The scales developed on the yttrium- and cerium-implanted surfaces were very adherent, with no loss of scale being observed. Figure 4(c) shows the interface between implanted and unimplanted surfaces and reveals the considerable differences in scale adhesion characteristics. In addition, the scale over the alloy grain boundaries of the implanted surfaces was not convoluted at all, but remained in close contact with the alloy across the boundary (Fig. 4(d)), suggesting a possible change in growth process. Analysis of these implanted surfaces indicated that the outer scale was NiO over the alloy grains but was largely Cr_2O_3 over the alloy grain boundaries. However, analysis of unimplanted surfaces where the scale was retained showed that the outer scale was NiO over the alloy grains and over the grain boundaries. The significant differences in scale thickness between the implanted and unimplanted

470

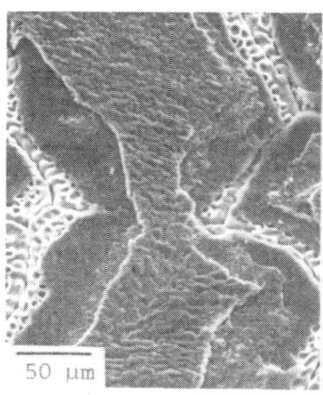

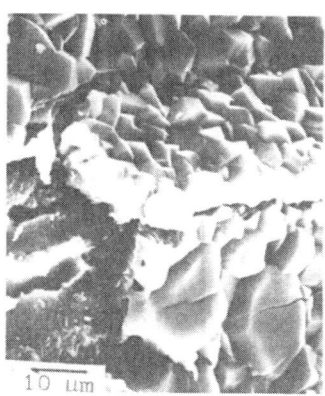

a) Unimplanted area, showing
   loss of scale from over alloy
   grain boundary.

b) Unimplanted area, showing convoluted scale over alloy grain
   boundary.

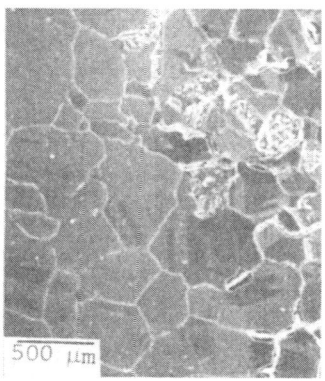

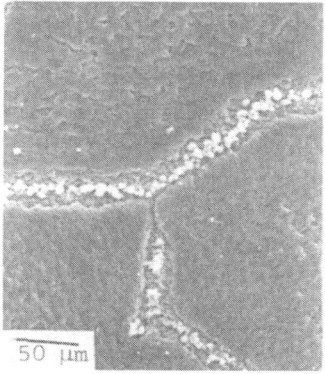

c) Cerium-implanted, showing
   interface between implanted
   (LHS) and unimplanted area(RHS)

d) Yttrium-implanted area.

Figure 4. Scanning electron micrographs of surface of Ni-20%Cr
          after oxidation for 24h at 1000°C.

areas are illustrated in figure 5(a). A thick NiO scale with a
healing Cr_2O_3-rich layer at the scale/alloy interface is present
on the unimplanted surface, while a significantly thinner scale is
present on the implanted surface. The latter scale consists of a
thin NiO outer layer and a thin Cr_2O_3 inner layer, both of which
are thinner than the corresponding layers on the unimplanted
surfaces (Fig. 5(b)). The scale developed on the chromium-
implanted surface was intermediate between those developed on the
unimplanted and the cerium-implanted surfaces, although the
thickness of the inner Cr_2O_3 layer was similar to that on the
unimplanted surface (Figs. 5(c)-(e)). Table 2 summarizes the
average thicknesses of the outer NiO- rich scale and the inner
healing Cr_2O_3-rich layer on various surfaces after 24h exposure

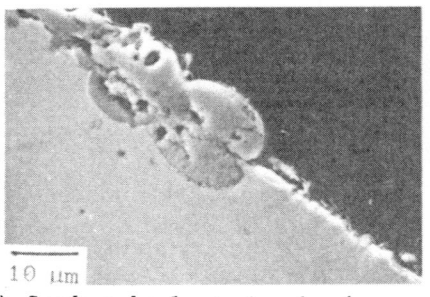

a) Cerium-implanted, showing interface between unimplanted (LHS) and implanted (RHS) areas.

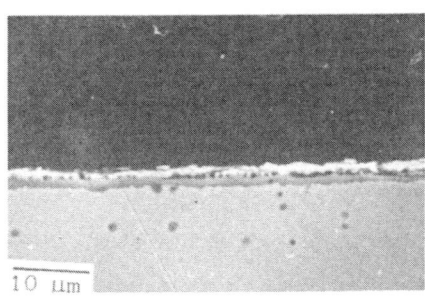

b) Cerium-implanted

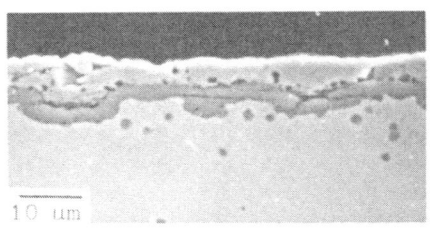

c) Chromium-implanted

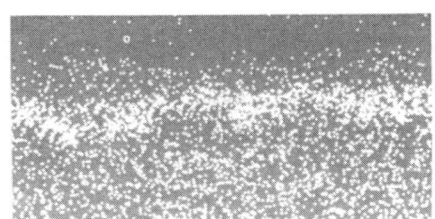

d) As c), showing chromium distribution.

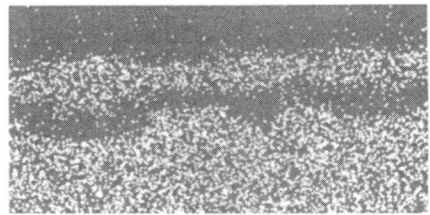

e) As c), showing nickel distribution.

Figure 5. Scanning electron micrographs of cross-sections of scale and alloy for Ni-20%Cr oxidized for 24h at 1000°C.

at 1000°C. (The measurements were made of scale over the alloy grains, not over the grain boundaries, where the NiO scale was invariably thinner for all specimens and non-existent for some of the implanted surfaces). These confirm that the scales on the yttrium-, cerium- and calcium-implanted surfaces are thinner than those on the unimplanted surfaces. Also, the inner Cr_2O_3 layers as well as the outer NiO-rich layers are thinner on these implanted surfaces. It is also interesting that there is a significant NiO-rich scale on the implanted surfaces, at least over the alloy grains, making it unlikely that significant loss of Cr_2O_3 (by volatilization of CrO_3) occurred during exposure at 1000°C.

Table 2

Average thicknesses of oxide scales on unimplanted and implanted Ni-20%Cr after oxidation at 1000°C in 1 atm oxygen.

Oxidation time	Type of Surface	Thickness of Cr_2O_3 layer, μm	Thickness of NiO layer, μm	Total scale thickness, μm
24h	Unimplanted	5.0	6.5	11.5
	Cr-implanted	4.9	5.5	10.5
	Ce-implanted	2.1	2.5	4.6
	Y-implanted	1.5	2.4	3.9

A particular effect of ion implantation of cerium or yttrium was to reduce quite considerably the grain size of the Cr_2O_3 scale (figure 6). The average sizes of the grains (Table 3) indicated that, for a given specimen, the grains were larger on the outer side of the Cr_2O_3 layer than on the inner side. Also, the grains were considerably larger for the unimplanted than for the implanted specimens, consistent with reported observations for reactive element alloying additions[2].

A major part of this research has been to locate the implanted element after oxidation. All the techniques used confirmed that the implanted yttrium or cerium was always incorporated into the scale and was not segregated at the scale/alloy interface. Also, the implanted species were always present in the Cr_2O_3 layer, not in the outer NiO-rich scale. Figure 7 shows that yttrium is associated with the Cr_2O_3 scale and not with the NiO scale. X-ray diffraction of the Cr_2O_3 scales, after stripping from the metal, indicated the presence of CeO_2 and Cr_2O_3 for the cerium-implanted surfaces.

Specimens for examination in the analytical transmission electron microscope were prepared by ion beam thinning in plan from both the inner and outer sides of the scale, enabling various locations across the scale to be analyzed. Energy dispersive analysis was able to detect yttrium (or cerium) in the Cr_2O_3 scale. However, it was not distributed uniformly, nor was it concentrated at the scale/metal interface or in the oxide grain boundaries. Regions enriched in the implanted species were detected within the oxide grains and at the grain boundaries. Conversely, other regions in the grains and grain boundaries did not contain any implanted species, at least not within the detection limit of the instrument. The implanted species-enriched regions occurred throughout the Cr_2O_3 after short exposure periods when the scale was thin. However, after longer periods (e.g. 24h at 1000°C),

Table 3

Average grain sizes of the Cr_2O_3 scales on unimplanted and implanted Ni-20%Cr, oxidized under various oxidation conditions.

Oxidation Conditions	Surface Conditions	Average grain size, nm	
		Top side of the Cr_2O_3 scale	Underside of the Cr_2O_3 scale
900°C for 24h	Unimplanted	980	250
	Ce-implanted	250	125
1000°C for 24h	Unimplanted	1400	500
	Y-implanted	450	136
1000°C for 6 min	Unimplanted	125	-
	Y-implanted	Not clear but < 125	-

when the scale was thicker than the original depth of implantation (100 nm), the implanted species-enriched regions were located towards the outside of the Cr_2O_3 scale only, consistent with inward growth of the scale. Electron diffraction indicated that these enriched regions were Y_2O_3 and possibly a mixed yttrium-chromium oxide ($YCrO_4$ or $YCrO_3$) for the yttrium-implanted surfaces and CeO_2 only for the cerium-implanted surfaces.

DISCUSSION

Implantation of yttrium or cerium into Ni-20%Cr produces similar effects as alloying additions of these elements on the oxidation behaviour, especially at 1000°C. In particular, a complete, healing Cr_2O_3 layer develops more rapidly on the implanted surfaces, resulting in a shorter transient oxidation period, the Cr_2O_3 layer thickens at a slower rate and the scale adhesion is improved considerably on cooling to room temperature. These effects persist for long periods, even when the scale is much thicker than the original depth of implantation. Also, the grain size of the Cr_2O_3 layer is much smaller on the implanted surfaces than on the unimplanted surfaces. Other implanted species, such as silicon and calcium, were partially effective, but less so than yttrium or cerium, while chromium, scandium and aluminium had relatively little effect. An important observation has been that the Cr_2O_3 healing layer on the implanted surfaces (notably cerium- and yttrium-) grows inwards while that on the unimplanted surfaces grows outwards. This is in agreement with other research on

a) Topside of Cr_2O_3 scale on unimplanted surface.

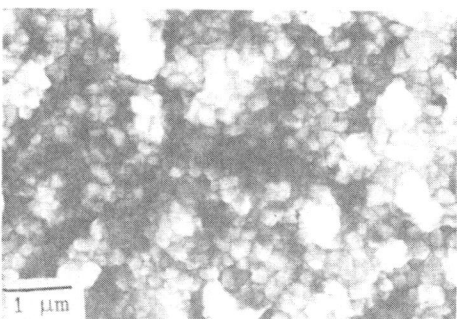

b) Underside of Cr_2O_3 scale on unimplanted surface.

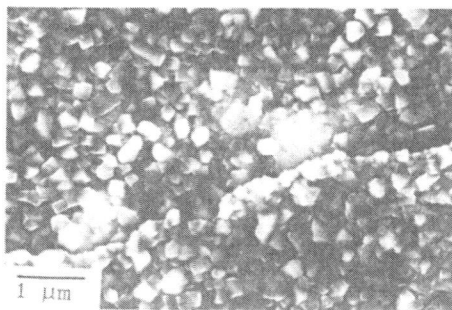

c) Topside of Cr_2O_3 scale on yttrium-implanted surface

d) Underside of Cr_2O_3 scale on yttrium-implanted surface.

Figure 6.   Scanning electron micrographs of inner Cr_2O_3 scale on Ni-20%Cr oxidized for 24h at 850°C.

Ni-20%Cr containing dispersed oxides of ThO_2[5,25] or Y_2O_3[2]. Similarly, ^{18}O tracer studies of cerium-coated 20/25 Nb stainless steel showed that the scale-forming reaction occurred at the scale/ alloy interface[24].

In the early stages of oxidation of the unimplanted surfaces, a considerable amount of NiO develops and thickens relatively rapidly while Cr_2O_3 forms as discrete precipitates in the sub-jacent alloy which are subsequently incorporated into the thick-ening NiO scale. The alloy grain boundaries provide easy diffusion paths for chromium from the bulk to the surface, enabling sufficient Cr_2O_3 to precipitate at the scale/alloy interface and coalesce to form, after some time, a continuous healing Cr_2O_3 layer which penetrates sideways from the intersection of the alloy grain boundaries with the interface. Eventually, the layer becomes complete and further oxidation involves thickening of this Cr_2O_3 scale. As this layer thickens, contact is lost with the alloy over the grain boundaries and a convoluted scale develops. The reasons for the development of these convolutions are not fully

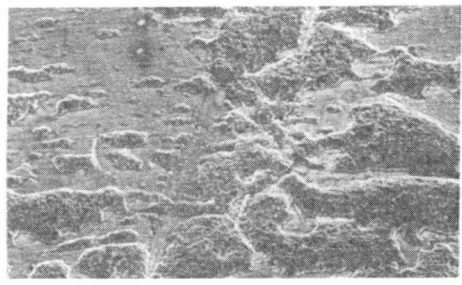 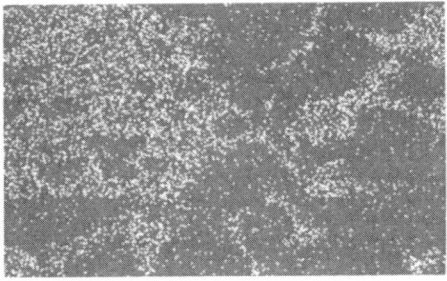

a) Showing NiO-rich nodules and   b) As a), yttrium distribution.

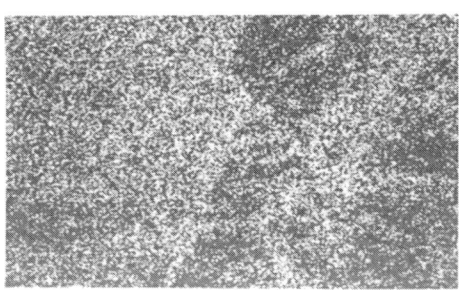 

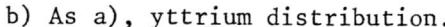

c) As a), chromium distribution   d) As a), nickel distribution.

Figure 7. Scanning electron micrograph of the surface of yttrium-
          implanted Ni-20%Cr after oxidation for 24h at 850°C.

understood, but may be associated with lateral growth of the scale
[21,24]. The resulting scale is stressed and out of contact with the
metal, leading to failure under the imposition of differential contraction stresses on cooling to room temperature.

    Scale growth on the implanted surfaces is similar except that
a Cr_2O_3 healing layer develops more rapidly and thus less transient
NiO is able to form. The fine grain size of the Cr_2O_3 scale
suggests that the more rapid development of the healing layer is
associated with an increased number of Cr_2O_3 nuclei on the alloy
surface in the initial stages. There is correlation between the
distribution of the small CeO_2 or Y_2O_3 precipitates formed in the
pre-internal oxidation treatment and the grain size of the Cr_2O_3
in the period immediately following its development as a complete
layer, indicating that these precipitates act as nucleation sites
for Cr_2O_3 [2]. Subsequently, the Cr_2O_3 layer thickens at a slower
rate than that on the unimplanted surfaces and grows inwards rather
than outwards. Hence, the implanted species influences the
diffusion processes in the Cr_2O_3 layer, suppressing the outward
transport of chromium without suppressing the inward transport of
oxygen.

At present, it is not possible to account fully for this effect on the diffusion processes, particularly as the transport paths for chromium in the scale are not really known. Chromium ions may diffuse via interstitial sites or via cation vacancies in the bulk lattice[27]. Alternatively, they may diffuse via short-circuit paths such as oxide grain boundaries or dislocation pipes[2]. Until these paths are identified, it is not possible to specify how the diffusion is inhibited by the implanted species. However, the implanted element apparently enters the scale as second-phase particles. There is some interaction between the Cr_2O_3 and the Y_2O_3 particles to form a mixed phase but apparently little between the Cr_2O_3 and CeO_2. Indeed, long-term sintering experiments of compacted CeO_2 and Cr_2O_3 powders at 1000°C and 1100°C have failed to produce any new phases when examined by diffraction[28]. Also, analysis was unable to detect any of the implanted species distributed uniformly in the oxide lattice, although they may have been present at levels below the detection limit of the instrument. There was no particular segregation of the second-phase particles to the oxide grain boundaries, so it is unlikely that the effect is due entirely to inhibition of oxide grain boundary transport. The fact that the implanted elements apparently occur as discrete second-phase precipitates in the oxide suggests that they may block short-circuit paths or even prevent their formation[2]. However, until the paths are identified, any further comment would be merely speculation.

It will be noted that, under the present conditions, there was no evidence for development of oxide pegs, changes in mechanical properties of the scale, formation of graded seals or of vacancy sinks for the reactive element implanted surfaces, although this does not rule out their being relevant to other situations.

CONCLUSIONS

1. Ion implantation of reactive elements has proved to be a useful technique for studying the influence of such elements on the high-temperature oxidation of Ni-20%Cr, since it enables the alloy surface to be modified and doped in a controlled manner.

2. Implantation of yttrium, cerium and, to a lesser extent, other reactive elements into Ni-20%Cr affects the oxidation behaviour in a similar manner as when the same elements are present as alloying additions. The development of a healing Cr_2O_3 layer occurs more rapidly. This layer thickens at a reduced rate and the adhesion of the scale to the alloy substrate is much improved.

3. Various techniques have shown that the Cr_2O_3 layer developed on the implanted surfaces has a much finer grain size than that on the unimplanted surfaces and apparently grows inwards rather than outwards as occurs on the unimplanted surfaces.

4. The change in growth direction of the Cr_2O_3 scale and its slower rate of thickening can be accounted for in terms of a suppression of chromium transport across the scale by the implanted species, enabling the slower-diffusing oxygen to predominate in the oxide growth process.

5. The implanted species, notably cerium and yttrium, are incorporated into the Cr_2O_3 layer during oxidation as second-phase oxide particles. These are retained as second-phase particles, any dissolution into the scale being below the detection limits of the techniques used in this investigation.

ACKNOWLEDGEMENTS

The authors are grateful to the Science and Engineering Research Council for a Research Studentship to one of them (JSP) and to the Atomic Energy Research Establishment, Harwell for financial support under the CASE scheme.

REFERENCES

1. L. B. Pfeil, U. K. Patent No. 459848 (1937).
2. J. Stringer, B. A. Wilcox and R. I. Jaffee, Oxidation Metals 5:11 (1972).
3. J. M. Francis and W. H. Whitlow, Corros. Science 5:701 (1965).
4. J. Stringer and I. G. Wright, Oxidation Metals 5:59 (1972).
5. C. S. Giggins and F. S. Pettit, Metall. Trans. 2:1071 (1971).
6. G. C. Wood and J. Boustead, Corros. Science 8:719 (1968).
7. I. G. Wright, B. A. Wilcox and R. I. Jaffee, Oxidation Metals 9:275 (1975).
8. B. Lustman, Trans. Met. Soc., AIME 188:995 (1950).
9. E. J. Felton, J. Electrochem. Soc. 108:490 (1961).
10. C. S. Wukusick and J. F. Collins, Mater. Res. Stand. 4:637 (1964).
11. J. E. Antill and K. A. Peakall, J. Iron Steel Inst. 205, 1136 (1967).
12. G. C. Wood, Werkst. u. Korros. 22:491 (1971).
13. D. P. Whittle and J. Stringer, Phil. Trans. Roy. Soc. A. 309 (1979).
14. J. M. Francis and J. A. Jutson, Corros. Science 8:445 (1968).
15. H. Pfeiffer, Werkst. u. Korros. 8:574 (1957).
16. J. Stringer, Oxidation Metals 5:49 (1972).
17. J. K. Tien and F. S. Pettit, Metall. Trans. 3:1587 (1972).
18. I. A. Kvernes, Oxidation Metals 6:45 (1973).

19. C. S. Giggins, B. H. Kear, F. S. Pettit and J. K. Tien, Metall. Trans. 5:1685 (1974).
20. G. C. Wood, J. A. Richardson, M. G. Hobby and J. Boustead, Corros. Science 9:659 (1969).
21. F. A. Golightly, F. H. Stott and G. C. Wood, Oxidation Metals 10:163 (1976).
22. G. C. Wood and F. H. Stott, Development and Growth of Protective α-Al$_2$O$_3$ Scales on Alloys, in: "High-Temperature Corrosion", ed. R. A. Rapp, NACE 6:227 (1983).
23. D. Mortimer and M. L. Post, Corros. Science 8:499 (1968).
24. D. Caplan and G. I. Sproule, Oxidation Metals 9:459 (1975).
25. G. R. Wallwork and A. Z. Hed, Oxidation Metals 3:229 (1971).
26. P. Skeldon, J. M. Calvert and D. G. Lees, Phil. Trans. Roy. Soc. London A296:557 (1980).
27. F. A. Kruger, Defects and Transport in SiO$_2$, Al$_2$O$_3$ and Cr$_2$O$_3$, in: "High-Temperature Corrosion", ed. R. A. Rapp, NACE-6:89 (1983).
28. J. S. Punni, Ph.D. Thesis, University of Manchester (1983).

DIFFUSIONAL GROWTH OF Al_2O_3 SCALES AND SUBSCALES ON Ni-Al, Fe-Al

AND Fe-Mn-Al ALLOYS

W. W. Smeltzer, P. R. S. Jackson and H. A. Ahmed

Institute for Materials Research
McMaster University
Hamilton, Ontario, Canada, L8S 4M1

ABSTRACT

Oxidation of Ni-Al, Fe-Al and Fe-Mn-Al alloys at high
temperature exhibits transitions from scale-subscale growth to
scale growth dependent upon alloy aluminum contents. Alloy
compositions presently can be specified leading to growth of
scales containing an Al_2O_3 layer which acts as the most protective
barrier presently available for imparting high temperature
corrosion resistance to these alloys. It is demonstrated that
this Al_2O_3 layer can exist either as an external film or beneath
oxide layers composed of solvent alloy metal oxides, dependent
upon alloy composition and structure.

INTRODUCTION

Nickel and iron alloys with aluminum at concentrations
sufficient for formation of oxide scales comprising a continuous
alumina layer are becoming of significantly more importance as
materials for technological applications where high temperature
corrosion resistance is required. Binary Ni-Al alloys are FCC
austenitic (γ) solid solutions up to an aluminum solubility of
$\sim$ 6 wt %; in contrast, Fe-Al alloys are of BCC ferritic (α)
structure up to much larger aluminum contents of $\sim$ 34 wt %. Iron
alloys based on the ternary Fe-Mn-Al series, moreover, can be
wholly austenitic, ferritic on duplex ($\alpha + \gamma$) dependent upon the
concentrations of the alloying elements and temperature. The
purpose of this work, accordingly, is to demonstrate the influence
of alloy composition and structure on the morphological develop-

ment and transport properties of oxidation resistant scales containing alumina found on γ Ni-Al, α Fe-Al and duplex $(\alpha + \gamma)$ Fe-Mn-Al alloys.

OXIDATION OF Ni-Al ALLOYS

The oxidation kinetics, compositions and structures of the scales and subscales formed on the γ Ni-Al solid solution phase have been subjects of a recent survey paper.[1] In the case of this alloy approaching its limiting aluminum solubility, oxidation ultimately leads to growth of a multilayered $NiO/NiAl_2O_4/Al_2O_3$* scale as depicted schematically by the models in Fig. 1, (a) to (d).

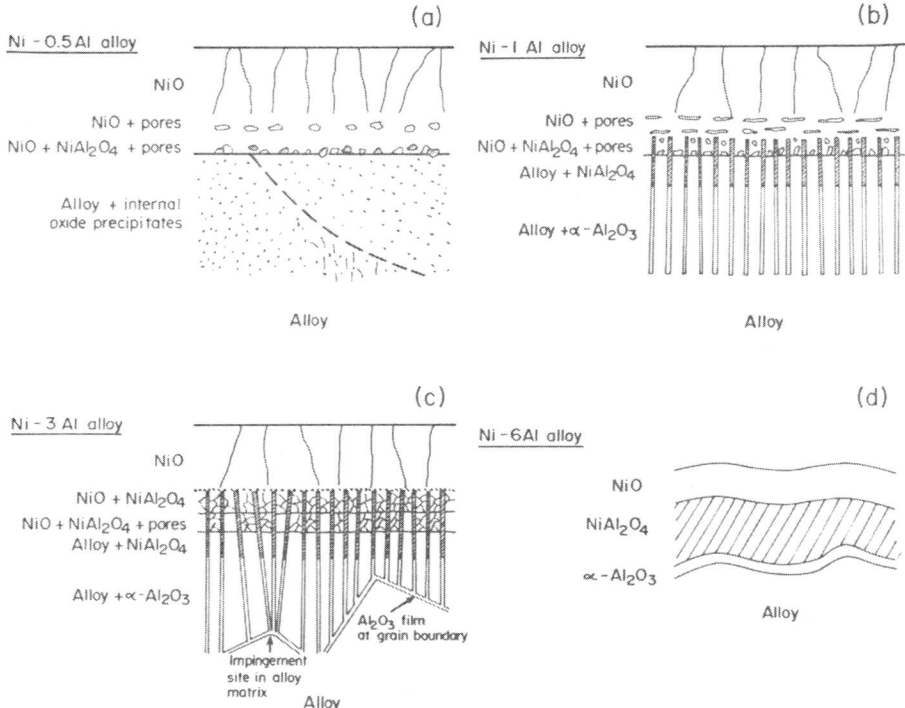

Fig. 1. Schematic models for scale and subscale growth on the Ni-Al γ-solid solution alloys.

---

*Al_2O_3 or alumina is used in the subsequent text for α-Al_2O_3.

Alloys containing up to 3 wt. % Al exhibited growth of scales
comprised of an external columnar NiO layer and an inner
NiO + $NiAl_2O_4$ layer and subscales of $NiAl_2O_4$ + Al_2O_3 composed of
either small dispersed precipitates or rods within the alloy
matrix, Fig. la, b. Oxidation kinetics were parabolic at temperatures in the range 1273-1473 K; the growth of the external scale
was largely controlled by the outward diffusion of metal and
inward penetration of the subscale was dependent upon inward
oxygen diffusion in the alloy. Oxygen required for growth of
Al_2O_3 as small particles or rods at the internal oxidation front
arose from the dissociation of NiO according to the reaction

$$3 \ NiO_{(scale)} + 2 \ Al_{(alloy)} = Al_2O_{3(alloy)} + 3 \ Ni_{(scale)}$$

with the resultant nickel diffusing through the scale to react
with atmospheric oxygen. Simultaneous reaction of Al_2O_3 with
nickel and oxygen in the alloy at intermediate depths within the
subscale caused inward penetration of $NiAl_2O_4$. Outward growth of
this oxide phase also occurred in the inner scale layer by
diffusion of aluminum through $NiAl_2O_4$ and release of nickel which
diffused through the NiO layer.

Transition from scale-subscale growth to growth of an
external scale having a continuous layer of Al_2O_3 at the alloy
interface occurred with aluminum contents of $\sim$ 3-6 wt. %, Fig. ld;
lateral growth of the alumina film was initiated at the internal
oxidation front from Al_2O_3 precipitates located at appropriately
oriented grain boundaries and at blockage sites for aluminum and
oxygen diffusion caused by impingement of Al_2O_3 rods in the alloy
matrix, Fig. lc. Complete development of this irregular Al_2O_3
film by the above two modes finally isolated the internal
oxidation zone and eventually led to conversion of the
scale/subscale layers to a multilayered $NiO/NiAl_2O_4/Al_2O_3$ scale.

It is apparent that predominant diffusion species during
growth of the scales and subscales were nickel in the scales and
oxygen in the alloy, the fluxes of which were drastically reduced
upon formation of the continuous Al_2O_3 layer. This behavior
reduced the alloy oxidation rate. It was possible to demonstrate
that this oxidation mechanism is consistent with diffusional
properties of the oxides and alloy phases and the thermodynamic
constraints of the Ni-Al-O system.[1,2]

OXIDATION OF Fe-Al ALLOYS

The beneficial effects of aluminum additions to the high
temperature oxidation resistance of iron have been the topic of
reviews.[3,4] Recent research has given a fairly detailed

classification of scale morphologies based on alloy composition.[5]
At 1073 K, stratified scales composed of Fe_2O_3 and Fe_3O_4 layers
with $FeAl_2O_4$ and Al_2O_3 dispersed near the metal interface are
formed on alloys with aluminum contents up to ~ 2.4 wt. %. Alloys
containing from 2.4 to 6.9 wt. % Al form an Al_2O_3 scale
interspersed with oxide nodules that penetrate to the alloy
surface. Continuous protective Al_2O_3 scales tend to form on
alloys containing more than 6.9 wt. % Al. The purpose of our
research has been to obtain an understanding of the transitions
leading to growth of the continuous Al_2O_3 layers on these alloys
as a function of alloy composition.[6]

    Alloys with aluminum contents ranging from 0.7 to 10 wt. % Al
were oxidized in oxygen at 1173 K and 10^5 Pa pressure. Scale and
subscale growth occurred on alloys of aluminum contents ranging up
to 1.5 wt. %. In the composition range 0.7-1.5 wt. % Al, a thick
striated scale composed of Fe_2O_3 and $(Fe,Al)_3O_4$ layers was formed
above an alloy-oxide subscale as shown in Fig. 1. The oxide of
the subscale which was composed of $FeAl_2O_4$ and Al_2O_3 extended into
the alloy as fine rectangular rods, Fig. 1b. At a higher alloy
aluminum content of 2.5 wt. %, an Al_2O_3 film grew and spread
laterally from the tips of the oxide rods to eventually isolate
the subscale from the alloy as shown in Fig. 3. Continued
oxidation up to long exposures of 60 days led ultimately to growth
of a three layered $Al_2O_3/(Fe,Al)_3O_4/Fe_2O_3$ scale.

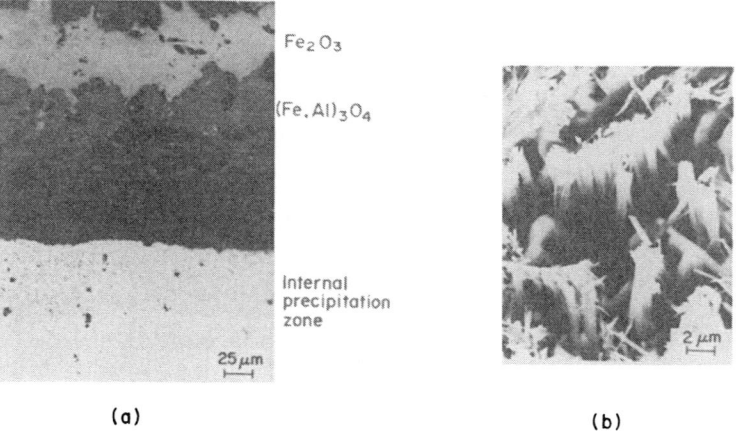

(a)   (b)

Fig. 2. Scale and subscale formed on Fe-(0.7-1.5) wt. % Al alloy:
(a) cross-section; (b) deep-etched lateral section of
subscale.

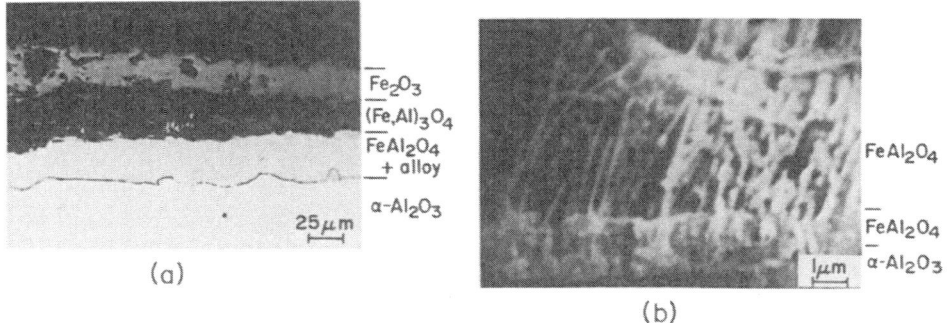

Fig. 3. Scale and subscale formed on Fe-2.5 wt. % Al alloy:
(a) cross-section; (b) deep-etched cross-section.

Subscale growth did not occur in alloys of aluminum contents greater than 3.5 wt. %. As illustrated in Fig. 4, an Fe-3.5 wt. % Al alloy formed a thin duplex Al_2O_3/Fe_2O_3 scale and an Fe-6 wt. % Al alloy formed a thin protective Al_2O_3 scale with undulating surface features. The scales which developed continuous Al_2O_3 layers grew rapidly during early exposure stages but ultimately attained high degrees of protection and very slow growth rates.

As shown schematically in Fig. 5, the stages in the development of the scales on α Fe-Al alloys exhibited a close correspondence to those stages for scale growth on the γ Ni-Al alloys. In

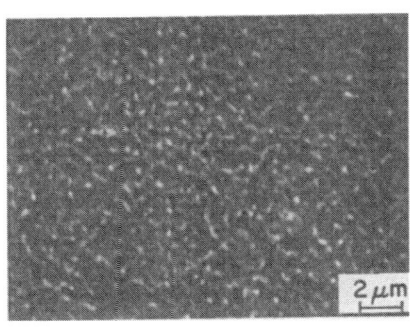 

Fig. 4. Scales formed on Fe-Al alloys: (a) Al_2O_3/Fe_2O_3 duplex scale on Fe-3.5 wt. % Al alloy; (b) Al_2O_3 scale formed on Fe-6 wt. % Al alloy.

the case of the Fe-Al alloys, oxygen required for growth of the $FeAl_2O_4 - Al_2O_3$ rectangular rods at the internal oxidation front was supplied by dissociation of the external scale inner layer according to the reaction

$$\frac{3}{4}(Fe,Al)_3O_{4(scale)} + 2\,Al_{(alloy)} = Al_2O_{3(alloy)}$$
$$+ \frac{9}{4}(Fe,Al)_{(scale)}$$

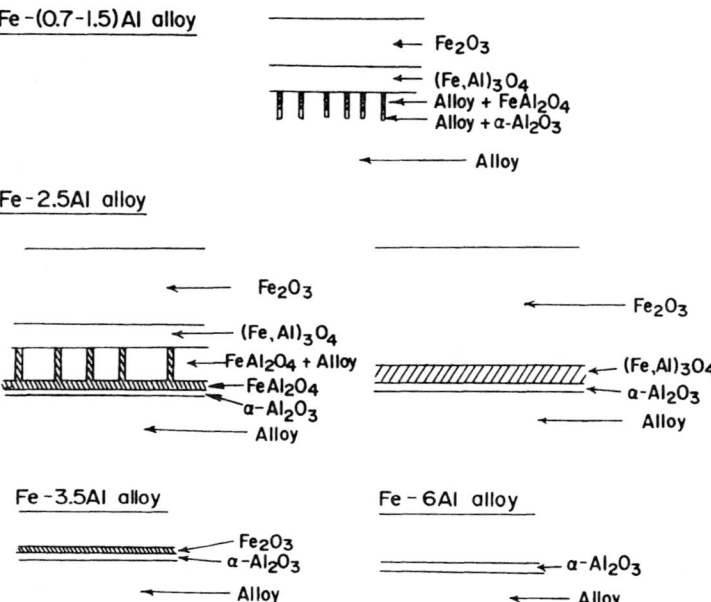

Fig. 5. Schematic models for scale and subscale growth of Fe-Al α-solid solution alloys.

with the resultant metals diffusing in and through the scale. Growth of scales on the α Fe-Al alloys, however, differed significantly from those on γ Ni-Al alloys in that the former alloys exhibited sufficient aluminum solubility for growth of an Al_2O_3 film directly on the alloy solid solution phase in the absence of subscale formation.

Oxidation behavior of these alloys is of particular interest due to the relationship between alloy structure and morphological development of the resulting scales.[7-9] This is represented in Fig. 6 by means of an oxide map which depicts compositional regions of the ternary alloy phase diagram over which similar scales occur. These alloys exhibit structures that are austenitic (γ-FCC), ferritic (α-BCC) or duplex ($\alpha + \gamma$) depending upon composition and temperature. Consequently the oxidation behavior of the Fe-Mn-Al alloys is classified according to their phase structures.

Austenitic alloys found to have optimum oxidation resistance have compositions within the range Fe-(15-35)Mn-(6-12)Al-(0-1)C in wt. % often balanced with additions of silicon, chromium and nickel.[9] The most protective scales are formed at temperatures up to 1073 K and consist of a continuous Al_2O_3 layer or $(FeMn)Al_2O_4$ which forms and grows beneath a manganese rich, iron-manganese oxide scale. At lower alloy aluminum concentrations or at higher temperatures, rapid internal oxidation prevents growth of the continuous Al_2O_3 layer and scales are composed of internally oxidized aluminum present as $(Fe,Mn)Al_2O_4$ and/or alumina and iron-manganese oxides.[9,10] Preferential depletion of carbon and/or manganese from the alloy during oxidation results in formation of ferrite adjacent to the scale.

Scales formed on ferritic alloys containing less than 10 wt. % Al are in many respects similar to those formed on Fe-Al alloys. Approximately 7-8 wt. % Al is required to form continuous

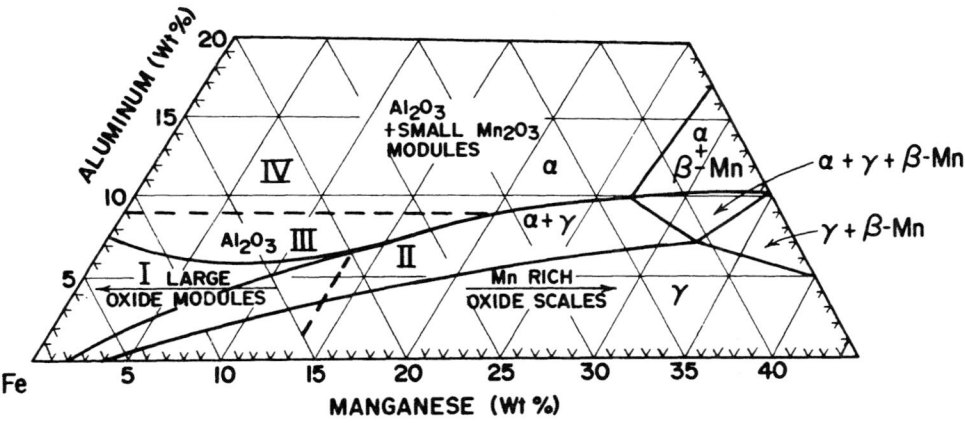

Fig. 6. Oxide map for the Fe-Mn-Al system at 1073 K.[8]

alumina scales (regions III and IV in Fig. 6) and, in contrast to austenites, such alloys exhibit extremely slow reaction kinetics at temperatures far exceeding 1073 K. Formation of austenite in a ferritic alloy by aluminum depletion during oxidation, results in the breakdown of pre-existing Al_2O_3 scales and growth of stratified scales as outlined above for austenitic alloys.[10]

Scales formed on duplex ($\alpha + \gamma$) alloys with compositions close to the upper boundary of region II in Fig. 6 consist of an external Al_2O_3 layer on the ferrite phase and stratified multicomponent oxide layers on the austenitic phase. Definitive statements concerning the factors which govern the development of these scale morphologies are difficult to make. An investigation, accordingly, using a duplex Fe-15 Mn-6 Al alloy was designed and carried out to determine the controlling factors governing alumina formation and growth on ferritic and austenitic alloy phases.[11] The structure of the annealed alloy containing grain boundary and grain interior austenite precipitates, of sizes and interprecipitate spacings generally < 10 μm is shown in Fig. 7. It is considered that the treatment given to the alloy was sufficient for equilibrium and attainment of equivalent elemental metal activities in both alloy phases.

Alloy Structure and Scale Morphology

Oxidation kinetics of this duplex ($\alpha + \gamma$) Fe-15 wt. % Mn-6 wt. % Al alloy in pure oxygen at 1023°K and 2.7×10^4 Pa pressure were non-parabolic and rapid for the first three hours followed by a declining rate to a very low value. The micrographs in Fig. 7 reveal the development of the scale. There was growth of a thick multiphase $\alpha - Mn_2O_3 + \alpha - Fe_2O_3$ layer on the austenite precipitates, Fig. 7b, which resulted in partial conversion of austenite to ferrite by preferential manganese depletion during oxidation.[7,9] Iron, manganese and aluminum X-ray images demonstrated manganese enrichment within the multiphase oxide layers. Aluminum enrichment occurred within the inner layer particularly at the alloy/scale interface, Fig. 7c and d, and elemental distributions were consistent with the suggestion that the inner oxide layer was cubic $(Fe,Mn)_3O_4$ and either Al_2O_3 or $(Fe,Mn)Al_2O_4$ at the metal interface. In view of the very low oxidation rates at long times it is believed that formation of the aluminum rich oxide layer at the alloy/scale interface reduced the further outward diffusion of metal. Elemental and phase distributions within the multiphase scale formed on austenite were consistent with previous results for nodular growths on ferritic alloys.[8] The thin protective scale formed on the ferrite phase was composed of an Al_2O_3 layer on the alloy covered by small Fe_2O_3 protrusions typically less than 1 μm in size. Stages in the development of the final scale on the duplex ($\alpha + \gamma$) alloy exhibiting a morphology of an innermost

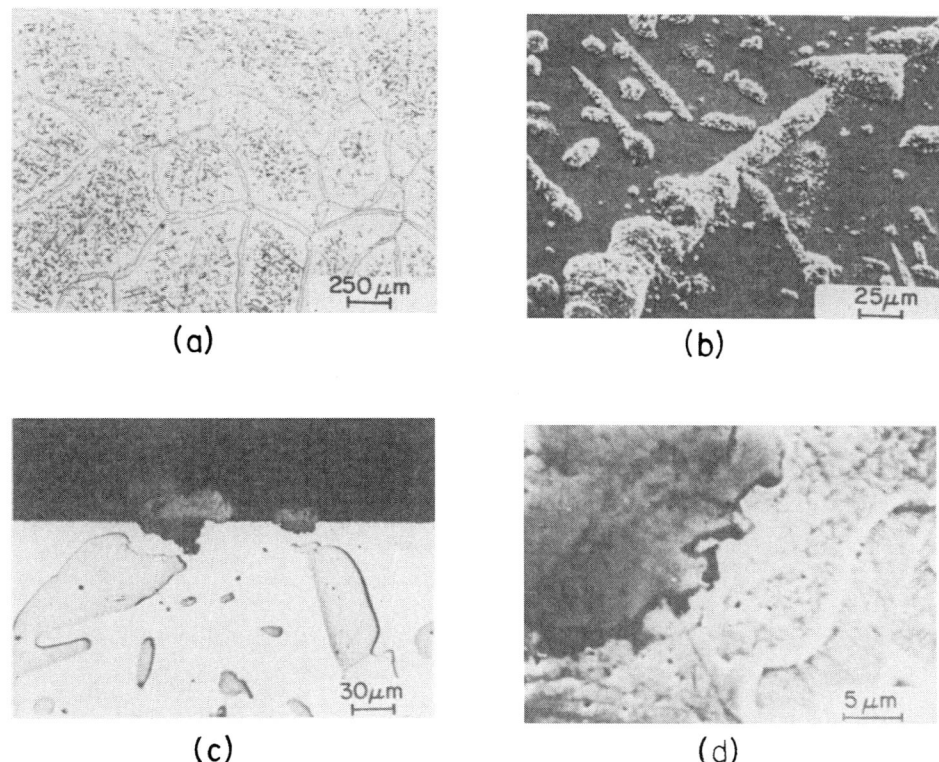

Fig. 7. Growth of oxides on Fe-15 wt. % Mn-6 Al wt. % alloy at 1023 K: (a) structure of unoxidized α + γ alloy showing austenite precipitates in ferrite matrix; (b) SEM micrograph showing thick oxides on the austenite precipitates; (c) optical cross-section of thick oxides of (b); (d) SEM micrograph showing distribution of Al as rich particles and as a thin layer adjacent to the alloy.

continuous alumina layer on the evolved ferrite alloy surface are shown schematically in Fig. 8.

DISCUSSION AND CONCLUSIONS

Factors which are of paramount importance in determining the type of scale containing alumina on the Ni-Al, Fe-Al and Fe-Mn-Al alloys under a given set of experimental conditions, whether it be as a thin external film, a continuous layer beneath solvent metal oxides or as a subscale are: (a) alloy aluminum activity,

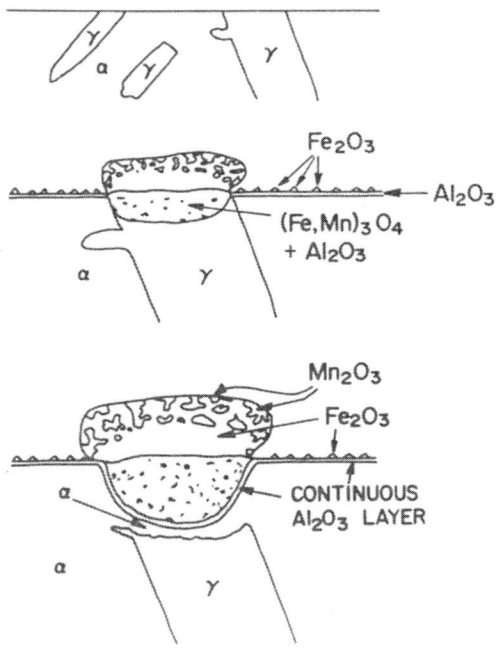

Fig. 8. Schematic models showing stages of scale and subscale
growth on the duplex ($\alpha+\gamma$) Fe-Mn-Al alloys.

(b) alloy diffusion coefficients, particularly D_{Al}, and (c) oxygen
flux into the alloy which is a function of oxygen diffusivity and
solubility. Growth of thin external alumina scales in preference
to the other morphologies described above is favorable for high
values of (a) and (b) and low values of (c).

Possibly the major difference between continuous Al_2O_3 scale
formation on α Fe-Al alloys and γ Ni-Al alloys is the ability of
the ferritic alloys to directly form this type of scale. The
comparative ease of Al_2O_3 growth on ferritic alloys as opposed to
γ Ni-Al alloys was considered previously as due to the differences
in aluminum diffusion rates and oxygen solubilities in the alloy
phases.[4] Diffusion,[12-16] aluminum activities[17,18] and
oxygen solubilities[19] are presented in Tables 1-3. At present,
similar data for ternary Fe-Mn-Al alloys are unknown.

490

In comparison to γ Ni-Al alloys, it is seen from Tables 1 and 2 that aluminum diffusivities and activites in α Fe-Al alloys are several orders of magnitude larger. Oxygen diffusion in nickel is several orders of magnitude lower than in iron; however, its solubility is somewhat greater. Since there is conjecture concerning the oxygen solubility data it is difficult to assess the influence of oxygen flux for both systems. Notwithstanding, one can conclude that high aluminum diffusivity and activity in the alloy favor formation of thin external alumina scales in ferritic Fe-Al alloys at lower alloy aluminum concentrations.

Table 1. Diffusion Coefficients (cm^2s^{-1}) in Metals and Binary Alloys

Phase	System	Diffusivity	1023 K	1173 K	1273 K	Ref.
α	Fe	D_O	7×10^{-7}	1×10^{-6}		12
γ	Fe	D_O		2×10^{-7}	7×10^{-6}	12
γ	Ni	D_O	8×10^{-12}	7×10^{-10}	8×10^{-9}	12
α	Fe	D_{Fe}	3×10^{-13}	4×10^{-11}		13
γ	Fe	D_{Fe}		3×10^{-13}	3×10^{-12}	13
γ	Ni	D_{Ni}				
α	Fe-(5-20) at. % Al	$\hat{D}$	$0.3\text{-}\times10^{-11}$	$0.8\text{-}5\times10^{-10}$	$0.8\text{-}3\times10^{-9}$	14,15
γ	Ni-Al	D_{Al}	5×10^{-14}	3×10^{-12}	2×10^{-11}	16

Table 2. Activity of Al in Binary γ Ni-Al and α Fe-Al Alloys at 1273 K.[17,18]

Ni- at. % Al	a_{Al}	Fe- at. % Al	a_{Al}
5	3×10^{-7}	5	9×10^{-4}
10	3×10^{-6}	10	2×10^{-3}
14	2×10^{-5}	15	3×10^{-3}

Table 3. Oxygen Solubility in Iron and Nickel.[19]

α Fe, γ Fe	$\sim 10^{-3}$ at. %
Ni	$\sim 10^{-1}\text{-}10^{-2}$ at. %

With respect to ternary duplex (α + γ) Fe-Mn-Al alloys, fixed and equal metal activities in both phases of the annealed alloy ingot rule out differences in metal activity as being responsible for the two observed scale morphologies. Oxygen diffusion is known to be higher in α iron than in γ iron, Table 1, and the scant oxygen solubility data, Table 3, indicate that oxygen solubilities in the metal phases are similar. It thus appears that the oxygen flux may be greater for ferritic alloys which would favor growth of alumina scales on austenitic alloys. This is not the case, which leads one to conclude that the two scale morphologies observed on the duplex (α + γ) Fe-15 wt. % Mn-6 wt. % Al alloy at 1023°K arise from the higher rate of aluminum diffusion in ferrite, favoring alumina scale growth on this alloy phase.

Transition from scale-subscale to external scale growth for these three alloys was associated with formation of a continuous protective Al_2O_3 layer at the alloy interface. During this period the oxidation kinetics decreased from early stage rapid rates to very slow rates at long times. It is only after this transitory period of oxidation that the more general criterion for the transition from internal oxidation to external scale formation could be expected to apply based on exceeding the solubility product for oxide formation at the scale/alloy interface as governed by diffusion and thermodynamics of the metal species and oxygen in the oxide and alloy phases.[20] The growth and morphological development of the Al_2O_3 layer, moreover, is determined by aluminum and oxygen transport and possibly reaction in the scale by mechanisms presently incompletely understood.[4]

ACKNOWLEDGEMENT

The authors express their appreciation to the Natural Sciences and Engineering Research Council of Canada for financial support of this research.

REFERENCES

1. W. W. Smeltzer, H. M. Hindam and F. A. Elrefaie, Growth and microstructures of oxide scales and subscales on Ni-Al γ-solid solution alloys, in: "High Temperature Corrosion," NACE 6, R. A. Rapp, ed., NACE, Houston (1983).

2. H. M. Hindam and D. P. Whittle, J. Matls. Sc. 18:1389 (1983).
3. P. Tomaszewicz and R. R. Wallwork, Rev. High Temp. Mats. 4:75 (1978).
4. G. C. Wood and F. H. Stott, The Development and Growth of Protective α-Al$_2$O$_3$ Scales on Alloys, in: "High Temperature Corrosion," NACE 6, R. A. Rapp, ed., NACE, Houston (1983).
5. P. Tomaszewicz and R. R. Wallwork, Oxid, Met. 19:165 (1983).
6. H. A. Ahmed and W. W. Smeltzer, to be published.
7. P. R. S. Jackson, The Oxidation of Iron-Manganese-Aluminum Based Alloys, Ph.D. Thesis, University of New South Wales, Sydney (1982).
8. P. R. S. Jackson and G. R. Wallwork, Oxid. Met., in press.
9. P. R. S. Jackson G. R. Wallwork, J. High Temp. Tech. 1:259 (1983).
10. J. P. Sauer, R. A. Rapp and J. P. Hirth, Oxid. Met. 18:285 (1982).
11. P. R. S. Jackson and W. W. Smeltzer, to be published.
12. D. Bergner, Diffusion of C, N and O in Metals, in: "DIMETA-82, Diffusion in Metals and Alloys," Proceedings of an International Conference, Tihany, Hungary, 30 Aug. - 3 Sept. 1982, F. J. Kedves and D. L. Beke, ed., Trans. Tech. Publication, Switzerland (1983).
13. S. Mrowec, "Defects and Diffusion in Solids," Elsevier Scientific Publishing Co., Amsterdam (1980).
14. H. C. Akuezue and D. P. Whittle, Metal. Sci. 17:27 (1983).
15. K. Nishida, T. Yamamoto and T. Nogata, Trans. Japan Inst. Metals 12:310 (1971).
16. W. Gust, M. B. Hintz, A. Lodding, H. Odelius and B. Predel, p. 82 in: "Diffusion and Defect Data," Vol. 25, F. H. Wohlbier, ed., Trans. Tech., Rockport (1981).
17. R. Hultgren, P. Desai, D. Hawkins, M. Gleiser and K. Kelly, "Selected Values of the Thermodynamic Properties of Binary Alloys," ASM, Metals Park (1973).
18. J. Eldridge and K. L. Komarek, Trans. Met. Soc. AIME 230:226 (1964).
19. M. Hansen, "Constitution of Binary Alloys," 2nd ed., McGraw-Hill, New York (1958).
20. W. W. Smeltzer and D. P. Whittle, J. Electrochem. Soc. 125:1116 (1978).

POINT DEFECT CONSIDERATIONS

THE BEHAVIOUR OF PROTONS IN OXIDES

A.N. Cormack, P. Saul, and C.R.A. Catlow

Department of Chemistry
University College London
20 Gordon Street, London WC1H OAJ, UK

INTRODUCTION

It is becoming increasingly apparent that there is an important need
to understand the behaviour of protons in oxides. It has been
suggested, for example, that all oxides, even of nominal high purity,
will contain some residual hydrogen that is, for practical purposes,
impossible to remove: this has been termed 'unextractable water'(1),
although its exact nature is still under debate. There are, conflicting reports in the literature: some features of the ESR spectra of
rutile, for example, have been ascribed to OH^- groups (2), whereas
IR spectra have been interpreted in terms of both OH groups (3) and
H^- ions trapped at oxygen vacancies(4).

Additionally, many interesting processes in oxides involve the
participation of protons: dehydroxylation reactions in the calcination
of refractory oxides, and, of topical technological interest,
'wetside'corrosion. In the former case, for example, the mechanism
for the reaction

$$Mg(OH)_2 ----> MgO + H_2O$$

is still not understood at the atomic level, there being two models
for the removal of water. Firstly there is the Goodman(5) or
homogeneous model which entails condensation between the brucite
layers to form water molecules which may then migrate out of the
crystal, and secondly, there is the inhomogeneous model(6) involving
'donor' and 'acceptor' regions, between which there is a counter
migration of cations and protons. MgO would then be formed in the
acceptor regions, whilst in the donor regions the protons combine with

OH groups to form water molecules which then escape, leaving
intercrystalline pores.

For 'wetside' corrosion, understanding the mechanism of dissolution
of metals and oxide films is of central importance. For example,
oxide films such as Al_2O_3 and Fe_2O_3 which may be thermodynamically
unstable with respect to dissolution in acid solutions are, nevertheless, kinetically stable. Here one is concerned with such processes
as the protonation of a surface oxide and its subsequent loss to the
aqueous medium.

Another area of current interest is that of fast ion proton conductors. Materials such as $SrCeO_3$ and $KTaO_3$ which have been doped with
a divalent cation and reoxidised or compensated with OH are used
industrially as water vapour sensors.

The common themes underlying these different areas are questions
relating to the presence of hyrogen in oxides: what is the nature of
the protonic species - e.g. $H, H_2, H^{\pm}, OH^-, H_3O^{\pm}$ etc.; how is it
incorporated into the structure; and what reactions between these
various species and between the protonic species and the host lattice
are possible.

We are currently undertaking a project to address these questions,
using computer simulation techniques. The contribution to our
understanding of defect solid state chemistry that these methods can
provide is now well established through previous studies (7).

It will be recognised that a central problem in the simulation of
proton behaviour is that a classical approach alone, as adopted in
studies of other materials, will not be adequate and thus the major
new feature of this work is the treatment of the O......H potential
interaction, using quantum mechanical techniques. This is in contrast
to the earlier work of Mackrodt and Stewart (8) who used an electron-
gas potential.

In the next section, we discuss our approach to the derivation of
these interactions, using the said quantum mechanical methods, and
then we will describe some preliminary results on the structure of
OH incorporation into MgO.

TREATMENT OF PROTON INTERACTIONS

Our treatment of this problem has been twofold, both of which are
based on a quantum mechanical approach. Firstly, we derive an
effective pair potential for the O....H interaction, for use directly
in the classical simulation codes such as HADES (9) and CASCADE (10).
This interaction is extracted from the potential surface calculated

from an ab initio quantum mechanical simulation. Secondly, we model the proton behaviour using an explicit quantum mechanical calculation of a cluster which is embedded in an array of point charges. The embedding is absolutely essential if the long range Madelung potentials found in crystalline materials are to be reproduced properly. It is our experience that a large array of point charges is needed if both the Madelung potential and its derivatives are to be correctly represented everywhere in the cluster: if this is not the case, then electrons may spill out into unphysical orbits, giving worthless results. Our calculations were performed at the ab initio Hartree-Fock SCF level using the ATMOL (11) program, with Gaussian basis sets. The effect of the choice of basis sets was carefully investigated for their influence on the O...H pair potential in order to justify our choice. The reliability of the 'effective' interaction derived in this manner was examined in a simulation of the structure of NaOH (12) This has an orthorhombic unit cell and its relatively low symmetry provided a good test of the potential. It was found that the calculated bond lengths matched the observed ones to within experimental error. Finally, we note that the transferability of the O...H potential to other systems was confirmed by the fact that it was found not to be strongly dependent on its environment. Greater details of this approach are available elsewhere (13).

INCORPORATION OF H IN OXIDES

As mentioned in the Introduction, Freund has suggested that there will always be a residual water content in refractory oxides, such as MgO, which arises in the following way:

$$H_2O \longrightarrow OH_o + V_{Mg} + 2OH$$

and the association of defects

$$V_{Mg} + OH_o \longrightarrow (V_{Mg}.OH)$$

$$(V_{Mg}.OH) + OH_o \longrightarrow (OH.V_{Mg} OH)$$

where the square brackets imply a complex (see structure in figure (1)) and the other symbols have their usual meaning. The presence of the complexes $(V_{Mg}.OH)$ and $(OH.V_{Mg}.OH)$, referred to as the V_{OH}^- centre (or partially compensated vacancy) and the V_{OH} centre (or fully compensated vacancy), respectively, have also been inferred from IR experiments. (1) Freund also proposed that the fully compensated vacancy dissociates into an H_2 molecule and an O_2^{2-} peroxy anion, as depicted in figure (2). This may be represent as

$$(OH.V_{Mg}.HO) \longrightarrow (O^{\cdot}(H_2)_{Mg})$$

Further, to account for his observation that atomic oxygen could be

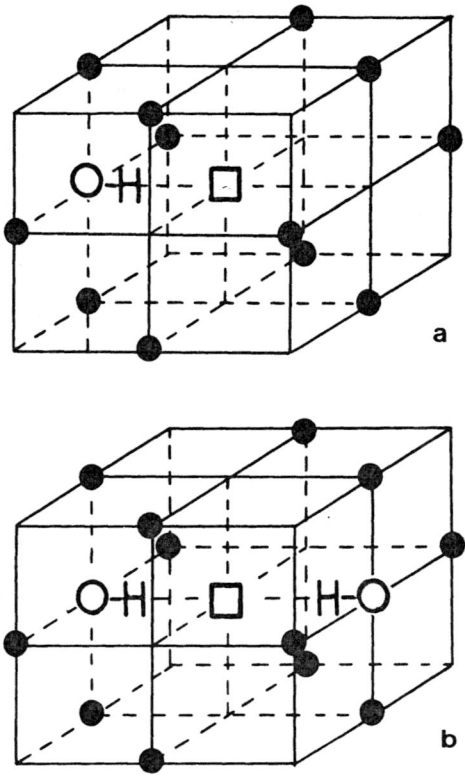

Figure 1   (a)   Structure of V_{OH}^- centre, or partially compensated cation vacancy in MgO.

           (b)   Structure of the V_{OH} centre, or fully compensated cation vacancy, in MgO. The linear structure is depicted here.

The filled circles are the cations but the anions are not shown except in OH groups.

TABLE   Binding Energies for OH to cation vacancies in MgO

Reaction	Binding Energy/eV
$OH_O + V_{Mg} \rightarrow [OH.V_{Mg}]$	3.67
$OH_O + [OH.V_{Mg}] \rightarrow [OH.V_{Mg}.HO]$	3.31

500

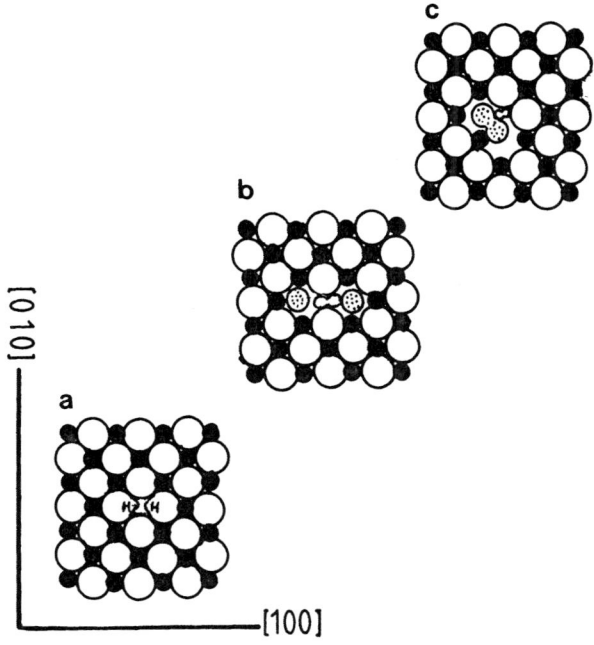

Figure 2   Freund's model for dissociation of fully compensated
vacancy into H_2 and O_2^{2-}.
The initial configuration in (a) reacts to form an H_2
molecule plus two O^- ions (stippled), (b), which then
react to form a peroxy anion O_2^{2-} (C).

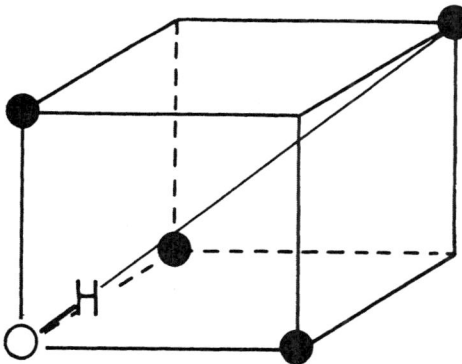

Figure 3   The minimum energy structure calculated for OH_0 in the
absence of any other defects.   The OH group is oriented
along (111).

evolved on heating (14), it was supposed that the peroxy anion decomposed into a lattice anion and an oxygen atom. We are using the techniques described above in order to ascertain the viability of these suggestions, and hope to report on these calculations shortly. Here, we present some initial results on the incorporation of OH groups in MgO, using the effective pair potential in the HADES code, using the potential model of Sangster and Stoneham for MgO (15). First we consider the stucture of the interstitial proton or OH group, in the absence of any cation vacancies. Figure (2) shows the minimum energy configuration for this species. We find that the proton clearly prefers to be associated with a lattice oxygen, as a distinct OH group, which will be oriented along (111) towards a next nearest neighour cation.

Secondly, we find that the energy of substitution is considerably lowered in the presence of a cation vacancy, to form a V_{OH}^- centre, or partially compensated vacancy. In addition the alignment of the OH group changes from (111) so that it is now oriented towards the vacancy along (100).

Thirdly, we predict that the fully compensated vacancy is bound with respect to V_{OH}^- and OH_O. An intriguing feature of our calculations here is that we find that there is virtually no difference between the energy of the non-linear structure for the fully compensated vacancy and the linear structure which is normally assumed to be the stable configuration. The former actually provides for a shorter H...,H separation which could presumably lead to enhanced H formation, and is of course compatible with the mechanism proposed by Freund - figure (2). A table of binding energies is provided for reference.

These results were obtained, as already indicated, from classical (static) simulations. To evaluate the energetics of the dissociation reaction postulated in eqn.(4), one needs to perform an explicit quantum mechanical embedded cluster calculation such as was described earlier . These calculations are currently in progress in our laboratory and will be reported in the near future.

SUMMARY AND CONCLUSIONS

In this presentation, we have described how important questions concerning the behaviour of protons in oxides may be addressed using computer simulation techniques that have been interfaced with a proper quantum mechanical treatment of the structure in the vicinity of the hydrogen ion. We have elucidated details of the structural incorporation of OH groups, both in the presence and absence of cation vacancies and shown how far these are consistent with the reaction mechanisms proposed by Freund.

ACKNOWLEDGEMENTS

We are very grateful to the Theoretical Physics Division AERE Harwell
for both financial support and the provision of computer facilities;
the SERC and ICI plc for a CASE studentship; and the University of
London for its computational resources.

REFERENCES

1. a) F. Freund, H. Wengeler, and R. Martens, Geochimica et
        Cosmochimica Acta 46:1821 (1982).
    b) F. Freund and H. Wengeler, J. Phys. Chem. Solids 43:129
        (1982).
2. L. Bursill and M. Blanchin, J. Solid State Chem. 51:321 (1984).
3. B. Henderson and W. Sibley, J. Chem. Phys. 55:1276 (1971).
4. Y. Chen, R. Gonzales, R. Schow, and E. Summer, Phys. Rev.
    B27:1276 (1982).
5. J. Goodman, Proc. Roy. Soc. London 247:346 (1958).
6. M. Ball and H. Taylor, Miner. Mag. 32:754 (1961).
7. C.R.A. Catlow and W. C. Mackrodt (eds), "Computer Simulation
    of Solids", Lecture Notes in Physics, 116, Springer-Verlag,
    Berlin (1982).
8. W. C. Mackrodt and R. F. Stewart, J. Phys. C12:5012 (1979).
9. C.R.A. Catlow, R. James, W. C. Mackrodt and R. F. Stewart,
    Phys. Rev. B25:1006 (1982).
10. M. Leslie, SERC Daresbury Laboratory Technical Memorandum
    DL/SCI/TM31T (1982).
11. V. R. Saunders and M. F. Guest, SERC Rutherford Laboratory
    Report: ATMOL3 (1976).
12. H. Stehr, Zeits. Kristallogr. 125:332 (1977).
13. P. Saul, C.R.A. Catlow and J. Kendrick, Philos. Mag. in press,
    (1984).
14. R. Martens, H. Gentsch and F. Freund, J. Catal. 44:366 (1976).
15. M. J. Sangster and A. M. Stoneham, Philos Mag. B43:597 (1981).

503

DEFECT STRUCTURE AND TRANSPORT PROPERTIES OF TITANATES

Nicholas G. Eror

Oregon Graduate Center
19600 N.W. Walker Road
Beaverton, OR 97006

The titanates are often described as nonstoichiometric compounds with oxygen vacancies as the principal ionic defects and
accompanying electrons as the compensating electronic defect.
This intrinsic defect model usually holds at low anion activities
but is not adequate in characterizing the defect properties at
higher anion activities (1-14). At all except the lowest anion
activities the intrinsic defect concentrations are very low and
are often less than 100 ppm in ambient air for oxides. This
means that exceptional purities, which are very seldom achieved,
are required in order to preserve intrinsic disorder at higher
oxygen activities. When potential impurities are considered
along with ionic size it becomes evident that the highest probability for aliovalent impurities are acceptors substituting for
titanium. The acceptor impurities will be compensated by the
introduction of an equivalent number of oxygen vacancies. For
the introduction of a two level acceptor on a titanium site (the
most probable) we may write.

$$[I_m''] \cong [V_o^{\cdot\cdot}] \tag{1}$$

This new electrical neutrality condition replaces the previous
intrinsic condition where doubly ionized oxygen vacancies were
compensated by electrons.

$$[n] \cong 2[V_o^{\cdot\cdot}] \tag{2}$$

which was derived from the oxygen extraction reaction

$$O_O = 1/2\ O_2 + V_O^{\cdot\cdot} + 2n \qquad (3)$$

which results in

$$[n]\ \alpha\ P_{O_2}^{-1/6} \qquad (4)$$

whereas the extrinsic disorder introduced by a fixed concentration of acceptor results in

$$[n]\ \alpha\ P_{O_2}^{-1/4} \qquad (5)$$

A Kroger-Vink diagram illustrates this transition from intrinsic to extrinsic disorder introduced by the presence of small concentrations of background single level acceptor impurities. See Figure 1.

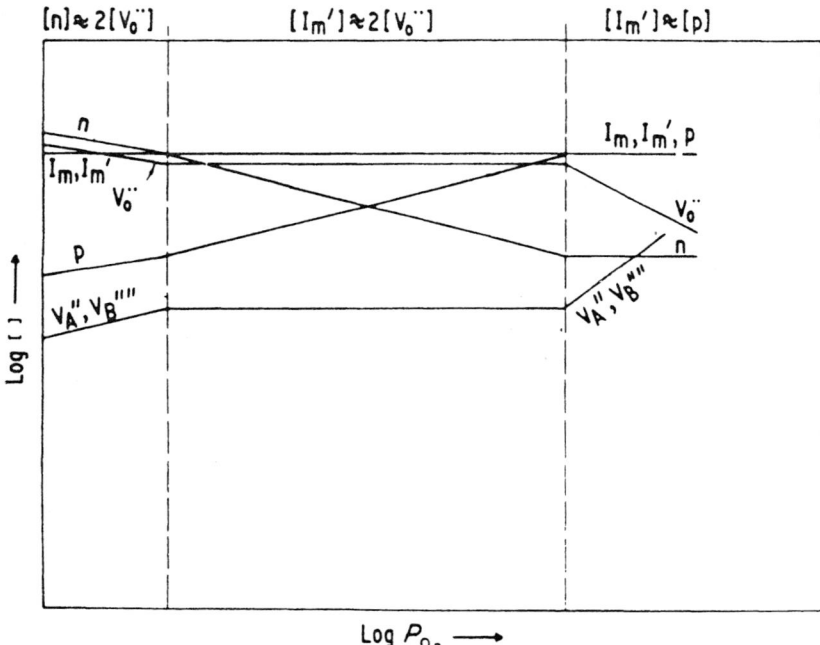

Fig. 1. Concentration of charged defects as a function of P_{O_2} for ternary oxide ABO_3 with Schottky-Wagner disorder and a fully ionized single level acceptor impurity.

Experimental observations (2-13) have shown the existence of a small intrinsic range for very lightly acceptor-doped ($\cong$ 100 ppm) compounds where

$$[n] \cong 2[V_o^{\cdot\cdot}] \tag{2}$$

and

$$[n] \ \alpha \ P_{o_2}^{-1/6} \tag{4}$$

followed by extrinsic ranges where

$$[I_m''] \cong [V_o^{\cdot\cdot}] \tag{1}$$

and

$$[n] \ \alpha \ P_{o_2}^{-1/4} \tag{5}$$

and

$$[p] \ \alpha \ P_{o_2}^{1/4} \tag{6}$$

Even for cases of very heavy (10,000 ppm) acceptor doping (15), there is no evidence for the range where

$$[I_m''] \ \alpha \ [p]^{1/2} \tag{7}$$

This means that for all levels of acceptor-doping the oxygen vacancy concentration will be dramatically enhanced, and since the oxide ions are the fast diffusing species, mass transport will be significantly increased at all levels of oxygen activity.

For the case of donor-doping, one would expect compensation of the donor by a controlled valency (16) or controlled electronic (17) imperfection mechanism.

$$[I_m^{\cdot}] \cong [Ti_{Ti}'] \cong [n] \tag{8}$$

A Kroger-Vink diagram illustrates the effect of a donor-dopant on the disorder in a compound such as a titanate. See Figure 2.

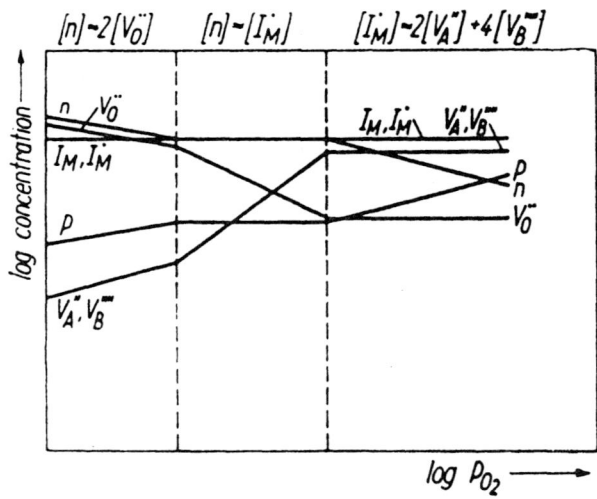

Fig. 2. Concentration of charged defects as a function of Po_2
for the ternary oxide ABO_3 with Schottky-Wagner disorder
and a fully ionized single level donor impurity.

As can be seen from the Kroger-Vink diagram, if the compound is
stable over a sufficiently wide range of anion activity, the compensation of the donor-dopant will change from the controlled
electronic imperfection mechanism to one of controlled atomic
imperfection (17). In this region, the donor-dopant is compensated by the formation of charged ionic point defects that are
also independent of equilibrium oxygen activity.

$$[I_m^\cdot] \cong 2[V_m''] \qquad (9)$$

This compensation mechanism has also been referred to as self
compensation (18-21) and stoichiometric compensation (22).

Self-compensation occurs because of the very rapid increase
in the metal deficiency point defect (proportional to the 1/2
power of the anion activity) in the controlled electronic imperfection range where

$$[I_m^\cdot] \cong [n] \qquad (8)$$

508

For the case of donor-dopants, while the n to p-type transition has been shifted to higher anion activities the metal excess to metal deficit transition has been shifted to lower anion activities. This means that there may be significant regions of anion activity where the dominant ionic disorder is metal deficit, while the electrical conductivity is n-type. It should also be noted that the ultimate concentration of metal deficiency ionic disorder will be linearly related to the donor-dopant concentration and may, therefore, be more than an order of magnitude larger in concentration than the maximum intrinsic disorder in the undoped compound for sufficient solid solubilities of suitable donor-dopants.

The implications of self-compensation on mass transport are obvious and follow directly from the preceding discussion of compensation mechanisms. While the concentration of the metal deficiency is increasing dramatically for the case of donor-doped compounds, the metal excess defect is correspondingly decreasing. Since the fast diffusing ion in the titanates is oxygen via oxygen vacancies, the rate controlling step for equilibration with the atmosphere is significantly decreased. For many compounds (22-30), the transition to the self-compensation range occurs at quite low anion activities ($\cong P_{O_2} < 10^{-8}$ atm.) so there may be a significant range of anion activity where the diffusion of the anion is very slow.

Further effects on the mass transport come from the consideration of second order contributions from defect complexes involving the aliovalent dopant.

The significance of neutral defect complexes was first pointed out by Mott and Gurney (31) when they considered the coulombic attraction of Schottky-Wagner defects in ionic crystals. Stasiw and Teltow (32) extended this further when they proposed that the cation vacancies that result when divalent cation impurities (donors) are added to alkali halides (33) form neutral divalent cation impurity-cation vacancy complexes because of strong coulombic forces.

$$I_m^{\cdot} + V_m' = (I_m^{\cdot}, V_m') \qquad (10)$$

Lidiard (34-36) considered the effect of the formation of these neutral defect complexes on the mass transport properties and prompted extensive verification in the alkali halides (37-56) of the theory.

Trivalent cation impurities (two-level donors) resulted in two cation vacancies for each trivalent-dopant which combined to form a charged defect complex with a single cation vacancy with a binding energy in excess of 1eV--which is close to the coulombic energy (51).

$$I_m^{..} + 2V_m' = (I_m^{..}, V_m')^. + V_m' \qquad (11)$$

The existence of Schottky-Wagner neutral vacancy pairs has also been demonstrated in the alkali halides (40,52). In this case, measurement of ionic conductivity and self-diffusion coefficients of intrinsic crystals demonstrated that the neutral Schottky-Wagner pairs gave a significant contribution to the self-diffusion of the ions at elevated temperatures but made no contribution to the ionic conduction. For the case of intrinsic KCl, there were equal concentrations of free vacancies and Schottky-Wagner vacancy pairs at 650°C with a pair binding energy of 0.8 eV (53) while for NaCl the concentrations were equal at 600°C with a pair binding energy of 0.7 eV (55).

$$V_m' + V_x^. = (V_m', V_x^.) \qquad (12)$$

Similar associations have been determined in the silver halides (57-60) where Frenkel-type disorder on the cation sublattice is the dominant intrinsic ionic defect.

The extension of these concepts to compounds such as oxides with more covalent bonding and predominantly electronic conduction has been more limited. Perkins and Rapp (61) studied the concentration dependent diffusion of chromium in nickel oxide and concluded that chromium diffuses by a mechanism involving a chromium ion-nickel vacancy complex

$$Cr_{Ni}^. + V_{Ni}'' = (Cr_{Ni}^., V_{Ni}'')' \qquad (13)$$

For the case of interactions between chromium impurities and magnesium vacancies in MgO, Glass (62) used the luminescence spectrum to measure the variation in concentration of chromium impurity ion-magnesium vacancy complexes

$$Cr_{Mg}^. + V_{Mg}'' = (Cr_{Mg}^., V_{Mg}'')' \qquad (14)$$

For $400°C \leqq T \leqq 900°C$, the defect complexes were accounted for by simple mass action theory with defect association energies of 0.5 - 0.8 eV.

Defect clusters or aggregates have been inferred in a number of oxides (63-85) and these are often associates of a single defect. Interaction of point defects and crystallographic shear planes have also been described (80-82). For the case of the fluorite structure oxides that are made anion deficient by the introduction of acceptor-dopants, association of the acceptor impurities and resulting anion vacancies, similar to what has been discussed for the halides, has been reported (83-85). Nowick and coworkers (83,84) have shown that for T < 1200 K and 28% acceptor-dopant concentrations, all of the oxygen vacancies introduced by the acceptor-dopant form complexes with the acceptor-dopant cations. For the case of trivalent impurities, one half of the dopant ions were found to be complexed with the oxygen vacancies with the other half free. The association energies were found to be in the 0.3 - 0.5 eV range with 50% association for T > 1000°C for acceptor-dopant concentrations > 1%.

The extensive data of Eror and Balachandran (2-5,7-11,86,87) and also by Chan et al. (6,12,13) point to a similar defect association in the titanates, niobates and tantalates—that of background acceptor impurity and the corresponding oxygen vacancy that was created to preserve charge neutrality. These compounds are all characterized by very small concentrations of intrinsic disorder at high anion activities and the typical background acceptor impurity concentration of the order of 100 ppm is more than sufficient to dominate the intrinsic disorder at all but the lowest anion activities.

$$x MO + TiO_2 = M''_x TiO_{2+x} + x V_O^{\cdot\cdot} \qquad (15)$$

Their data (2-13,86,87) show either a temperature dependence of the change of the -1/4 to -1/6 power of the oxygen activity dependence of the electrical conductivity; or the -1/6 power of oxygen activity to oxygen activity independent region of the oxygen vacancy concentration observed in thermogravimetric measurements. This is the transition point for the change in the intrinsic electrical neutrality condition

$$[n] \cong 2[V_O^{\cdot\cdot}] \qquad (2)$$

to the extrinsic

$$[M''_x] \cong x[V_O^{\cdot\cdot}] \qquad (16)$$

case. These data are consistent with the concept that there is some association between the acceptor impurity and its corresponding oxygen vacancy.

$$M''_x + xV''_o = x(M'', V''_o) \tag{17}$$

and

$$K_{17}(T) = \frac{[(M'', V''_o)]}{[M''][V''_o]} = C_{17}e^{-\Delta H_{17}/RT} \tag{18}$$

If the published electrical conductivity data in the literature for TiO_2, $BaTiO_3$, $SrTiO_3$, $CaTiO_3$ and Nb_2O_5 are examined carefully and the temperature dependence of the transition from the intrinsic to extrinsic disorder ($-1/4$ to $-1/6$ power of the oxygen activity dependence) are evaluated, ΔH_{17}, the binding enthalpy of the acceptor impurity-oxygen vacancy pair may be estimated. This is done for those data that are available for more than two temperatures. These data yield binding energies that range from 0.4 to 1.1 eV (88).

One may extend the above discussion of acceptor-dopant defect complexes to postulate what may occur when donor-dopants are added to transition metal oxides. Self-compensation of donor-dopants has been established in a number of compounds (22-28). And as mentioned above, the self-compensation occurs at quite low oxygen activities, so there is a wide range of anion activities where

$$[I^{\cdot}_m] \cong 2[V''_m] \tag{9}$$

What is likely to occur is the association of the compensating metal vacancies and the donor-dopant. For the case of a perovskite titanate, ABO_3, the defect association may be expressed.

$$V''_A + 2M^{\cdot} = (V''_A, M^{\cdot})' + M^{\cdot} \tag{19}$$

$$V''_A + 2M^{\cdot} = (M^{\cdot}, V''_A, M^{\cdot}) \tag{20}$$

of if the donor has two available ionization states

$$V''_A + M^{\cdot\cdot} = (V''_A, M^{\cdot\cdot}) \tag{21}$$

For the case of single level donors, reaction (19) will probably be the preferred association because of entropy effects. For this case, assuming complete association, only one half of the donor-dopants would be required to have the A-site vacancies actually associated with a donor-dopant. For the case of a two-level donor, all of the donors would be required to complex all of the A-site vacancies. Lewis (89) has calculated the association energy to be 0.90 eV for a $(La_{Ba}^{\cdot}, V_{Ti}^{''''})^{'''}$ complex for heavily doped $BaTiO_3$ (> 1%).

Study of the effect of acceptor and donor impurity induced defect complexes on mass transport properties has been very limited in oxides and has not been determined for the titanates.

ACKNOWLEDGEMENT

The author thanks the Gas Research Institute and the Office of Naval Research for financial support.

REFERENCES

1. N. G. Eror and D. M. Smyth, J. Solid State Chem., 24:235 (1978).
2. N. G. Eror and U. Balachandran, J. Solid State Chem., 42:227 (1982).
3. N. G. Eror and U. Balachandran, J. Solid State Chem., 43:196 (1982).
4. U. Balachandran and N. G. Eror, J. Mat. Sci., 17:2133 (1982).
5. U. Balachandran and N. G. Eror, Mat. Sci. and Engineering, 54:221 (1982).
6. N. H. Chan, R. K. Sharma and D. M. Smyth, J. Electrochem. Soc., 128:1762 (1981).
7. U. Balachandran and N. G. Eror, J. Solid State Chem., 39:351 (1981).
8. U. Balachandran, B. Odekirk and N. G. Eror, J. Solid State Chem., 41:185 (1982).
9. N. G. Eror and U. Balachandran, J. Amer. Ceram. Soc., 65:C73 (1982).
10. U. Balachandran and N. G. Eror, J. Phys. Chem. Solids, (in press).
11. N. G. Eror and U. Balachandran, to be published.
12. N. H. Chan and D. M. Smyth, J. Electrochem. Soc., 123:1584 (1976).
13. N. H. Chan, R. K. Sharma and D. M. Smyth, J. Amer. Ceram. Soc., 64:556 (1981).
14. U. Balachandran and N. G. Eror, J. Less-Common Metals, 85:111 (1982).
15. N. G. Eror and U. Balachandran, J. Amer. Cer. Soc., 61:426 (1982).
16. E. J. W. Verwey, et al., Philips Res. Rept., 5:173 (1950).

17. F. A. Kroger and H. J. Vink, in "Solid State Physics" (F. Seitz and D. Turnbull, Eds.), Vol. III, pp. 307-435, Academic Press, New York (1956).
18. R. F. Brebrick, "Progress in Solid State Chemistry," Vol. III, Pergamon, Oxford/New York (1966).
19. G. Mandel, Phys. Rev. A, 134:1073 (1964).
20. G. Mandel, Phys. Rev. A, 136:826 (1964).
21. F. A. Kroger, "Physical Chemistry: An Advanced Treatise" (H. Eyring, Ed.), Vol. X, pp. 229-259, Academic Press, New York (1970).
22. N. G. Eror and D. M. Smyth, "The Chemistry of Extended Defects in Non-Metallic Solids" (L. Eyring and M. O'Keefe, Eds.), pp. 62-74, North-Holland, Amsterdam (1970).
23. N. G. Eror, J. Solid State Chem., 37:281 (1981).
24. N. G. Eror and U. Balachandran, J. Solid State Chem., 40:85 (1981).
25. U. Balachandran and N. G. Eror, J. Mat. Sci., 17:1207 (1982).
26. U. Balachandran and N. G. Eror, J. Mat. Sci., **17:1795 (1982).**
27. U. Balachandran and N. G. Eror, J. Less-Common Metals, 85:11 (1982).
28. U. Balachandran and N. G. Eror, Phys. Stat. Solidi A, 71:179 (1982).
29. U. Balachandran and N. G. Eror, J. Electrochem. Soc., 129:1021 (1982).
30. U. Balachandran and N. G. Eror, J. Phys. Chem. Solids, 44:231 (1983).
31. N. F. Mott and R. W. Gurney, "Electronic Processes in Ionic Crystals", Clarendon Press, Oxford, England, p. 41 (1940).
32. O. Stasiw and J. Teltow, Ann. Phys. Lpz, 1:261 (1947).
33. E. Koch and C. Wagner, Z. Phys. Chem., B338:295 (1937).
34. A. B. Lidiard, Phil. Mag., 46:815 (1955).
35. A. B. Lidiard, Phil. Mag., 46:1215 (1955).
36. A. B. Lidiard, "Handbuch der Physik", Vol. XX Springer-Verlag, Berlin, p. 246 (1957).
37. N. Laurance, Phys. Rev., 120:57 (1960).
38. F. J. Keneshea and W. J. Fredericks, J. Chem. Phys., 38:1952 (1963).
39. F. J. Keneshea and W. J. Fredericks, J. Phys. Chem. Solids, 26:501 (1965).
40. R. G. Fuller, Phys. Rev., 142:524 (1966).
41. C. A. Allen, D. T. Ireland and W. J. Fredericks, J. Chem. Phys., 46:2000 (1967).
42. W. A. Mannion, C. A. Allen and W. J. Fredericks, J. Chem. Phys., 48:1537 (1968).
43. C. A. Allen and W. J. Fredericks, J. Solid State Chem., 1:205 (1970).
44. U. C. Nelson and R. J. Frianf, J. Phys. Chem. Solids, 31:825 (1970).

45. F. Beniere, M. Beniere and M. Chemla, J. Phys. Chem. Solids, 31:1205 (1970).
46. J. C. Krause and W. J. Fredericks, J. Phys. Chem. Solids, 32:2673 (1971).
47. R. E. Chaney and W. J. Fredericks, J. Solid State Chem., 6:240 (1972).
48. F. Beniere, M. Beniere and M. Chemla, J. Chem. Phys., 56:549 (1972).
49. C. A. Allen and W. J. Fredericks, Phys. Stat. Sol (b), 55:615 (1973).
50. J. L. Krause and W. J. Fredericks, J. Phys. (Paris) 34:C9-25 (1973).
51. F. Beniere and R. Rokbani, J. Phys. Chem. Solids, 36:1151 (1975).
52. H. Machida and W. J. Fredericks, J. Phys (Paris), 37:C7-385 (1976).
53. M. Beniere, M. Chemla and F. Beniere, J. Phys. Chem. Solids, 37:525 (1976).
54. J. S. Dryden and R. G. Heydon, J. Phys. C: Solid State Phys., 10:2333 (1977).
55. H. Machida and W. J. Fredericks, J. Phys. Chem. Solids, 39:797 (1978).
56. E. Lilley, J. Phys. (Paris) 41:C6-429 (1980).
57. A. P. Batra and L. M. Slifkin, J. Phys. C: Solid State Phys., 9:947 (1976).
58. A. P. Batra and L. M. Slifkin, J. Phys. C: Solid State Phys., 11:L317 (1978).
59. A. P. Batra and L. M. Slifkin, Phys. Stat. Sol. (b), 93:K77 (1979).
60. J. E. Hanlon, J. Chem. Phys., 32:1492 (1960).
61. R. A. Perkins and R. A. Rapp, Met. Trans., 4:193 (1973).
62. A. M. Glass, J. Chem. Phys., 46:2080 (1967).
63. W. L. Roth, Acta Cryst., 13:40 (1960).
64. F. Koch and J. B. Cohen, Acta Cryst., Sect. B, 25:275 (1969).
65. A. K. Cheetham, B. E. F. Fender and R. I. Taylor, J. Phys. C: Solid State Phys., 4:2160 (1971).
66. C. R. A. Catlow and B. E. F. Fender, J. Phys. C: Solid State Phys., 8:3267 (1975).
67. C. R. A. Catlow, B. E. F. Fender, and D. G. Muxworthy, J. Phys (Paris), C7:67 (1977).
68. P. D. Battle and A. K. Cheetham, J. Phys. C: Solid State Phys., 12:337 (1979).
69. M. Morinaga and J. B. Cohen, Acta. Cryst., A32:387 (1976).
70. B. T. M. Willis, Proc. Br. Ceram. Soc., 1:9 (1965).
71. B. T. M. Willis, J. Phys. (Paris), 25:431 (1964).
72. L. Manes and B. Manes-Pozzi, "Plutonium 1975 and Other Actinides", H. Blank and R. Lindner, Eds., North-Holland Publ., Amsterdam, p. 145 (1976).

73. M. DeFranko and J. P. Gatesoupe, "Plutonium 1975 and Other Actinides", H. Blank and R. Lindner, Eds., North-Holland Publ., Amsterdam, p. 133 (1976).

74. P. Chereau and J. P. Wadier, J. Nucl. Mater., 46:1 (1973).

75. B. E. F. Fender, "Chemical Applications of Thermal Neutron Scattering", B. T. M. Wills, Ed., Oxford Univ. Press, London, p. 250 (1973).

76. L. W. Hobbs, J. Phys. (Paris), C7:3 (1976).

77. M. Hayakawa, M. Morinaga and J. B. Cohen, "Defects and Transport in Oxides", M. S. Selzer and T. I. Jaffe, Eds., Plenum, New York, p. 177 (1974).

78. B. T. M. Willis, Proc. Roy. Soc. (London), A274:133 (1963).

79. B. T. M. Willis, Acta. Cryst., 18:75 (1965).

80. C. R. A. Catlow and R. James, Chem. Phys. Solids Their Surf., 8:108 (1980).

81. C. R. A. Catlow, "Physics and Chemistry of Refractory Oxides", Proc. NATO Summer School (1981).

82. A. N. Cormack, C. R. A. Catlow and P. W. Taskar, Radiat. Eff., 74:237 (1983).

83. A. S. Nowick and D. S. Park, "Superionic Conductors", G. Mahan and W. Roth, Eds., Plenum, New York, p. 395 (1976).

84. A. S. Nowick, D. Y. Wang, D. S. Park and J. Griffith, "Fast Ion Transport in Solids", P. Vashishta, J. N. Mundy and G. K. Shenoy, Eds., Elsevier/North-Holland, Amsterdam, p. 673 (1979).

85. K. El Adham, Thesis, Univ. Grenoble, Grenoble (1978).

86. U. Balachandran and N. G. Eror, Scripta Met., 16:275 (1982).

87. U. Balachandran and N. G. Eror, J. Less-Common Metals, 84:291 (1982).

88. N. G. Eror and U. Balachandran, Solid State Commun., 44:1117 (1982).

89. G. V. Lewis, Ph.D. Thesis, Univ. of London, London (1984).

Structure and Stoichiometry in Lithium Inserted Metal Oxides

R. J. Cava

AT&T Bell Laboratories
Murray Hill, New Jersey 07974

ABSTRACT

The stoichiometries of lithium inserted transition metal oxides synthesised at ambient temperature are strongly dependent on the structure of the host compound. Recent chemical and structural studies of insertion compounds are discussed in this paper as examples of the structural characteristics critical in the determination of lithium stoichiometries.

The recent interest in materials which can incorporate lithium atoms into their crystal structures at ambient temperature has been stimulated by their potential use as the positive electrode in rechargeable (secondary) batteries.[1] Lithium is the lightest and most electropositive metal, and has an ionic radius which facilities its incorporation into many crystal structures without severe distortion or bond breaking. The secondary cells generally consist of a lithium metal anode, a liquid or polymer electrolyte which passes lithium ions, and a cathode material which acts as a host for the Li on discharge of the cell. Most commonly, Li reacts with solid compounds in a displacement type reaction, e.g.:

$$y\,Li + MO_x \rightarrow \frac{y}{2}Li_2O + MO_{x-\frac{y}{2}}$$

These reactions involve extensive bond breaking and structural reorganization, and are therefore not useful in a rechargeable cell. For some transition metal compounds, however, Li is incorporated into the structure on cell discharge to form a single compound:

$$Li\,y + MO_x \rightarrow Li_y\,MO_x$$

with an accompanying chemical reduction of the host metal atom. This reaction will be easily reversed if the structural change on Li accommodation is not severe; that generally means M-O bonds in the host are not broken. For such a reaction to be fast enough to form the basis of a practical electrochemical cell, the diffusion coefficient of Li in MO_x must be unusually large at ambient temperature. In its most general form, the reaction of this type is known as an insertion reaction, if the host compound has a layered structure, the more specific term intercalation is used.

The EMF of an electrochemical cell as a function of discharge is a sensitive measure of the stoichiometry of the chemical phases produced during the insertion process. The cell potential is given by:

$$V_x = \frac{RT}{F} \ln \frac{\mu_x}{\mu_{Li}}$$

where x is the lithium stoichiometry, R, T and F are gas constant, temperature, and Faraday constant, μ_x is the chemical potential of Li in the host at stoichiometry x and μ_{Li} is the chemical potential of Li in the anode. The chemical potential of Li in the host depends on the reducibility of the host transition metal and is sensitive to the energy of the site Li occupies: Li in octahedral and tetrahedral sites in the same host will have different chemical potentials. Thus, if Li_xMO_y is a solid solution, V_x changes continuously with x, whereas if a 2 phase region occurs during discharge (in a particular range of x, 2 phases with different fixed x values are changing in proportion only) then V_x does not change over that range.

Insertion compounds form a broad class of materials in which both line phases and wide ranges of Li non-stoichiometry have been reported.[2] Common stoichiometries are in the range between 0.01 and 2.0 Li/ host metal. Aside from their technological use, insertion compounds are of great interest to the solid state chemist as they most often form metastable phases which cannot be synthesized by conventional routes. Such compounds may be synthesised either electrochemically, through discharge of Li/electrolyte/insertion compound cell, or chemically, by reaction of the host with a nonaqueous reagent which contains Li at a fixed chemical potential. Such chemical synthesis routes often make it possible to synthesize larger quantities of the pure insertion compound which are of interest for physical study, once the appropriate Li chemical potential for favorable reaction has been determined. Due to the significant volume changes associated with the insertion process, the compounds are generally available only in powder form.

Chalcogenides which have been studied extensively for lithium insertion reactions have had primarily layer type structures: the inserted lithium is accommodated in octahedral or tetrahedral sites between MX_n slabs otherwise loosely bonded to each other through chalcogen-chalcogen Van der Waals interactions.[2] In oxides, however, layer compounds are generally unstable, and most reversible Li insertion reactions have involved transition metal oxide hosts with close packed oxygen arrays, or crystallographic shear structures. There are four general structural requirements for such compounds: an interconnected network of sites in the structure allowing the fast transport of the Li ions, sites of appropriate geometry in large numbers which allow the inserted Li to be accommodated in favorable coordination polyhedra, stability of the host structure against bond bending, and strong host M-O bonds that will not break on reaction with lithium. Structural factors play a major role in determining the stoichiometries of the lithium insertion compounds, defining not only limiting and line phase stoichiometries, but also the ranges of composition within which solid solution phases form. Many electrochemical and structural studies have been performed on metal oxide insertion compounds in recent years. The results of those studies can be used to determine the important structural influences on stoichiometry in this class of compounds. I will use the results of studies on two classes of compounds, close packed structures, and crystallographic shear structures, to discuss and illustrate those structural influences.

Close Packed Structures

Lithium inserted rutile structure oxides have been studied extensively.[3] Rutile has an HCP oxygen array with the MO_6 octahedra sharing edges to form infinite chains, with interleaving one dimensional tunnels parallel to one crystallographic direction. The vacant octahedral sites in the tunnels have the same geometry as the MO_2 framework. The rutile structure is shown in Fig. 1a. The Lithium insertion compounds $LiMO_2$ can be generally prepared by chemical reactions with n-butyllithium (n-BuLi) at ambient temperature. This reagent has a chemical potential which approximates the discharge of a Li/electrolyte/cathode electrochemical cell to a potential of one volt.[4]

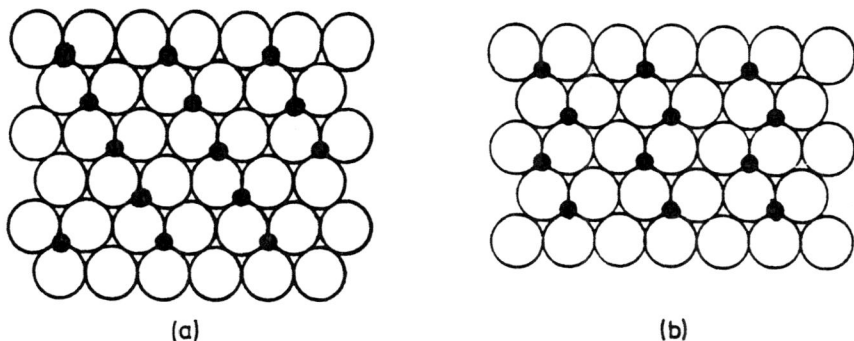

(a)   (b)

Figure 1.   a) Close packed plane in Rutile structure TiO_2 Vacant octahedral interstices form linear tunnels b) Close packed plane in Anatase structure TiO_2. Vacant octahedral interstices form zig-zag tunnels. Host metals solid circles, ocygen atoms open circles.

Figure 2 presents the EMF vs discharge for Li insertion into rutile structure RuO_2 and WO_2.[3] For the case of RuO_2 the EMF is relatively constant as a function of discharge for x between about 0.05 and 0.75, indicating a two phase mixture in that range. Non-stoichiometric solid solution phases apparently occur for $0.0 \leqslant x \leqslant 0.05$ and $0.75 \leqslant x \leqslant 1.0$, where the EMF changes quickly with x. The reversibility of the insertion is due to the small change in the host structure in going from RuO_2 to Li_1RuO_2. The case is significantly different for WO_2. The potential of the Li/WO_2 cell is about half that of the Li/RuO_2 cell because W^{4+} is more resistant to reduction than Ru^{4+}, and Li has a potential in WO_2 relatively close to that of metallic lithium. The data suggest a single phase non-stoichiometric solid solution for $0.0 \leqslant x \leqslant 0.5$, with a region of relatively large structural change between $0.4 \leqslant x \leqslant 0.5$. This may be due to an ordering of the Li ions at $x = 0.5$. A two phase region occurs between 0.5 and 0.8 and another single phase solid solution for $0.8 \leqslant x \leqslant 1.0$. Li_xMoO_2 also shows evidence for ordering at $x = 0.5$ in the EMF vs. discharge curves.

Crystallographic unit cell parameters have been obtained for rutile structure insertion compounds $Li_{0.8}CrO_2$, $LiMoO_2$, $LiWO_2$, $Li_{1.3}RuO_2$ $Li_{1.5}OsO_2$, $Li_{1.5}IrO_2$ and $Li(Mo_{.5}V_{.5})O_2$.[3] They indicate the rutile structure to remain stable to large inserted lithium stoichiometries, with a volume expansion of 10-20%. Lithium has been found to occupy tetrahedrally and octahedrally coordinated sites in many metal oxides. Neutron diffraction studies of $LiMoO_2$, $LiRuO_2$ and $LiIrO_2$ have found the Li to occupy all available octahedral sites.[5,6] Based on the number of available octahedral sites, a stoichiometry limit of 1 per host metal would be expected. For the compounds with $x > 1$, Li must begin to occupy tetrahedral sites which share faces with octrahedrally coordinated Li sites, resulting in very close Li-Li neighbors. This may be possible in metallic conducting compounds where conduction electrons screen the coulombic Li-Li repulsion. Rutile structure VO_2, MnO_2, NbO_2 and TiO_2 do not react in significant amounts with BuLi to form insertion compounds. The limit of zero Li solid solution has been attributed to the fact that these materials have either low crystallographic cell volumes, and therefore hindered Li diffusion, and/or low conductivity, which prevents screening of the repulsive coulombic Li-Li interactions by conduction electrons. The stoichiometies of Lithium inserted rutile structure oxides illustrate the importance of chemical, structural and electronic factors: they are affected by chemical potential, number and type of available sites, crystal structure volume, and electronic screening of repulsive Li-Li interactions.

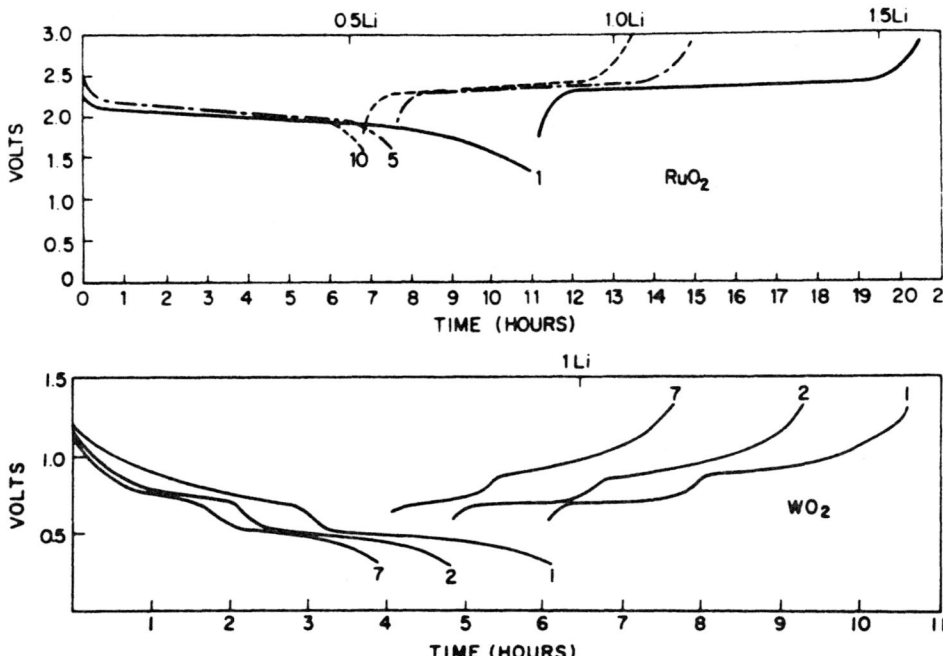

Figure 2.   EMF vs. discharge curves[3] for electrochemical cells employing cathodes of rutile structure a) RuO_2, b) WO_2

Of the CCP transition metal oxides, relatively few are known to undergo Li insertion reactions. Comparison of Li inserted TiO_2 anatase (CCP) and TiO_2 rutile (HCP) illustrates further several structural factors influencing Li stoichiometry. The behavior of these two electronically similar polymorphic materials toward Li is radically different: a stoichiometry limit of Li_7TiO_2 obtains for anatase, but virtually no Li can be inserted into rutile.[7] The CCP structure of TiO_2 anatase is illustrated in Figure 1b. The TiO_6 octahedra share edges to form zig-zag chains. Reaction of TiO_2 anatase with n-BuLi under normal conditions results in a compound Li_5TiO_2, with intermediate x values forming two phase $TiO_2 + Li_5TiO_2$ mixtures. The crystallographic unit cell changes symmetry from tetragonal to orthorhombic with a volume change of a few percent. Under highly forcing conditions of excess n-BuLi, Li stoichiometries up to Li_7TiO_2 are obtained, with a single phase nonstoichiometric solid solution between 0.5 and 0.7. The anatase structure contains 1 available octahedrally coordinated vacant site per transition metal, and so the 0.7 Li limiting stoichiometry is due to some hinderance other than the number of available sites. Neutron diffraction studies of Li_5TiO_2 have found the Li atoms to randomly occupy 1/2 of the available octahedral sites in the CCP anatase framework. On insertion of Li, Ti atoms form metal-metal bonded chains along one crystallographic direction. The bonding more

520

severely distorts the already distorted CCP oxygen array. The Li ions are actually 5 coordinate in the highly distorted octahedral interstices.[8] The limiting stoichiometry of 0.7 Li/Ti is reached when the Li-Li repulsive interaction becomes too strong. This interaction is mediated by the structural distortion of the TiO_2 host in 2 ways: (1) near-neighbor 5 coordinate Li sites, which must begin to fill in significant numbers for $X > 0.5$, are 2.54 Å distant, which is a small separation and (2) the Ti-Ti bonding, which becomes more pronounced as x increases (due to the donation of electrons to the bond) more and more severely distorts the CCP array, making the geometry of the interstitial sites more unfavorable for Li accommodation. One polymorph of $LiFeO_2$ has the same basic structure as Li_5TiO_2 anatase,[9] with the stoichiometry limited by the number of available sites. For this compound there are no metal-metal bonds, the CCP array is nearly regular, and Fe and Li occupy more regular octahetral sites with no extremely close neighbors. The stoichiometry limit of the Li_xTiO_2 anatase solid solution is determined by the extensive crystallographic strain present in the anatase structure, and the increasing strain with increasing Li content.

The case of $Li_{1+x}Ti_2O_4$[10] dramatically illustrates the effects of cation-cation interactions on the structure of the solid solutions. It also illustrates the high sensitivity of EMF vs. discharge curves to the structural changes. When Li_5TiO_2 anatase is heated above 450°C it forms the spinel $LiTi_2O_4$ with a nearly perfectly regular CCP oxygen array. The Li ions are in tetrahedral coordination. $LiTi_2O_4$ spinel can react with n-BuLi at ambient temperature to form $Li_2Ti_2O_4$, a 1:1 ratio of Li/host metal unattainable in the anatase form due to the crystallographic strain, further illustrating its influence on stoichiometry. Figure 3 presents the EMF vs. x for a $Li/electrolyte/Li_{1+x}Ti_2O_4$ spinel cell. A dramatic drop in potential occurs for $x > 0$ to a two phase region for $0.1 \leqslant x \leqslant 0.70$. A single phase solid solution occurs for $0.70 \leqslant x \leqslant 1.0$. A significant structural change must occur between $LiTi_2O_4$ and $Li_2Ti_2O_4$. The cubic unit cells of $Li Ti_2O_4$ and $Li_2Ti_2O_4$

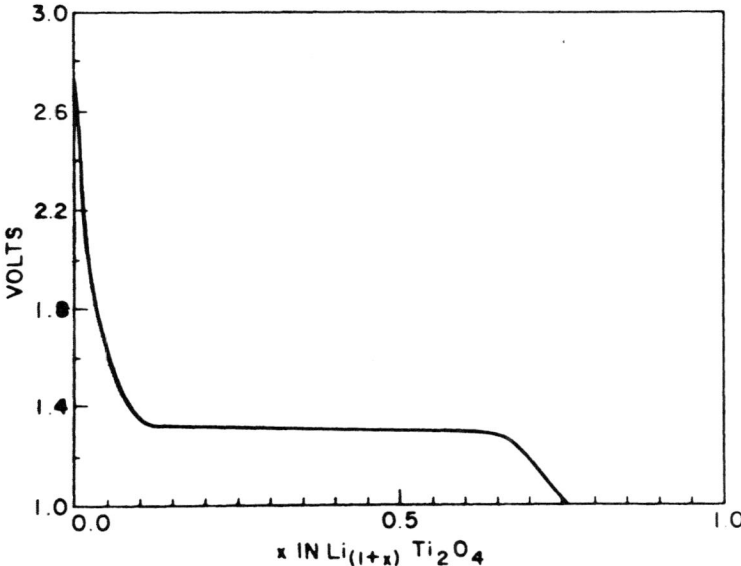

Figure 3.   EMF vs. discharge curve[10] for an electrochemical cell employing a cathode of Spinel structure $LiTi_2O_4$

differ by less than 1% in volume. The neutron diffraction measurements indicate that the Ti_2O_4 array is unchanged as Li goes from 1 to 2/formula unit, but that Li has moved from the tetrahedral interstitial sites at $X = 1$ to the octahedral interstitial sites at $X = 2$.[8] The structural transition on the Li sublattice occurs because Li-Li and Li-Ti polyhedra sharing is more energetically favorable for $X > 1$ if Li occupies solely octahedral sites. The high mobility of Li allows such a wholesale rearrangement of Li ions at ambient temperature, and the three dimensional interlocking of the Ti_2O_4 spinel host maintains its structural stability.

Comparison of TiO_2 anatase and rutile illustrates the importance of ion-ion interaction, electronic screening and cell volume on the stoichiometry of Li insertion compounds, as Li reacts with TiO_2 anatase to form Li_7TiO_2 but will not react to a significant extent with TiO_2 rutile. In the CCP structure (anatase) Li can occupy octahedral sites in which it shares only edges with the Ti atoms of the host. In rutile there are no tetrahedral or octahedral sites that do not share at least one face with a TiO_6 octahedra. Such face sharing is extremely unfavorable due to close Li-Ti approach.[10] It can only occur in cases where the coulombic interaction is screened by metallic conduction electrons or in phases where large crystallographic unit cells yield large near-neighbor separation.

Crystallographic Shear Structures: Wadsley-Roth Phases

Wadsley-Roth phases are a special class of crystallographic shear structures based on blocks of MO_6 octahedra sharing corners as in ReO_3.[11] The blocks of octahedra are generally of finite size in 2 dimensions and infinite in the other. Neighboring blocks of ReO_3 type are joined either through edge sharing, or a combination of edge sharing and tetrahedrally coordinated metal atoms at the corners of the blocks. Compounds of this type occur in many chemical systems based on niobium-oxygen and vanadium-oxygen octahedra, and occur at stoichiometries between MO_3 and MO_2. As the oxygen to metal ratio decreases from 3:1 to 2:1, the amount of crystallographic shear increases and the fraction of ReO_3 like volume decreases. Oxygen vacancies do not occur: small differences in stoichiometry result in different amounts of crystallographic shear; many distinct but similar crystal structures are observed. Figure 4 presents the structures of ReO_3, Nb_2O_5 and $WV_2O_{7.5}$.

Crystallographic shear structures generally display 3 important structural characteristics necessary for good Li insertion reactions: they are mechanically stable against the accommodation of a large number of Li ions, they have many sites of the correct geometry for Li ion accommodation, and there are tunnels or open regions in the structures which allow more rapid diffusion of Li than is possible in close packed oxides. ReO_3 itself has been found to undergo Li insertion reactions to a stoichiometry of 2 Li/Re. Stoichiometries of Li_xReO_3 solid solutions are determined by the instability to bond bending of the exclusively corner shared ReO_6 octahedra. For values of $x \leqslant 0.3$, the basic ReO_3 structure is maintained, but the octahedra twist somewhat about their shared corners to move 4 oxygens toward the center of the normally 12 coordinate cavity to create 4-coordinate sites for Li.[12] Stoichiometries up to $Li_{.75}ReO_3$ could be accommodated in sites of the appropriate geometry in the cubic phase, but this is never reached. The EMF vs. discharge curves indicate a large drop in potential for $X > 0.3$. At this stoichiometry the 3/4 CCP ReO_3 framework collapses to an HCP oxygen array through a 60° twist about one cubic [111] direction.[13] The twist creates two face sharing octahedral interstices from the original single large cavity. The twisted structure has stoichiometry $LiReO_3$. Stoichiometries between $0.3 \leqslant X \leqslant 1.0$ are not allowed (a 2 phase region). For $LiReO_3$, the Li ions fully occupy one of the two available octahedral interstices in an ordered

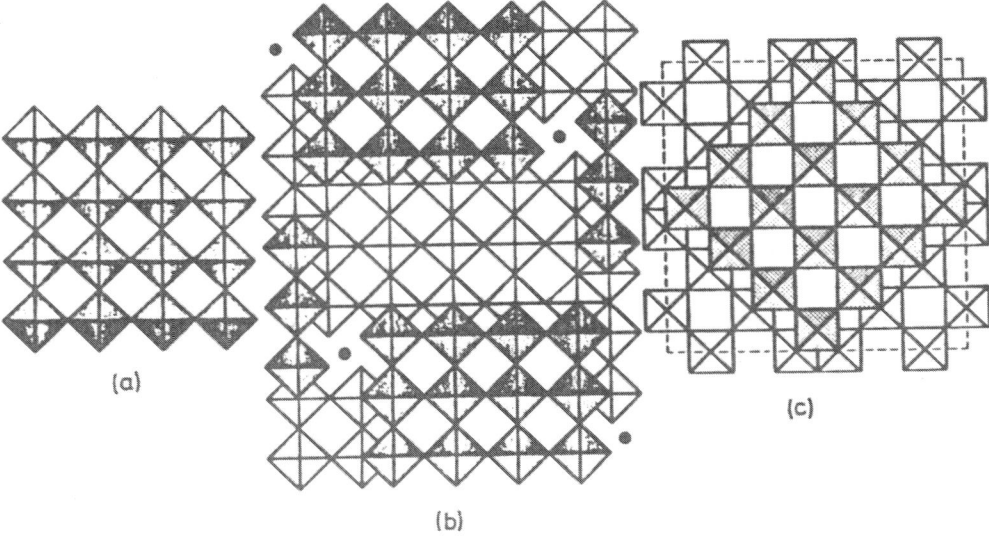

Figure 4.   Crystal structures of ReO_3 (a) and two representative crystallographic shear structures: (b) Nb_2O_5, (c) $WV_2O_{7.5}$. MO_6 octahedra are shown, those at the same level have same shading.

manner. The lithium ion is significantly displaced from the center of the octahedron away from the face shared Re octahedron and towards the face shared vacant octahedron, as in $LiNbO_3$. The structure cannot accommodate Li concentrations greater than 1/Re, and a 2 phase region exists for $1.0 \leqslant X \leq 1.8$ where the lithium rich phase begins to form. Once twisted to the HCP oxygen array (x > 1), the ReO_3 host is structurally stable and does not change further. The fact that a continuous solid solution does not form for $1.0 \leqslant x \leqslant 2.0$ is surprising. The structure of Li_2ReO_3 has both available octahedral sites filled, with Li ons near the centers of the octahedra. We believe that the solid solution does not form because the single Li ion in $LiReO_3$ precludes occupancy of the vacant octahedron without a high chemical driving force - one sufficient to drive the Li toward the Re ions. This chemical driving force establishes the minimum allowed stoichiometry of the high lithium phase at $Li_{1.8}ReO_3$; for fewer than 0.8 Li in excess of $LiReO_2$ the repulsion between neighboring Li is apparently insufficient to allow any occupancy of the second octahedron. The maximum stoichiometry of the second solid solution is Li_2ReO_3, limited by the number of available octahedral sites.

Although the structural instability of ReO_3 on Li insertion increases the number of potential sites for Li, resulting in 2 inserted Li/host metal, the twist to an HCP oxygen array is accompanied by a significant decrease in the potential of the electrochemical cell, and a significant decrease in the rate of lithium diffusion. For the crystallographic shear structures, the finite size ReO_3 type blocks bordered by regions of edge sharing are structurally stable against bond bending. We have found Wadsley-Roth crystallographic shear structures to be structurally stable on Li insertion if they contain intersecting crystallographic shear planes.[14] Nb_3O_7F and V_2O_5, for instance, which are $3 \times \infty \times \infty$ and $2 \times \infty \times \infty$ block structures, respectively, (with shear in only on direction) distort significantly Li insertion. For the structurally stable compounds, the general aspects of the

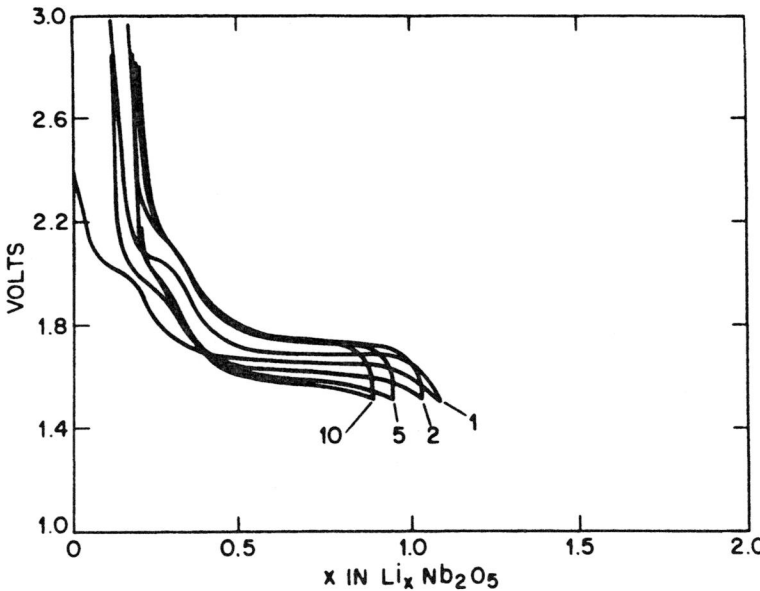

Figure 5.   EMF vs. discharge curve for electrochemical cell with Nb_2O_5 cathode[14].

stoichiometry of the Li inserted solid solutions are determined by two factors: the number of available sites for Li accommodation, and the number of reducible host cations. If lithium is postulated to occupy 5 and 4 coordinate sites the type found to be occupied in $Li_2FeV_3O_8$[15] and $Li_{.35}WO_3$,[16] the number of sites available is between approximately 0.7-0.9/host metal. Thus the increased structural stability and Li diffusion rate has been obtained in these compounds at the expense of the number of total available Li sites. Chemical limitations to the Li solid solution stoichiometries involve the reducibility of the host cations. Vanadium, for instance, which is easily reducible, can be reduced from V^{5+} to V^{3+} in an insertion reaction, thus allowing the possible incorporation of 2 Li/V^{5+} if other limiting factors are not operative. Similarly, Nb^{5+} can reduce to Nb^{4+}, allowing the incorporation of 1 Li/Nb^{5+}. When analyzed in this manner, the Wadsley-Roth phases present between 0.95 and 1.5 reducible ion equivalents/host cation. The number of potentially reducible ions is generally larger than the number of available sites, and is therefore the absolute limiting factor for the maximum allowed inserted lithium ion stoichiometry. In many cases, however, the chemically determined limits are not reached in the insertion compounds, and the limitation may be the number of available sites. On the other hand, the number of Li inserted is often larger than the number of available sites determined by simple site counting and indicates that we do not completely understand the structural limitations to the stoichiometry: Li may occupy sites of unusual geometry in these insertion compounds. Finally, in several titanium-niobium oxides, the maximum number of inserted Li ions is only half that expected from site or reducible ion counting, suggesting there may be another limiting mechanism, perhaps Li-Li interaction, operative.

Two examples of the kinds of non-stoichiometry which may occur in this class of compounds are illustrated in Figures 5 and 6, the EMF vs. lithium concentration curves for $Li_xNb_2O_5$ and $Li_xWV_2O_{7.5}$. For $Li_xNb_2O_5$ there appear to be three non-stoichiometric single phase solid solutions, $0.0 \leqslant x \leqslant 0.1$, $0.2 \leqslant x \leqslant 0.5$ and $x \geqslant 0.9$: the host Nb_2O_5

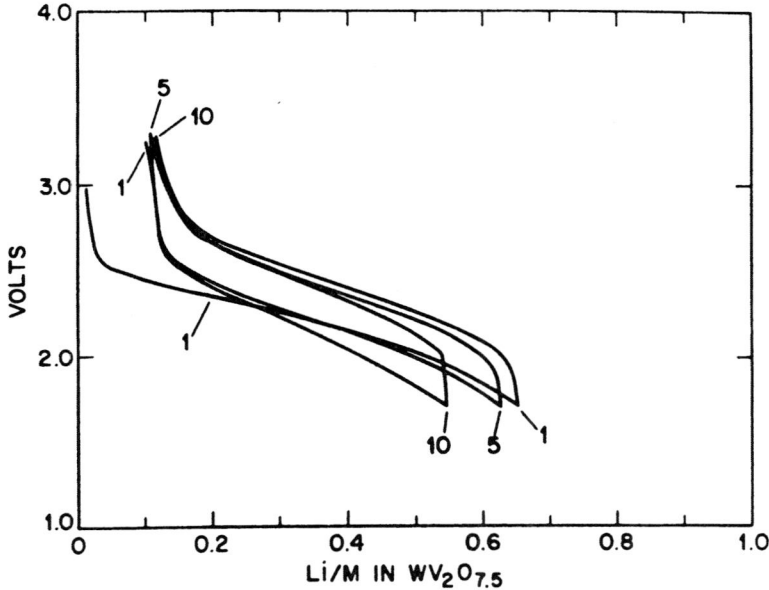

Figure 6.   EMF vs. discharge curve for electrochemical cell with $WV_2O_{7.5}$ cathode.[17]

structure is virtually unchanged during insertion (the volume increases approximately 6%) and thus the different phases must involve the filling of sites of different energy by the Li ions. The drops in EMF indicate the order in which the sites are filled: there are four types of simple Li sites in Nb_2O_5 distinguishable by their oxygen and cation neighbor host environments; non-stoichiometric solid solution regions may occur over the compositional ranges in which a particular site goes from empty to filled. Somewhat different behavior is observed for $Li_xWV_2O_{7.5}$, where there are no flat regions in the discharge curve. This suggests the occurrence of a continuous non-stoichiometric solid solution for $0.0 \leqslant x \leqslant 2.1$. The changes in slope of EMF vs x nonetheless suggest different energies of the Li as they fill the three different types of simple available cavities in the solid solution. Very slow discharge rates might show plateaus between distinct phases as for $Li_xNb_2O_5$, however a continuous EMF vs discharge type curve is observed for many insertion compounds, indicating the formation of very broad range solid solutions. It is not understood why a series like $Li_xNb_2O_5$ might form distinct phases and a series like $Li_xWV_2O_{7.5}$ might form a continuous solid solution.

Conclusions

Through the synthesis of compounds by lithium insertion reactions at ambient temperatures, a great number of new line phases and non-stoichiometric solid solutions have been discovered. Such compounds are generally metastable and decompose to stable compounds on heating in some cases to only a few degrees above ambient temperature, and therefore cannot readily be synthesized in any other manner. Many factors are operant in determining the lithium stoichiometries, depending strongly on the structure of the host metal oxide. Among these factors are: number and type of available sites, fast diffusion pathways for diffusion of lithium, crystallographic cell volume, number of reducible host metal atoms, electrical conductivity, and ion-ion interactions on the lithium ion sublattice (including short range order). Examples of the effects of these factors can be observed in many close packed and crystallographic shear type transition metal oxides. Although the chemistry of these materials has in many cases been studied in great detail, careful physical property characterization is generally lacking. This is due to the unavailability of single crystals because of the large stresses imposed on the host during the insertion process. Due to the wide ranges of non-stoichiometry often available in many of the solid solution phases synthesised such physical property measurements might be of great interest.

Acknowledgements

I would like to acknowledge fruitful collaborations and discussions on the structure and chemistry of lithium inserted metal oxides with D. W. Murphy, A. Santoro and R. S. Roth. This report includes many of their ideas.

REFERENCES

1. See, for instance, K. M. Abraham and S. B. Rummmer, in *Lithium Batteries*, J. P. Gabano, ed., Academic press Inc., New York 1983, p. 371.

2. M. S. Whittingham, Prog. Sol. St. Chem. *12* 41 (1978).

3. D. W. Murphy, F. J. DiSalvo, J. N. Carides and J. V. Waszczak Mat. Res. Bull *13* 1395 (1978).

4. D. W. Murphy and P. A. Christian, Science *205* 651 (1979).

5. I. J. Davidson and J. E. Greedan, J. Sol. St. Chem. *51*, 104 (1984).

6. D. E. Cox, R. J. Cava, D. B. McWhan and D. W. Murphy J. Phys. Chem. Sol. *43* 657 (1982).

7. D. W. Murphy, M. Greenblatt, S. M. Zahurak, R. J. Cava, J. V. Waszczak, G. W. Hull, Jr. and R. S. Hutton, Rev. Chim. Min. *19* 441 (1982).

8. R. J. Cava, D. W. Murphy, S. Zahurak, A. Santoro and R. S. Roth, J. Sol. St. Chem. 1984 to be published.

9. D. E. Cox, G. Shirane, P. A. Flynn, S. L. Ruby, and W. J. Takei, Phys. Rev. *132* 1547 (1963).

10. D. W. Murphy, R. J. Cava, S. M. Zahurak and A. Santoro, Sol. St. Ionics. *9/10* 413 (1984).

11. A. D. Wadsley and S. Andersson in *Perspectives in Structural Chemistry, 3* Dunitz and Ikers, eds., John Wiley and Sons, New York (1970) p. 1.

12. R. J. Cava, A. Santoro, D. W. Murphy, S. M. Zahurak and R. S. Roth, J. Sol St. Chem. *50* 121 (1983).

13. R. J. Cava, A. Santoro, D. W. Murphy, S. M. Zahurak and R. S. Roth, J. Sol. St. Chem. *42* 251 (1982).

14. R. J. Cava, D. W. Murphy and S. M. Zahurak, J. Electrochem. Soc. *130* 2347 (1983).

15. R. J. Cava, A. Santoro, D. W. Murphy, S. Zahurak and R. S. Roth, J. Sol. St. Chem. *48* 309 (1983).

16. P. J. Wiseman and P. G. Dickens, J. Sol. St. Chem. *17* 91 (1976).

17. R. J. Cava, D. W. Murphy and S. M. Zahurak J. Electrochem. Soc. *130* 244 (1983).

A PREDICTIVE MOLECULAR ORBITAL THEORY APPLIED TO DEFECTS AND

STRUCTURES OF TRANSITION METAL OXIDES[†]

Alfred B. Anderson,* Robin W. Grimes and Arthur H. Heuer

Department of Metallurgy and Materials Science
Case Western Reserve University
Cleveland, Ohio 44106

ABSTRACT

A quantum chemical approach for determining defect structures in cation-deficient transition metal monoxides is described. The method employs molecular orbital electronic energies for cations and anions in nearest-neighbor coordinations and adds to them interatomic pair-wise repulsion energies. Using this approach, zincblende structure extended defect clusters composed of 4:1 cluster building blocks are found to be most stable in $Fe_{1-x}O$. These clusters account for the observed P' and P" phases in the iron oxide. The theory predicts that isolated cation vacancies with no clustering are most stable in $Ni_{1-x}O$, in agreement with conductivity and diffusion data in the literature. For $Co_{1-x}O$ the experimental situation is unclear, and our theory suggests that small, but not extended, 4:1 defect clusters may form.

1. ASED-MO THEORY

The quantum mechanical treatment of structure and diffusion questions in non-stoichiometric compounds demands simplified models. The purpose of this paper is to explain one such model and show how it may be used to predict defect structures in the cation-deficient sodium chloride structure iron, cobalt, and nickel monoxides.

Our approach is based on the atom superposition and electron

[†]Supported by a grant to the CWRU/MRL from the NSF.
*Contact this author at the Department of Chemistry.

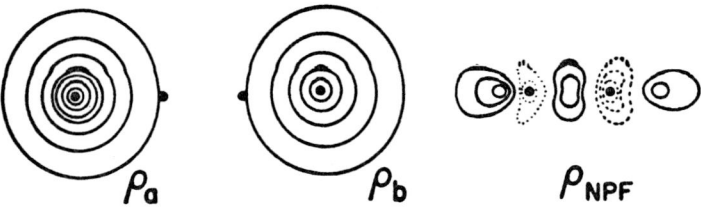

Fig. 1. Schematic representation of electronic charge density components for a diatomic molecule according to Eq. (1).

delocalization molecular orbital (ASED-MO) theory.[1] Though already used for numerous studies, this name was used first in Ref. 1. In this approach the electronic charge density distribution of a molecule, or a solid, is partitioned into atomic parts and the bond charge part. The energy of the system is calculated as the non-interacting separated atoms are brought into a molecular configuration so that the atomic density distributions begin to overlap and some charge density is subtracted from the atoms to form the bond charge distribution. Thus, the molecular electronic charge density distribution function, ρ_{mol}, is a sum of rigid or perfectly-following atomic components ρ_a and a structure-dependent non-perfectly-following bond charge component ρ_{npf}:

$$\rho_{mol} = \sum_{all\ atoms} \rho_a + \rho_{npf} . \tag{1}$$

It may be noted that $\int \rho_a\, d\underset{\sim}{r} = Z_a$, the atomic number, and $\int \rho_{npf} d\underset{\sim}{r} = 0$. For a diatomic molecule at a given internuclear distance, the density components are shown schematically in Fig. 1. The bond formation energy is determined by integrating the electrostatic force on one of the nuclei as the atoms are brought into a molecular configuration. The atom supposition energy, E_r, is repulsive because the nuclear repulsion force is greater than the nuclear-electron attraction force (the attraction of the nucleus to the electronic density function centered on it is constant and set equal to zero). The integral of the force due to the formation of the bond charge, ρ_{npf}, yields an attractive energy, E_{npf}. The two energy components for a diatomic molecule are shown schematically in Fig. 2, along with their sum which is exactly the molecular binding energy, E_{mol}[2]:

$$E_{mol} = E_r + E_{npf} \tag{2}$$

where

528

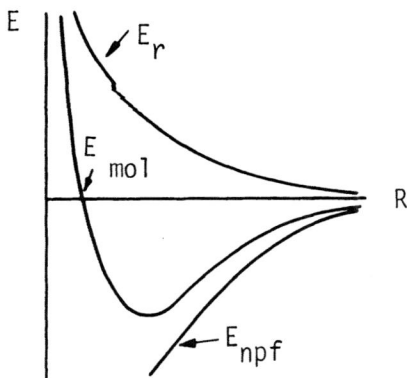

Fig. 2.   Schematic representation of energy components for a
diatomic molecule according to Eq. (2).

$$E_r = \frac{Z_a Z_b}{R} - Z_a \int \rho_b(r)|R-\underset{\sim}{r}|^{-1} d\underset{\sim}{r} \ , \tag{3}$$

$$E_{npf} = -Z_a \int_{\infty}^{R} [\int \rho_{npf}(R',\underset{\sim}{r})\frac{d}{dR'}|R'-\underset{\sim}{r}|^{-1} d\underset{\sim}{r}]dR' \ . \tag{4}$$

To construct a polyatomic molecule or a solid one would repeat the
process, adding one atom after the other and summing the pairwise
repulsion energies and the attraction energies due to bond formation.

Because the bond charge ρ_{npf} is generally well-approximated by
point charges, it may be shown that harmonic force constants, k_e, are
well-approximated by the Poisson equation[2]

$$k_e = 4\pi Z_a \rho_b(R_e) \tag{5}$$

where R_e is the equilibrium internuclear distance. This formula has
been applied to numerous molecular systems, for example various
states of the C_2 molecule and diamond.[3] The molecular charge density
partitioning model, Eq. (1), and force constant formula, Eq. (5),
have provided the basis for defining useful atomic radii and empirical
relationships relating force constants to bond lengths.[4]

From the above it may be concluded that molecular density partitioning of Eq. (1) has physical significance. We now address the
question of determining the molecular energy using Eqns. (2)-(4).
The pairwise repulsion energy contributions to E_r in a polyatomic
molecule are readily calculable using available atomic orbital
electronic wave functions which have been tabulated.[5] The attractive
contributions to E_{npf} can be calculated only when ρ_{npf} is known for

all internuclear distances, but ρ_{npf} is unknown. However, it has been found that a molecular orbital energy E_{MO} from diagonalizing a hamiltonian similar to extended Hückel is often a sufficiently good approximation to E_{npf} so as to allow the calculation of E_{mol}, leading to predictions of molecular structures and stabilities.[6] Then we have

$$E_{mol} \simeq E_r + E_{MO} \; . \tag{6}$$

Equation 6 is the basis for the ASED-MO method. This theory retains the flexibility and ease of interpretation of the well-known extended Hückel molecular orbital theory, but has the important advantage of allowing structure predictions. It is widely used in all areas of chemistry. In extended Hückel theory model structures with fixed bond lengths are necessarily used because E_R is not included. It may be noted that E_R plays the role, somewhat crudely speaking, of the nuclear repulsion energy minus the electron repulsion energy in Hartree-Fock theory and E_{MO} plays the role of the orbital energy in Hartree-Fock theory. In many instances, the ASED-MO theory describes bonding in molecules more accurately than Hartree-Fock theory, as for example predicting bonding in F_2, which Hartree-Fock theory fails to do. However, the E_{MO} approximation to E_{npf} is not exact, and the ASED-MO theory must be used with care.

2. MODEL FOR DEFECTS IN OXIDES

The massive non-stoichiometry in wustite, $Fe_{1-x}O$, where x ranges from 0.05 to 0.15, involves defect clusters whose size and periodicity result in unit cells that have too many atoms to allow tight-binding solid state theoretical treatments with the ASED-MO or any other theory. Because of this, we have developed an approach where anion and cation energies in defect clusters and in the surrounding matrix depend on local coordination. Applications to wustite, including defect cluster structure predictions, will soon be published.[7] The purpose of this section is to summarize the model and our results for wustite and introduce new predictions for cation-deficient cobaltous oxide, $Co_{1-x}O$ and bunsenite, $Ni_{1-x}O$.

The success of our model is a direct result of the nature of the hamiltonian used to obtain the E_{MO} approximation to E_{npf}. This hamiltonian excludes electron repulsion integrals and its matrix elements depend only on anion and cation ionization potentials and orbital overlaps. Diagonal matrix elements, H_{ii}, are measured ionization potentials. Off-diagonal elements, H_{ij}, are averages of these.[6]:

$$H_{ij} = 1.125 \; (H_{ii} + H_{jj}) \; S_{ij} e^{-0.13R} \tag{5}$$

where $S_{ij} = \langle i | j \rangle$ and R is the internuclear distance. Thus, one may calculate Fe^{II} cation orbital energies in octahedral sites in wustite

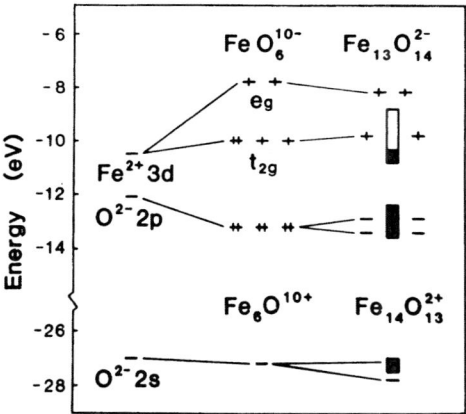

Fig. 3. Molecular orbital energy levels for octahedral Fe^{II} and O^{2-} coordination. The $Fe_{13}O_{14}^{2-}$ cluster is a cube with Fe^{II} in the center; similarly, $Fe_{14}O_{13}^{2+}$ has O^{2-} in the center; heavy bands represent doubly-occupied energy levels; others are singly-occupied.

using FeO_6^{10-} clusters. Similarly, anion energies in octahedral sites are determined using OFe_6^{10+} clusters. When designing defect clusters, various types of cation and anion coordinations occur, and orbital energies for each are determined using corresponding $FeO_n^{(2n-2)-}$ ferrous clusters, $FeO_n^{(2n-3)-}$ ferric clusters and $OFe_{a+b}^{(2a+3b-2)+}$ anion clusters where there are a ferrous and b ferric nearest neighbors, respectively. We call these the "normalized ion" energies.

The relationship between the occupied molecular orbital energy levels for Fe^{II} and O^{2-} ions in the octahedral environment of wustite and in larger clusters is shown in Fig. 3. There it is evident that the iron d energy levels in FeO_6^{10-} shift only slightly in the larger $Fe_{13}O_{14}^{2-}$ cluster, and similarly for the oxygen 2s and 2p levels on comparing OFe_6^{10+} to $Fe_{14}O_{13}^{2+}$. Thus, the minimal local coordination clusters yield normalized ion energies representing average bulk band energies. Similar comparisons have been published for hematite, Fe_2O_3.[8]

The cation normalized ion energies are equivalent to the familiar cation crystal field energies. The anion normalized ion energies are anion crystal field energies, but prior to our work they do not seem to have been considered in attempting to understand

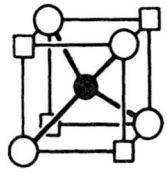

4 : 1

Fig. 4. A 4:1 cluster. Circles represent octahedral O^{2-} anions,
the dot represents a ferric tetrahedral interstitial, and
the squares represent octahedral ferrous vacancies.

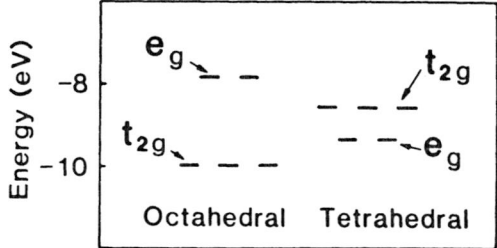

Fig. 5. The d molecular orbital energy levels for iron in octahedral lattice sites and tetrahedral interstices in
wustite.

structures in transition metal oxides.

In modeling defect clusters and predicting their stabilities in
the Fe, Co, and Ni monoxides, we set up a defect structure within a
constant oxygen anion sublattice, count the number of each type of
ion and sum the normalized ion energies. To this sum is added the
total of all repulsive two-body energies. Then the total energies
of different defect structures are compared. The neglect of lattice
relaxations around defects is a part of our modeling approach using
normalized ion energies. Such relaxations will be ∿0.1Å, the difference between Fe^{III}–O and Fe^{II}–O bond lengths. Given the small
force constraints in these materials, corrections for relaxations
are not expected to change our quantitative conclusions.

3. RESULTS FOR $Fe_{1-x}O$, $Co_{1-x}O$, AND $Ni_{1-x}O$

Defect clusters in wustite are generally believed to have 4:1
defect clusters as building blocks.[9,10,11] A 4:1 cluster consists

532

of a tetrahedral Fe^{III} interstitial surrounded by a tetrahedron of octahedral Fe^{II} vacancy sites, as shown in Fig. 4. It has been proposed on the basis of x-ray,[12] neutron,[13] and electron diffraction studies[14,15] that the 4:1 clusters share edges or corners in extended defect clusters.

In our recent study[7] we employed the normalized ion energy approach described in the previous section to estimate relative stabilities for a variety of defect clusters. Four types of cation coordinations, tetrahedral and octahedral ferrous and ferric, and eleven types of anion coordinations are possible. For both the ferrous and ferric oxidation states, octahedral coordination within the oxygen sublattice is more stable because of the greater destabilization of the 3d levels due to the shorter Fe-0 bond distances in the tetrahedral interstitial cases. Relative positions are given in Fig. 5. The metal d levels are shifted up in energy in Figs. 3 and 5 because the d orbitals are antibonding to the surrounding oxygen anion orbitals, which mix with them a small amount. Conversely, anion levels are most stable for coordinations with several tetrahedral OFe bonds because these orbitals mix in iron orbitals by a small amount in a bonding way. The shorter the bonds and the greater the anion-cation orbital overlap, the greater the bonding stabilization and antibonding destabilization, a well-known result of molecular orbital theory. It is evident that defect cluster structure stabilities depend on a balance of anion stabilization, cation destabilization, and two-body repulsion energies.

Defect cluster formation energies were calculated assuming isolated point defects as the initial state. The formation of a m:n cluster (m ferrous octahedral vacancies and n tetrahedral ferric interstitials) is given by the reaction

$$2(m-n)Fe^{\cdot}_{Fe} + (m-n)V''_{Fe} \rightarrow nFe^{\cdots}_{i} + (2m-3n)Fe^{\cdot}_{Fe} + mV''_{Fe} . \quad (8)$$

The energies for such reactions are divided by $(m-n)$ to yield cluster formation energies per vacancy. These numbers are approximate measures of the defect cluster stabilities.

As may be seen in Fig. 6, zinc-blende structure defect clusters, with numerous tetrahedral OFe type bonds, are determined by the theory to be especially stable. Of these, the 13:5 and 16:7 zinc-blende clusters fit the diffraction properties of the ordered P" phase of wustite, such as the next nearest neighbor vacancy-vacancy vectors from x-ray work[12] and structural data from electron diffraction studies.[15] Both clusters are consistent with the poorly ordered P' phase of wustite.[15] These clusters also fit the requirement that $m/n = 2.4 \pm 0.5$ as determined from neutron diffraction studies.[13] It may be noted in support of the theoretical model that it correctly predicts that the zinc-blende structure is less stable than the sodium chloride structure for FeO and that inverse spinel

Fig. 6. Binding energies (B.E.) per vacancy determined using the normalized ion energy plus two-body energy sum as outlined in the text.

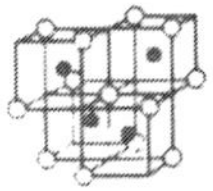

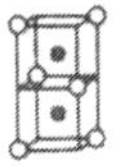

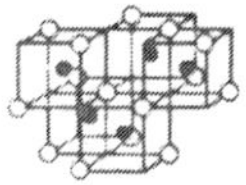

Cluster Type	B.E./vacancy (eV)		
	FeO	CoO	NiO
4:1	0.51	0.26	-0.06
6:2	0.29	0.17	-0.28
7:2 <110>	0.53	0.21	-0.16
7:2 <111>	0.62	0.31	-0.07
8:2	0.40	0.10	-0.20

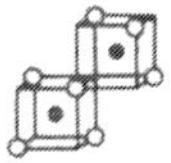

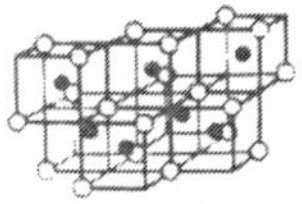

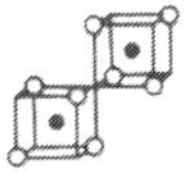

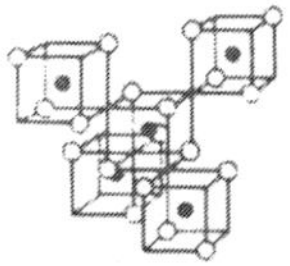

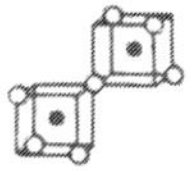

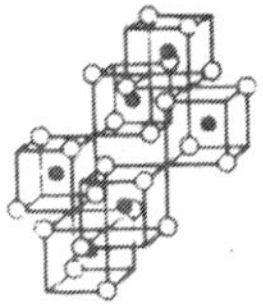

Cluster Type	B.E./vacancy (eV)		
	FeO	CoO	NiO
10:4	0.86	0.33	-0.29
13:5	0.75	0.24	-0.33
16:7	0.91	0.28	-0.43
16:5	0.70	0.35	-0.08
18:6	0.77	0.39	-0.09

is more stable than normal spinel magnetite, Fe_3O_4. Furthermore, the extended magnetite structure is more stable than any $Fe_{0.75}O$ wustite which can be constructed from these clusters. Finally, it should be mentioned that the zinc-blende clusters are limited in size by the instability associated with charge separation in and around the clusters, which increases with size. In the center, a net positive charge builds up due to the tetrahedral ferric cations. At the surface there is an excess of compensating negative charge due to the deficiency in cations. Surrounding the cluster is a cloud of (2m-3n) electron holes consisting of octahedral ferric cations which may be an influence on the development of the cluster spacing and ordering.

The oxides $Co_{1-x}O$ and $Ni_{1-x}O$ are much closer to ideal MO stoich- iometry than wustite. In cobaltous oxide[16] x ranges from $\sim 10^{-2}$ to $\sim 10^{-4}$ and in bunsenite[17] from $\sim 10^{-3}$ to $\sim 10^{-5}$. There is no clear evidence concerning defect clustering in these oxides. Isolated Ni^{II} vacancy point defects evidently account for the non-stoichiometry in bunsenite[18]. Some believe isolated Co^{II} vacancy point defects dominate in cobaltous oxide[16], while others argue for the presence of Frenkel defects as well[19].

We have modeled defect clustering in cobaltous oxide and bun- senite. In the absence of evidence for 4:1 clustering, we tried bringing octahedral M^{II} vacancies to adjacent sites. For the cobalt and nickel oxides, as well as wustite, such arrangements are always repulsive when the vacancies lie along <100> and <110> directions and there is no interaction along <111> directions. The repulsions are due to the low coordinations of oxygen anions bridging the vacancies. For <100> Fe, Co, and Ni oxides the energies per vacancy are 0.15, 0.05, and 0.16 eV, respectively, and for the <110> direc- tion the respective energies are 0.09, 0.16, and 0.13 eV.

While low-levels of cation vacancy concentrations in $Co_{1-x}O$ and $Ni_{1-x}O$ would appear, from a consideration of configurational entropy, to preclude the formation of 4:1 defect clusters and their aggrega- tion, we have examined their formation energies nevertheless. As seen in Fig. 6, we predict 4:1 defect clusters will never form in bunsenite, and neither will aggregates of them form. This result is in agreement with other workers[18]. In cobaltous oxide, the 4:1 cluster is stable, but less so than in wustite. Larger aggregates show, in Fig. 6a, a relatively slow rate of increase in binding energy per vacancy; for example, the 13:5 and 16:7 cluster are not much more stable than the 4:1 cluster. Given the small value of x, configurational entropy would probably prevent aggregation of 4:1 defect clusters in cobaltous oxide. However, our results suggest some 4:1 clusters may form and may be responsible for the observa- tion of interstitial tetrahedral cations discussed in Ref. 19.

The instability of 4:1 clusters in bunsenite is easy to understand. As shown in Fig. 5, the center of gravity of metal d levels is raised in the tetrahedral environment. Consequently, high-spin d^7 Ni^{III} is, in our calculations, 3.32 eV less stable in the tetrahedral site. Anion stabilizations are insufficient to overcome this in bunsenite. For cobaltous oxide and wustite the destabilizations of high-spin d^6 Co^{III} and d^5 Fe^{III} in the tetrahedral sites are, respectively, 2.15 and 1.30 eV, and anion stabilizations can overcome cation instabilities in 4:1 cluster formation in these oxides. It may be noted that variations in lattice constants and d orbital sizes cause small relative shifts in levels in Fig. 5 for the cobaltous oxide and bunsenite and small shifts in the oxygen anion energies. These shifts do not detract from our qualitative explanation.

Other workers have estimated stabilities of some of the defect clusters considered here by using pairwise ion interaction potentials and a polarization energy.[20] That technique favors the formation of large aggregates of 4:1 clusters in cobaltous oxide and bunsenite as well as in wustite. There is some doubt that such methods can include the important consequences of variations in d orbital occupation which we see in our work, and which we believe prevent the formation of large clusters in $Co_{1-x}O$ and $Ni_{1-x}O$.

SUMMARY

By working within a molecular orbital model wherein cation and anion packing energies in cation-deficient iron, cobalt, and nickel monoxides are assumed to depend predominantly on local coordination, it has been possible to make a number of predictions in accord with experiment. Extended zinc-blende structure defect clusters are found to be stable in wustite and appear to account for the available experimental data on the P' and P'' phases. We predict 4:1 clusters and possibly small aggregates may form in cobaltous oxide and that in nickel oxide isolated cation vacancies are most stable. The loss of clustering ability on going from the iron to cobalt to nickel oxides is caused by the addition of d electrons across the series. These additional electrons strongly favor octahedral lattice sites for the cations over the tetrahedral interstices which must be occupied for defect clustering to occur.

REFERENCES

1. A. B. Anderson, Chem. Phys. Lett. 76:155 (1980).
2. A. B. Anderson, J. Chem. Phys. 60:2477 (1974).
3. A. B. Anderson, J. Chem. Phys. 63:4430 (1975).
4. A. B. Anderson and R. G. Parr, Chem. Phys. Lett. 10:293 (1971).
5. E. Clementi and C. Roetti, "Atomic Data and Nuclear Data Tables", Academic, New York (1974).
6. A. B. Anderson, J. Chem. Phys. 62:1187 (1975).

7. A. B. Anderson, R. W. Grimes, and A. H. Heuer, <u>J. Sol. St. Chem.</u> (to be published).
8. N. C. Debnath and A. B. Anderson, <u>J. Electrochem. Soc.</u> 129:2170 (1982).
9. P. Tarte, J. Prendhomme, F. Jeannot, and O. Ewrard, <u>C. R. Acad. Sc. Fr.</u> C269:1529 (1969).
10. A. K. Cheetham, B. E. F. Fender, and R. I. Taylor, <u>J. Phys. C: Solid St. Phys.</u> 4:2160 (1971).
11. N. N. Greenwood and A. T. Howe, <u>J. C. S. Dalton</u> 1:110 (1972).
12. F. Koch and J. B. Cohen, <u>Acta Cryst.</u> B25:275 (1969).
13. J. R. Gavarri, C. Carel, and D. Weigel, <u>J. Sol. St. Chem.</u> 29:81 (1979).
14. B. Andersson and J. O. Sletnes, <u>Acta Cryst.</u> A33:268 (1977).
15. C. Lebreton and L. W. Hobbs, <u>Rad. Effects</u> 74:227 (1983).
16. R. Dieckmann, <u>Z. Physik. Chem. Neue Folge</u> 107:189 (1977).
17. G. Dhalenne, J. C. Rouchaud, G. Revel, A. REvcolevscki and R. Collongues, <u>C. R. Acad. Sci.</u> 272:538 (1971).
18. N. L. Peterson and C. L. Wiley, <u>J. Phys. Chem. Solids</u> (submitted).
19. E. Fryt, <u>Oxid. Met.</u> 10:311 (1976).
20. C. R. A. Catlow and A. M. Stoneham, <u>J. Am. Ceram. Soc.</u> 64:234 (1981).

DEFECT CLUSTERING IN ROCK-SALT STRUCTURED TRANSITION METAL OXIDES

S.M. Tomlinson*, C.R.A. Catlow*, and J.H. Harding**

*Department of Chemistry, University College London
20 Gordon Street, London WC1H OAJ, UK
**Theoretical Physics Division, AERE, Harwell, Didcot
Oxfordshire, OX11 ORA, UK

ABSTRACT

We present results of a survey, based on theoretical methods, of defect aggregation in non-stoichiometric, rock salt structured transition metal oxides, with special emphasis on the widely studied $Fe_{1-x}O$ phase. Calculations of the energies and entropies of defect formation and aggregation are presented. We consider both heavily defective phases, where we concentrate on the nature of the dominant type of cluster structure, and near stoichiometric compositions where our main concern is with the magnitude of the deviation from stoichiometry required for clustering to occur.

(1) INTRODUCTION

The nature and extent of defect aggregation in non-stoichiometric rock-salt structured transition metal oxides has led to much controversy in recent years. The nature of the basic defect structure is clearly established: transport, thermodynamic and structural studies show that oxidised lattice cations in the non-stoichiometric phases $M_{1-x}O$ are compensated by the creation of cation vacancies. In addition, it is clear from both diffraction[1,2,3,4,5] and microscopy studies[6,7] on heavily defective wustite, $Fe_{1-x}O$, that there is extensive vacancy aggregation, and that the formation of vacancy clusters is accompanied by cation interstitial formation. However, the precise nature of these clusters remains uncertain as does the question of the extent of the deviation from stoichiometry (the value of x in the formula $M_{1-x}O$) that is necessary for clustering to occur.

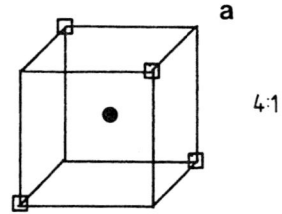

a

4:1

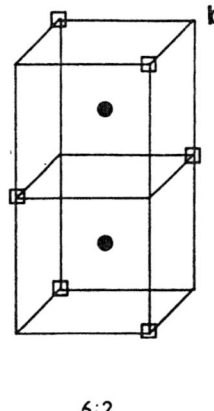

b

6:2

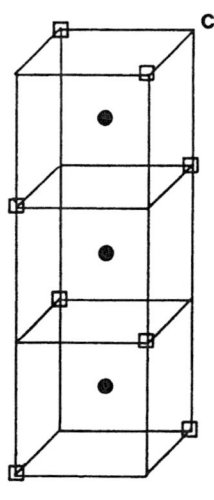

c

8:3

Figure 1. Clusters in
$Fe_{1-x}O$

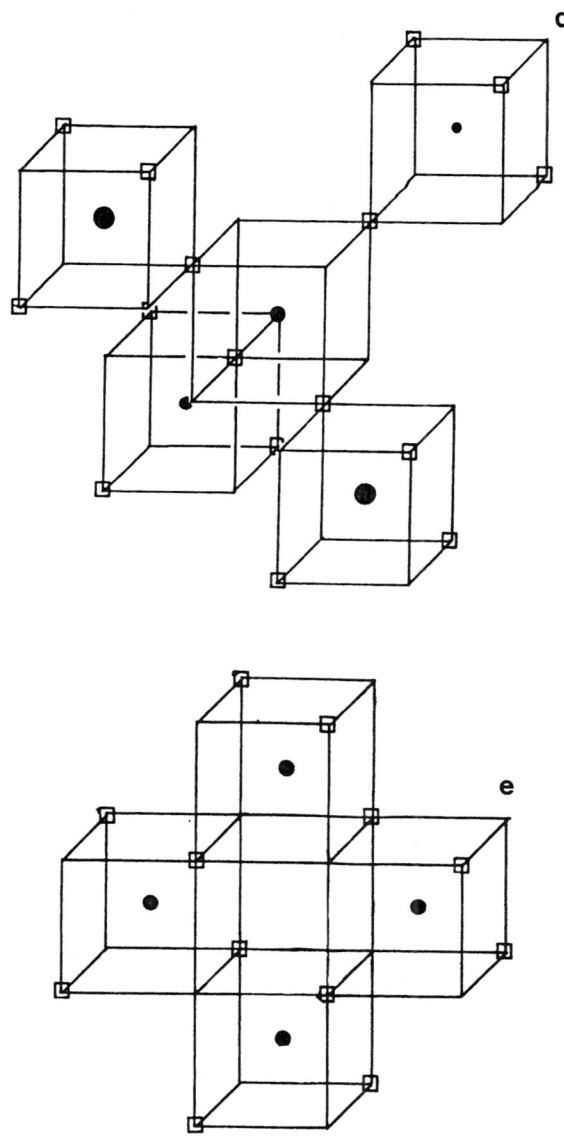

Fig. 1 (Continued)

Theoretical methods have proved to be particularly useful in examining the stability of defect clusters in strongly ionic materials. An application of these techniques to $Fe_{1-x}O$ was reported by Catlow and Fender[8]. The work was subsequently extended[9] to include the other rock salt structured oxides; $Mn_{1-x}O$, $Co_{1-x}O$ and $Ni_{1-x}O$. The calculations found high stability for a tetrahedral aggregate in which four cation vacancies surround an Fe^{3+} interstitial i.e. the 4:1 cluster shown in Fig. 1a. The calculated energies also indicated subsequent modes of aggregation in which these tetrahedral clusters share edges to give the 6:2 and 8:3 clusters shown in Figs. 1b and 1c. Large binding energies were also found for the corner shared 16:5 cluster (Fig. 1d) which is an element of the inverse spinel structure adopted by Fe_3O_4, while the binding energy of an alternative corner shared aggregate, the 13:4 cluster originally proposed by Koch and Cohen[2], was found to be somewhat smaller.

The cluster models proposed by Catlow and Fender seemed to accord reasonably well with the available experimental data. Bragg diffraction studies[1,2,3,4,5] indicated the ratio of vacancies to interstitials was about 2.5-3.0 in accordance with the formation of 6:2 and 8:3 clusters, while the results of a study of the magnetic structure of the phase by Battle and Cheetham[4] were compatible with the formation of these clusters. Recently, however, the microscopy studies of Lebreton and Hobbs[7,10] on quenched Wustite have found evidence for the formation of larger clusters; in particular, these workers suggested the formation of a new type of edge shared aggregate, the 12:4 cluster shown in Fig. 1e. Furthermore, Anderson et al.[11] have applied approximate quantum mechanical methods to the study of cluster stabilities; they proposed a new range of 'Zincblende' related structures of which the simplest is the 10:4 cluster.

The first aim of this paper is therefore to update our theoretical survey of defect clustering by reporting calculations on these recently proposed clusters. Our second concern will be the important point raised above, namely the magnitude of the deviation from stoichiometry at which clustering becomes important. In an earlier study, Catlow and Stoneham[12] proposed that, at 1000 K in $Mn_{1-x}O$, 4:1 and 6:2 clusters would begin to dominate the defect structure at stoichiometry deviations (i.e. values of x) as low as approximately 10^{-3}. The simple mass action treatment which led to this prediction, however, ignored vibrational entropy contributions to the free energy of clustering. This deficiency is removed in the present study which reports calculated vibrational entropies. We also examine a wider range of temperatures than in the earlier study. A final point considered in this paper concerns the charge state of the vacancy aggregates. In earlier work it was assumed that the effective charge of the vacancies will be neutralised by a surrounding distribution of Fe^{3+} ions on octahedral sites. To test this assumption we have performed calculations on charged clusters.

Due to restrictions of space we cannot consider all four rock-salt oxides. The results we report pertain to $Fe_{1-x}O$. We will describe the extent to which these results are generally applicable, and in a subsequent publication will present results for all four oxides.

(2) THEORY AND COMPUTATIONAL TECHNIQUES

(2.1) Defect Energy Calculations

As in the earlier work of Catlow and Fender[8], all defect energy calculations were performed using the generalised Mott-Littleton procedure available in the HADES codes; in the present calculations the HADES3[13] code was employed. A general discussion of the techniques used by the program is given by Catlow and Mackrodt[14]. In the present study we took advantage of the increase in computer power since the original work of Catlow and Fender[8], in order to perform calculations with a larger size of the explicitly simulated region (region I); between 300 and 400 ions were included in all the calculations.

As in our earlier study, calculations were performed on clusters which were, as nearly as possible, electroneutral (except for the calculations performed specifically to investigate the stability of charged clusters as discussed in the Introduction). Electroneutrality is achieved by surrounding the cluster by M^{3+} ions (on the octahedral, lattice sites). In many cases, a zero net charge cannot be achieved as the resulting cluster structure would be of too low a symmetry to permit the use of a region I of adequate size. The net effective charges are given in Table 1 where the calculated energies are reported. All calculations were performed with the lattice parameter of 2.1554 Å as in the earlier study of Catlow and Fender.

The calculations require the specification of interatomic potentials. Results were obtained both for the original potentials reported by Catlow and Fender, and for a more recent potential model developed by Lewis and Catlow[15]. As the difference between the results were found to be insignificant, we report only results obtained using the earlier Catlow-Fender potential model.

(2.2) Entropy Calculations

The SHEOL program[16] uses the large crystallite method to calculate entropies of formation at constant volume, S_V. The crystal is divided into two regions. In the inner region, containing the defect and surrounding ions, the ions are allowed to vibrate. In the outer region the ions are held fixed and contribute only through the diagonal elements of the force constant matrix. We may write the vibrational entropy of formation, ΔS_d, in the high temperature limit as

$$\Delta S_d = S \text{ (defect crystal)} - S \text{ (perfect crystal)}$$

$$= -\tfrac{1}{2} \ln \frac{\det (\underline{\underline{D}})}{\det (\underline{\underline{D}}')} + 3k(N' - N)(1 - \ln \frac{\hbar}{kT})$$

where primes refer to the defect crystal and N,N' are the number of ions in the inner region. D, the dynamical matrix is given by

$$\underline{\underline{D}} = \underline{\underline{M}}^{-\tfrac{1}{2}} \underline{\underline{\Phi}} \underline{\underline{M}}^{-\tfrac{1}{2}}$$

where $\underline{\underline{M}}$ is the mass matrix and $\underline{\underline{\Phi}}$ the force-constant matrix.

In the calculation of entropies of charged defects we must apply a correction arising from the long range distortion field. This is discussed by Gillan and Jacobs[17] and an expression for the correction ΔS_{corr} given. The entropy we require:

$$S_v = \Delta S_d - \Delta S_{corr}$$

The program uses symmetry to reduce the diagonal matrix to block diagonal form. The details are given in reference 18.

(2.3) Mass Action Calculations

This well known procedure requires the specification of a defect reaction to describe the formation of each cluster. For example, the appropriate equation for the 4:1 cluster is given by:

$$3V_m'' + 6h^\bullet \overset{K}{=} (4:1) \tag{1}$$

in which we assume that a neutral cluster is being formed - an assumption also made in our treatment of the 6:2 cluster. (Larger clusters have not yet been included in our full treatment). Corresponding to reaction 1 there is a mass action equation, $x_v^3 x_h^6 = K$, where K is the equilibrium constant for the reaction which can be expressed using standard chemical thermodynamics in terms of the free energy, g_p, of the reaction:

$$K = \exp(-g_p/kT) \tag{2}$$

However, as shown by Gillan[19], for defect free energies we have:

$$g_p = f_u \tag{3}$$

i.e. the Gibbs and Helmholtz free energies are equal.

Thus we can write:

$$K = \exp[-(\Delta u_v - T\Delta s_v)/kT] \tag{4}$$

where Δu_v and Δs_v are, respectively, the constant volume energies and entropies calculated for our defect reactions by the techniques summarised in sections (2.1) and (2.2) above.

Having obtained the equilibrium constant for the defect reactions, using equation 4, the procedure is then straightforward: the coupled set of mass action equations are solved subject to the linear restraints describing conservation of stoichiometry (i.e. the value of x is fixed for a given calculation) and the electroneutrality condition. This enables us to calculate the concentrations of all the defect species included in the mass action analysis, as a function both of x and T.

The simple mass action treatment assumes that all species have activity coefficients of unity; that is an ideal solution theory is employed with interactions between different defect species being ignored. A possible improvement would be the inclusion of Debye-Huckel activity coefficients. We consider, however, that the errors caused by the omission of these terms are probably small compared with other uncertainties in the calculations.

(3) RESULTS

(3.1) Cluster Binding Energies

Results of our calculations are reported in Table 1; following Catlow and Fender we report our results in terms of binding energy, E_B, per net vacancy in the cluster. For example for the 4:1 cluster, the energy reported is the total binding energy divided by three. The purpose of presenting results in this fashion is to ensure that all the results for different clusters are strictly comparable.

Three features of the results deserve emphasis: firstly, we found greatest stability for the edge-shared aggregates, that is the 6:2, 8:3 and 12:4 clusters (note that the 12:5 cluster, which is formed from the 12:4 by placing an interstitial in the central tetrahedron is considerably less stable). Secondly, we note that the 10:4 'zinc blende' like cluster discussed by Anderson et al.[11] appears to be appreciably less stable than the edge shared aggregates. Thirdly, for clusters considered by Catlow and Fender[8] we obtain very similar results to theirs.

In order to investigate the question referred to in the Introduction, of the charge states of the clusters, we have performed calculations on the 4:1 cluster with no surrounding holes (i.e. octa-

Table 1

Cluster	Effective Charge	Total Defect Energy/eV	Binding Energy Per Net Vacancy/eV
Vacancy/Hole	1-	-7.895	-
Vacancy/2 Hole	0	-38.101	-
4:1	1-	-91.966	-1.752
6:2	0	-164.943	-2.043
8:3	1+	-238.050	-2.244
10:4	4-	-121.770	-1.675
12:4	0	-331.329	-2.250
16:5	1-	-421.919	-2.129

hedral Fe^{3+} ions). Interestingly, we find that the cluster is still bound, with a binding energy per net vacancy of 0.502 eV. This corresponds to an <u>average</u> dissociation energy per hole from the 4:1 cluster of approximately 0.937 eV. With a value of this magnitude for the average energy we consider it highly probable that most clusters will exist in charged states, and that net charges of -3 and -4 may be common at elevated temperatures.

(3.2) Calculated Vibrational Entropies

Table 2 gives the calculated values of S_V for isolated defects and simple clusters. Clusters more complicated than the 6:2 aggregate have not been considered due to the excessive computational requirement. Also reported in the table are the resulting values of the changes in S_V (ΔS_V) for the various reactions.

Table 2a

Defect Cluster	S_V/kB
Cation Vacancy	-6.463
Hole	2.326
Vacancy/Hole	-4.781
Vacancy/2 Hole	-3.826
4:1	-12.849
6:2	-22.968

Table 2b

Cluster Reaction	$\Delta S_V/kB$
$V_m'' + h^\bullet = (Vac/Hole)$	-0.644
$V_m'' + 2h^\bullet = (Vac/2\ Hole)$	-2.015
$3V_m'' + 5h^\bullet = 4{:}1$	-5.090
$4V_m'' + 8h^\bullet = 6{:}2$	-15.724

546

We should stress that the calculated entropies are very sensitive to the interatomic potentials. We are continuing to investigate these effects, and so the results we have presented should be considered as preliminary values. It is clear, however, that the energy changes make by for the greater contribution to the free energy of clustering, and it is unlikely that these terms will drastically change our predictions concerning the variation of cluster concentration with stoichiometry.

(3.3) Mass Action Calculations

Using the defect energies obtained from the energies and entropies summarised in Tables 1 and 2, we have calculated defect concentrations as a function of x and T using the mass action formalism described in section (2.3). Results are summarised in Figs. 2a, 2b and 2c for three temperatures, 1200 K, 1400 K and 1600 K; the figures show, as a function of x, the fractions of the total vacancy concentration present in each cluster.

The results clearly show a transition from a defect structure dominated by isolated vacancies and holes to one in which the 4:1 cluster dominates; the cross over point moves to higher x as temperature increases, being at approximately $x = 10^{-2}$ for 1600 K. The concentration of 6:2 clusters is suppressed by its more significant entropy term, but the calculated isolated hole entropy suggests this would be less apparent if we were not assuming clusters exist in an as near as possible electroneutral state. Finally, a notable feature of the calculations is the absence of the simple aggregates of a single vacancy and one or two holes (i.e. the singly and uncharged vacancies) in high concentration for any value of x.

(4) DISCUSSION AND CONCLUSIONS

On the first theme outlined in the Introduction, namely the nature of clustering in heavily defective oxides, two new results have emerged in this study: firstly the stability of the 12:4 edge shared cluster proposed by Lebreton and Hobbs, and indeed of edge shared clusters in general; secondly, the contrasting result for the 10:4, 'zinc-blende' like aggregate which appears to be appreciably less stable. In addition, the preliminary calculation we have presented on the energetics of hole dissociation from aggregates suggest that this could be important at higher temperatures. Clearly, further work is needed on this latter topic.

Our second theme concerned the clustering of defects in the near-stoichiometric oxides. Here we have suggested that clustering is to be expected for $x \geq 10^{-2}$ at 1600 K. Again, however, more precise predictions require the inclusion of the effects of hole dissociation from the vacancy aggregates.

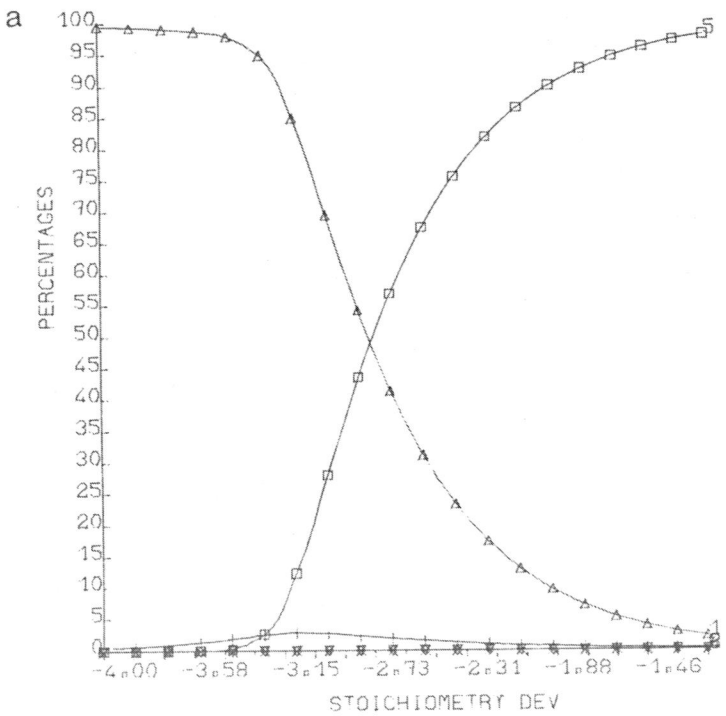

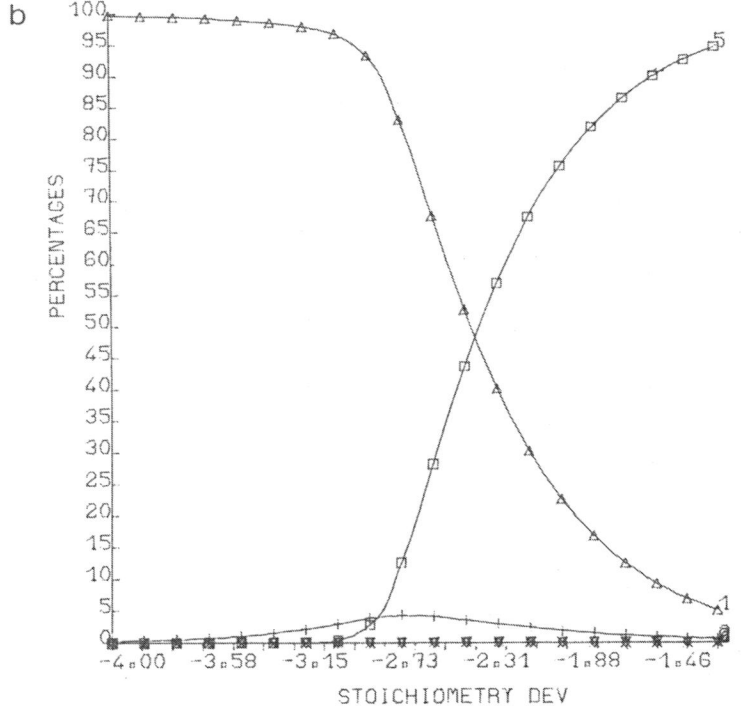

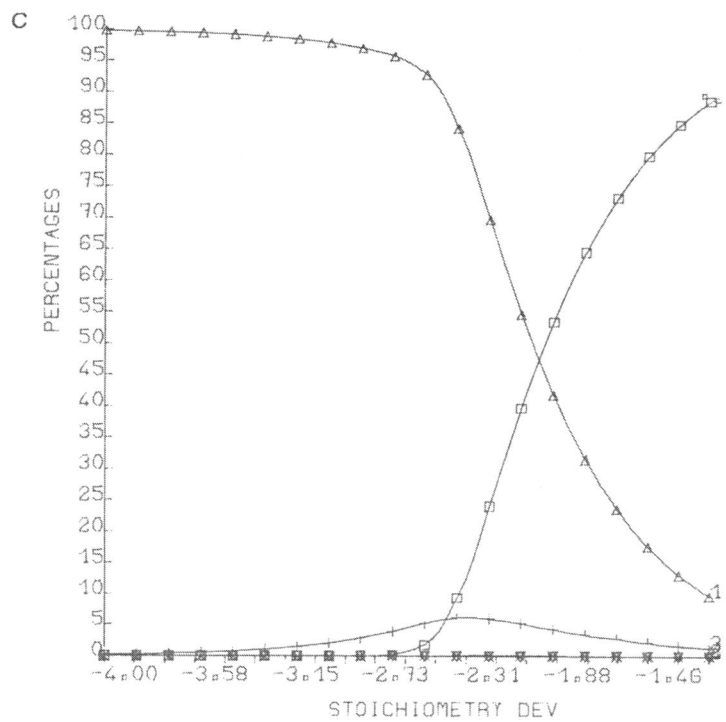

Figure 2. Variation of fraction of total vacancy
concentration present in clusters for
various temperatures: a) 1200 K
b) 1400 K,   c) 1600 K.

Key: △ cation vacancy; + singly charged
vacancy; □ 4:1 clusters. (Other
defects not present in appreciable
concentrations)

A final point concerns the difference between the results presented here for $Fe_{1-x}O$ and those for other rock salt structured
oxides. Greater details will be presented in a future publication;
but in general we find relatively small differences between the
binding energies calculated for the various clusters in the different oxides. However, we should note that our calculations do not
include a detailed representation of ligand field effects (although
they may be partly absorbed in the parameterisations of the pair
potentials). As discussed by Anderson et.al.[11], such terms may be
appreciable and could give rise to significant differences between

cluster energetics for the different oxides.

We consider, however, that the calculations we have presented provide a good basis for an understanding of non-stoichiometry in these materials. Future work should be directed towards confirming our predictions.

ACKNOWLEDGEMENTS

We are grateful to AERE Harwell and SERC for supporting this work. We should also like to thank A.M. Stoneham for several useful discussions.

REFERENCES

1. W. L. Roth, Acta. Cryst. 13:140 (1960).
2. F. Koch and J. B. Cohen, Acta. Cryst. B25:275 (1969).
3. A. K. Cheetham, B. E. F. Fender and R. I. Taylor, J. Phys. C. 4:2160 (1970).
4. P. D. Battle and A. K. Cheetham, J. Phys. C. 12:337 (1979).
5. J. R. Gavarri, C. Carel and D. Wiegler, J. Solid State Chem. 29:81 (1979).
6. S. Iijima, Proc. Int. Conf. Real Atoms Real Cryst. (Melbourne) p135 (1974).
7. C. Lebreton and L. W. Hobbs, Radiation Effects 74:227 (1983).
8. C. R. A. Catlow and B. E. F. Fender, J. Phys. C. 8:3267 (1975).
9. C. R. A. Catlow, B. E. F. Fender and D. G. Muxworthy, J. Phys (Paris) C7:67 (1977).
10. L. W. Hobbs - these proceedings.
11. A. B. Anderson - these proceedings.
12. C. R. A. Catlow and A. M. Stoneham, J. Amer. Ceram. Soc. 64:234 (1981).
13. C. R. A. Catlow, R. James, W. C. Mackrodt and R. F. Stewart, Phys. Rev. B25:1006 (1982).
14. C. R. A. Catlow and W. C. Mackrodt, eds., Computer Simulation of Solids, in "Lecture Notes in Physics" Vol.166, Springer-Berlin (1982).
15. G. V. Lewis and C. R. A. Catlow, J. Phys. C., in press.
16. J. H. Harding, Physica B, in press.
17. M. J. Gillan and P. W. M. Jacobs, Phys. Rev. B28:759 (1983).
18. R. Ball and J. H. Harding, UKAEA report, AERE-M3294.
19. M. J. Gillan, Phil. Mag. A43, 291 (1981).

THIRD INTERNATIONAL CONFERENCE ON
TRANSPORT IN NONSTOICHIOMETRIC COMPOUNDS
June 10-15, 1984
University Park, Pennsylvania

PARTICIPANTS

Dr. A. B. Anderson
   Chemistry Department, Case Western Reserve
   University, Cleveland, OH 44106 USA
Dr. Harlan U. Anderson
   Dept. of Ceramic Engineering, 222 Fulton Hall, Univ. of
   Missouri - Rolla, Rolla, MO 65401 USA
Dr. Ken Ando
   Dept. of Nuclear Engineering, Kyusku University
   Fukuoka, 812, Japan
Dr. A. Atkinson
   Materials Development Div. AERE, Harwell, Oxfordshire,
   OX11 ORA United Kingdom
Dr. H. V. Atkinson
   Materials Development Div. AERE, Harwell, Oxfordshire,
   OX11 ORA United Kingdom
Dr. N. I. Birks
   Dept. of Metal and Mats. Eng., University of Pittsburgh,
   Pittsburgh, PA 15261 USA
Dr. R. A. Catlow
   Dept. of Chemistry, Univ. College London, 20 Gordon
   Street, London WC1, United Kingdom
Dr. Robert Cava
   Bell Laboratories, 600 Mountain Ave., Murray Hill, NJ
   07974 USA
Dr. Alfred R. Cooper
   Case Western Reserve University, Dept. of Metal. and Mats.
   Science, 10900 Euclid Avenue, Cleveland, OH 44106 USA
Dr. Alastair N. Cormack
   Dept. of Chemistry, Univ. College London, 20 Gordon
   Street, London WC1H OAJ, United Kingdom
Dr. -Ing. R. Dieckmann
   Institut fur Phy. Chemie and Electrochemie der Universitat
   Hannover, Collinstrasse 3-3A, 3000 Hanover 1, Federal Republic
   of Germany
Dr. Nicklaus Eror
   Oregon Graduate Center, 19600 Northwest Walk Road,
   Beaverton, Oregon 97006

Dr. F. Gesmundo
   Instituto di Chimica Fisica Applicata dei Materiali,
   Lungobisagno Istria 34, 16141 Genova, Italy
Dr. John Halloran
   Case Western Reserve University, Dept. Metal. and
   Materials Science, Cleveland, OH 44106 USA
Dr. A. H. Heuer
   Case Western Reserve University, Dept. Metal. and
   Materials Science, Cleveland, OH 44106 USA
Dr. L. W. Hobbs
   Ceramic Division, Mass. Inst. of Tech., Cambridge, MA
   02139 USA
Dr. P. R. S. Jackson
   Institute for Materials Research, McMaster University
   Hamilton, Ontario, L85-4MI, Canada
Dr. J. Janowski
   Institute of Metallurgy, Academy of Mining and
   Metallurgy, Krakow, Poland
Dr. H. M. Kizilyalli
   Dept. of Physics, Orta Dogu Teknik Universitesi, Middle
   East Technical University, Ankara, Turkey
Dr. F. A. Kröger
   University of Southern California, Dept. of Materials
   Science, Los Angeles, CA 90007 USA
Dr. W. C. Mackrodt
   I.C.I. Corporate Laboratory, P.O. Box 11, The Heath
   Runcorn, Cheshire WA7 4QE, United Kingdom
Dr. C. M. Mari
   Instituto di Electrochimica e Metalurgica, Universita di
   Milano, Via Golgi 19, 20133 Milano, Italy
Dr. T. O. Mason
   Dept. of Materials Science and Engineering,
   Northwestern University, Evanston, IL 60201 USA
Dr. H. J. Matzke
   European Institute for Transuranium Elements, Postfach
   2266, D-7500 Karlsruhe, Republic of Germany
Dr. A. E. McHale
   Imperial College of Science & Technology, Dept. of
   Metallurgy, Royal School of Mines, Prince Consort Road,
   London SW 7 2BP, United Kingdom
Dr. C. Monty
   C.N.R.S. Laboratoire de Physique des Materiaux, 1, Place
   Aristide Briaud, 92190 Meudon, France
Dr. A. S. Nowick
   Div. of Metal & Chem. Metal., Henry Krumb School of Mines,
   Columbia University, New York, NY 10027 USA
Dr. J. Nowotny
   Institute of Catalysis and Surface Chemistry, Polish Academy of
   Sciences, UI, Niezapominajek, 30-239 Krakow, Poland
Dr. Y. Oishi
   NGK Insulators, Ltd., 2-56 Suda-cho, Mizuho-ku, Magoya
   467 Japan

Dr. M. Onillon
  C.N.R.S. Laboratoire de Chimie du Solide, University de
  Boreaux 1, 351 Cours de la Liberation, 33405 Talance-
  Cedex France
Dr. N. L. Peterson
  Materials Science Division, Argonne National Laboratory,
  9700 South Cass Avenue, Argonne, IL 60439 USA
Dr. G. Petot-Ervas
  C.N.R.S. Laboratoire P.M.T.M., Avenue J. B. Clement,
  93430 Villetaneuse, France
Dr. F. S. Pettit, Head
  Dept. of Metal. and Materials Science, University of
  Pittsburgh, Pittsburgh, PA 15261 USA
Dr. J. Philibert
  C.N.R.S. Laboratoire de Physique des Materiaux, 1, Place
  Aristide Briaud, 92190 Meudon, France
Dr. A. Revcolevschi
  Universite de Paris - Sud, Laboratoire de Chimie Appliquee,
  Batiment 44, 91405 Orsay, France
Dr. G. Simkovich
  206 Steidle Building, The Pennsylvania State University,
  University Park, PA 16802 USA
Dr. W. W. Smeltzer
  Institute for Materials Research, McMaster University,
  Hamilton, Ontario, L85-4ML Canada
Dr. D. M. Smyth, Director
  Materials Research Center, Lehigh University, Coxe
  Laboratory, Bethlehem, PA 18015 USA
Dr. A. Steinbrunn
  Laboratoire de Recherches sur la Reactivite des Solides,
  Associé au C.N.R.S., Faculté des Sciences, Mirande, B.P.
  138, 21004 Dijon, France
Dr. F. H. Stott
  UMIST, P.O. Box 88, Manchester M60 1QD, United Kingdom
Dr. V. S. Stubican
  328 Steidle Building, The Pennsylvania State University,
  University Park, PA 16802 USA
Dr. D. S. Tannhauser
  Dept. of Physics, Technion, Haifa, Israel
Dr. H. L. Tuller
  Mass. Institute of Technology, Dept. of Materials Science &
  Engineering, Room 13-4022, Cambridge, MA 02139 USA
Dr. J. B. Wagner
  Arizona State University, Center for Solid State Science, Tempe,
  AZ 85281 USA
Dr. W. L. Worrell
  Dept. of Materials Science & Engineering, University of
  Pennsylvania, Philadelphia, PA 19104 USA
Dr. B. J. Wuensch
  Ceramics Division, Rm 13-4037, Mass. Institute of
  Technology, Cambridge, MA 02139 USA

COMPOUND INDEX

INDEX

ERRATUM

The following information was inadvertently omitted from this volume:

CONFERENCE ORGANIZATION

ORGANIZING COMMITTEE - 1984

    F. A. Kröger - University Southern California, U.S.A.
    G. Simkovich - The Pennsylvania State University, U.S.A.
    V. S. Stubican - The Pennsylvania State University, U.S.A.
    J. B. Wagner, Jr. - Arizona State University, U.S.A.

INTERNATIONAL ADVISORY BOARD

    J. C. Colson - France
    P. Kofstad - Norway
    H. J. Matzke - West Germany
    J. Nowotny - Poland
    G. Simkovich - U.S.A.
    V. S. Stubican - U.S.A.
    G. C. Petot-Ervas - France
    F. Gesmundo - Italy

SPONSORS

    NATO - ARW 861-83
    Army Research Office - Durham, N.C., U.S.A.
    Office of Naval Research - Washington, D.C., U.S.A.

TRANSPORT IN NONSTOICHIOMETRIC COMPOUNDS
G. Simkovich and V.S. Stubican, editors   0-306-42086-4